**Immunology**

*Edited by*
*Robert A. Meyers*

## 1807–2007 Knowledge for Generations

Each generation has its unique needs and aspirations. When Charles Wiley first opened his small printing shop in lower Manhattan in 1807, it was a generation of boundless potential searching for an identity. And we were there, helping to define a new American literary tradition. Over half a century later, in the midst of the Second Industrial Revolution, it was a generation focused on building the future. Once again, we were there, supplying the critical scientific, technical, and engineering knowledge that helped frame the world. Throughout the 20th Century, and into the new millennium, nations began to reach out beyond their own borders and a new international community was born. Wiley was there, expanding its operations around the world to enable a global exchange of ideas, opinions, and know-how.

For 200 years, Wiley has been an integral part of each generation's journey, enabling the flow of information and understanding necessary to meet their needs and fulfill their aspirations. Today, bold new technologies are changing the way we live and learn. Wiley will be there, providing you the must-have knowledge you need to imagine new worlds, new possibilities, and new opportunities.

Generations come and go, but you can always count on Wiley to provide you the knowledge you need, when and where you need it!

*William J. Pesce*
President and Chief Executive Officer

*Peter Booth Wiley*
Chairman of the Board

# Immunology

From Cell Biology to Disease

*Edited by*

*Robert A. Meyers*

WILEY-VCH Verlag GmbH & Co. KGaA

**The Editor**

***Dr. Robert A. Meyers***
RAMTECH LIMITED
122 Escalle Lane
Larkspur, CA 94039
USA



**Library of Congress Card No.**: applied for

**British Library Cataloguing-in-Publication Data**: A catalogue record for this book is available from the British Library.

**Bibliographic information published by the Deutsche Nationalbibliothek**
Die Deutsche Nationalbibliothek lists this publication in the Deutsche Nationalbibliografie; detailed bibliographic data are available in the Internet at http://dnb.d-nb.de.



**Composition:** Laserwords Private Ltd, Chennai, India
**Printing:** Strauss GmbH, Mörlenbach
**Bookbinding:** Litges & Dopf GmbH, Heppenheim
**Cover Design:** Adam Design, Weinheim
**Wiley Bicentennial Logo:** Richard J. Pacifico

Printed in the Federal Republic of Germany.
Printed on acid-free paper.

**ISBN** 978-3-527-31770-7

# Contents

**Preface** vii

**Color Plates** xiii

**Part I Innate and Adaptive Immunity** 1

1 Immunology 3
*Cindy Takeuchi and Paul Wentworth*

2 Superantigens 27
*Matthew D. Baker and K. Ravi Acharya*

3 Antigen Presenting Cells (APCs) 43
*Harald Kropshofer and Anne B. Vogt*

4 Cell Mediated Immune Defence 93
*Martin F. Bachmann and Thomas M. Kundig*

5 Immunologic Memory 113
*Alexander Ploss and Eric G. Pamer*

**Part II Signaling in the Immune System** 135

6 Molecular Mediators: Cytokines 137
*Jean-Marc Cavaillon*

7 Interleukins 167
*Anthony Meager*

8 Viral Inhibitors and Immune Response Mediators: The Interferons 205
*Anthony Meager*

9 Signaling Through JAKs and STATs: Interferons Lead the Way 241
*Christian Schindler and Jessica Melillo*

*Immunology. From Cell Biology to Disease.* Edited by Robert A. Meyers.

ISBN: 978-3-527-31770-7

**Part III Techniques** 269

10 Flow Cytometry 271
*Michael G. Ormerod*

11 Immunoassays 295
*James P. Gosling*

12 Genetic Engineering of Antibody Molecules 319
*Manuel L. Penichet and Sherie L. Morrison*

**Part IV Immunological Disorders** 343

13 Autoantibodies and Autoimmunity 345
*Kenneth Michael Pollard*

14 Synovial Mast Cells in Inflammatory Arthritis 367
*Theoharis C. Theoharides*

15 Molecular and Cell Biology of AIDS/HIV 395
*Andrew M.L. Lever*

Index 421

# Preface

This treatise on molecular immunology and molecular medicine approaches to understanding and treatment of immune diseases was compiled from a selection of key articles from the recently published 16 volume *Encyclopedia of Molecular Cell Biology and Molecular Medicine* (ISBN 978-3-527-30542-1, http://www.meyers-emcbmm.de/). This volume is comprised of 15 detailed articles arranged in four sections covering innate and adaptive immunity, signaling in the immune system, techniques and immunological disorders. The articles were prepared by eminent researchers from many of the major global molecular cell immunology research institutions including The Scripps Research Institute; University of Oxford; University of Cambridge; Institut Pasteur; University of Zürich; National University of Ireland; Roche Center for Medical Genomics; Sloan-Kettering Institute; Cornell University; National Institute for Biological Standards and Control, UK; Columbia University; and University of California, Los Angeles.

Each article begins with a concise definition of the subject and its importance, followed by the body of the article and extensive references for further reading. The references are divided into secondary references (books and review articles) and primary research papers. Each subject is presented on a first-principle basis, including detailed figures, tables and drawings. Because of the self-contained nature of each article, some overlap among articles on related topics occurs. Extensive cross-referencing is provided to help the reader expand his or her range of inquiry.

The master publication, which is the basis of the Proteins set, is the *Encyclopedia of Molecular Cell Biology and Molecular Medicine,* which is the successor and second edition of the *VCH Encyclopedia of Molecular Biology and Molecular Medicine,* covering the molecular and cellular basis of life at a university and professional researcher level. The First Edition, published in 1996–97 was utilized in libraries around the world. This second edition is double the first edition in length and comprises the most detailed treatment of both molecular and cell biology available today. The Board with eleven Nobel laureates and I believe that there is a serious need for this publication, even in view of the vast amount of information available on the World Wide Web and in text books and monographs. We feel that there is no substitute for our tightly organized and integrated approach to selection of articles and authors and implementation of

*Immunology. From Cell Biology to Disease.* Edited by Robert A. Meyers.

ISBN: 978-3-527-31770-7

peer review standards for providing an authoritative single-source reference for undergraduate and graduate students, faculty, librarians and researchers in industry and government.

Our purpose is to provide a comprehensive foundation for the expanding number of molecular biologists, cell biologists, pharmacologists, biophysicists, biotechnologists, biochemists and physicians as well as for those entering molecular cell biology and molecular medicine from majors or careers in physics, chemistry, mathematics, computer science and engineering. For example there is an unprecedented demand for physicists, chemists and computer scientists who will work with biologists to define the genome, proteome and interactome through experimental and computational biology.

The Board and I first divided all of molecular cell biology and molecular medicine into primary topical categories and each of these was further defined into subtopics. The following is a summary of the topics and subtopics:

- Nucleic Acids: amplification, disease genetics overview, DNA structure, evolution, general genetics, nucleic acid processes, oligonucleotides, RNA structure, RNA replication and transcription.
- Structure Determination Technologies Applicable to Biomolecules: chromatography, labeling, large structures, mapping, mass spectrometry, microscopy, magnetic resonance, sequencing, spectroscopy, x-ray diffraction.
- Proteins, Peptides and Amino Acids: analysis, enzymes, folding, mechanisms, modeling, peptides, structural genomics (proteomics), structure, types.
- Biomolecular Interactions: cell properties, charge transfer, immunology, recognition, senses.
- Molecular Cell Biology of Specific Organisms: algae, amoeba, birds, fish, insects, mammals, microbes, nematodes, parasites, plants, viruses, yeasts.
- Molecular Cell Biology of Specific Organs or Systems: excretory, lymphatic, muscular, neurobiology, reproductive, skin.
- Molecular Cell Biology of Specific Diseases: cancer, circulatory, endocrine, environmental stress, immune, infectious diseases, neurological, radiation.
- Biotechnology: applications, diagnostics, gene altered animals, bacteria and fungi, laboratory techniques, legal, materials, process engineering, nanotechnology, production of classes or specific molecules, sensors, vaccine production.
- Biochemistry: carbohydrates, chirality, energetics, enzymes, biochemical genetics, inorganics, lipids, mechanisms, metabolism, neurology, vitamins.
- Pharmacology: chemistry, disease therapy, gene therapy, general molecular medicine, synthesis, toxicology.
- Cellular Biology: developmental cell biology, diseases, dynamics, fertilization, immunology, organelles and structures, senses, structural biology, techniques.

We then selected some 340 article titles and author or author teams to cover the above topics. Each article is designed as a self-contained treatise. Each article begins with a key word section, including definitions, to assist the scientist or student who is unfamiliar with the specific subject area. The Encyclopedia includes more than 3000 key words, each defined within the context of the particular scientific field covered by the article. In addition to these definitions, the glossary of basic terms found at the back of each volume, defines the most commonly used terms in molecular and cell biology. These definitions should allow most readers to understand articles in the Encyclopedia without referring to a dictionary, textbook or other reference work.

Larkspur, May 2007

**Robert A. Meyers**
Editor-in-Chief

# List of Contributors

**K. Ravi Acharya**
University of Bath, Claverton Down,
Bath, UK

**Martin F. Bachmann**
Cytos Biotechnology AG,
Zürich-Schlieren,
Switzerland

**Matthew D. Baker**
University of Bath, Claverton Down,
Bath, UK

**Jean-Marc Cavaillon**
Institut Pasteur, Paris, France

**James P. Gosling**
National University of Ireland, Galway,
Ireland

**Harald Kropshofer**
Roche Center for Medical Genomics,
F. Hoffman-La Roche Ltd, Basel,
Switzerland

**Thomas M. Kundig**
University of Zürich, Zürich, Switzerland

**Andrew M. L. Lever**
University of Cambridge, Cambridge, UK

**Anthony Meager**
National Institute for Biological Standards
and Control, South Mimms, Herts, UK

**Jessica Melillo**
Columbia University, New York, USA

**Sherie L. Morrison**
University of California, Los Angeles,
CA, USA

**Michael G. Ormerod**
34 Wray Park Road, Reigate, UK

**Eric G. Pamer**
Sloan-Kettering Institute, New York, USA

**Manuel L. Penichet**
University of California, Los Angeles,
CA, USA

**Alexander Ploss**
Sloan-Kettering Institute, New York, USA

**Kenneth Michael Pollard**
The Scripps Research Institute, La Jolla,
CA, USA

**Christian Schindler**
Columbia University, New York, USA

**Cindy Takeuchi**
The Scripps Research Institute, La Jolla,
CA, USA

**Theoharis C. Theoharides**
Tufts University School of Medicine,
Boston, MA, USA

*Immunology. From Cell Biology to Disease.* Edited by Robert A. Meyers.

ISBN: 978-3-527-31770-7

**Anne B. Vogt**
Roche Center for Medical Genomics, F. Hoffman-La Roche Ltd, Basel, Switzerland

**Paul Wentworth**
University of Oxford, Oxford, UK

# Color Plates

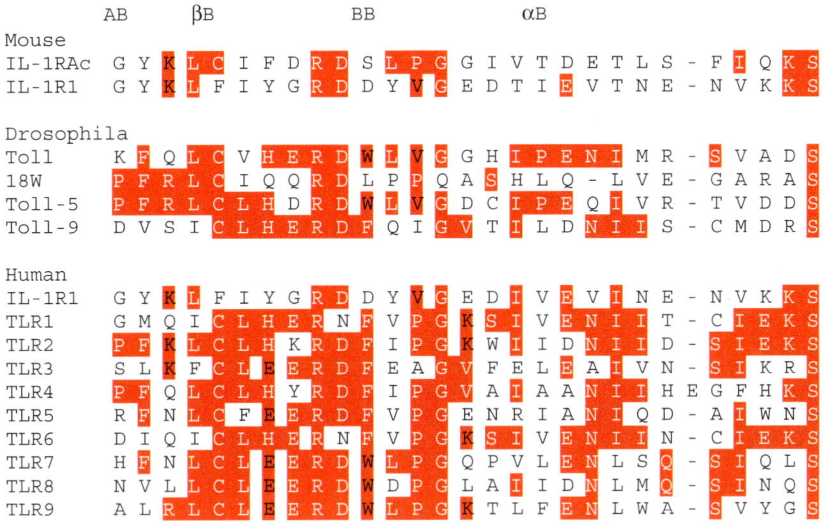


**Fig. 2 (p. 13)** Sequence alignment of the TLR domain-signaling region of mouse and human IL-1R proteins, *Drosophila* Toll proteins, and human TLRs. Residues conserved in at least four sequences are shaded. The greatest homology between *Drosophila* and human proteins occurs between the cytoplasmic domains of Toll-9 and TLRs 1, 2, 4, and 6.

*Immunology. From Cell Biology to Disease.* Edited by Robert A. Meyers.

ISBN: 978-3-527-31770-7

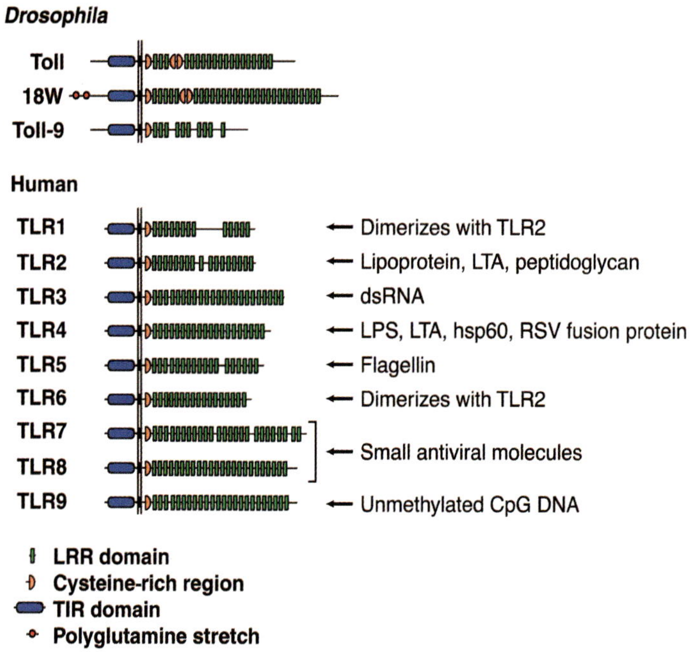


**Fig. 3 (p. 15)** Schematic representation of human TLR domains and their stimuli. Adapted from Imler, J.-L., Hoffmann, J.A. (2001) Toll receptors in innate immunity, *Trends Cell Biol.* **11**, 304–311.

**Fig. 1 (p. 30)** Ribbon diagram of staphylococcal enterotoxin A (SEA), representative of the common structural features of the staphylococcal and streptococcal SAg family. Blue spheres represent the positions of the two zinc sites. The cysteine residues that form the disulfide loop are shown in stick representation (yellow), the surfaces defining the TCR binding region (green), and MHC class II binding region (red) are shown.

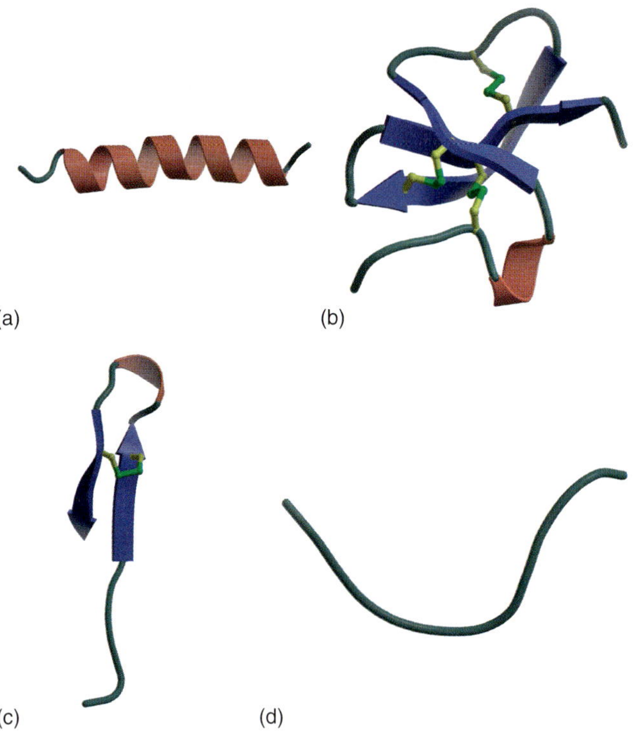


**Fig. 7 (p. 23)** Antimicrobial peptides. Representative member of each structural class: (a) magainin 2, alpha-helix; (b) $\beta$-defensin, beta-sheet; (c) thanatin, loop; (d) indolicidin, extended. Pictures generated by Matt Kelker, using bobscript and raster3D. Coordinates obtained from the RCSB Protein Data Bank under the accession codes 1BNB (defensin), 1G89 (indolicidin), 2MAG (magainin), and 8TFV (thanatin). Raster3D: Merritt & Bacon (1997) *Meth. Enzymol.* **277**, 505–524. bobscript: Esnouf, RM (1997) *J. Mol. Graph. Model.* **132–134**, 112–113.

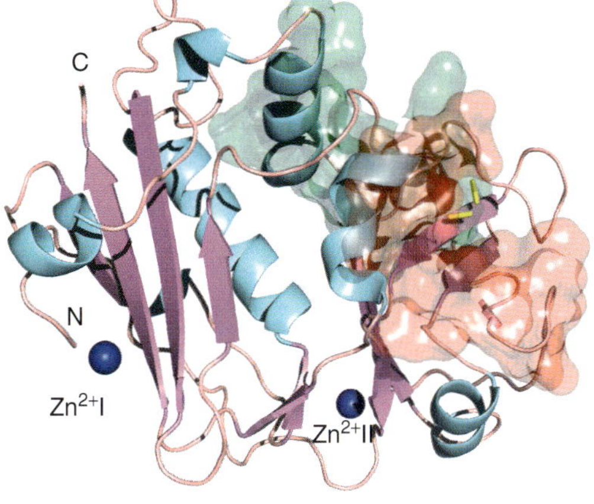

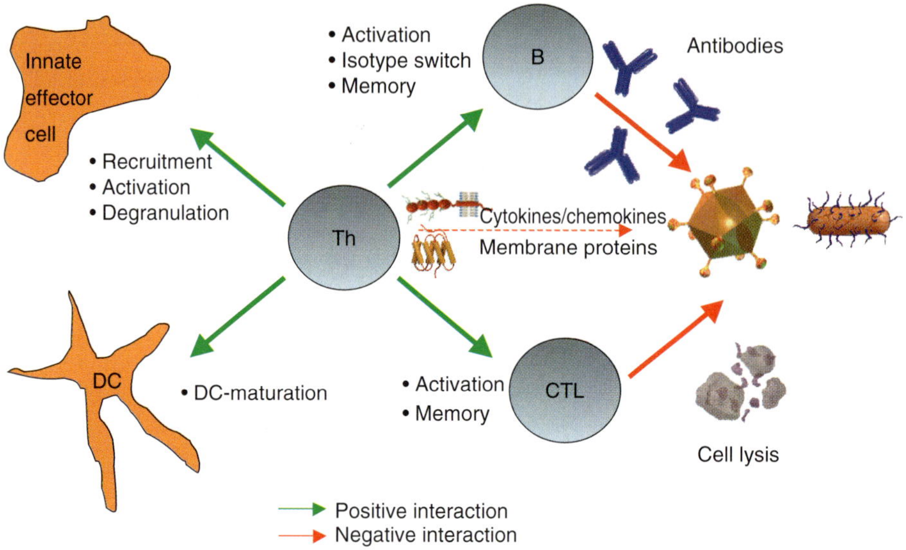


**Fig. 3 (p. 107)** Cross-talk between the innate and adaptive immune system and the critical role of Th cells in orchestrating the immune response.

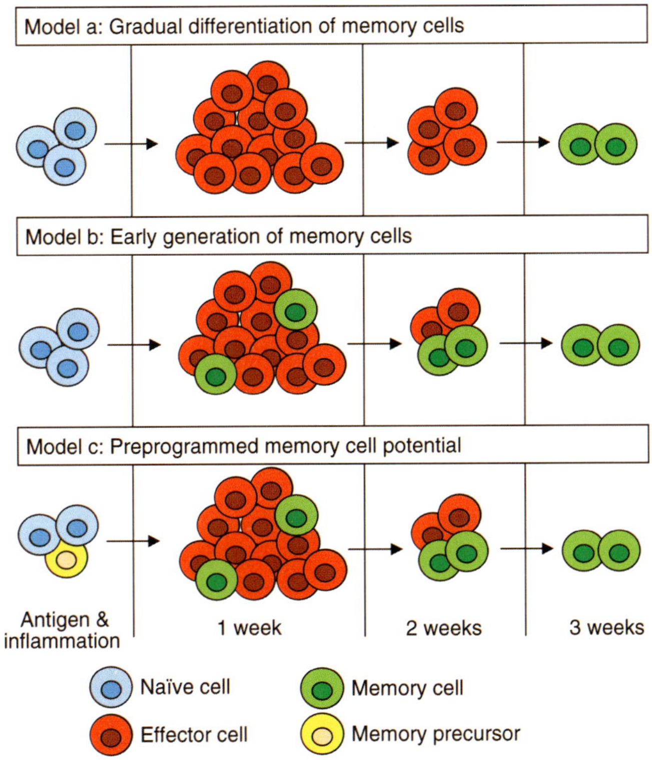


**Fig. 1 (p. 124)** Models of programmed memory T-cell generation.

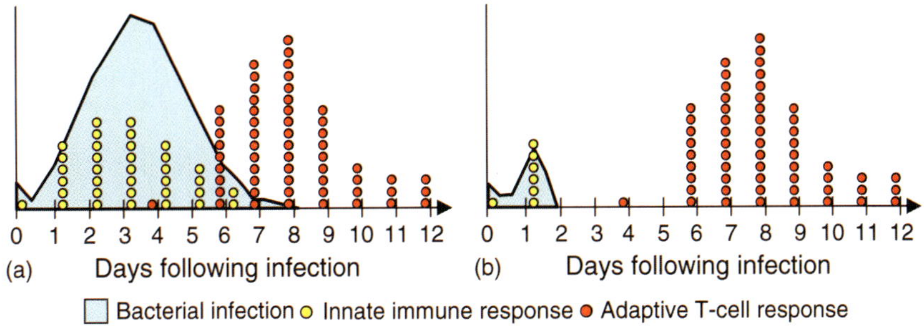


**Fig. 2 (p. 125)** Antigen-independent proliferation and memory generation.

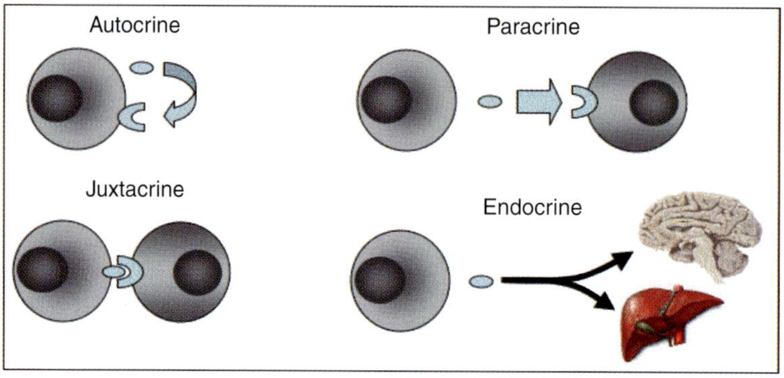


**Fig. 1 (p. 142)** Mode of action of cytokines.

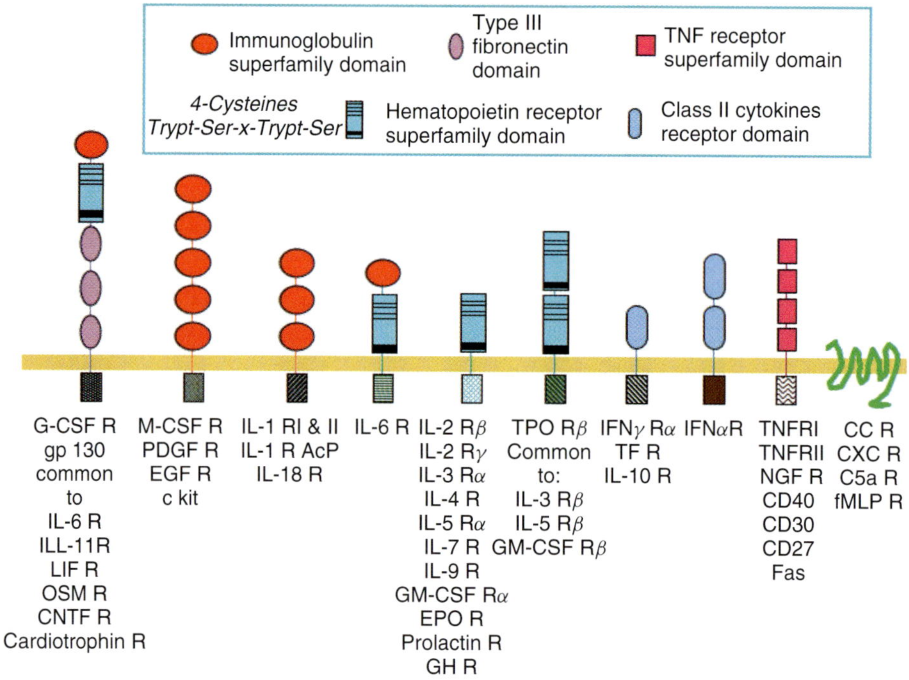


**Fig. 2 (p. 145)** Families of cytokine receptors.

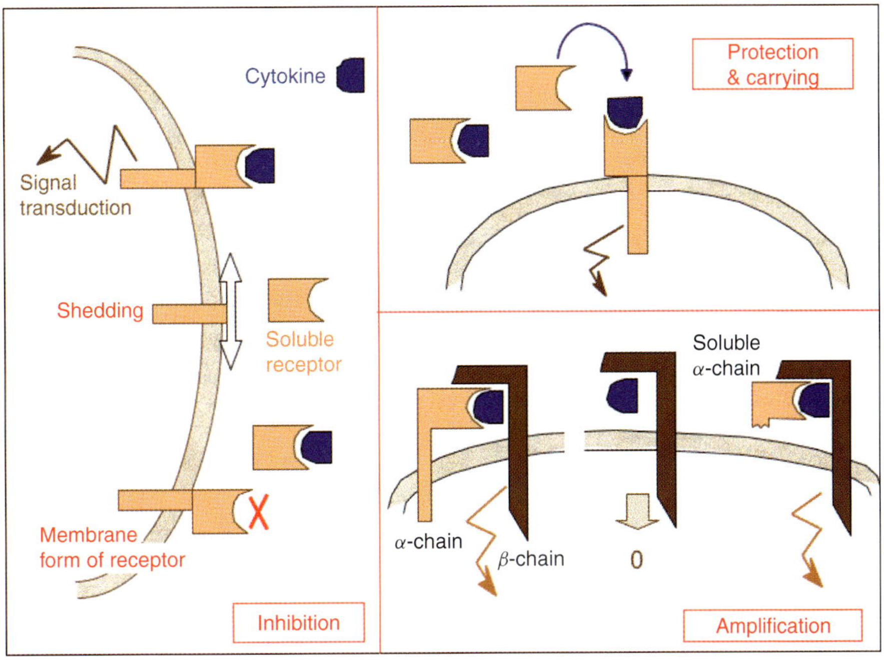


**Fig. 3 (p. 146)** Different properties of soluble receptors.

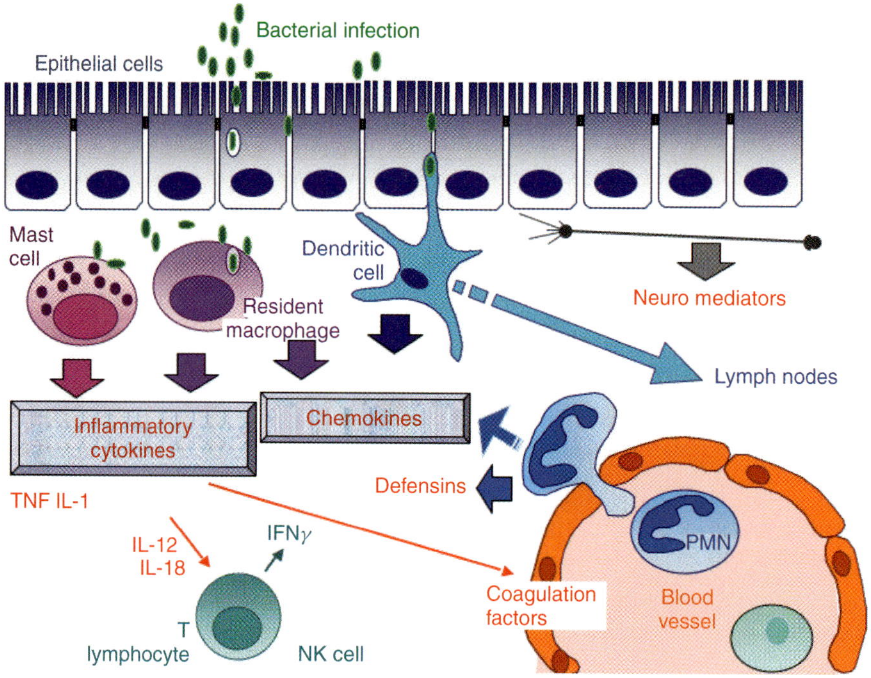


**Fig. 4 (p. 148)** Cytokines and soluble mediators as coordinators of regulated processes taking place in a bacteria-loaded sites.

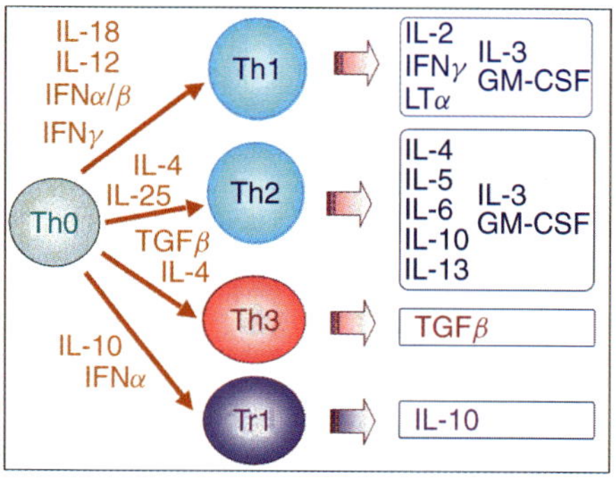


**Fig. 5 (p. 149)** T-cell subpopulations and the nature of environmental cytokines required for their differentiation.

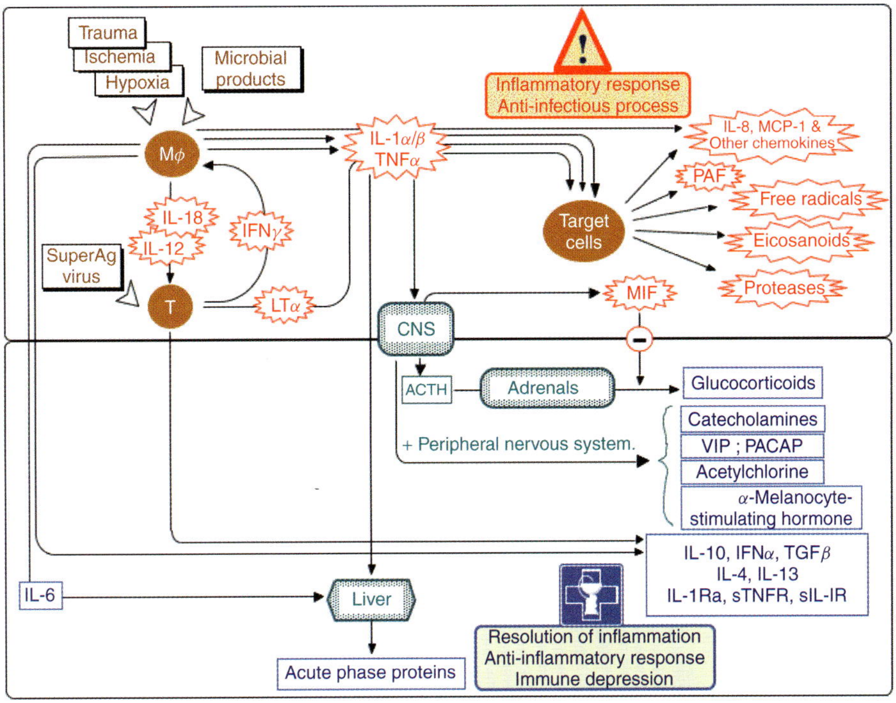


**Fig. 6 (p. 152)** Inflammation is the consequence of a cascade of events initiated by IL-1 and/or TNF. The action of these cytokines on various target cells leads to the release of numerous inflammatory mediators. Anti-inflammatory cytokines, the activation of the neuroendocrine pathway and the effects of glucocorticoids as well as the enhanced production of acute-phase proteins control negatively the inflammatory process (ACTH: adrenocorticotropic hormone; CNS: central nervous system; PACAP: pituitary adenylate cyclase-activating polypeptide; PAF: platelet activating factor; VIP: vasoactive intestinal peptide).

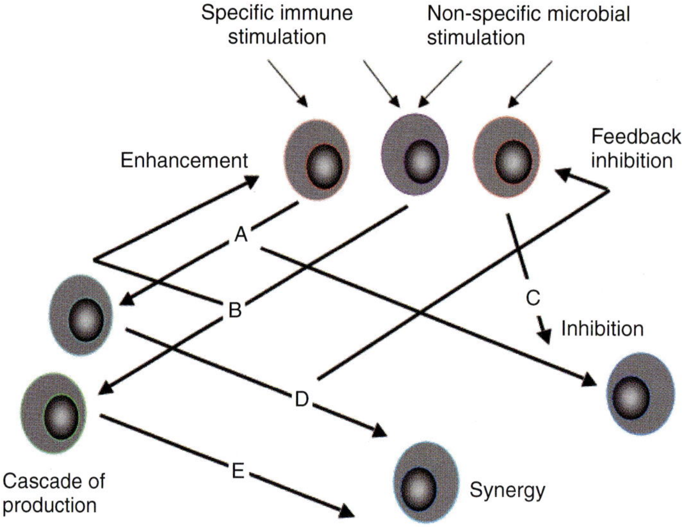


**Fig. 7 (p. 157)** The cytokine network (see Sect. 7).

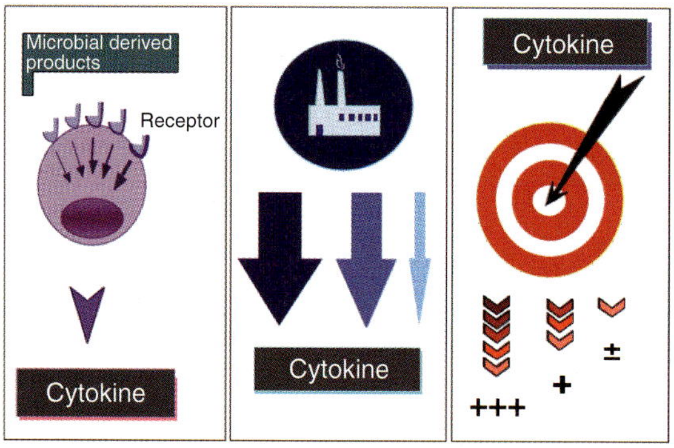


**Fig. 8 (p. 158)** Individual heterogeneity for cytokine production and responsiveness.

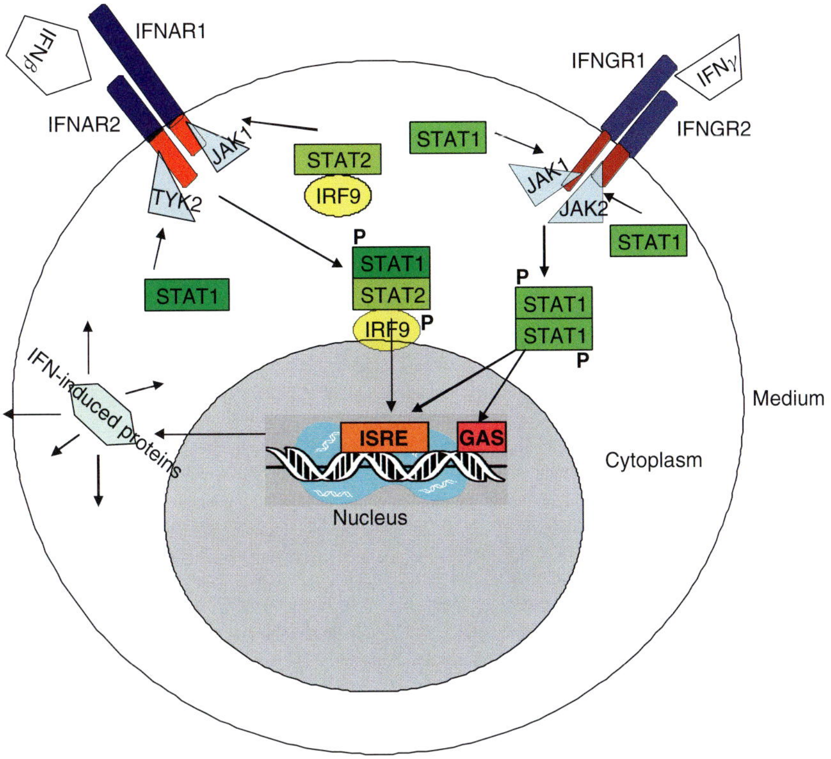


**Fig. 5 (p. 221)** Schematic drawing depicting signal transduction pathways from type I and type II IFN receptors at the cell membrane to the cell nucleus. ISGF3, interferon-stimulated gene factor-3. The ISGF3α complex contains three structurally related proteins (p84), p91[STAT1] and p113 [STAT2], and the ISGF3γ subunit is composed of a single protein, p48/IRF-9. ISRE, interferon-stimulated response element, present in IFN-inducible genes. GAS, γ-activated sequences, present in IFN-γ-inducible genes. JAK1, JAK2 and TYK2, nonreceptor protein tyrosine kinases involved in the phosphorylation of ISGF3α proteins, p91 [STAT1] and p113 [STAT2].

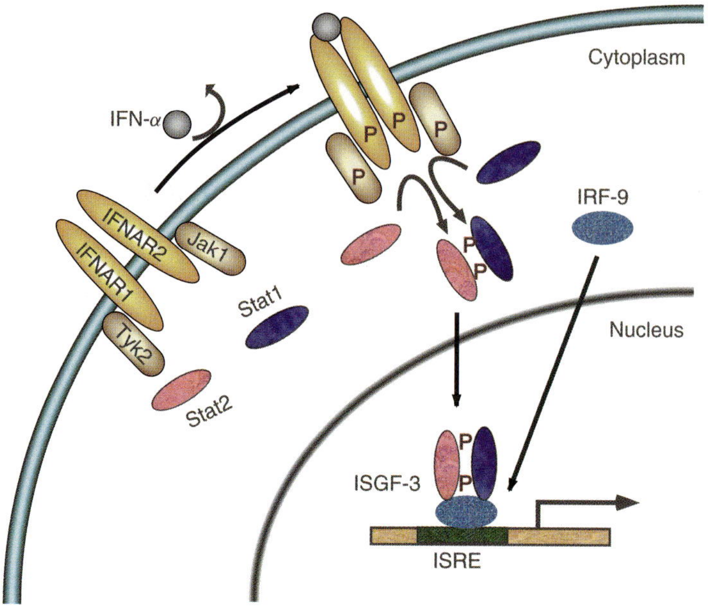


**Fig. 1 (p. 247)** The IFN-I signaling paradigm. Upon binding to its dimeric receptor, type I IFN promotes the apposition of two receptor associated JAKs, which then activate each other by transphosphorylation. The activated JAKs in turn phosphorylate receptor tyrosine(s), directing the SH2 domain dependent recruitment of Stat1 and Stat2. At the receptor, Stat1 and Stat2 are activated by phosphorylation, whereupon they heterodimerize, translocate into the nucleus, and associate with IRF-9 to bind to the ISRE enhancers. This DNA binding complex, referred to as ISGF-3, directs the expression of ISRE target genes. IFN-I also promotes the formation of Stat1 homodimers, which directs the expression of GAS-driven genes as outlined in Fig. 2.

**Fig. 3 (p. 250)** Structural models of the JAK and STAT family of signaling proteins. (a) The four members of the JAK family share seven homology domains regions, JH1–JH7. JH1 serves as the catalytic domain, whereas JH2 represents a pseudokinase domain, which is a characteristic feature of this family of tyrosine kinases. JH3–JH7 comprise a FERM domain that is responsible for association with cytokine receptors. (b) The seven members of the STAT family of transcription factors share seven functionally conserved domains. This includes the amino-terminal domain ($NH_2$), the coiled-coiled domain (coiled coil), the DNA binding domain (DBD), the Linker domain (LK), the SH2 domain, the tyrosine activation domain, and the transcriptional activation domain (TAD), which is conserved in function but not in sequence. Domain colors correspond with those in Stat1 Crystal structure in Fig. 4.

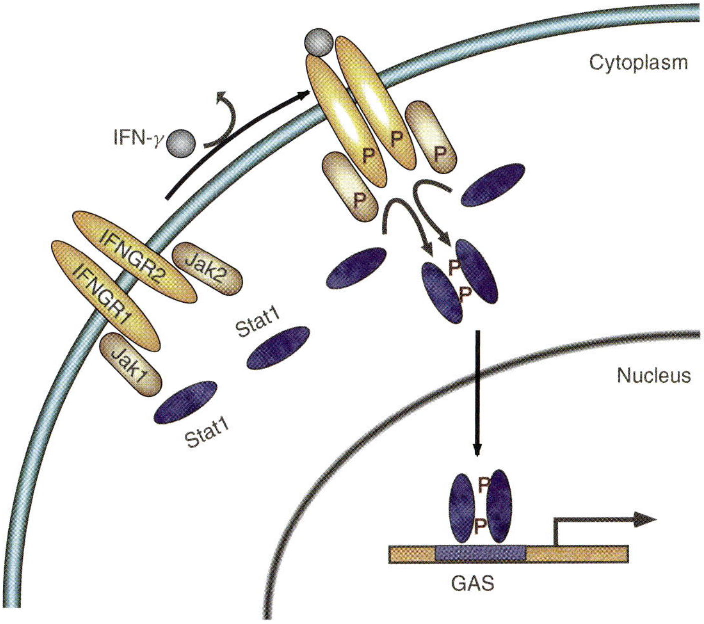


**Fig. 2 (p. 248)** The IFN-II signaling paradigm. Upon binding to its dimeric receptor, IFN-$\gamma$ promotes the activation of two receptor associated JAKs. The activated JAKs in turn activate the receptor. Stat1 is recruited to the receptor, whereupon it is activated and forms homodimers. Stat1 homodimers translocate to the nucleus and bind to the GAS family of enhancers, culminating in the expression of a distinct set of genes.

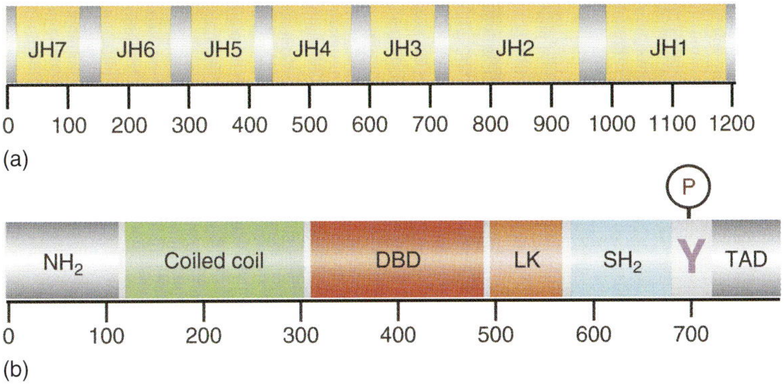

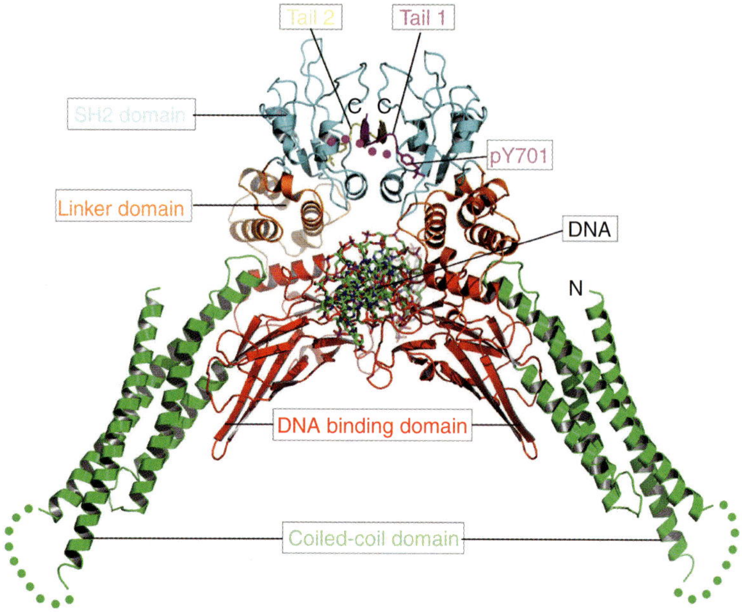


**Fig. 4 (p. 252)** Stat1 crystal structure. The structure represents two activated Stat1 fragments (amino acids 135–710) that are dimerized and bound to a GAS palindrome. Each domain is shown in a distinct color and labeled (see Fig. 3). Note that the CCD forms a four-$\alpha$-helix bundle and the DBD forms a $\beta$-barrel with an immunoglobulin fold.

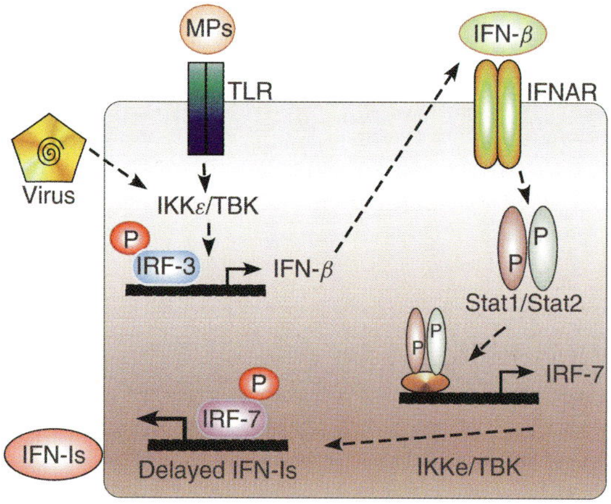


**Fig. 5 (p. 257)** IFN-I autocrine/paracrine loop. Viral and bacterial pathogens include molecular patterns (MPs) that are recognized by toll-like receptors (TLRs). Once bound to these receptors, they promote the activation of IKK$\varepsilon$/TBK1 kinases, which in turn phosphorylate IRF-3. Activated IRF-3 then drives the expression and secretion of IFN-$\beta$, an immediate early IFN-I. IFN-$\beta$ binds the IFN-$\alpha$ receptor (IFNAR) initiating an important autocrine loop. This leads to the expression of IRF-7, an important ISGF-3 target gene, which is responsible for driving the expression/secretion of the delayed IFN-Is.

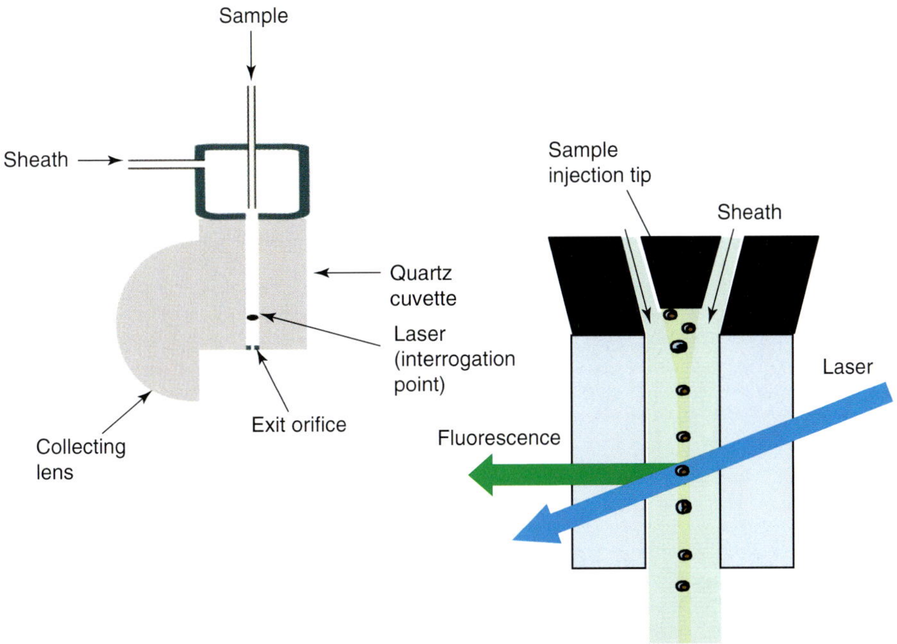


**Fig. 1 (p. 274)** Diagrammatic representation of a flow chamber.

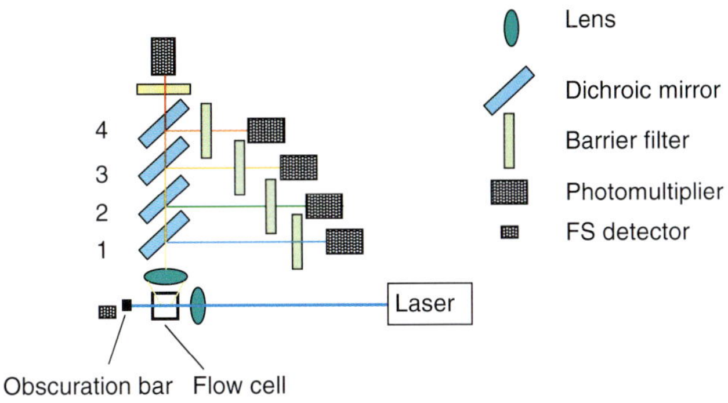


**Fig. 2 (p. 275)** The optical layout of a typical flow cytometer. Using an argon-ion laser emitting light at 488 nm, the optical components could be as follows: (1) Dichroic selecting light <500 nm; blue bandpass (488 nm). (2) Dichroic selecting light <540 nm; green bandpass (520 nm). (3) Dichroic selecting light <595 nm; orange bandpass (575 nm). (4) Dichroic selecting light <640 nm; red bandpass (620 nm) Final filter: long pass >650 nm.

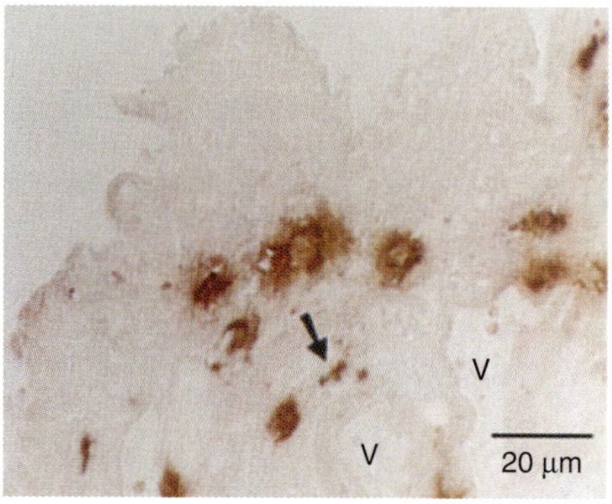


**Fig. 1 (p. 370)** Tryptase-positive mast cells (brown color) adjacent to blood vessels (V) from the joint of a patient with rheumatoid arthritis.

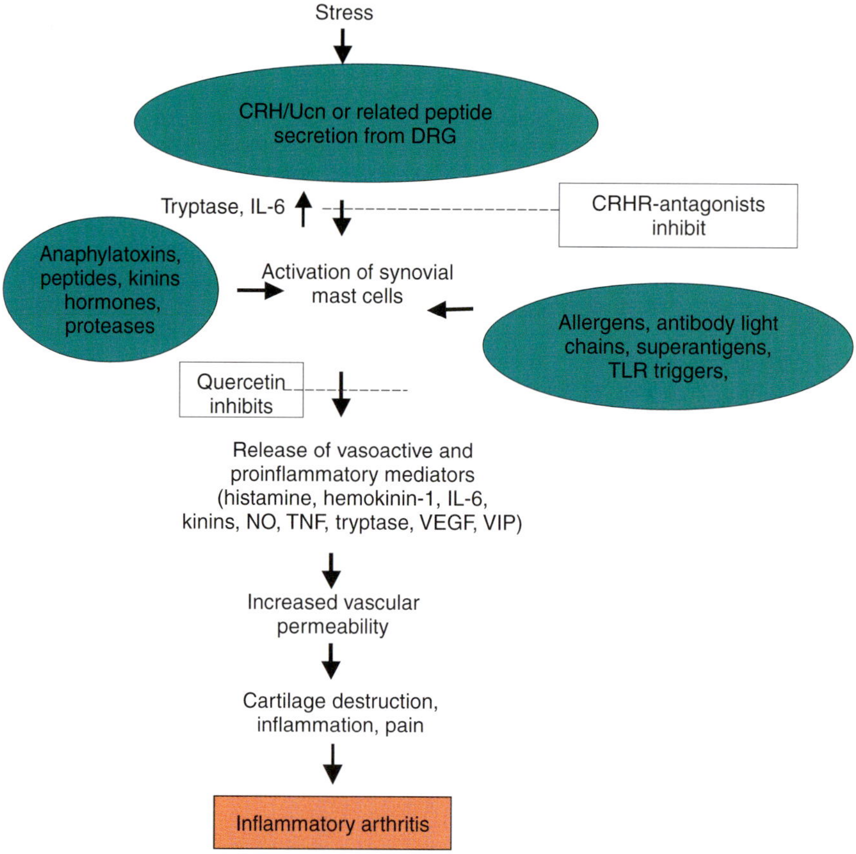


**Fig. 3 (p. 373)** Schematic representation of the possible role of mast cells in inflammatory arthritis. DRG = dorsal root ganglia; NO = nitric oxide; VIP = vasoactive intestinal peptide; VEGF = vascular endothelial growth factor.

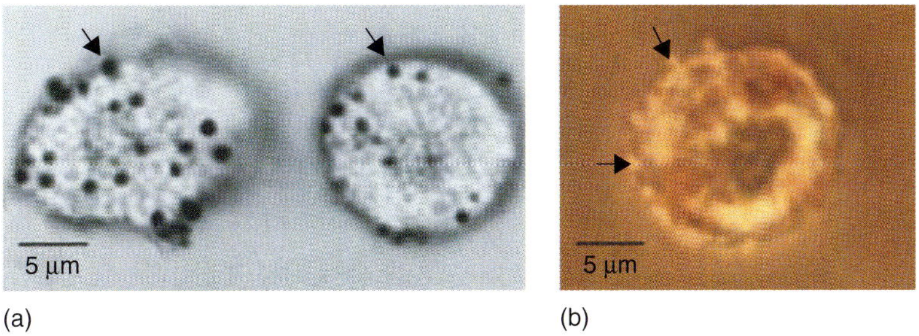


**Fig. 4 (p. 376)** Punctate release of contents of individual secretory granules from purified rat peritoneal mast cells incubated with calcium containing liposomes. (a) Content of individual granules (arrow) visualized by ruthenium red staining (with a complementary filter to show black) using Nomarski optics; (b) individual granules in the process of exocytosis from a mast cell visualized by ruthenium red staining.

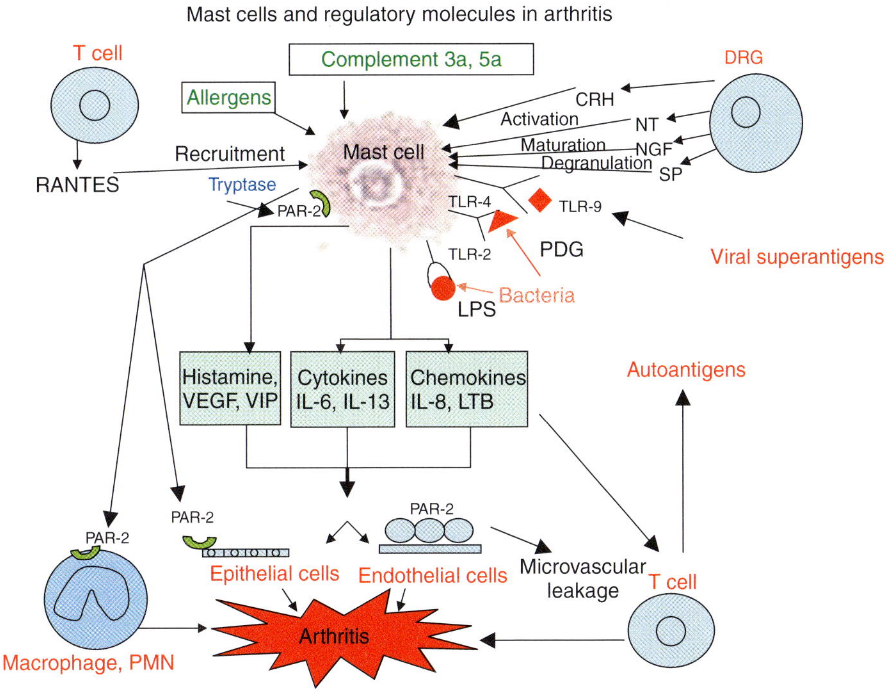


**Fig. 7 (p. 381)** Diagrammatic representation of synovial mast cells interaction with other cells in the initiation of synovial inflammation and arthritis.

# Part I
# Innate and Adaptive Immunity

*Immunology. From Cell Biology to Disease.* Edited by Robert A. Meyers.

ISBN: 978-3-527-31770-7

# 1
# Immunology

*Cindy Takeuchi*[1] *and Paul Wentworth*[1,2]
[1] *The Scripps Research Institute, La Jolla, California, USA*
[2] *University of Oxford, Oxford, UK*

**1 Essential Components and Functions of the Immune System 5**
1.1 Innate and Adaptive Immunity 5
1.2 Constitutive and Inducible Defenses 6

**2 Molecular Basis of Innate Immunity 8**
2.1 Evolution 8
2.2 Affinity and Specificity 9

**3 Innate Immune Recognition 10**
3.1 Pathogen-associated Molecular Patterns 10
3.2 Pattern-recognition Receptors 11
3.3 Toll-like Receptors 13

**4 Cellular Effector Mechanisms 17**
4.1 Cell-mediated Inflammation 17
4.2 Oxygen-independent Destruction of Pathogens 18
4.3 Oxygen-dependent Killing 19

**5 Humoral Effector Mechanisms 20**
5.1 Complement 20
5.2 Antimicrobial Peptides 22
5.3 Cytokines 25

**Bibliography 25**
Books and Reviews 25
Primary Literature 26

*Immunology. From Cell Biology to Disease.* Edited by Robert A. Meyers.

ISBN: 978-3-527-31770-7

## Keywords

**Cytokines**
Low molecular weight, soluble proteins that mediate signaling and affect cellular behavior by specific binding to receptors on cell surfaces. Cytokines are pleiotropic, redundant, and multifunctional.

**Chemokines**
A class of cytokines with chemoattractant properties, which enables migration of leukocytes to sites of inflammation.

**Inflammation**
Recruitment of body defenses against an invading organism or tissue damage and ensuing repair of damage.

**Leukocytes**
White blood cells including the lymphocytes and most myeloid cells, which perform various significant immune functions.

**Phagocytosis**
The process of engulfment of invading particles within an invagination of the phagocyte cell membrane.

Prior to the late-1960s, most studies on vertebrate immune systems focused on the recognition and killing mechanisms of adaptive immunity. The evolutionarily ancient mechanisms of innate immunity, generally thought to be nonspecific and obsolete, were, for the most part, overlooked. However, the relevance of innate systems has become apparent in recent decades, especially with the discovery of Toll-like receptors and with the ability of pattern-recognition receptors to stimulate effective immune responses via the induction of genes and cytokine expression. This review focuses on the complex interplay at the molecular level, between innate immune activation and adaptive immune responses and on how this relationship is essential for host defense.

Current molecular studies of innate immunity seek to clarify the interconnectivity between innate and adaptive phenomena and to understand the mechanisms of innate defense in their own right. These include examinations of the ligand specificities and structural interactions of the innate immune receptors, gene expression regulation during the maturation of immune responses, and the interplay of signal transduction pathways that culminate in induced innate responses to various pathogens. The perception of the innate immune system as a highly evolved biological design rather than a relic from ancestral species is allowing significant advances to be made in modern immunology.

## 1
## Essential Components and Functions of the Immune System

### 1.1
### Innate and Adaptive Immunity

The principal responsibilities of an immune system are to recognize a wide variety of pathogens and to destroy those pathogens while tolerating the tissues of the host organism. Innate immunity is the most universal and expeditious type of immunity, having evolved over the last hundred million years or so to accomplish these tasks. It includes physical barriers, such as the skin and epithelia that protect our internal organs, and the cells that bear innate immune receptors, such as macrophages, dendritic cells (DCs), neutrophils, basophils, eosinophils, mast cells, and natural killer (NK) cells. These white blood cells derive from myeloid and lymphoid precursors in the bone marrow. An encounter with a pathogen leads to the prompt removal of the agent or recruitment, activation, and differentiation of short-lived effector cells, which have further capacity to rid the host of most infections. Innate immune mechanisms work effectively against a broad range of pathogens during the first encounter with a pathogen. The response does not require an extended period of time to develop, due to the fact that the complete protein receptors are encoded in the germ-line DNA. Expression of innate immune receptors thus does not seem to involve either gene rearrangement or clonal expansion.

While the majority of organisms survive utilizing only innate immune mechanisms, vertebrates have developed secondary means for pathogen detection and eradication. When innate immunity is circumvented or overwhelmed, an adaptive system is utilized. Adaptive immunity relies on activated lymphocytes to generate highly variable antigen receptors, which provide specific recognition of protein, carbohydrate, lipid, and nucleic acid moieties of pathogens and foreign material. These recombinant receptors, the immunoglobulins and T-cell receptors (TCR), are capable of recognizing virtually any pathogen. Each lymphocyte generates a unique antigen receptor by rearrangement of multiple gene segments. Activated cells undergo selection and clonal expansion for four to five days in order to produce large quantities of these antigen-specific receptors. Binding events by these molecules activate downstream humoral and cellular effectors, many of which are components of the innate defenses. The adaptive immune system is thus dependent on the same effector mechanisms as innate immunity but imparts activation and antigen specificity to the process. Adaptive immunity also bestows immunological memory to a host, providing enhanced protection against reinfection by the same pathogen.

Host defense by innate immune mechanisms controls infection by pathogens with common molecular surface motifs and by those that induce interferons and nonspecific defenses. But innate defenses do not have the ability to recognize and fight a broad range of pathogens. Prior to infection, viruses are rarely recognizable by macrophages as the viruses do not have many invariant structures. Other pathogenic organisms can produce a capsule that conceals the otherwise susceptible surface molecules to avoid detection. Significantly, innate defenses are also unable to elicit long-lasting memory of specific pathogens. This ability is limited to the adaptive immune processes found exclusively in vertebrates. Nevertheless, the large number of invertebrate species that exist on earth (much greater than vertebrates) attests to the success of the innate immune system.

The beneficial features unique to the adaptive immune responses include, great diversity for antigen recognition, specificity for a particular antigen or pathogen, and the ability to confer immunological memory after an encounter. However, adaptive immune responses do not become activated until approximately three days postinfection and the production of their primary components is delayed by five days. During this delay, the innate immune system limits the proliferation and spread of pathogens until the mechanisms of adaptive immunity achieve maximum potential. In the absence of innate immunity, acquired immune responses are inadequate for host protection. In contrast, the potency and pathogenicity of antibody-mediated effector functions can be seen in autoimmune diseases, allergic responses, and allograft rejection, which all result from the misregulation of adaptive immunity. This inopportune recognition of nonpathogenic material or self is generally confined to the adaptive immune response; autoreactivity has not been observed in organisms that accommodate only the components of innate immunity.

Immunological memory, preserved in the form of long-lived cells that persist in tissues and bone marrow, can last for months, years, or a lifetime and accounts for the efficacy of vaccines. These memory cells are readily activated by a subsequent encounter with antigen and can rapidly differentiate into effector cells. Unfortunately, this memory is confined to the individual challenged by the pathogen. Apart from passive immunization, such as transfer of antibody from mother to child, there is little carryover to future generations.

Despite the limitations of the individual systems, the benefits of incorporating two systems for immune defense are obvious. The strength of the adaptive immune response lies in its great diversity and specificity, providing complexity to the innate systems in the ability to recognize exquisite details of pathogenic organisms. Meanwhile, the innate defenses allow rapid discrimination of self from nonself and delay the need for adaptive immunity during the time required to generate effector lymphocytes. Chemokine secretion recruits leukocytes to appropriate sites, and cytokine-mediated signals provide optimal interplay between the divisions. In this manner, the innate immune system can direct the adaptive immune system in combating the infection. Thus, the synergy between the innate and adaptive mechanisms creates successful host defenses in vertebrates.

## 1.2 Constitutive and Inducible Defenses

The components of innate immune systems, found in most multicellular organisms, can be divided conceptually by function into the afferent and efferent limbs (Table 1). These defense mechanisms may also be discussed in terms of the point at which they are active during the course of infection (Table 2). The constitutive defense mechanisms include physical, mechanical, and chemical barriers that prevent initial invasion

**Tab. 1** Afferent and efferent components of the innate immune system.

| | *Afferent function* | *Efferent function* |
|---|---|---|
| Humoral components | Soluble innate immune receptors such as LPS-binding protein, CD14, collectins, properdin, complement component C3b, and C-reactive protein (pentraxin) | Secreted and cytosolic cytokines, antimicrobial peptides, enzymes, late components of complement, lactoferrin, and acute-phase proteins |
| Cellular components | Surface-bound receptors including toll-like receptors, MSR, MMR, and NOD proteins | Cellular enzymes, cell adhesion molecules, reactive oxygen species, and other antimicrobial substances |

**Tab. 2** Constitutive and inducible innate defense mechanisms.

| *Component* | *Location* | *Function* |
|---|---|---|
| *Physical factors* | | |
| Intact skin | External | Physical barrier to the entrance of microbes. |
| Mucous membranes | Internal cavities | Physical barrier to the entrance of microbes |
| Mucus | Respiratory and digestive tracts | Trap microbes |
| Epiglottis | Respiratory tract | Prevent microbes and particles from entering trachea |
| Hair | Nose | Filter microbes and dust |
| *Mechanical factors* | | |
| Cilia | Upper respiratory tract | Trap and remove microbes and dust |
| Tears | Eyes | Dilute and remove irritating substances and microbes |
| Saliva | Oral cavity | Wash microbes from mucosal surfaces |
| Urine | Urogenital tract | Wash microbes from urethra |
| *Chemical factors* | | |
| Gastric juice | Digestive tract | Destroy bacteria and toxins in stomach |
| Acid pH | External | Discourage growth of microbes |
| Unsaturated fatty acids | Various | Antibacterial in sebum |
| *Other defenses* | | |
| Lysozyme | Various | Antimicrobial substance in perspiration, tears, saliva, nasal secretions, and tissue fluids. |
| Interferon (IFN) | Circulation | Protect host cells from viral infection |
| Complement | Circulation | Lyse microbes, promote phagocytosis, contribute to inflammation (leukocyte recruitment) |
| Phagocytosis | Peripheral tissues and circulation | Ingestion and destruction of foreign particles |
| Inflammation | Peripheral tissues | Confine and destroy microbes, repair tissues |
| Fever | General | Inhibit microbial growth, aid repair |

by potential pathogens. Rigid junctions between epithelial cells effectively separate the interior milieu from the external environment. Production of a glycoprotein-rich mucus layer prevents adhesion of the microorganism to the epithelium, and movement of hairlike cilia aids in foreign-particle clearance. Antimicrobial enzymes and peptides (see Sect. 5.2) are also continuously expressed at mucosal epithelia, which have frequent exposure to microbes. The destructive capability of these substances affects microorganisms but is not directed at host cells and tissues. This distinction between self and nonself is accomplished by compartmentalization, or specific activity against microbes (e.g. susceptibility of bacterial cell walls, but not host cell membranes, to lysozyme and defensin). Yet another mechanism of defense is the colonization of epithelial surfaces with commensal microbes. These normally nonpathogenic bacteria compete with harmful microorganisms for nutrients and sites of attachment. Some bacteria even produce their own antimicrobial substances, or bacteriocins, to help outcompete other species.

If a focus of infection is established, inducible host defense mechanisms including local inflammatory responses are initiated at sites of infection. Heat, pain, redness, and swelling associated with inflammation are due to increased vascular diameter, increased endothelial adhesion, and increased vascular permeability mediated by cytokines. Pathogens that have penetrated host tissue are recognized by macrophages, which can phagocytose the offending microbes. Macrophages may also release cytokines, chemokines, and inflammatory mediators, which include tumor necrosis factor-$\alpha$ (TNF-$\alpha$), prostaglandins, leukotrienes, and platelet activating factor. Another set of inflammatory mediators is produced via the activation of the complement cascade (discussed in Sec. 5.1), which also occurs in response to pathogen binding. These molecules can stimulate an acute-phase response, activate natural killer (NK) cells, and trigger interferon production and nonspecific defenses by regulating transcriptional activation. Cytokine production is also responsible for instructive cross talk to antigen-presenting cells and adaptive immune mechanisms.

## 2
## Molecular Basis of Innate Immunity

### 2.1
### Evolution

The innate immune system has its origins in the bilateria approximately one billion years ago. Today, nearly all multicellular organisms rely on derivative innate strategies for protection from infection. For example, mammals utilize skin as a barrier to pathogen invasion, whereas insects have an exoskeleton that performs the same function. Following this first tier of immune defense, specialized immune cells and tissues are induced to contain infections. White blood cells and liver of mammals function analogously to the hemocytes and fat body of certain insects in this respect. Diverse organisms have also evolved similar mechanisms for the recognition of infection and for swift responses to infection. Animals and plants alike have acquired the ability to recognize invariant patterns associated with pathogenic microbial organisms. In plants, these patterns are termed *general elicitors of plant defense*, while in animals they are designated *pathogen-associated molecular patterns*. Amazingly, plant receptors recognize patterns similar to those reported to activate innate defense mechanisms in mammals and insects, and some of the specific receptors and signaling molecules used appear to share homology. Genomic evidence also exists for common pathways for transcriptional activation in plants, vertebrates, and invertebrates. But while *Drosophila* produce many different antimicrobial peptides to combat fungal and bacterial infection, mammalian systems produce fewer of these types of peptides and rely more on cytokine-mediated inflammatory responses. It has not been conclusively determined whether similarities between diverse animal and plant defense pathways are due to conservation through a billion years of evolution or whether the systems converged independently.

The mechanisms of adaptive immunity are thought to have developed much later on the evolutionary timeline in the context of a functioning innate immune system. The genes of the TCR and antibody first appeared in the cartilaginous fish, such as shark, approximately 400 million years ago. The origin of gene rearrangement and junctional diversity in immune receptors is thought to be an invasion by a retrotransposon. The recombinase-activating genes (RAGs), required to generate diversity in the receptors, are encoded in opposite orientations in the genome, and lack introns, similar to the bacterial transposases. Additionally, recognition signal sequences (RSSs) are present at the 3′-end of the V-gene segments and 5′-end of the J-gene segments. These properties support the notion of the integration of RAGs and RSSs into a gene for an ancient antigen receptor.

## 2.2
## Affinity and Specificity

The innate and adaptive arms of the immune system differ in their modes of recognition. While antibodies are known for their exquisite specificity for antigens, the innate immune system recognizes a few highly conserved structures found in a large variety of microorganisms. The molecules targeted are vital components of microbes, so they are not often altered by mutation. Central to our understanding of the molecular basis of innate immune recognition are investigations of the receptor complexes that recognize these molecules. Binding of these receptors mediates ingestion of pathogens, allows cells to recognize chemotactic substances and migrate toward sites of infection, and induces effector molecules that regulate further immune responses. The affinities of innate immune receptors for their targets can run in the nanomolar range, for example, soluble CD14 binds Lipopolysaccharide (LPS) from a *Salmonella* strain with dissociation constants of $4.1 \times 10^{-8}$ M and $7.1 \times 10^{-9}$ M, in the absence and in the presence of LPS-binding protein (LBP) respectively. In comparison, affinity constants for antibody–ligand binding usually range from $10^{-8}$ M to $10^{-10}$ M, and TCR MHC–antigen affinities run in the range of $10^{-4}$ M to $10^{-5}$ M.

Importantly, innate receptors can distinguish nonself from self. While the receptors of T- and B lymphocytes allow great variability in adaptive immune recognition, they cannot determine the nature of the antigen. Although usually rare, adaptive immune responses to nonpathogen antigens can potentiate autoimmunity, allergy, or tissue-graft rejection.

## 3
## Innate Immune Recognition

### 3.1
### Pathogen-associated Molecular Patterns

Pathogen-associated molecular patterns (PAMPs) are the main targets of innate immune recognition. PAMPs are usually formed by constitutively expressed structures essential to the survival of a microbe and are often shared among members of large groups of microorganisms. Significantly, PAMPs are motifs that are not found on host cells. Host immune systems have evolved to take advantage of the following three characteristics. First, constitutive expression of the PAMP allows constant recognition regardless of pathogen life cycle, and because the structure is indispensable, it is unlikely to acquire mutations. Second, the host can employ a limited number of germ-line-encoded receptors to recognize a very broad range of microorganisms and can customize its responses to them because the target molecules are class specific. And third, the presence of a PAMP allows the host to distinguish itself from the foreign microbe. Lipopolysaccharide, peptidoglycan, and lipoteichoic acids (LTA) are examples of PAMPs that are fundamental components of bacterial cell walls and are not present on eukaryotic cells. Other PAMPs include the CpG motif present in bacterial DNA, components of flagella, and viral components. The detection of these PAMPs provides some information on the identity of the pathogen.

Lipopolysaccharide is the principal outer-membrane constituent of gram-negative bacteria. It is an amphipathic molecule consisting of a hydrophilic polysaccharide core and O-antigen and a hydrophobic glycolipid anchor termed *lipid A* (Fig. 1). The core region contains Kdo (3-deoxy-D-*manno*-octulopyranosonic acid) and heptose-sugar residues, while O-antigen is composed of a sequence of repeating units of oligosaccharide, which varies between bacterial species or strain. Lipid A is composed of a phosphorylated diglucosamine carrying acyl residues and is the primary immunostimulatory component of LPS. The primary effects of LPS were first reported over a century ago, but properties of this molecule continue to be studied today. LPS, the source of bacterial endotoxin and lipid A, is, to a degree, responsible for systemic inflammatory response syndrome (SIRS) and septic shock associated with gram-negative bacteremia. LPS stimulates the secretion of cytokines from leukocytes, namely, TNF-$\alpha$, interleukin (IL)-1, and IL-6, and causes changes in body temperature, heart rate, respiratory

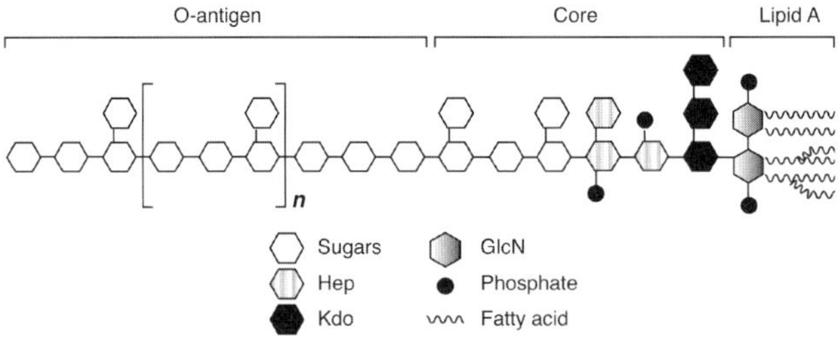


**Fig. 1** General structure of LPS.
Kdo = 3-deoxy-D-manno-octulopyranosonic acid; GlcN = *N*-acetylglucosamine; Hep = heptose.

function, and leukocyte circulation. The resultant hypotension, systemic organ failure, and other significant complications lead to hundreds of thousands of deaths per year from this syndrome. However, most LPS-activated responses resolve infections quickly by initiating killing of the microbe.

Peptidoglycan and LTA may be the causative reagents in gram-positive bacterial septic shock. Peptidoglycan is a network of carbohydrate polymers covalently cross-linked by proteins, which composes the structural core of the gram-positive bacterial cell wall. Linear macromolecules are made of alternating *N*-acetylglucosamine (GlcNAc) and *N*-acetylmuramic acid (MurNAc) residues joined by $\beta$ – (1–4) linkages. Tetrapeptide side chains attached to MurNAc form covalent bonds with adjacent polymers directly through its side chains or through peptide interbridges. Teichoic acids, also found in the cell wall of gram-positive organisms, are polymers of glycerol and ribitol, linked by phosphodiester groups. Teichoic acids are connected to either the peptidoglycan itself or to plasma membrane lipids (LTA). The biological functions of LTA are known to include binding of metal cations. The LTA structure consists of two fatty acid tails and a long charged hydrophilic chain. The classic *Staphylococcal* LTA is an unbranched 1,3-linked glycerophosphate chain bound covalently to a membrane diacylglyceroglycolipid. LTAs are structurally distinct between bacterial species, varying in fatty acid composition, chain length, glycosyl substituents, glycolipid structures, spatial distributions, and charge densities. One or more of these features may explain the variable immunological properties of LTA from different sources. The fatty acid-containing lipid anchor and additional structural components on the polyglycerophosphate backbone contribute to the immunostimulatory activity of LTA. Recent studies have suggested a key role of D-alanine substituents in inducing cytokine secretion. Although the effects of LTA alone are not particularly severe, it can act synergistically with peptidoglycan, leading to systemic inflammation and multiple systems failure via induction of TNF-$\alpha$ and IFN$\gamma$.

### 3.2 Pattern-recognition Receptors

Pattern-recognition receptors (PRRs) detect conserved structures on microorganisms during innate immune defense. PRRs are structurally and functionally heterogeneous. They may contain distinct motifs such as C-type lectin domains, scavenger receptor cysteine-rich domains, and leucine-rich repeat (LRR) domains. The receptors may be secreted, or expressed on the cell surface, within intracellular compartments or in the cytoplasm. PRRs function through opsonization, complement activation, proinflammatory signaling pathways, phagocytosis, and the induction of apoptosis.

Secreted recognition molecules include the proteins synthesized in the liver during the acute-phase response, such as mannan-binding lectin (MBL), C-reactive protein (CRP), and serum amyloid P component (SAP). MBL is a collectin (contains both collagen-like and C-type lectin domains) that binds specifically to the carbohydrate arrays accessible on many microorganisms. MBL–ligand binding may be inhibited by mannose, fucose, *N*-acetylglucosamine, and DNA, but avidity is greatest for the terminal mannose residues typically present on microbial surfaces. By comparison, on vertebrate cells, such carbohydrate patterns are usually masked by sialic acid residues, so for the most part, MBL recognizes only pathogens. MBL can effectively neutralize bacteria by binding or can opsonize bacteria and activate complement via an associated serum protease, MASP-2, initiating the lectin pathway by cleaving complement proteins C2 and C4. CRP and SAP are pentraxins that opsonize bacterial surfaces containing phosphorylcholine (PC). Both proteins display calcium-ion-dependent binding, but SAP has a slightly lower avidity for PC than does CRP. The distribution of PC in polysaccharides of foreign molecules and damaged host membranes allows CRP and SAP to recognize pathogens and altered self. CRP is found in trace amounts in human serum, but its concentration can elevate dramatically in response to inflammation and tissue necrosis. Upregulation occurs at the transcriptional level in response to cytokines IL-1$\beta$ and IL-6. Multivalent binding of CRP activates complement by initiating assembly of a C3 convertase. Additionally, the protein interacts with Fc$\gamma$RI receptors, mediating phagocytosis and leading to the generation of proinflammatory cytokines. The topology and chemical composition of the CRP ligand-binding site has been determined by crystallography. SAP shares many features with CRP, including remarkable structural similarity, 59% sequence homology, and the ability to activate complement. Plasma LBP mediates the binding of minute amounts of LPS to glycosylphosphatidylinositol-anchored membrane CD14, triggering a proinflammatory response in macrophages, which is crucial to control infection by gram-negative pathogens. The CD14 LPS receptor complex may also involve molecules of TLR-4 and MD-2.

Several cell surface receptors function as PRRs to facilitate direct phagocytosis of microorganisms. The macrophage mannose receptor (MMR) is a C-type lectin found on macrophages but not on monocytes or neutrophils. The protein contains eight carbohydrate-recognition domains, the fourth of which is central to ligand binding, and like MBL, it can bind mannose, fucose, and *N*-acetylglucosamine. MMR is closely related to the dendritic cell PRR, DEC205, which is also a mediator of phagocytosis. Macrophage scavenger receptors (MSRs) bind a variety of charged ligands such as double-stranded RNA (dsRNA), LPS, and LTA, coordinating responses to a variety of viral and bacterial pathogens. MSRs also participate in lipid homeostasis by endocytosing acetylated and oxidized low-density lipoprotein. PRRs for a broad range of ligands have also been identified in invertebrates. These include the broad range $\beta$2integrin CR3, peptidoglycan recognition protein (PGRP), and gram-negative binding protein (GNBP), which have counterparts in the mammalian Toll-like receptors (TLRs).

Recognition systems also exist in the cytosol to prevent replication of intracellular pathogens. Activation of dsRNA-dependent protein kinase (PKR) inhibits initiation of viral and cellular protein synthesis by phosphorylating the translation initiation factor eIF2$\alpha$. Interferon-inducible 2′,5′-oligodenylate synthase (OAS) also binds viral dsRNA within cells. Resulting synthesis of ppp(A2'p)nA oligoadenylates activate the RNaseL endonuclease, which cleaves RNA. This OAS/RNaseL pathway is particularly important in the antiviral action against picornaviruses. Members of the NOD (nucleotide-binding oligomerization domain) family are proteins containing a CARD, NOD, and LRR region. After PAMP binding, protein kinase RIP2 associates with CARD domains and activates MAP kinase and NFκB signaling pathways. NOD proteins have been shown to bind LPS, but the full extent of ligand specificity for these molecules has not been elucidated. Downstream transcriptional activation by PKR (in TLR signaling) and NOD induces IFN genes.

## 3.3
## Toll-like Receptors

Multiple innate immune receptors are able to distinguish nonself from self, as well as recognize individual classes of pathogens. Among diverse organisms, there exist genetic and structural homologies between the protein receptors and similarities in the pathways of immune-mechanism activation. The identification of the mammalian Toll-like receptor (TLR) family, homologs of the *Drosophila* Toll (Toll-1) protein, provided the primary evidence for evolutionary conservation within eukaryotic immunity. Both Toll and the TLRs are single-transmembrane spanning receptors possessing a variable number of extracellular amino-terminal LRR motifs, followed by a cysteine-rich domain. At the carboxy-end of the Toll/TLR transmembrane domain is an intracellular signaling domain, homologous to that of the cytoplasmic interleukin-1 receptor (IL-1R). It is therefore called the Toll/IL-1R (TLR) domain (Fig. 2).

The original function identified for Toll was the development of dorsoventral polarity in the fruit-fly embryo. It has since been discovered that Toll is required to combat systemic fungal infection in *Drosophila* by inducing the expression of drosomycin, an antifungal peptide. In addition to Toll, there are at least eight related homologs (Tolls-2–9), which appear to have functions in *Drosophila* development.

```
            AB     βB            BB             αB
Mouse
IL-1RAc   G Y K L C I F D R D S L P G G I V T D E T L S - F I Q K S
IL-1R1    G Y K L F I Y G R D D Y V G E D T I E V T N E - N V K K S

Drosophila
Toll      K F Q L C V H E R D W L V G G H I P E N I M R - S V A D S
18W       P F R L C I Q Q R D L P P Q A S H L Q - L V E - G A R A S
Toll-5    P F R L C L H D R D W L V G D C I P E Q I V R - T V D D S
Toll-9    D V S I C L H E R D F Q I G V T I L D N I I S - C M D R S

Human
IL-1R1    G Y K L F I Y G R D D Y V G E D I V E V I N E - N V K K S
TLR1      G M Q I C L H E R N F V P G K S I V E N I I T - C I E K S
TLR2      P F K L C L H K R D F I P G K W I I D N I I D - S I E K S
TLR3      S L K F C L E E R D F E A G V F E L E A I V N - S I K R S
TLR4      P F Q L C L H Y R D F I P G V A I A A N I I H E G F H K S
TLR5      R F N L C F E E R D F V P G E N R I A N I Q D - A I W N S
TLR6      D I Q I C L H E R N F V P G K S I V E N I I N - C I E K S
TLR7      H F N L C L E E R D W L P G Q P V L E N L S Q - S I Q L S
TLR8      N V L L C L E E R D W D P G L A I I D N L M Q - S I N Q S
TLR9      A L R L C L E E R D W L P G K T L F E N L W A - S V Y G S
```

**Fig. 2** Sequence alignment of the TLR domain-signaling region of mouse and human IL-1R proteins, *Drosophila* Toll proteins, and human TLRs. Residues conserved in at least four sequences are shaded. The greatest homology between *Drosophila* and human proteins occurs between the cytoplasmic domains of Toll-9 and TLRs 1, 2, 4, and 6 (see color plate p. xiii).

Of these proteins, Toll-2 (or 18 Wheeler, 18 W) and Toll-5 have been found to have involvement in immune responses. Mutagenesis has shown that 18 W facilitates the recognition of gram-negative bacterial pathogens and the subsequent production of attacin, an anti-gram-negative bacterial peptide. Likewise, it has been suggested that Toll-5 acts as a pattern-recognition receptor mediating local antifungal responses in the absence of Toll-1/Spätzle signaling. These Toll family members respond to different pathogen classes and induce appropriate effector responses, presumably on the basis of PAMP recognition, that is, antibacterial and antifungal peptides are differentially expressed in response to the presence of bacteria or fungi. Significantly, these findings demonstrated that an organism lacking an adaptive immune system can detect infection as well as determine the type of pathogen present, entirely on the basis of innate immune mechanisms.

Mammalian TLRs may also be able to distinguish between pathogen classes. The TLRs function as PRRs in mammals, binding to microbial products to activate host defense responses. To date, 11 identified mammalian TLRs have been described, differing in sites of expression, ligand specificity, and downstream effector activation. Ten TLR genes (TLRs 1–10) are dispersed throughout the human genome, whereas mice encode nine of these, plus two additional paralogs not present in humans. Surface expression is usually low in monocytes ($\sim 10^3$ per cell) and DCs ($\sim 10^2$ per cell) but may be enhanced in response to various stimuli. TLR expression is also observed in adipocytes, myocytes, vascular endothelial cells, and intestinal epithelial cells. TLR4, the first mammalian TLR identified, when activated induces the expression of inflammatory cytokines and costimulatory molecules in macrophages. The murine TLR4 protein responds to the class of gram-negative bacteria. The molecular signature detected in this case is LPS. Although other PRRs were previously known to initiate signals in response to LPS, the existence of an additional coreceptor was hypothesized. The cellular response to LPS occurs through interaction with a circulating LBP and a GPI-linked surface receptor, CD14, followed by activation of TLR4. Point mutations in the TLR4 TIR domain led to defective LPS signal transduction, as did the absence or deletion of TLR4. These results demonstrated the necessity for a functional TLR4 in the recognition of this essential component of gram-negative bacteria. TLR4 has also been shown to mediate signals initiated by other endogenous and exogenous compounds (e.g. Taxol), respiratory syncytial virus protein F, heat shock protein 60, and domain A of fibronectin. Other conserved PAMPs have been shown to signal immune responses via other members of the TLR family, such as peptidoglycan by TLR2 and flagellin by TLR5. It seems likely that elucidation of newly discovered mammalian TLR homologs will elucidate the specificity of one or more of the homologs for mannan or glucan, and of fungal and mycobacterial pathogens for lipoarabinomannan or muramyldipeptide PAMPs, respectively.

Structural variation among TLRs is mainly due to differences in the numbers of LRRs, which affect the proteins' sequence length and molecular weight (Fig. 3). LRRs are found in a wide array of molecules, including proteoglycans, adhesion molecules, enzymes, tyrosine kinase receptors, and G-protein-coupled receptors. Toll and 18 Wheeler have 22 and 31 LRR ectodomains respectively, comparable to the 18, 19, 24, and 22 LRR-containing

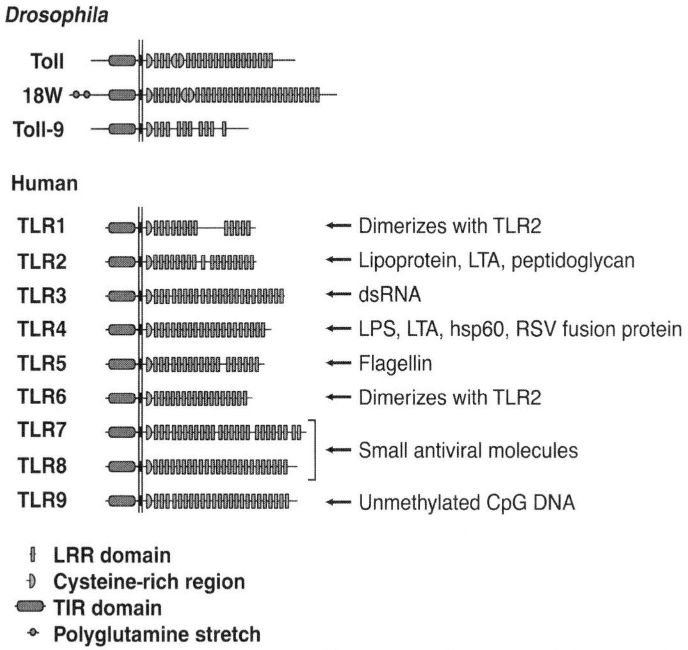


**Fig. 3** Schematic representation of human TLR domains and their stimuli. Adapted from Imler, J.-L., Hoffmann, J.A. (2001) Toll receptors in innate immunity, *Trends Cell Biol.* **11**, 304–311. (See color plate p. xiv)

TLRs 1–4. The LRRs are terminated by a domain of cysteine residues; however, the human TLR chains have lost an approximately 90-residue cysteine-rich region found variably in the *Drosophila* proteins. The extracellular domain of 18 W contains 16 putative N-glycosylation sites and a single glycosaminoglycan site. The cysteine clusters allow the receptors to form a compact module with paired termini that can be inserted between any pair of LRRs without altering the overall fold of the extracellular region. Two hydrophobic regions are apparent in the protein structure of 18 W. One corresponds to a signal peptide and the second to a transmembrane domain. The intracellular moiety of 18 W is about 380-aa long and consists of two domains, the first of which shows significant homology to the cytoplasmic domain of IL-1R. Interleukin-1 receptor, Toll, and 18 W share six conserved stretches of amino acids. The second domain consists of a glutamine-rich repeat.

Upon ligand binding, TLRs activate expression of host defense genes. Some of the intracellular signaling pathways induced are similar to those employed by IL-1R (Fig. 4). These pathways are coupled to the nuclear translocation of Rel-type transcription factors. An adaptor, such as MyD88, is first recruited to the receptor complex, mediating an interaction with the receptor via its amino-terminal TIR domain and associating with protein kinase IRAK1 through a carboxy-terminal death domain. IRAK1 undergoes autophosphorylation, which is thought to be followed by IKK-$\beta$ activation via TRAF6 and by degradation of I$\kappa$B. As a result, NF$\kappa$B may translocate into the nucleus

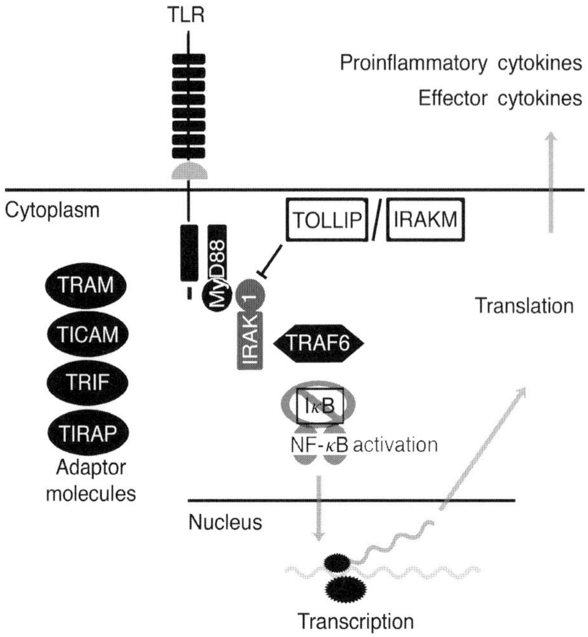


**Fig. 4** MyD88-dependent and -independent TIR domain signaling. The MyD88-dependent pathway is common to all TLRs. Upon binding, TLRs induce the recruitment of MyD88 via its TIR domain, which activates IRAK1 by phosphorylation. IRAK1 then associates temporarily with TRAF6, eliciting downstream signaling, NFκB activation, and resultant production of proinflammatory cytokines such as TNF-$\alpha$, IL-1, and IL-12. IRAK1 activity is negatively regulated by Tollip and IRAK-M. Some differences between signaling pathways induced by individual TLRs may be attributed to distinct adaptor protein usage. TLR4 and TLR2 MyD88-dependent signaling requires the adaptor TIRAP/Mal. IFN-$\beta$ production in response to double-stranded RNA recognition by TLR3 is mediated by TRIF/TICAM-1 in a MyD88-independent manner. TRAM/TICAM2 is another adaptor pertinent to the TLR4 MyD88-independent pathway.

and allow transcription of appropriate targets. Although some TLRs are dependent on this conserved Toll signaling pathway (TLR2 and TLR9), MyD88 is not required for other TLR signaling events, such as the MyD88-independent pathway associated with the stimulation of IFN-$\beta$ and the maturation of DCs. Additional regulatory mechanisms and downstream signals must also be in place to effect responses appropriate to the initiating stimulus. In particular, TLR4 activates transcription of NF-kappaB- and interferon regulatory factor (IRF)-3/IFN-inducible genes in macrophages and dendritic cells in a MyD88-independent manner, potentially by another TIR domain-containing adaptor protein, TIRAP. Data has suggested that TLRs 1, 2, and 6 can also use this alternative adaptor molecule. A third TIR-containing adaptor molecule (TICAM-1), which binds TLR3, inducing interferon-$\beta$

has also been described. Suppression of signaling may also occur by association of either of the proteins Tollip or IRAK-M with TRAF6. TLRs have been found to heterodimerize with one another, which may also contribute to the repertoire of specificities. It remains to be determined whether differences in adaptor usage and involvement of suppressor molecules can account for the activation of distinct immune responses among the TLRs. It is plausible other unidentified adaptors play a role in TLR signaling or that combinations of adaptor usage provide more distinct signals than those involving individual TIR proteins.

## 4
## Cellular Effector Mechanisms

### 4.1
### Cell-mediated Inflammation

Myeloid cells are critical to the defenses of the innate immune system. These include the monocyte-derived macrophages and dendritic cells and the granulocytic polymorphonuclear neutrophils (PMN), basophils, and eosinophils. These are the first cells to counter infectious processes by internalizing and degrading foreign material. The lymphoid NK cells also play a role in innate immune responses to viral infection.

Macrophages are most concentrated in lymphoreticular organs (the lymph nodes and spleen) but are present throughout the body, forming morphologically distinct populations in different tissues. Tissue histiocytes, osteoclasts, hepatic Küpffer cells, as well as microglial cells of the central nervous system are all considered macrophages. These phagocytic cells survey the blood and connective tissue for potential pathogens. Macrophages are usually the first cell type to encounter pathogen in tissues. Upon encountering a microbe, the cells can engulf and kill it directly, in the process of phagocytosis, or can recruit other myeloid cells, especially PMNs, to the sites of damage or infection. Cytokines secreted by macrophages, including IL-6, 8, and 12 and TNF-$\alpha$, are all inducers of fever and inflammation associated with an acute-phase response. Secretion of TNF-$\alpha$ and IL-1 alter expression of vascular adhesion molecules such as E-selectin and VCAM-1. The leukocyte CD18 complex integrins allow adhesion to the vascular endothelium. These include CD11a/CD18, LFA-1 and CD11b/CD18, or Mac-1. Interactions mediated by these molecules between the leukocytes and vessel walls allow firm adhesion and diapedesis. Activation of the macrophage by IFN$\gamma$ and TNF-$\alpha$ from NK cells allow it to kill most intracellular pathogens. Macrophages are also able to present antigens to T-lymphocytes to activate adaptive immune responses and to modulate the type of response by the production of other cytokines. These include secretion of IL-12 for Th1 and IL-10 for Th2 responses. Consequently, macrophages provide one link between innate and specific immunity.

Dendritic cells are specialized phagocytes that are particularly efficient at presenting antigen to T-lymphocytes. Dendritic cells regularly ingest bits of their immediate environment by endocytosis and macropinocytosis. When a pathogen is detected, TLR stimulation and signaling initiates maturation, enabling the DCs to produce costimulatory molecules and enhancing the capacity to present antigen. MHC molecule expression is upregulated and cells proceed to the lymph nodes for T-cell priming. In this way, DCs are poised for activation of adaptive immune responses. Like the macrophages, DCs take on multiple identities, such as the Langerhans cells of the skin and plasmacytoid cells of the spleen. However, DCs are less numerous than the macrophage.

Polymorphonuclear neutrophils are the most abundant white blood cells in circulation that play a critical role in the destruction of circulating pathogens. These cells are not normally found in tissues but will migrate in response to the release of chemotactic factors by injured tissues or in response to the activation of other cells. At these inflammatory sites, PMNs may become activated and produce large amounts of reactive oxygen species, and release a variety of other toxic products from cytosolic granules. These cells have a short lifetime, spending approximately six hours in circulation, before undergoing apoptosis.

Eosinophils are responsive to cytokines, and themselves produce a variety of chemokines that mediate inflammation. Under normal conditions, these cell types are found in limited numbers, but counts are elevated during parasitic infection and during an allergic adaptive immune response. Most reside in the subepithelium of respiratory, digestive, and urinary tracts. Mast cells reside in nearly all tissues of the body and play major roles in the defense against parasites and in allergic inflammation. Basophils are very similar to mast cells, except that they reside in blood. These cells are distinguishable by large cytoplasmic granules and high-affinity IgE receptors. Degranulation of mast cells and basophils releases histamine, other vasoactive mediators, and certain proinflammatory cytokines, for example, TNF-$\alpha$ and IL-5. These compounds trigger vasodilation and increase vascular permeability, mucus secretion, bronchiolar smooth muscle constriction, leukocyte recruitment, platelet activation, tissue destruction, fever, and pain.

Natural Killer cells migrate in response to chemotactic factors and kill virally infected cells and tumor cells. Binding to target cells is followed by release of perforins into the cell membrane and lysis or stimulation of programmed cell death.

## 4.2 Oxygen-independent Destruction of Pathogens

Phagocytosis of pathogens occurs following ligation of certain surface receptors on the cell (Fig. 5). Macrophages and neutrophils are highly efficient at phagocytosis, utilizing the external receptors FcR, complement receptors, mannose receptors, and TLRs to mediate binding to foreign or opsonized material. It is an active process involving actin and microtubule restructuring, which surrounds the pathogen with the phagocyte membrane and internalizes it within a phagosome. Intracellular killing takes place when this vesicle is acidified and introduced to lysosomal granules in the macrophage or to azurophil granules in neutrophils, which contain toxic enzymes, proteins, and peptides. Lysosomal proteases and glycolases will disturb parasitic, bacterial, and viral membrane function. Lysozyme does so by cleaving the $\beta-(1-4)$ glycosidic bond within the peptidoglycan of gram-positive bacterial cell walls. Insertion of defensins into pathogen membranes also disrupts membrane permeability. Acid hydrolases and neutral proteinases digest multiple components of bacteria. The antibacterials transferrin and lactoferrin sequesters free iron, thereby preventing the growth of ingested microorganisms.

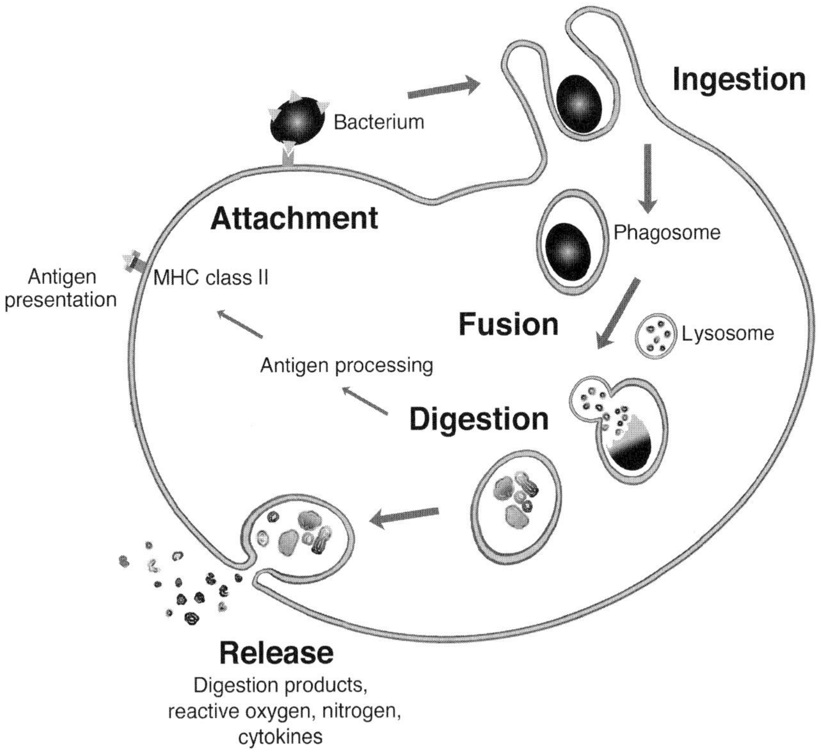


**Fig. 5** Processes of phagocytosis. Attachment of the bacterium via FcRs, CRs, TLRs, and other receptors is followed by ingestion of the bacterium, forming a phagosome. Acidification of the phagosome and fusion with the lysosome allows toxic products to kill microbes and proteolytic enzymes to digest the ingested material. Release of digestion products from the cell is accompanied by antigen presentation and release of effectors.

While azurophil granules function predominantly intracellularly within the phagolysosomal vacuole, neutrophil-specific granules are prone to release their contents to the extracellular milieu and appear to have a significant role in the initiation of inflammation.

### 4.3 Oxygen-dependent Killing

Phagocytes may also undergo a respiratory burst to help kill engulfed microorganisms. This involves a reduction and metabolism of oxygen by cellular enzymes to superoxide ($O_2^{\bullet -}$) and hydrogen peroxide ($H_2O_2$), ultimately forming toxic products such as hypohalous acid (HOCl/HOBr) and hydroxyl radical (•OH). Subunits of the enzyme NADPH oxidase assemble in the phagosomal and cellular membranes in response to neutrophil activation. This activation arises from the interaction between opsonized particles or pathogens and neutrophil receptors. The phagocyte oxidase transfers electrons from NADPH to oxygen, forming superoxide anion (Eq. 1). Superoxide may spontaneously dismutate or the enzyme superoxide dismutase may catalyze the formation of $H_2O_2$ and ground state $O_2$ (Eq. 2). Myeloperoxidase in neutrophil azurophilic granules catalyzes the formation of HOCl from $H_2O_2$ and chloride ions (Eq. 3). In the same manner, brominated species may also be formed, however, at much lower concentrations.

$$2O_2 + NADPH \longrightarrow 2O_2{\bullet}^- + NADP^+ + H^+ \quad (1)$$

$$2H^+ + 2O_2{\bullet}^- \longrightarrow 2H_2O_2 + O_2 \quad (2)$$

$$X^- + H_2O_2 + H^+ \longrightarrow HOX + H_2O \quad X = Cl, Br \quad (3)$$

Superoxide ion, hydrogen peroxide, and hypochlorite are all efficient oxidizing agents that can kill bacteria and fungi by oxidizing membrane lipids and proteins. Further reactions of these oxidants catalyzed by transition metals can produce •OH (Eq. 4). Hydroxyl radical is the most biologically reactive of the aforementioned oxidants. Singlet oxygen ($^1O_2$) is an excited state of dioxygen that debatably forms downstream of HOCl or •OH, which has deleterious effects on macromolecules (Eqs. 5 and 6).

$$O_2{\bullet}^- + HOCl \longrightarrow O_2{\bullet}^- + {\bullet}OH + Cl^- \quad (4)$$

$$OCl^- + H_2O_2 \longrightarrow {}^1O_2 + Cl^- + H_2O \quad (5)$$

$$O_2{\bullet}^- + {\bullet}OH \longrightarrow {}^1O_2 + HO^- \quad (6)$$

Singlet oxygen is also thought to be formed during the dismutation of peroxyl radical and during the decomposition of 1,2-hydroperoxides. The presence of $^1O_2$ during the oxidative burst provides a means by which the proposed oxidation of water by antibodies that has been postulated leads to trioxygen species generation (Eq. 7).

$$x\,{}^1O_2 + yH_2O \longrightarrow [H_2O_3(y-1)H_2O] \longrightarrow O_3 + H_2O_2 + (x-1)O_2 \quad (7)$$

While this reaction has mainly been characterized as a function of antibodies, other immunoglobulin fold-containing proteins may also catalyze the process. Its contribution to innate defenses and immunity as a whole has not yet been fully elucidated. The reactive oxygen species produced by activated phagocytes act in concert to effect microbial killing. Reactive nitrogen species such as nitric oxide (NO) and peroxynitrite ($OONO^-$) are also important for immune defense in that they mediate intercellular signaling and are also toxic to bacteria.

## 5 Humoral Effector Mechanisms

### 5.1 Complement

The complement system is a component of innate immunity concerned with the lysis and clearance of foreign particles and host debris. Activation of this system involves binding and cleavage of multiple proteins and glycoproteins, both soluble and membrane bound. The formation of active enzyme complexes and deposition of proteolytic fragments in this process recruits phagocytes to sites of infection, prompts direct cell lysis, and opsonizes pathogens for phagocytosis. Induction of inflammatory mediators also occurs as a result of complement activation (Fig. 6).

Complement enzyme precursors are activated locally and initiate inflammatory events in response to specific stimuli such

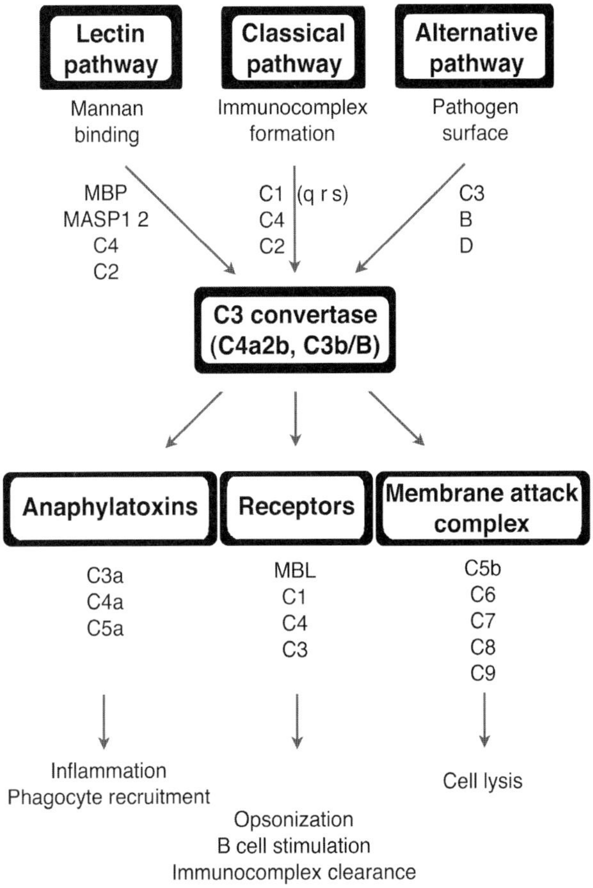


**Fig. 6** Complement pathways and proteins. Complement regulatory proteins not shown include C1INH, C4bp, CR1, MCP, DAF, H, I, P, and CD59.

as pathogen binding and immunocomplex formation. This may occur by three pathways. In the classical activation pathway, the pentameric protein complex C1, made up of a C1q subunit and two subunits each of C1r and C1s, binds directly to pathogens or to antibodies that are in complex with pathogens. In the latter case, C1 links adaptive immunity to the innate complement system. Proteolytic cleavage of C1r and C1s activates C1 to sequentially cleave other complement proteins (C4 then C2) to form an active C3 convertase (C4b2a). The mannan-binding lectin pathway uses a similar route to form an identical C3 convertase: serum PRRs bind mannose residues on bacterial or viral carbohydrates, forming an MBL/MASP complex, to activate complement proteins C4 and C2 and to form an active C3 convertase. Finally, alternative activation occurs when complement component C3 becomes spontaneously activated, for example, by microbial compounds LPS or

LTA, and gets deposited onto the surface of a pathogen. Properdin is a positive regulator of the alternative pathway of complement activation, released by peripheral blood cells and the endothelium. Binding and cleavage of plasma factor B produces a fluid-phase C3 convertase.

The C3 convertases cleave C3 to produce C3a, an anaphylactic by-product that mediates inflammatory responses, and C3b, which acts as an opsonin for phagocytes with complement receptors. C3b also activates a C5 convertase, thereby producing C5a, an important inflammatory mediator, and C5b, which initiates late events of complement. These entail polymerization of the terminal complement components, C6 to C9, to form porelike membrane attack complexes (MACs), leading to loss of bacterial cell integrity and lysis. Phagocyte ingestion of complement-opsonized pathogens is mediated by specific complement receptors. There are five known types of complement receptors. The receptor for C3 (CR1) is present on neutrophils and macrophages and can promote binding and phagocytosis.

The complement system is efficient in the defense against many viral and bacterial infections as well as in the removal of cellular debris. However, if improperly regulated, complement can cause severe host tissue damage, as in rheumatoid arthritis, acute hemolytic anemia, or glomerulonephritis. Other conditions including mechanical injury and myocardial infarction can stimulate inopportune complement-mediated damage as well.

## 5.2 Antimicrobial Peptides

Cationic antimicrobial peptides produced by vertebrates, arthropods, plants, insects, and bacteria are small stretches of less than 50 amino acids that possess activity against bacteria, fungi, and protozoa. Of the more than 800 antimicrobial peptides submitted in the Antimicrobial Sequences Database (http://www.bbcm.univ.trieste.it/~tossi/amsdb.html), over 175 are produced by insects. These peptides are particularly important in the immune defenses of the well-studied fruit fly, *Drosophila melanogaster*. Proteins include the antibacterials Andropin, Attacin, Cecropin, Diptericin (gram negative), Defensin (gram positive), and Drosocin; and the antifungals Drosomycin and Metchnikowin, among many others. The genes encoding most of these protein effectors are induced in response to infection and the peptides secreted from *Drosophila* fat bodies. Andropin is the exception to the rule in that it is constitutively expressed in males. In contrast, most mammalian antimicrobial proteins, including the cryptidins or $\alpha$-defensins present in the intestines and $\beta$-defensins produced in the skin and respiratory tract, are ubiquitously produced as part of innate constitutive defenses.

Two general mechanisms of action of these molecules have been indicated and segregated on the basis of being either membrane disruptive or nonmembrane disruptive. Initial associations of these peptides with microbes occur through electrostatic interactions, for example, between the peptide and negatively charged LPS in bacterial cell membranes. Destabilization of the outer membrane creates sites on the cytoplasmic membrane that are more susceptible to translocation or membrane disruption. It is postulated that transient pores are formed by the peptide that span the membrane, resulting in membrane depolarization and disintegration. Alternatively, peptides may form micellar aggregates in the outer leaflet, which collapse allowing translocation into the cytoplasm. Inside the bacterium, peptides interact with nucleic acids and cellular proteins, inducing transcriptional changes and inhibiting further synthesis of macromolecules.

The peptides are classified on the basis of secondary structure into four major classes: $\beta$-sheet, $\alpha$-helix, loop, or extended peptides (Fig. 7). The majority of known peptides belong to the first two groups, according to circular dichroism spectroscopy. High-resolution structures of these peptides, obtained by 2-d $^1$H NMR, exist for about 50 of these molecules. The $\alpha$-helical antimicrobials are characterized by an $\alpha$-helical conformation containing a slight bend in the center of the molecule. This bend may be crucial for suppressing undesirable hemolytic activity. Structure–activity studies of magainin 2, a typical $\alpha$-helical peptide, suggest no

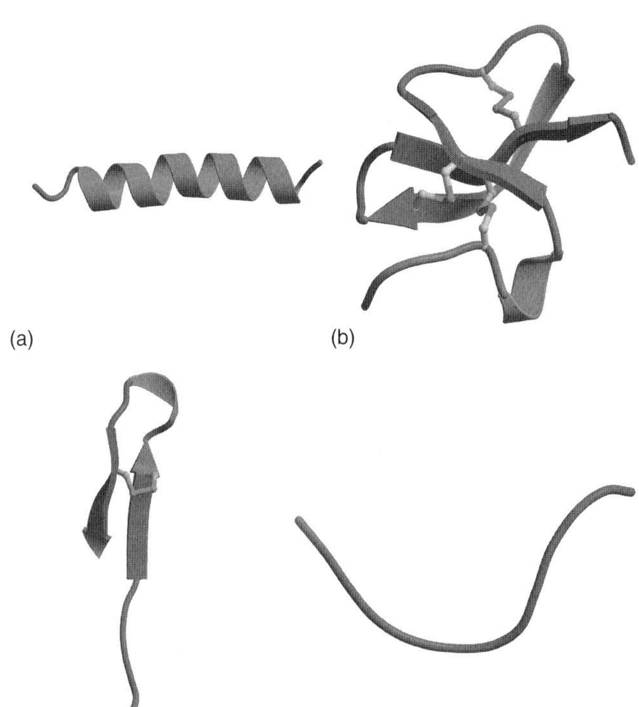


**Fig. 7** Antimicrobial peptides. Representative member of each structural class: (a) magainin 2, alpha-helix; (b) $\beta$-defensin, beta-sheet; (c) thanatin, loop; (d) indolicidin, extended. Pictures generated by Matt Kelker, using bobscript and raster3D. Coordinates obtained from the RCSB Protein Data Bank under the accession codes 1BNB (defensin), 1G89 (indolicidin), 2MAG (magainin), and 8TFV (thanatin). Raster3D: Merritt & Bacon (1997) *Meth. Enzymol.* **277**, 505–524. bobscript: Esnouf, RM (1997) *J. Mol. Graph. Model.* **132–134**, 112–113. (See color plate p. xv).

difference in D- and L-enantiomers, ruling out chiral receptors or enzyme targets as mechanisms of action. Furthermore, an increase in cationic charge, up to +5, leads to an increase in antimicrobial activity. This is presumably due to enhanced association with the negatively charged cell surface. Peptides containing antiparallel $\beta$-sheets, such as tachyplesin, disrupt lipid organization and undergo membrane translocation. Stabilizing disulfide bonds within the structure suggest that flexibility is pertinent for translocation to occur. The class of peptides with extended conformations are high in proline or glycine content and hence lack overall secondary structure. These extended peptides tend to form hydrogen bonds and Van der Waals interactions with membrane lipids. Evidence exists that extended peptides, such as indolicidin, translocate into the cytoplasm to inhibit DNA synthesis; however, the mechanism has not been explicitly determined. Finally, loop peptides such as the insect antimicrobial thanatin contain a single disulfide, amide, or isopeptide bond. The exact mechanism of killing by thanatin is unknown, but the molecule evidently

**Tab. 3** Effects of cytokines on the innate immune system.

| *Cytokine* | *Origin* | *Function* |
|---|---|---|
| IL-1$\alpha$ and -1$\beta$ | Monocytes, macrophages, epithelium, endothelium, fibroblasts | costimulation of APCs and T cells, inflammation, adhesion and fever, acute-phase response |
| IL-2 | NK cells, activated $TH_1$ cells | proliferation of B cells and activated T cells, NK functions |
| IL-4 | Mast cells, basophils $TH_2$ cells | B-cell proliferation, eosinophil and mast-cell growth and function, IgE and class-II MHC expression on B cells, inhibition of monokine production |
| IL-5 | $TH_2$ and mast cells | B-cell and eosinophil growth and function |
| IL-6 | Activated $TH_2$ cells, monocytes, macrophages, epithelium, endothelium | acute-phase response, B-cell proliferation, thrombopoiesis, synergistic with IL-1 and TNF-$\alpha$ |
| IL-8 | Monocytes, macrophages, epithelium | chemoattractant for neutrophils and T cells |
| IL-10 | Macrophages, activated $TH_2$ cells, CD8+ B cells | inhibits cytokine production (IL-1, -2 and TNF-$\alpha$), promotes B-cell proliferation and antibody production, suppresses cellular immunity, mast-cell growth |
| IL-12 | Monocytes, macrophages | NK and $TH_1$ cell activation |
| IFN-$\alpha$ and -$\beta$ | DCs, epithelium, fibroblasts | Antiviral, increases MHC I expression, suppresses proliferation |
| IFN-$\gamma$ | Macrophages, NK cells, activated $TH_1$ cells | Antiviral, increases MHC I &II expression, suppresses proliferation |
| TNF-$\alpha$ | Macrophages, NK cells, mast cells, epithelium | Phagocyte activation, fever, antitumor activity |
| G-CSF, M-CSF, GM-CSF | Various | Growth and maturation of PMNs and monocytes |

does not perturb membrane permeability. The antimicrobial mechanism appears to be target specific, as only L-thanatin is effective against gram-negative bacteria but both enantiomeric forms of the peptides possess activity against gram-positive and fungal species.

Cationic charge and amphipathic conformation appear to be the greatest requirements for antimicrobial activity. Significantly, components of the target membrane may also determine the efficacy of these peptides. Further study of peptide–membrane interactions may explain activity of these molecules with individual organisms.

### 5.3 Cytokines

Whereas peptides seem to predominate in the immune defenses of insects, cytokine-mediated inflammation is highly significant in mammals. Cytokines are primarily secreted by leukocytes, stimulating both humoral and cellular immune responses as well as activation of phagocytosis (Table 3). The proinflammatory cytokines TNF-$\alpha$, IL-1, and IL-6, and chemokines such as IL-8 are released in response to encounters with pathogens, specifically by PAMP/PRR interactions. TNF-$\alpha$ is the primary mediator of acute inflammation, and the interleukins enhance this action by increasing endothelial adhesion, increasing chemokine secretion, activating cell-mediated immunity, and stimulating hepatocytes to secrete acute-phase proteins. Type-I interferons produced by virus-infected cells inhibit viral replication and kill infected cells by degrading mRNA. Interferons also activate $TH_1$ cells, which activate cytotoxic T cells and NK cells, which kill pathogens.

## Bibliography

### Books and Reviews

Aderem, A., Ulevitch, R.J. (2000) Toll-like receptors in the induction of the innate immune response, *Nature* **406**, 782–787.

Akira, S., Takeda, K., Kaisho, T. (2001) Toll-like receptors: critical proteins linking innate and acquired immunity, *Nat. Immunol.* **2**, 675–680.

Beutler, B. (2004) Innate immunity: an overview, *Mol. Immunol.* **40**, 845–849.

Cook, D.N., Hollingsworth II, J.W., Schwartz, D.A. (2003) Toll-like receptors and the genetics of innate immunity, *Curr. Opin. Allergy Clin. Immunol.* **3**, 523–529.

Coutinho, A., Poltorack, A. (2003) Innate immunity: from lymphocyte mitogens to toll-like receptors and back, *Curr. Opin. Immunol.* **15**, 599–602.

Fearon, D.T. (1999) Innate immunity and the biological relevance of the acquired immune response, *Q. J. Med.* **92**, 235–237.

Imler, J.-L., Hoffmann, J.A. (2001) Toll receptors in innate immunity, *Trends Cell Biol.* **11**, 304–311.

Kimbrell, D.A., Beutler, B. (2001) The evolution and genetics of innate immunity, *Nat. Rev. Genet.* **2**, 255–267.

Means, T.K., Golenbock, D.T., Fenton, M.J. (2000) Structure and function of toll-like receptor proteins, *Life. Sci.* **68**, 241–258.

Medzhitov, R., Janeway Jr., C. (2000) Innate immune recognition: mechanisms and pathways, *Immunol. Rev.* **173**, 89–97.

Medzhitov, R., Janeway Jr., C. (2002) Innate immune recognition, *Annu. Rev. Immunol.* **20**, 197–216.

O'Neill, L.A. (2002) Signal transduction pathways activated by the IL-1 receptor/toll-like receptor superfamily, *Curr. Top. Microbiol. Immunol.* **270**, 47–61.

### Primary Literature

Adachi, O., et al. (1998) Targeted disruption of the MyD88 gene results in loss of IL-1- and IL-18-mediated function, *Immunity* **9**, 143–150.

Dziarski, R., Tapping, R.I., Tobias, P.S. (1998) Binding of bacterial peptidoglycan to CD14, *J. Biol. Chem.* **273**, 8680–8690.

Horng, T., et al. (2002) The adaptor molecule TIRAP provides signaling specificity for toll-like receptors, *Nature* **420**, 329–333.

Jurk, M., et al. (2002) Human TLR7 or TLR8 independently confer responsiveness to the antiviral compound R-848, *Nat. Immunol.* **3**, 499.

Kobayashi, K., et al. (2002) IRAK-M is a negative regulator of toll-like receptor signaling, *Cell* **110**, 191–202.

Medzhitov, R., et al. (1997) A human homologue of the Drosophila Toll protein signals activation of adaptive immunity, *Nature* **388**, 394–397.

Miyake, K. (2003) Innate recognition of lipopolysaccharide by CD14 and toll-like receptor 4-MD-2: unique roles for MD-2, *Int. Immunopharmacol.* **3**, 119–128.

Ozinsky, A., et al. (2000) The repertoire for pattern recognition of pathogens by the innate immune system is defined by cooperation between toll-like receptors, *Proc. Natl. Acad. Sci. U.S.A.* **97**, 13766–13771.

Palaniyar, N., Nadesalingam, J., Clark, H., Shih, M.J., Dodds, A.W., Reid, K.B. (2004) Nucleic acid is a novel ligand for innate immune pattern recognition collectins surfactant proteins A and D and mannose-binding lectin, *J. Biol. Chem* Published online before print.

Powers, J.S., Hancock, R.E.W. (2003) The relationship between peptide structure and antibacterial activity, *Peptides* **24**, 1681–1691.

Radons, J., Dove, S., Neumann, D., Altmann, R., Botzik, A., Martin, M.U., Falk, W. (2003) The interleukin 1 (IL-1) receptor accessory protein Toll/IL-1 receptor domain, *J. Biol. Chem.* **278**, 49145–49153.

Tabeta, K., et al. (2004) Toll-like receptors 9 and 3 as essential components of innate immune defense against mouse cytomegalovirus infection, *Proc. Natl. Acad. Sci. U.S.A.* **101**, 3516–3521.

Underhill, D.M., Ozinsky, A. (2002) Toll-like receptors: key mediators of microbe detection, *Curr. Opin. Immunol.* **14**, 103–110.

Welch, R.A., et al. (2002) Extensive mosaic structure revealed by the complete genome sequence of uropathogenic *Escherichia coli*, *Proc. Natl. Acad. Sci. U.S.A.* **99**, 17020–17024.

Wentworth Jr., P., McDunn, J.E., Wentworth, A.D., Takeuchi, C., Nieva, J., Jones, T., Bautista, C., Ruedi, J.M., Gutierrez, A., Janda, K.D., Babior, B.M., Eschenmoser, A., Lerner, R.A. (2002) Evidence for antibody-catalyzed ozone formation in bacterial killing and inflammation, *Science* **298**, 2195–2199.

Yamamoto, M., et al. (2002) Cutting edge: a novel Toll/IL-1 receptor domain-containing adaptor that preferentially activates the IFN-beta promoter in the toll-like receptor signaling, *J. Immunol.* **169**, 6668–6672.

Yamamoto, M., et al. (2003) TRAM is specifically involved in the toll-like receptor 4-mediated MyD88-independent signaling pathway, *Nat. Immunol.* **4**, 1144–1150.

Zhang, G., Ghosh, S. (2002) Negative regulation of toll-like receptor-mediated signaling by Tollip, *J. Biol. Chem.* **277**, 7059–7065.

Zhang, D., et al. (2004) A toll-like receptor that prevents infection by uropathogenic bacteria, *Science* **303**, 1522–1526.

# 2
# Superantigens

*Matthew D. Baker and K. Ravi Acharya*
*University of Bath, Claverton Down, Bath, UK*

1 **Introduction 29**

2 **The Superantigen Family 29**
2.1 Molecular Architecture 30
2.2 Nonclassical Superantigens 31
2.3 The Immune Response to Bacterial Superantigens 31

3 **Binding to MHC Class II Molecules 31**
3.1 The Generic Binding Site and the MHC Class II $\alpha$-chain 32
3.2 Zinc-dependent Binding to MHC Class II $\beta$-chain 33

4 **Binding to the T-Cell Receptor 35**
4.1 General Mechanism of Binding for Classical Superantigens to T-Cell Receptor $V_\beta$ Elements 35
4.2 Binding to the T-Cell Receptor V$\alpha$ Element 36

5 **Formation of the Trimeric Complex for Signal Transduction 37**

6 **Roles of Superantigens in Disease 37**
6.1 Toxic Shock Syndrome 38
6.2 Staphylococcal Food Poisoning 38
6.3 Autoimmunity 39

**Bibliography 39**
Books and Reviews 39
Primary Literature 39

*Immunology. From Cell Biology to Disease.* Edited by Robert A. Meyers.

ISBN: 978-3-527-31770-7

## Keywords

**Antigen Presenting Cell**
Cells, including macrophages, B cells, and dendritic cells that take up and process an antigen into a peptide fragment capable of being presented by an MHC class II molecule on the cell surface.

**Major Histocompatibility Complex Class II**
Present peptide antigen to T cells.

**Major Histocompatibility Complex Class II Binding Site**
The region of a superantigen that recognizes and binds to MHC class II molecules.

**Pyrogenic Exotoxins**
A group of superantigens produced by the gram-negative bacteria *Streptococcus pyogenes.*

**Staphylococcal Enterotoxins**
A group of superantigens produced by the gram-negative bacteria *Staphylococcus aureus.*

**Superantigen**
A substance (as an enterotoxin) that acts as an antigen capable of stimulating much larger numbers of T cells than ordinary antigen.

**T Cell**
T lymphocyte; a cell bearing a T-cell receptor, which can control and mediate humoral immunity.

**T-Cell Receptor**
The receptor that enables the T cell to recognize and react to specific antigens.

**T-Cell Receptor Binding Site**
The region of a superantigen that recognizes and binds to the $V_\beta$ elements of T-cell receptors.

Superantigens (SAgs) are a family of proteins produced primarily by *Staphylococcus aureus* and *Streptococcus pyogenes.* These toxins acquired the name *Superantigen* due to their ability to hyperstimulate the immune system. SAgs are not recognized by the immune system in the same way as conventional foreign proteins, and interact in a unique fashion with major histocompatibility complex (MHC) class II molecules and T-cell receptors (TCRs). Once formed, the trimolecular complex induces vast

T-cell proliferation resulting in massive cytokine release, epithelial damage, capillary leak, and hypotension. Recent additions to the family include toxins produced by *Mycoplasma arthritidis* and *Yersinia Pestis*. SAgs are implicated in several diseases including toxic shock syndrome (TSS), scarlet fever, and food poisoning; their function appears to be primarily to weaken the host sufficiently to allow disease to take hold.

## 1 Introduction

Conventional antigens: peptide fragments derived from exogenous proteins are displayed on the surface of antigen presenting cells (APC) by major histocompatibility complex II (MHC) molecules. These peptides are then recognized by circulating T cells and an immune response targeted toward that particular peptide is initiated. In contrast, superantigens (SAgs) are not processed by APC and remain as fully intact proteins. Structural evidence shows SAgs to bind to APCs on the external face of the MHC class II molecule and concomitantly to T cells via the external face of the T-cell receptor (TCR) $V_\beta$ element. In contrast with the specificity that results from TCR binding to one particular type of conventional peptide antigen, interaction between SAg and the $V_\beta$ element of the TCR supplies a much broader range of specificity, as whole subsets of T cells displaying the same $V_\beta$ types are able to recognize the SAg.

Superantigenic toxins are very stable proteins. They are highly resistant to proteases, denaturation from heat and extreme pH, and are often found in food that are high in protein, salt, and sugar. In order to cause food poisoning, they must be able to survive not only cooking and preserving procedures, but also the digestive process.

## 2 The Superantigen Family

In order to be classified as an SAg, a toxin must demonstrate the ability to stimulate T cells in a $V_\beta$ specific manner. The number of bacterial SAgs or proteins highly homologous to known SAgs has grown considerably over the last decade so much so that an international nomenclature committee for staphylococcal SAgs have described an international procedure for the naming of newly described SAgs and putative SAgs.

The staphylococcal enterotoxins (SEs) A, B, C1–3, D, E, H, I, and J toxic shock syndrome toxin-1 (TSST-1) were originally designated enterotoxins because of their emetic property when ingested. Since 2000, several new "SEs" have been discovered on the basis of sequence alignments, though it is proposed that none should be classified as SEs until it is demonstrated that they induce emesis after oral administration.

Along with the staphylococcal enterotoxins, the streptococcal pyrogenic exotoxins (SPEs) A, C, and H; streptococcal mitogenic exotoxin (SME)-$Z_2$, and streptococcal superantigen (SSA) produced by *Streptococcus pyogenes* represent the most well-characterized bacterial SAgs. Other pathogens, such as *Mycoplasma arthritidis* and *Yersinia pseudotuberculosis* have also been shown to secrete superantigenic proteins. For the most part, this review will focus on the staphylococcal and streptococcal SAgs, with some discussion on the emerging SAgs from *M. arthritidis* and *Y. pseudotuberculosis* (*Mycoplasma arthritidis* mitogen – MAM and *Yersinia pseudotuberculosis* mitogen – YPM), as well as the wealth of putative SAgs uncovered by genomic initiatives. Broadly speaking, staphylococcal and streptococcal SAgs can be grouped into four subfamilies on the basis of amino acid sequence and three-dimensional structure: The first group is made up of the staphylococcal enterotoxins SEA, SED, SEE, SEH, SEI, and SEJ; both staphylococcal and streptococcal toxins form the second group, which consists of SEB, SEC1-3, SpeA1-3, SSA, and SEG; the third group contains streptococcal pyrogenic- and mitogenic toxins SpeC, SpeJ, SpeG, SpeH, SME-Z, SME-$Z_2$, and streptococcus dysgalactiae-derived mitogen (SDM); the SAg-like proteins SSL and TSST-1 form the fourth group. MAM and YPM have no sequence or structural homology to the rest of the bacterial SAgs and cannot be grouped with any of these subfamilies.

## 2.1 Molecular Architecture

Comparison of the three-dimensional structures of staphylococcal and streptococcal SAgs reveal a conserved two-domain architecture (Fig. 1). The two domains, known as the N- and C-terminal domains, are separated by a long solvent-exposed $\alpha$-helix, spanning the center of the molecule. Typically, the N-terminal domain contains many hydrophobic residues in its solvent-exposed regions and structurally it has a significant similarity to the oligosaccharide/oligonucleotide binding fold (OB-fold) found in other proteins. In spite of this similarity, SAgs have not been seen to bind to either DNA or carbohydrates. The C-terminal domain resembles the $\beta$-grasp motif and is composed of a four-stranded $\beta$-sheet that is centrally capped by a long $\alpha$-helix. Several SAgs also contain a highly flexible disulfide loop. Located in the N-terminal domain (Fig. 1), this flexible loop is implicated in the emetic properties of the staphylococcal and streptococcal toxins, as mutations of the two cysteine residues that form at the base

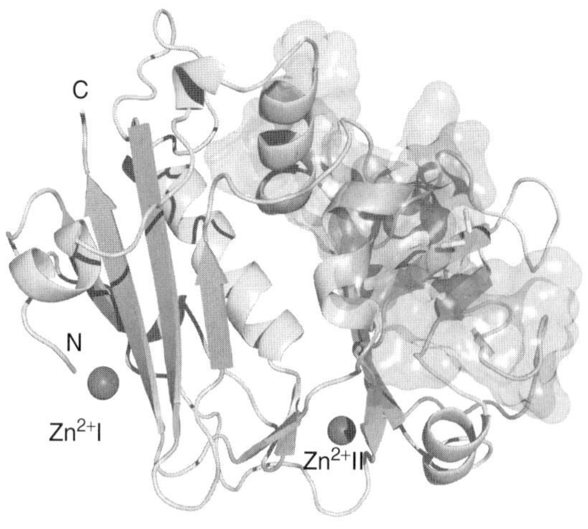


**Fig. 1** Ribbon diagram of staphylococcal enterotoxin A (SEA), representative of the common structural features of the staphylococcal and streptococcal SAg family. Blue spheres represent the positions of the two zinc sites. The cysteine residues that form the disulfide loop are shown in stick representation (yellow), the surfaces defining the TCR binding region (green), and MHC class II binding region (red) are shown (see color plate p. xv).

of this disulfide loop abolishes the ability of these toxins to induce vomiting in primates. Additionally, this loop has also been implicated in T-cell stimulation for several members of the staphylococcal and streptococcal SAg family.

## 2.2 Nonclassical Superantigens

The staphylococcal SAg-like proteins SSL5 and SSL7 also show the common architecture preserved among these related groups of proteins. This group of proteins is most closely related to TSST-1, although none of the SSLs have demonstratable superantigenic activity, and as such are linked to the SAg family purely due to sequence identity and structural homology. Differences between the SSLs and classical SAgs include extended secondary structure elements, extended loop regions, and differences in charge distribution over the protein surface. At present, it is not known whether these features have any biological significance.

The discovery and characterization of SAgs from bacterial sources such as *M. arthritidis* and *Y. pseudotuberculosis* have extended the topology of the family even further. MAM and YPMa show no structural homology to the rest of the SAg family whatsoever. MAM is composed entirely of $\alpha$-helices, arranged in two domains to form an L shape. The N-terminal domain consists of four $\alpha$-helices and a 25-residue N-terminal loop that wraps around the four-helix bundle, whereas The C-terminal domain contains six $\alpha$-helices. YPMa is much smaller than the classical staphylococcal and streptococcal toxins (14 kDa), and consists of a jelly-roll fold comprising two $\beta$-sheets, each containing four antiparallel strands. Similarly, the structure of YPMa is, unlike any other known SAgs, structurally similar to viral capsid proteins.

## 2.3 The Immune Response to Bacterial Superantigens

Superantigens primarily target $CD4^+$ T cells, leading to a T-helper type 1 (Th1) response and no Th2 response. A dominant Th1 response causes the suppression of antibody production, which in turn reduces the clearance of the invading microbe, allowing disease progression. The resultant release of a variety of cytokines in high quantities from macrophages and T cells cause many of the symptoms associated with Sag-related disease, and eventually leads to hypotension through capillary leak. The array of cytokines produced include IL-1 and TNF-$\alpha$ from macrophages and TNF-$\beta$, IL-2, and IFN-$\gamma$ from T cells. Most cytokine expression is transient and occurs within 12 h of SAg exposure, whereas signals involved in proliferation such as DNA synthesis and IL-2 receptor expression do not occur until over 24 h after toxin exposure. Once cells have undergone DNA synthesis, they are capable of re-expressing cytokines prior to a second round of proliferation.

# 3 Binding to MHC Class II Molecules

MHC class II molecules possess two separate binding sites for the classical staphylococcal and streptococcal SAgs. The first, a low-affinity binding site that is also known as the *generic site* is located on the $\alpha$-chain of the MHC class II molecule (Fig. 2), whereas the second, a high-affinity zinc-dependent site is located

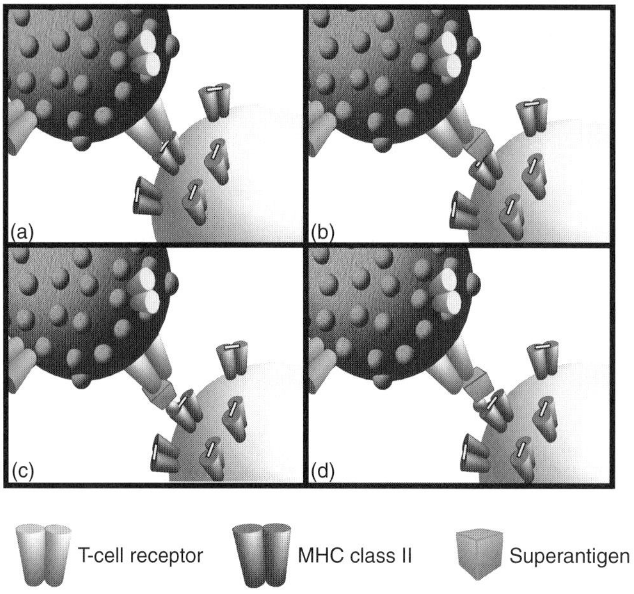


**Fig. 2** Schematic representation illustrating the differences between conventional peptide antigen presentation and SAg presentation to MHC class II and T-cell receptors. (a) Conventional antigen is processed by the antigen presenting cell and displayed as discrete peptide fragments within the peptide binding groove of MHC class II molecules, direct interaction occurs between TCR and MHC II molecule; (b) SAgs bind to the solvent-exposed face of the MHC class II molecule ($\alpha$-1) via its generic site, forming a bridge between TCR (V$\beta$) and MHC class II; (c) SAg binds to MHC class II ($\beta$1) via a bridging zinc atom in a high affinity, zinc-dependent manner; and interacts with TCR V$\alpha$ as predicted for SEH; (d) SAg binds to MHC class II ($\beta$1) via a high-affinity zinc site, and interacts with TCR V$\beta$ as is postulated for the remainder of the zinc containing SAgs. In both cases, the MHC class II molecule associated antigenic peptide has been shown to influence T-cell recognition of SAg/MHC class II molecule.

on the $\beta$-chain. SAgs bind via this high-affinity site approximately 100 times more strongly than to the generic site. A great deal of structural and mutational data has revealed that a range of diversity exists within the SAg family for binding to MHC class II molecules. Each SAg binds to different alleles of class II molecules to varying extent. While most of the SAgs, including TSST-1, SEB, and MAM bind preferentially to HLA-DR alleles, SAgs, such as SEC, SpeA, and SSA bind predominantly to HLA-DQ alleles.

### 3.1 The Generic Binding Site and the MHC Class II $\alpha$-chain

X-ray crystallographic studies on SEB and TSST-1 in complex with the MHC class

II molecule HLA-DR1 via the low-affinity generic site, SpeC in complex with HLA-DR2, and SEH in complex with HLA-DR1 via the high-affinity zinc site have yielded a great deal of structural information about the binding of SAgs to MHC class II molecules. The SEB– and TSST-1–HLA-DR1 complexes demonstrate that both SEB and TSST-1 have similar binding modes to HLA-DR1. Interaction between HLA-DR1 and these SAgs mainly involves the $\alpha$-chain of the DR1 molecule and the solvent-exposed, hydrophobic core at the N-terminal domain of the toxin. Similar hydrophobic ridge regions exist in several other SAgs and form the generic MHC class II binding site. One feature seen in the TSST-1–HLA-DR1 complex, (but not in the SEB–HLA-DR1 complex) is the number of additional contacts between TSST-1 and the peptide antigen displayed by the HLA-DR1. Indeed, truncating the C-terminal end of the antigenic peptide significantly affects TSST-1 binding to MHC class II molecules in mice.

### 3.2
### Zinc-dependent Binding to MHC Class II $\beta$-chain

Several of the bacterial SAgs, with the exception of SEB, TSST-1, SSA, SSL1-, and YPM, have one or more zinc binding sites (Fig. 1). Zinc ions have been shown to play two distinct roles in SAg function. First, they act as dimerization sites, second, they have been shown to be important for the recognition of the MHC class II $\beta$-chain. SEA possesses a high-affinity zinc binding site in its C-terminal domain; this site has a substantially higher affinity for DR1 ($K_d = 100$ nM) compared with the generic binding site ($K_d = 10$ μM), and if the two binding sites coexist, a $K_d$ of 13 nM is observed. The importance of both of these sites is demonstrated by the fact that mutation of residues in either of these sites abolishes the toxin's ability to induce cytokine expression. The coexistence of these two distinct MHC class II binding sites may enable it to form a trimeric SEA–MHC–SEA complex, a phenomenon that has been observed in solution. Several other SAgs including SEE possess identical zinc ligands, and a similar case for SAg–MHC–SAg complex formation can be argued. Both SED and SpeC possess analogous zinc binding sites to SEA. In SED, the zinc ion acts as a means for toxin dimerization; SED can form zinc-dependent homodimers, whereas SpeC forms zinc-independent homodimers. Both SED and SpeC bind to the $\beta$-chain of MHC class II molecule by a zinc-mediated mechanism similar to that of SEA, and because of the ability of these toxins to dimerize, formation of trimers and/ or tetramers is a possibility. A similar binding mechanism has been proposed for SEH, SME-Z, SME-$Z_2$, SpeG, and SpeH, all of which lack a generic MHC class II binding site. This high-affinity zinc site, present in SEA is not seen in SEC or SpeA. Instead, a separate zinc binding site (with an estimated $K_d$ of approximately 1 μM) is seen at the N-terminal domain of these toxins. In a similar fashion to the SEC-like zinc binding site, this secondary zinc binding site is also important for MHC class II binding (Figs. 1 and 2). Direct observation of zinc-mediated SAg–MHC class II complexes such as those seen in the crystal structures of SEH–HLA-DR1 and SpeC–HLA-DR2, have enabled a detailed analysis of these interactions. A great deal of similarity exists between the two complexes. The interactions between both SAgs and their MHC class II molecules are governed by a tetrahedrally coordinated zinc ion. For SpeC, three of the coordinating ligands are from the toxin, and for SEH, two are from the toxin and one is from a water molecule. In both complexes, the fourth ligand is provided by residue His81 of the $\beta$1 chain of the class II molecule. In a similar fashion to the TSST-1-MHC class II complex, interaction between the SAg and the antigenic peptide is observed. Up to one-third of the contact area between SAg and MHC class II is taken up by antigenic peptide. In both cases, a majority of the interactions occur with the backbone atoms of the peptide, enabling both SpeC and SEH to have similar interactions with the antigenic peptide even though the composition of the peptide differs in both complexes. This would suggest that although the antigenic peptide plays an important role in the complex interaction, MHC class II binding is not entirely peptide specific.

To confuse matters further, another zinc binding site has recently been discovered in SEC2. This zinc ion is located near the generic MHC class II binding site and is thought to act as a site of dimerization for SEC2. If this zinc ion does indeed allow dimerization of SEC2, the generic MHC class II site would be blocked by the dimer interface; thus, MHC class II binding would be governed by the SEC2 primary zinc binding site. As a result, SEC2 would be able to bind to MHC class II molecules as a zinc-mediated dimer. SpeA1 and SpeC are also capable of forming homodimers, although in this case the mechanism is zinc independent. SpeA1 has recently been shown to exist in a disulfide linked dimeric form via cysteine residues located within the flexible disulfide loop, whereas the SpeC dimer is formed by direct interaction of the surfaces that usually form the generic MHC class II binding site.

Zinc was also thought to play a similar role in MAM, namely, binding to MHC class II molecules and toxin dimerization. However, the absence of zinc in the crystal structure of dimeric MAM in complex with HLA-DR1 suggests that zinc fulfills neither role. Clearly, the exact role of zinc in the action of this novel SAg requires further investigation.

Despite the $MAM_2$–HLA-DR1 complex bearing no immediate likeness to that of other bacterial SAgs, there are some similarities in the overall mechanism of interaction. Complex formation does not induce any major structural changes at the binding interface, which is consistent with other SAg–MHC class II complexes. The $MAM_2$–HLA-DR1 complex is formed through interaction of the N-terminal domain of MAM and a number of regions of HLA-DR1. Contact with the antigenic peptide (hemagglutinin peptide) in the peptide binding groove of the HLA-DR1 molecule is again observed. Interestingly, the MAM binding site on HLA-DR1 also overlaps the generic binding site for staphylococcal and streptococcal SAgs. In support of this, both SEB and TSST-1 have been shown to block MAM binding to MHC class II molecules. Moreover, MAM binding is also disrupted by those SAgs that bind to the high-affinity zinc site, as the MAM binding site on MHC class II also includes the region on the MHC $\beta$-chain, which contains His81. Although the crystal structure reveals MAM to bind to MHC class II molecules as a dimer, both monomer and homodimers are shown to be present in solution, indicating that MAM may be able to act on MHC class II molecules as both monomer and dimer.

Current understanding of SAg structure and function suggests that there is an increasing range of mechanisms by which SAgs can interact with MHC class II molecules. Although differences exist between SAgs, three main binding modes are observed: (1) through zinc-mediated interaction; (2) via the generic site; or (3) as a homodimer, which can be either zinc or nonzinc mediated. In addition, it is also possible that a combination of any of these three mechanisms may be employed.

## 4 Binding to the T-Cell Receptor

In comparison to their interaction with MHC class II molecules, the binding of SAgs to T-cell receptors is at first sight a much more localized interaction. SAgs were thought to bind exclusively TCR via the TCR $V_\beta$ element through a universally similar mechanism, where the binding is mediated by interactions between the side chains of the SAg and the backbone atoms from the $V_\beta$ region. This results in each SAg selectively expanding T-cell subsets bearing certain $V_\beta$ elements, while excluding others.

### 4.1 General Mechanism of Binding for Classical Superantigens to T-Cell Receptor $V_\beta$ Elements

The TCR binding sites of the staphylococcal and streptococcal SAgs consist of comparable scaffold regions within which specific amino acids affect the $V_\beta$ specificity. The binding site is formed by a shallow groove between the two domains of the molecule (Fig. 1). The structures of SEB, SEC2, and SEC3 in complex with TCR $V_\beta$ seem to support this simple binding mechanism, a majority of the contacts between SAg and TCR are between the amino acid side chains within this groove and backbone atoms of the TCR $V_\beta$ elements. The main interactions are between the side-chain atoms of the SAg and complementarity determining regions 1 and 2 (CDRs 1 and 2), and hypervariable region 4 (HV4) of the $V_\beta$ chain. Close examination of the TCR binding sites of SEC2/3, SEA, and SEB indicate that an invariant asparagine residue (Asn23 in SEB/SECs; Asn25 in SEA) is essential for direct interactions with the TCR, and if this residue is mutated, T-cell stimulation is lost. This residue is located on the surface of the toxin, and is thought to have similar interactions in all the SEs. Mutagenesis studies have revealed that further interaction exists between SAg and TCR, which are conserved among many of the staphylococcal and streptococcal SAgs, including SEC1-3, SEB, SpeA, and SSA. In addition to this "common core" of residues, the differences in TCR affinity among the SAgs can be accounted for by several residues unique to each toxin. For example, the type of residue at position 26 of SEC2 confers specificity between SEC1, SEC2, SEA, and SEB via its interaction with Gly53 from the $V_\beta$ chain. Similarly, the residue at position 91 of SEC2 is also implicated in TCR binding and is not conserved in SEA or SEB. By switching several residues for a dissimilar group of amino acids, a decreased affinity for certain $V_\beta$ chains and an increase in affinity for others may occur. This is likely to be as a direct result of an increase in the amount of unfavorable interactions for one $V_\beta$ element followed by an increase of favorable interactions for another $V_\beta$ element. Experiments with SEA support this idea. SEA specificity for TCR is thought to be governed by three residues at position 21, 206, and 207. Exchange of these three residues for the homologous residues in SEE causes a change in the profile of the responding

$V_\beta$ elements to that of the profile normally seen for T cells stimulated by SEE. It has also been shown that those residues that define the specificity of an SAg for particular $V_\beta$ elements make the greatest energetic contribution to the overall stability of the SAg–TCR complex, and as the above evidence shows, these residues are comparatively few in number.

More recently, the two SAg–TCR $V_\beta$ complexes SpeA1–$V_\beta$ 8.2 and SpeC–$V_\beta$ 2.1 have indicated that TCR binding, for at least some SAgs is not as simple as first thought. The interaction of SpeA1 with the TCR $\beta$-chain is similar to that of SEB, with several of the important contacts preserved. In addition, SpeA1 has an extra five hydrogen bonds with the TCR $V_\beta$ compared to SEB. Thus, the mechanism of TCR binding employed by SpeA1 is slightly more intricate than those employed by SEB and SEC2/3. The complexity increases still further with SpeC; this toxin was found to bind to significantly more $V_\beta$ residues than SEC2/3, SEB, or SpeA1, including numerous interactions with both main-chain and side-chain atoms of the TCR $\beta$-chain. The increased amount of contacts are most likely due to the deeper and broader cleft between the two domains of SpeC compared with other SAgs, as this is where the TCR binding site is located. Structural similarity between SpeC and TSST-1 mean that although the TCR binding site of TSST-1 is not yet fully characterized, it is likely that it shares similar binding characteristics. The available mutagenesis data suggests that the TCR binding site of TSST-1 is located in a similar position to SpeC, between the two domains of the toxin.

Both MAM and YPMa have been shown to bind to the TCR $\beta$-chain, although in the absence of a crystal structure of their complexes, the exact location and composition of their TCR binding sites are speculative. Clearly, more work needs to be carried out to determine the binding mechanisms of these novel SAgs before a direct comparison can be made to the staphylococcal and streptococcal toxins.

The characterization of SAg–TCR complexes have demonstrated that multiple modes of TCR interaction exist. The first, a high specificity binding mode, involves many contacts by both backbone atoms and side-chain atoms over an increased area as demonstrated by SpeC. This mode has high affinity for relatively few TCR $V_\beta$ elements. The second, a moderate specificity mode, has fewer interactions over a reduced contact area. This mode has a modest affinity for a larger group of TCR $V_\beta$ elements, such as seen in SpeA1. The final mode, a promiscuous binding mode, such as is seen in SEB and SEC2/3, suggests that binding to TCR $V_\beta$ chains for these SAgs is a matter of simple conformational dependence.

### 4.2
### Binding to the T-Cell Receptor $V\alpha$ Element

The discovery that SEH, in contrast to all other SAgs, stimulates T cells in a $V_\alpha$ specific fashion, completely lacking any $V_\beta$ expansion, adds further complication to the scheme. It is suggested that SEH may bind to TCR in a $V_\alpha$ specific fashion due to its presentation by MHC class II molecules via a zinc atom. Whether other SAgs that are presented to TCR by MHC class II in the same way can bind to $V_\alpha$ in the same way is yet to be established.

## 5
## Formation of the Trimeric Complex for Signal Transduction

The complex course of events at the cell surface that lead to the formation of the MHC II–SAg–TCR complex is still ambiguous. It is possible for SAg to exist as monomer and dimer, and many SAgs have multiple MHC class II binding sites enabling them to form complexes with more than one MHC class II molecule or TCR. In terms of trimeric complex formation, this means that even further diversity exists in the ability of these toxins to interact with the immune system.

When considering complex formation, it is important to first consider the nature of membrane-bound receptors. In order for a complex to form, cell membranes need to be in close proximity to each other, and the receptors must diffuse to the site of interaction. It has been shown that in order for superantigenic T-cell activation to occur, a comparatively low percentage (<0.3%) of the MHC class II molecules must be occupied by the SAg. Higher concentrations of bound toxin result in an aborted T-cell response after a few cell divisions due to apoptosis. Thus, a low local concentration of MHC class II molecules on the cell interface is preferable for optimum superantigenicity. It is thought that the binding of a SAg to TCR induces clustering of the TCRs on the cell surface and the acquirement of the intracellular components required to transduce a signal. This is thought to occur in a manner that mimics peptide antigens with regard to receptor clustering, either through direct clustering events as proposed by the TCR oligomerization model, or by the binding of SAg homodimers to multiple MHC class II molecules, which in turn would promote T-cell clustering. Molecular modeling suggests that signal transduction stimulated by SEA through large-scale assembly is limited to four or five TCR–(DR1$\beta$-SEA-DR1$\alpha$) tetramers, and requires the dimerization of MHC class II molecules. While TCRs would be clustered together in this model, TCR dimerization is thought unlikely. SEA is not unique in its ability to form zinc-mediated dimers. SED and SEC2 can form zinc-dependent homodimers, whereas SpeA1 forms a nonzinc mediated disulfide linked dimer; a SpeC dimer is also formed in the absence of zinc. MAM has also been shown to be able to form zinc-dependent and zinc-independent dimers, and YPM is thought to be able to form zinc-independent trimers. Clearly, SAgs have evolved slightly different ways of inducing receptor clustering. SAgs that act as monomers and possess only a single MHC class II binding site appear to rely on the interactions of the TCR $V_\alpha$ and MHC class II-$\beta$1, which increases the stability of the ternary complex to within the range seen for conventional antigen. A stable MHC–SAg–TCR complex with an extended half-life may therefore assist receptor clustering.

## 6
## Roles of Superantigens in Disease

It should be noted that the association of an SAg with a certain disease does not mean that it necessarily causes the disease. In some cases, the presence of a microbe known to produce an SAg has been correlated with the disease. In other cases, expansion of a certain $V_\beta$ subset is interpreted as evidence of the involvement of SAg in a disease. A definitive confirmation of the involvement of an SAg requires the cloning of the putative SAg and its subsequent testing in an animal model. Thus, the involvement of SAgs in disease is established for relatively few conditions: these include toxic shock syndrome (TSS), food poisoning, Kawasaki disease, and scarlet fever.

### 6.1
### Toxic Shock Syndrome

TSST-1 is the main causative agent for TSS and induces most of the symptoms associated with the disease in experimental animals. TSST-1 is responsible for nearly all cases of menstrual TSS, and approximately 60% of nonmenstrual staphylococcal TSS. The remainder of the cases can be attributed primarily to other SEs including SEB, SEC, and SEA, whereas SpeA is associated with most cases of streptococcal TSS. Although the exact nature of the disease is not fully understood, it is clear that toxic shock is in part a result of the superantigenic nature of these toxins. Superantigenicity and massive T-cell proliferation affect the cardiovascular system by causing extensive epithelial damage, capillary leak, and a decrease in peripheral vascular resistance resulting in shock as well as affecting kidney and liver function. The ability of TSST-1 to cause both localized or systemic symptoms in this manner dictates that it must be able to transverse the epithelial barrier. At present, the mechanism by which it is able to do this is unknown, but could include either passive diffusion, or cellular receptors. The ability of TSST-1 to cross epithelial barriers, if it exists, would make it unique among the SAgs. The use of a cellular receptor would require some kind of specific interaction with TSST-1, and therefore a receptor binding site. Structural evidence for such a binding site on TSST-1 is limited, and is confined to the observable structural differences between TSST-1 and the other SAgs. These differences include the lack of an $\alpha$-helix in the C-terminal domain, the long N-terminal extension, and the absence of a disulfide loop. TSST-1 also has unique patches of hydrophobic and neutral residues on the front and rear of the $\beta$-barrel at the N-terminal domain. It is possible that a combination of these features could produce a receptor specific binding site in order for TSST-1 to traverse epithelial cells and allow systemic shock.

### 6.2
### Staphylococcal Food Poisoning

*Staphylococcus aureus* is the leading cause of microbial food-borne disease in the world. Symptoms of staphylococcal food poisoning (SFP) include abdominal pain, nausea, vomiting, and diarrhea and are usually experienced within 2 to 6 h of ingestion. Little is known about how the physical properties of SAgs relate to the symptoms of emesis and diarrhea. It has been suggested that the symptoms of food poisoning are a result of high cytokine levels. Indeed, the side effects of IL-2 therapy often mimic those of SFP. More recent work indicates that the emetic properties of these toxins are not completely correlated with their superantigenicity. Several areas in the N-terminal region of SEA have been identified that are important for both emetic and superantigenic function. Carboxymethylation of histidine residues in SEB was found to abrogate emetic activity but still induce T-cell proliferation, indicating that the two activities are separate. The areas of the toxin that induce emesis and diarrhea are also thought to be separate, because intravenous administration of a C-terminal fragment of SEC1 was found to induce diarrhea but not emesis in primates. The disulfide loop is considered to be one of the main regions responsible for emesis. Investigation has shown that the disulfide bond itself is not an absolute requirement for emetic activity, but the conformation within or adjacent to the loop is important for emesis. Further analysis of these toxins is required in order to assess the contribution of certain amino acids to the overall pathology of SAgs in the context of the diseases they cause.

### 6.3 Autoimmunity

The action of SAgs as immunomodulators has identified them as possible candidates for autoimmune disease. Since they do not discriminate between autoimmune and normal lymphocytes, SAgs would be able to stimulate autoreactive T cells, both locally and systemically. Although SAgs have not yet been directly implicated in human autoimmune disease, there is evidence to suggest their involvement. For example, SED-reactive T-cell lines have been shown to stimulate autoantibody production by B cells. Superantigens have been shown to be associated with a growing list of autoimmune conditions, including multiple sclerosis, arthritis, diabetes, and Kawasaki disease. Whether SAgs act as the initial trigger for autoimmunity or as external factors changing an ordinary controllable illness into an autoimmune condition for susceptible individuals, both remain possible, and clearly further work is needed to assess the roles of these toxins in this type of disease.

## Bibliography

### Books and Reviews

Baker, M., Acharya, K.R. (2002) Superantigens: Structure Function and Diversity, in: Krakauer, T. (Ed.) *Superantigen Protocols*, Humana Press Totowa, New Jersey.

Krakauer, T. (Ed.) (2002) *Superantigen Protocols*, Humana press, Totowa, New Jersey.

Papageorgiou, A.C., Acharya, K.R. (2000) Microbial superantigens: from structure to function, *Trends Microbiol.* **8**, 369–375.

Petersson, K., Forsberg, G., Walse, B. (2004) Interplay between superantigens and immunoreceptors, *Scand. J. Immunol.* **59**, 345–355.

Thibodeau, J., Sekaly, R.P. (Eds.) (1995) *Superantigens: Structure Function and Therapeutic Potential*, R.G. Landes company, Austin, Texas.

### Primary Literature

Arcus, V.L., Langley, R., Proft, T., Fraser, J.D., Baker, E.N. (2002) The three-dimensional structure of a superantigen-like protein, SET3, from a pathogenicity island of the *Staphylococcus aureus* genome, *J. Biol. Chem.* **277**, 32274–32281.

Cole, B.C. (1991) The immunobiology of *Mycoplasma arthritidis* and its superantigen MAM, *Curr. Top. Microbiol. Immunol.* **174**, 107–119.

Deringer, J.R., Ely, R.J., Stauffacher, C.V., Bohach, G.A. (1996) Subtype-specific interactions of type C staphylococcal enterotoxins with the T-cell receptor, *Mol. Microbiol.* **22**, 523–534.

Donadini, R., Liew, C.W., Kwan, A.H., Mackay, J.P., Fields, B.A. (2004) Crystal and solution structures of a superantigen from *Yersinia pseudotuberculosis* reveal a jelly-roll fold, *Structure (Camb.)* **12**, 145–156.

Fields, B.A., Malchiodi, E.L., Li, H., Ysern, X., Stauffacher, C.V., Schlievert, P.M., Karjalainen, K., Mariuzza, R.A. (1996) Crystal structure of a T-cell receptor $\beta$-chain complexed with a superantigen, *Nature* **384**, 188–192.

Germain, R.N. (1997) T-cell signaling: the importance of receptor clustering, *Curr. Biol.* **7**, 640–644.

Jardetzky, T.S., Brown, J.H., Gorga, J.C., Stern, L.J., Urban, R.G., Chi, Y.I., Stauffacher, C., Strominger, J.L., Wiley, D.C. (1994) Three-dimensional structure of a human class II histocompatibility molecule complexed with superantigen, *Nature* **368**, 711–718.

Kim, J., Urban, R.G., Strominger, J.L., Wiley, D.C. (1994) Toxic shock syndrome toxin-1 complexed with a class II major histocompatibility molecule HLA-DR1, *Science* **266**, 1870–1874.

Kline, J.B., Collins, C.M. (1997) Analysis of the interaction between the bacterial superantigen streptococcal pyrogenic exotoxin A (SpeA) and the human T-cell receptor, *Mol. Microbiol.* **24**, 191–202.

Lavoie, P.M., McGrath, H., Shoukry, N.H., Cazenave, P.A., Sekaly, R.P., Thibodeau, J. (2001) Quantitative relationship between MHC class II-superantigen complexes and the balance of T cell activation versus death, *J. Immunol.* **166**, 7229–7237.

Leder, L., Llera, A., Lavoie, P.M., Lebedeva, M.I., Li, H., Sekaly, R.P., Bohach, G.A., Gahr, P.J., Schlievert, P.M., Karjalainen, K., Mariuzza, R.A. (1998) A mutational analysis of the binding of staphylococcal enterotoxins B and C3 to the T cell receptor *β* chain and major histocompatibility complex class II, *J. Exp. Med.* **187**, 823–833.

Li, H., Llera, A., Mariuzza, R.A. (1998a) Structure-function studies of T-cell receptor-superantigen interactions, *Immunol. Rev.* **163**, 177–186.

Li, H., Llera, A., Tsuchiya, D., Leder, L., Ysern, X., Schlievert, P.M., Karjalainen, K., Mariuzza, R.A. (1998b) Three-dimensional structure of the complex between a T cell receptor *β* chain and the superantigen staphylococcal enterotoxin B, *Immunity* **9**, 807–816.

Li, Y., Li, H., Dimasi, N., McCormick, J.K., Martin, R., Schuck, P., Schlievert, P.M., Mariuzza, R.A. (2001) Crystal structure of a superantigen bound to the high-affinity, zinc-dependent site on MHC class II, *Immunity* **14**, 93–104.

Lina, G., Bohach, G.A., Nair, S.P., Hiramatsu, K., Jouvin-Marche, E., Mariuzza, R. (2004) Standard nomenclature for the superantigens expressed by Staphylococcus: international nomenclature committee for staphylococcal superantigens, *J. Infect. Dis.* **189**, 2334–2336.

Marrack, P., Kappler, J. (1990) The staphylococcal enterotoxins and their relatives, *Science* **248**, 705–711.

Petersson, K., Hakansson, M., Nilsson, H., Forsberg, G., Svensson, L.A., Liljas, A., Walse, B. (2001) Crystal structure of a superantigen bound to MHC class II displays zinc and peptide dependence, *EMBO. J.* **20**, 3306–3312.

Petersson, K., Pettersson, H., Skartved, N.J., Walse, B., Forsberg, G. (2003) Staphylococcal enterotoxin H induces V *α*-specific expansion of T cells, *J Immunol.* **170**, 4148–4154.

Roussel, A., Anderson, B.F., Baker, H.M., Fraser, J.D., Baker, E.N. (1997) Crystal structure of the streptococcal superantigen SPE-C: dimerization and zinc binding suggest a novel mode of interaction with MHC class II molecules, *Nat. Struct. Biol.* **4**, 635–643.

Schlievert, P.M., Jablonski, L.M., Roggiani, M., Sadler, I., Callantine, S., Mitchell, D.T., Ohlendorf, D.H., Bohach, G.A. (2000) Pyrogenic toxin superantigen site specificity in toxic shock syndrome and food poisoning in animals, *Infect. Immun.* **68**, 3630–3634.

Sundberg, E., Jardetzky, T.S. (1999) Structural basis for HLA-DQ binding by the streptococcal superantigen SSA., *Nat. Struct. Biol.* **6**, 123–129.

Sundberg, E.J., Li, H., Llera, A.S., McCormick, J.K., Tormo, J., Schlievert, P.M., Karjalainen, K., Mariuzza, R.A. (2002) Structures of two streptococcal superantigens bound to TCR *β* chains reveal diversity in the architecture of T cell signaling complexes, *Structure (Camb.)* **10**, 687–699.

Swaminathan, S., Furey, W., Pletcher, J., Sax, M. (1992) Crystal structure of staphylococcal enterotoxin B, a superantigen., *Nature* **359**, 801–806.

Swaminathan, S., Furey, W., Pletcher, J., Sax, M. (1995) Residues defining V*β* specificity in staphylococcal enterotoxins, *Nat. Struct. Biol.* **2**, 680–686.

Woodland, D.L., Wen, R., Blackman, M.A. (1997) Why do superantigens care about peptides?, *Immunol. Today* **18**, 18–22.

Zhao, Y., Li, Z., Drozd, S.J., Guo, Y., Mourad, W., Li, H. (2004) Crystal structure of Mycoplasma arthritidis mitogen complexed with HLA-DR1 reveals a novel superantigen fold and a dimerized superantigen-MHC complex, *Structure (Camb.)* **12**, 277–288.

# 3
# Antigen Presenting Cells (APCs)

*Harald Kropshofer and Anne B. Vogt*
*Roche Center for Medical Genomics, F. Hoffman-La Roche Ltd, Basel, Switzerland*

**1 Antigen-specific Immunity 47**

**2 Antigen Presenting Cells (APCs) 48**
2.1 The Discovery of Langerhans 48
2.2 Professional APCs 48
2.3 Nonprofessional APCs 49

**3 Antigens Presented by APCs 50**
3.1 Antigen Types 50
3.2 Biochemical Nature of Antigens 51
3.3 Origin of Antigens 52

**4 Antigen Entry Sites 52**
4.1 APCs in the Skin 52
4.2 APCs in the Mucosa 53
4.3 APCs in the Gut 53
4.4 Antigen Uptake in the Respiratory Tract 54

**5 Antigen Presenting Molecules 54**
5.1 MHC Class I Molecules 54
5.2 MHC Class II Molecules 56
5.3 CD1 Molecules 56

**6 Antigen Processing by APCs 57**
6.1 MHC Class I Processing Pathways 58
6.1.1 Generation of Peptide Ligands 58
6.1.2 TAP 58
6.1.3 The MHC Class I Loading Complex 59

*Immunology. From Cell Biology to Disease.* Edited by Robert A. Meyers.

ISBN: 978-3-527-31770-7

6.2 MHC Class II Processing Pathways 60
6.2.1 Endosomal Generation of Peptides 60
6.2.2 Invariant Chain and CLIP 60
6.2.3 HLA-DM 60
6.2.4 Tetraspan Network 61

**7 Activation of T Cells by APCs 61**
7.1 Cell Adhesion Molecules 62
7.2 Costimulatory Molecules 63
7.3 The Immunological Synapse 64

**8 Macrophages 65**
8.1 Evolutionary and Cellular Origin 65
8.2 Antigen Recognition Receptors 66
8.3 Macrophage Activation 68
8.4 Immunological and Nonimmunological Effector Functions 69
8.4.1 Microbial Killing 69
8.4.2 Tissue Remodeling 70
8.5 Deactivated and Alternatively Activated Macrophages 70

**9 B lymphocytes 71**
9.1 Antigen-induced B-Cell Activation 71
9.2 Antigen Presentation Outside the Follicular Center 72
9.2.1 B-cell Epitope versus T-cell Epitope 72
9.2.2 CD40–CD40L Engagement 74
9.3 Antigen Presentation in the Germinal Center 74
9.3.1 B Cell/Follicular DC Contacts 75
9.3.2 HLA-DO 75

**10 Dendritic Cells (DCs) 76**
10.1 DC Subsets 77
10.1.1 Myeloid Versus Lymphoid DCs 77
10.1.2 Human DC Subtypes 78
10.2 Antigen Uptake Mechanisms 79
10.2.1 Macropinocytosis Versus Phagocytosis 80
10.2.2 Receptors for Endocytosis 80
10.2.3 The Adhesion Receptor DC-SIGN 81
10.3 Maturation of DCs 81
10.3.1 Maturation Stimuli 81
10.3.2 The Danger Hypothesis 82
10.4 Antigen Processing and Presentation 84
10.4.1 Classical and Cross-presentation via MHC Class I 84
10.4.2 MHC Class II–restricted Antigen Presentation 84

10.5 T Cell Priming and Polarization 85
10.5.1 Priming of $CD8^+$ and $CD4^+$ T Cells 85
10.5.2 T Helper Cell Polarization 86
10.6 Immune Evasion and Plasticity of DCs 87
10.7 DCs in Immunotherapy 88

**Bibliography 89**
Books and Reviews 89
Primary Literature 90

## Keywords

**Antigen**
Any piece of a substance that can induce a specific immune response or reacts to a specific antibody or T cell.

**Adhesion Molecules**
Proteins, such as integrins, selectins, members of the Ig superfamily, or CD44, that mediate the binding of one cell to other cells or to the extracellular matrix.

**Adjuvant**
A substance that enhances the immune response to antigens and is therefore used in immunizations.

**Activation**
A process by which the immune system (e.g. T cells, B cells) is switched on to deal with an infection.

**Costimulation**
Secondary signals provided to T cells by costimulatory molecules (e.g. CD80, CD86) on antigen presenting cells. Regulated expression of costimulatory molecules allows T-cell activation in inflammation but not under healthy conditions.

**CpG**
Denotes a deoxynucleotide motif in bacterial DNA that differs in sequence and methylation state from mammalian DNA. It binds to and activates certain subsets of dendritic cells.

**Danger**
A term popularized by Matzinger, that covers the various signals from damaged tissues or from microbial products that trigger activation of dendritic cells.

**dsRNA**
Denotes double-stranded RNA, known as the genome of retro-viruses.

**Endotoxin**
Lipopolysaccharides (LPS) that are an integral part of the outer cell envelope membrane of gram-negative bacteria. They evoke toxic and pyrogenic effects and cause sepsis.

**Epitope**
Specific amino acid sequence of foreign or self-antigens that an antibody (B-cell epitope) or a T-cell receptor (T-cell epitope) recognizes, binds to, and reacts against.

**Germinal Center**
Sites in secondary lymphoid organs where activated B cells undergo proliferation, selection, maturation, and apoptosis.

**HLA**
Acronym for human leukocyte antigen, denoting the human MHC molecules.

**Humoral Immunity**
Immune response involving B cells that produce antibodies.

**Immunoglobulin**
A protein produced by a B cell that binds to a specific antigen leading to attack by the immune system.

**LPS**
Lipopolysaccharide (cf. endotoxin).

**Lymphatic System**
A network of vessels that is separate from the blood circulation. It includes the lymph nodes that are the command centers of the immune system.

**MHC**
Major histocompatibility complex – a gene cluster encoding MHC molecules, originally described as antigens responsible for transplant rejection. MHC molecules bind antigenic peptides in order to activate T cells in an antigen-specific manner.

**Peptide**
Short fragment of a protein. On binding to MHC molecules, it can activate T cells and thereby becomes antigenic.

**Processing**
Set of activities of an antigen presenting cell encompassing proteolysis of an antigen, generation of antigen peptides, peptide loading onto MHC molecules, and transport of MHC-peptide complexes to the cell surface.

**$T_{H}1$**
Helper T cells that produce high levels of IFN-$\gamma$ and promote activation of macrophages.

**$T_{H}2$**
Helper T cells that produce IL-4, IL-5, and IL-10 and promote eosinophil, B-cell, and mast-cell functions.

**Tolerance**
The failure of the immune system to respond to antigens.

**Vaccination**
Administering an antigen to generate long-term antibody or cellular immune responses or both.

Antigen presenting cells play a key role in the immune system as they are the only cells of the body that are able to trigger antigen-specific immune responses upon infection with foreign invaders. Three differentially specialized types of cells are in charge of this vital task: dendritic cells, macrophages, and to a certain extent, B lymphocytes. We are beginning to understand the molecular mechanisms of how these cells take up, process, and present antigens and how this translates into T-cell activation or silencing. This knowledge is fundamental for both our current concepts about how the immune system discriminates between self and nonself and for the development of novel vaccination strategies against diseases still threatening mankind.

## 1 Antigen-specific Immunity

Our environment is rich in microbes and parasites, a multitude of them being pathogenic to humans and/or to other mammals or vertebrates. The necessity of means to combat microbial invaders is evident and is accomplished through the presence of a functional immune system. One type of weapon was discovered by Behring and Kitasato at the end of the nineteenth century in the serum of rabbits immunized with diphtheria bacilli: the cell-free serum was capable of destroying diphtheria toxins, it was specific for diphtheria and remained effective in the organism of other animals – the breakthrough in immunizing against diphtheria. Behring and his colleagues named the serum components of this monumental discovery *antitoxins.* Later on they were named *antibodies,* and the agents able to induce the generation of antibodies became known as *antigens.*

An immune response can be antigen-specific, as in Behring's immunization experiments with diphtheria and tetanus toxins, but also antigen-nonspecific (Fig. 1).

| Innate immunity | Acquired immunity |
|---|---|
| Eosinophils | B lymphocytes |
| Neutrophils | Macrophages |
| Macrophages | Dendritic cells |
| NK cells | T lymphocytes |

**Fig. 1** Cells of the innate immune system versus cells of the acquired immune system. Eosinophils, neutrophils, macrophages, and natural killer (NK) cells confer innate immunity. B lymphocytes, macrophages, and dendritic cells are professional antigen presenting cells, often referred to as "APCs." They are the key players in facilitating acquired immunity.

Nonspecific immunity is based on specialized cells, such as natural killer (NK) cells, eosinophils and neutrophils, and phagocytic cells, such as microglial cells in the brain or Kupffer cells in the liver. None of these elements require specific antigen recognition. These cells belong to the innate immune system. However, most pathogens cannot be eliminated by innate immunity alone but only through the involvement of elements of the adaptive or acquired immune system, which is antigen-specific. This system relies on antigen presenting cells that do not simply destroy antigens but expose their antigens on the surface for recognition by T cells. In particular, macrophages link the innate with the acquired immune system being both phagocytes and antigen presenters. Antigen exposure to effector T lymphocytes leads to production of antigen-specific antibodies, to cytolytic elimination of infected cells in an antigen-specific manner and/or to pathogen killing by phagocytosis. Thus, antigen presenting cells govern the initiation of both humoral and cellular immunity.

## 2 Antigen Presenting Cells (APCs)

### 2.1 The Discovery of Langerhans

In the second half of the nineteenth century, the histologist Langerhans was the first to detect a cell that immunologists nowadays call an *antigen presenting cell.* He used a gold impregnation method to stain skin sections and saw irregularly shaped cells with long protrusions reminiscent of dendrites of cutaneous nervous system cells. These cells formed a network of dendrites within the spinal layer of the epidermis and were called *Langerhans cells* (LCs). At the ultrastructural level, LCs can be identified by unique organelles in their cytoplasma, termed *Birbeck granules.* Birbeck granules are unique to LCs and allow us to distinguish them from macrophages and other dendritic cells. Almost a century later, LCs were rediscovered by immunologists and shown to function as APCs. Today, it is obvious that LCs are representatives of the dendritic cell family, specialized in trapping antigens that have penetrated the epidermis of the skin. Hence, they function as the sentinels of the antigen-specific skin immune response.

### 2.2 Professional APCs

The term APC refers to not only LCs but to all types of leukocytes that are responsible for taking up antigens at the entry sites of the body, transporting them to lymphoid organs where they recruit antigen-specific T lymphocytes (short: T cells). APCs thus act at the very beginning of an adaptive immune response. In order to mount such a response against an antigen that enters the body for the first time, APCs must be able to activate so-called *naive* T cells that have not been stimulated before and "see" the respective antigen for the first time.

APCs that are able to prime naive T cells are termed *professional APCs* and include three sets of cells (Fig. 2): mononuclear phagocytes or macrophages, dendritic cells (DCs), and B lymphocytes (B cells). These three classes of APCs can be distinguished from nonprofessional APCs in that they express a specific combination of surface proteins that are critical for priming naive T cells. These surface proteins are (1) the gene products of the major histocompatibility complex (MHC): MHC class I and class II molecules bind peptides derived from antigenic proteins and present them to either cytotoxic T cells, characterized by the coreceptor CD8, or helper T cells expressing the coreceptor CD4, respectively; (2) the costimulatory molecules CD80 (B7.1) and CD86 (B7.2), which promote growth and differentiation of T cells upon engaging APCs; and (3) adhesion molecules such as ICAM-1 or LFA-3 that facilitate the primary contact between T cells and APCs.

DCs, although being highly diverse in origin and shape, appear to be the most potent professional APCs: DCs localize to all putative antigen entry sites of the skin, mucosa, and airways, they express high amounts of both classes of MHC molecules and, as members of the phagocyte family, they have an enormously high capacity to pino- and phagocytose antigenic material. Owing to their superior impact on the functionality of the acquired immune system, numerous attempts are ongoing to exploit DCs in immunotherapy of cancer and other diseases. Interestingly, about 100 years ago, the writer George Bernard Shaw stated in his famous book The Doctor's Dilemma: "There is only one genuinely scientific treatment for all diseases and that is to stimulate the phagocytes."

## 2.3 Nonprofessional APCs

One major prerequisite of bone marrow–derived professional APCs is their concerted expression of MHC class II and

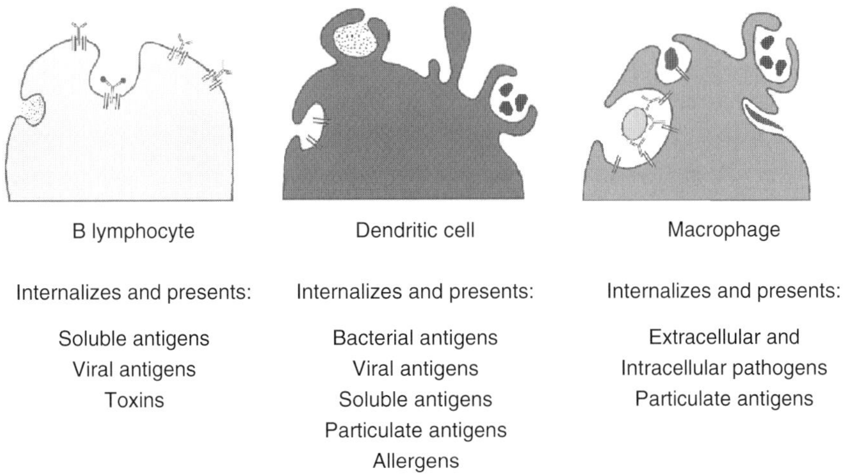


**Fig. 2** The three types of antigen presenting cells (APCs). They have overlapping but distinct repertoires of antigens that they internalize, process and present to T cells.

**Tab. 1** Nonprofessional APCs of the human immune system.

| *Cell type* | *Constitutive MHC class I* | *Constitutive MHC class II* | *Costimulators CD80/CD86* |
|---|---|---|---|
| Thymic cortical epithelial cells | + | + | − |
| Thymic medullary epithelial cells | + | + | − |
| Keratinocytes | + | − | − |
| Myoblasts | + | − | − |
| T lymphocytes | + | − | − |

costimulatory molecules CD80 and CD86. Apart from a very few exceptions, for example, vascular endothelial cells, only B cells, macrophages, and DCs are able to express reasonable amounts of CD80 and CD86 that are essential for activation of naive T cells. However, a number of other cell types can express MHC class II molecules (Table 1).

Thymic cortical epithelial cells (cTECs) that are critical for positive selection of $CD4^+$ thymocytes, and thymic epithelial cells of the medulla, which are responsible for deletion of self-reactive $CD4^+$ thymocytes, express MHC class II molecules in a constitutive fashion. The importance of MHC class II expression on cTECs is demonstrated by MHC class II–negative mutant mice that lack mature $CD4^+$ T cells. Following cytokine stimulation *in vitro* or during inflammatory states *in vivo*, MHC class II molecules are induced on a variety of epithelial and endothelial cells. IFN-$\gamma$ induces MHC class II molecules in keratinocytes of the skin and even in myoblasts of muscles. Likewise, activated T lymphocytes in humans can become MHC class II positive.

It has been shown that cells becoming MHC class II–positive in the presence of proinflammatory cytokines, such as IFN-$\gamma$, are able to present peptides to $CD4^+$ T cells and to activate them, provided that the T cells do not need costimulation. Neo-MHC class II–expressing cells that have the potential to present antigen, are termed *nonprofessional* APCs.

## 3 Antigens Presented by APCs

### 3.1 Antigen Types

According to the initial definition brought up by the discoveries of Behring and Ehrlich, "antigens" are substances that cause the production of antibodies. However, this historic definition only covers humoral immune responses. Nowadays, molecules that are recognized by the antigen receptor of a B lymphocyte (B-cell receptor) and/or the antigen receptor of a T lymphocyte (T-cell receptor) are termed *antigens*. Antigens that initiate an immune response are named *immunogens*, whereas others may induce tolerance and are named *tolerogens*. Another subset of antigens that have a low molecular weight (<4000 Da) and are immunogenic only if they are covalently bound to a carrier protein, is termed *haptens*. Examples of haptens are substances, such as poison ivy or penicillins, that bind covalently to proteins before or after take-up by APCs and then can cause allergy.

## 3.2 Biochemical Nature of Antigens

APCs are able to successfully present a large variety of organic substances (mainly consisting of carbon, hydrogen, oxygen, and nitrogen atoms); however, not a single inorganic substance is known to be able to trigger an antigen-specific immune response without the involvement of proteins. This is also true for large crystals, such as renal calculi. This clear difference in the immunogenicity indicates that the adaptive immune system evolved to circumvent the danger originating from organic and not inorganic material.

The most potent antigens are proteins and polypeptides. This is also true for polypeptides that consist of D-amino acids instead of naturally occurring L-amino acids. The reason for the strong immunostimulatory capacity is that proteins are easily recognized by the B-cell receptor and by (soluble) antibodies in the blood as well. In addition, proteins give rise to peptides during processing by APCs and peptides are the substances of choice to activate cellular immune responses through recognition by the T-cell receptor of T lymphocytes (Table 2).

Polysaccharides and glycans are by far weaker antigens than proteins. They can activate B cells but not T cells because APCs cannot present polysaccharides to T cells. Nevertheless, *Saccharomyces glycan* stimulates macrophages to secrete IL-1 and TNF-$\alpha$, thereby supporting antitumor immunity in mice. Polysaccharides from bacteria, such as multiple branched glucose polymers (dextranes), induce strong antibody responses in humans and mice, however, not in rabbits or guinea pigs.

Most lipids are too small to act as antigens that induce antibody production. Similar to other low molecular weight substances, a few lipids can function as haptens, such as cardiolipin. Cardiolipin is a phospholipid derived from the mitochondrial membrane. It is released from cells upon bacterial infection with *Treponema pallidum*, binds to carrier proteins and, in this context, initiates formation of anticardiolipin antibodies (Table 2).

However, DCs such as LCs express CD1 molecules that bind and present certain lipids or glycolipids in order to induce T-lymphocyte activation. Nonglycosylated and glycosylated mycolates,

**Tab. 2** Antigens presented by APCs via MHC and CD1 molecules, versus haptens.

| ***Restriction*** | ***Antigen*** | ***Hapten*** |
|---|---|---|
| MHC class I/MHC class II | Proteins<br>Polypeptides<br>Peptides<br>Glycopeptides | Polysaccharides<br>DNA<br>Penicillin<br>Cardiolipin<br>Organic substances |
| CD1 | Mycolates<br>Diacylglycerols<br>Sphingolipids<br>Polyisoprenoids | |

diacylglycerols, sphingolipids, and polyisoprenoids derived from endogenous and microbial sources have been described (Table 2).

Similar to pure lipids, only DNA or RNA is widely unable to trigger generation of antibodies. Moreover, APCs obviously lack specialized receptors for DNA or RNA necessary for activating T lymphocytes. Nucleoproteins, however, are suitable to generate anti-DNA antibodies. Thus, DNA can serve as a hapten. Anti-DNA antibodies are actually found in the serum of patients suffering from the autoimmune disease systemic lupus erythematosus (SLE). Some of these antibodies react with denatured DNA, others with intact double-strand DNA.

### 3.3 Origin of Antigens

Under normal circumstances, the immune system relying on APCs reacts only against foreign antigens, but not against self-antigens. In cases in which a foreign antigen is a homolog of a self-antigen, the extent of the immune response decreases with increasing similarity between both antigens. To take this rule into account, transplantation immunologists differentiate between the donor and the recipient of a transplanted organ: antigens derived from a recipient of the same species are called *alloantigens*. When the antigen is derived from another species, it is named *xenoantigen*. In contrast to that, *autoantigens* are derived from the same organism and may cause autoimmune diseases.

## 4 Antigen Entry Sites

Vertebrates offer microbes a large interface to enter the body. Consequently, the lymphatic system is organized in such a way that lymphoid tissues are located in close proximity to all putative entry sites. Afferent lymphatic vessels drain fluid and APCs from the skin and from other peripheral tissues to the lymph node. The spleen filters antigen mainly from the blood stream. The gut-associated lymphoid tissues (GALT), which include the tonsils, adenoids, appendix, and the Peyer's patches (PP) in the small intestine, collect antigen from the epithelial surfaces of the gastrointestinal tract.

Likewise, the bronchial-associated lymphoid tissues (BALT) and the mucosal-associated lymphoid tissues (MALT) protect the respiratory epithelia and other mucosa respectively. Each of these lymphatic tissues traps foreign antigens through the activity of migratory APCs that present antigenic peptides to T lymphocytes, thereby initiating antigen-specific immunity. Importantly, even in the absence of foreign antigen, lymphoid tissues provide signals to lymphocytes via presentation of self-antigens. This is important for the survival of naive and silent T lymphocytes, thereby contributing to the maintenance of T-cell homeostasis.

### 4.1 APCs in the Skin

Antigen that has penetrated the epidermis is primarily picked up by LCs. LCs are derived from monocytes that leave the blood stream and migrate into the epidermis where they adopt the shape of typical DCs. Antigen-fed LCs leave the skin and migrate as so-called *veiled cells* via the afferent lymph vessels to the draining lymph nodes. There, they accumulate as interdigitating dendritic cells (IDCs) in the T cell–rich areas.

Under pathological conditions, for example, in skin suffering from the autoimmune disease psoriasis, not only LCs but also keratinocytes express high amounts of MHC class II molecules and become capable of functioning as APCs. In other squamous epithelia, DCs different from LCs do occur. This is true of the oral and nasal mucosa, the trachea and bronchus, and the esophagus and the tonsils. In the epithelia lining the lung, the stomach, and the gut, DCs have not been found, as yet.

### 4.2 APCs in the Mucosa

Although tight junctions and the mucous layer of mucosal surfaces widely exclude microbes and large particulate antigen(s), antigenic proteins are known to surpass this barrier through endo- and transcytosis. Granulocytes, as representatives of the innate immune system, and macrophages functioning as APCs take up these antigens in the lamina propria. Furthermore, in the gut and bronchus, an antigen transfer system exists that transports antigens from the lumen to the GALT and BALT respectively. A prerequisite of this system is the presence of a specialized type of epithelial cell, the so-called *M cell* (see below).

### 4.3 APCs in the Gut

The uptake of bacteria, proteins, and abiotic substances (e.g. latex) by the Peyer's patches and by lymphoid follicles in other parts of the gut, for example, the appendix, is well known. The GALT is separated from the gut lumen by a single layer of epithelium, containing M cells (Fig. 3). In ultrathin sections, an M cell is seen as a rim of apical cytoplasm that bridges the space between two adjacent epithelial cells. An M cell forms a kind of an umbrella above a space surrounded by epithelial cells and filled with all sorts of APCs including DCs, macrophages, and B cells. Thus, antigen from the gut lumen is transcytosed through the epithelial monolayer and

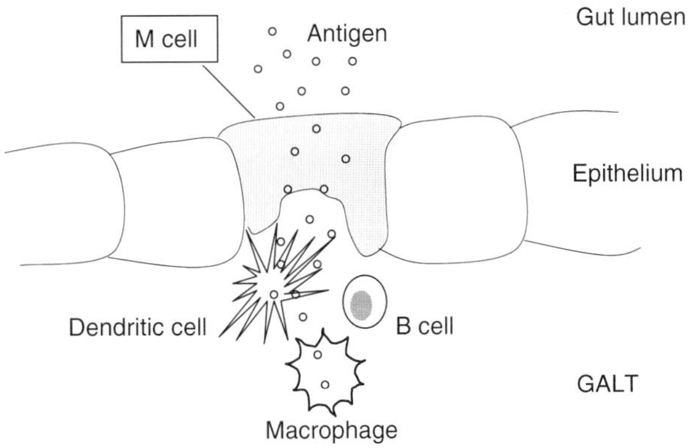


**Fig. 3** The gut-associated lymphoid tissue (GALT). Antigens from the gut lumen traverse the single layer of epithelia via M cell–mediated transcytosis whereupon they get in contact with APCs (dendritic cells, macrophages, B cells) of the GALT. APCs are very similarly organized in the lymphoid tissues of the bronchus and Waldeyer's ring.

reaches APCs, which take up and process the antigen and present it to T lymphocytes in the PP or in mesenteric lymph nodes.

## 4.4
## Antigen Uptake in the Respiratory Tract

In the upper respiratory tract, antigen presenting LCs occur in the transitional zones between the keratinizing epithelium of the skin and the mucosal surface of the nasal cavity and the nasopharyngeal region, for example, the lips. Moreover, APCs fulfill their sentinel function in and around the lymphoid tissue of Waldeyer's ring. The epithelium above the ring of Waldeyer contains, just as BALT and GALT do, antigen transporting M cells. Furthermore, DCs are found in the epithelium of the trachea. The rest of the mucosal surface of the upper respiratory tract serves to exclude rather than take up antigen.

Antigen exclusion is also dominant in the lower respiratory tract, in particular, the lung. Antigen clearance mainly depends on mucociliary activity of the epithelia and/or alveolar macrophages. Macrophages in the alveoli act as both phagocytic cells of the innate defense system and as APCs giving rise to acquired immunity. Some of them simply phagocytose and digest foreign antigens involving the lung; others stimulate a local immune response. For this purpose, alveolar macrophages loaded with antigen cross the lining of the alveoli and migrate into lymph nodes located in the lung tissue. Thus, similar to DCs, alveolar macrophages form a lung-specific antigen handling system.

# 5
# Antigen Presenting Molecules

Both types of the classical molecules of the MHC, class I and class II, are peptide receptors and encoded by genes that display the highest degree of polymorphism in the genome. Within the MHC molecules, the polymorphism is mainly restricted to the regions that constitute the antigenic peptide-binding cleft. The advantage of the multitude of MHC alleles encoding for a multitude of distinct peptide-binding specificities is that there will be hardly any foreign peptide that cannot be bound by at least a few individuals of a population. In contrast, CD1 molecules are MHC class I–like receptors for mycobacterial lipids or glycolipids. They are not encoded in the MHC and are monomorphic. Their physiological role is, as yet, ill defined.

## 5.1
## MHC Class I Molecules

MHC class I molecules are glycosylated transmembrane proteins that consist of two subunits: the larger $\alpha$-chain encoded in the MHC genetic locus is polymorphic and forms the peptide-binding site. It associates with the smaller subunit, $\beta_2$-microglobulin ($\beta_{2m}$), which is nonpolymorphic and encoded outside the MHC. $\beta_{2m}$ and the membrane-proximal $\alpha_3$-domain of the $\alpha$-chain fold as immunoglobulin domains, whereas the membrane-proximal (membrane-distal) $\alpha_1$- and $\alpha_2$-domain constitute the peptide-binding cleft (Fig. 4).

Quite in contrast to other peptide receptors described in the hormone system, the binding cleft of MHC molecules is highly promiscuous so that a large variety of antigenic peptides can be presented. In the case of MHC class I molecules, this peculiarity is accomplished through three structural criteria shared by peptide ligands: (1) the peptide length is limited to 8 to 10 amino acids, with the majority

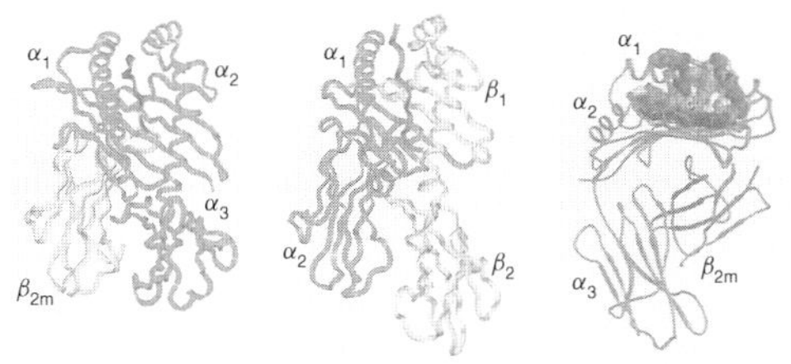


**Fig. 4** X-ray structure of the MHC class I molecule HLA-Aw68, the MHC class II molecule HLA-DR1, and the CD1b molecule, which are key surface molecules of APCs. Adapted from Guo, H. C. et al. (1992) *Nature* **26**, 300–301; Murthy, V. L., Stern, L. J. (1997) *Structure* **5**, 1385–1396; Gadola, S. D., Zaccai, N. R., Harlos, K., Shepherd, D., Castro-Palomino, J. C., Ritter, G., Schmidt, R. R., Jones, E. Y. Cerundola, V. (2002) Structure of human CD1b with bound ligands at 2.3 Å, a maze for alkyl chains, *Nat. Immunol.* **3**, 721–726.

of peptides being 9-mers; (2i) 1 to 2 side chains of the peptide, called *anchors*, fit into corresponding specificity pockets in the binding cleft. Both the position and identity of these anchor residues vary, depending on the particular MHC class I molecule (allele-specific anchor motifs); and (3) the amino- and carboxyterminni of peptides are fixed in the binding cleft through a hydrogen-bonding network.

To illustrate this in more detail, the sequences of self- and foreign peptides restricted by the human MHC class I molecule HLA-A2 are compared (Table 3): 70% of these peptides are 9-mers with a leucine or methionine at the anchor position P2 and a valine at the anchor position P9. Owing to the fact that the sequences at the nonanchor positions are widely irrelevant for binding, more than

**Tab. 3** Peptide anchor motif of antigenic peptides binding to the MHC class I molecules HLA-A1 and HLA-A2.

| ***Antigen*** | | | | | ***Sequence*** | | | | | | ***Allele*** |
|---|---|---|---|---|---|---|---|---|---|---|---|
| | | **P2** | | | | | | | **P9** | **P10** | |
| Tyrosinase (369–377) | Y | **M** | N | G | T | M | S | Q | **V** | | |
| EBV LMP2 (426–434) | C | **L** | G | G | L | L | T | M | **V** | | |
| HIV RT(476–484) | I | **L** | K | E | P | V | H | G | **V** | | |
| HTLV-1 Tax(11–19) | I | **L** | F | G | Y | P | V | Y | **V** | | HLA-A2 |
| Hepatitis B sAg(335–343) | W | **L** | S | L | L | V | P | F | **V** | | |
| Tyrosinase (1–9) | M | **M** | N | G | T | M | S | Q | **V** | | |
| pmel 17/gp100 | I | **L** | D | G | T | A | T | L | R | **L** | |
| Influenza B NP(85–94) | K | **L** | G | E | F | Y | N | Q | M | **M** | |
| Influenza MP(59–68) | I | **L** | G | F | V | F | T | L | T | **V** | |
| HPV11 E7 (4–12) | R | **L** | V | T | L | K | D | I | **V** | | |
| | | | | | Anchor motif | | | | | | |
| | | L | | | | | | | V | | |
| | | M | | | | | | | L | | |
| | | | | | | | | | M | | |
| | | | | | Anchor motif | | | | | | |
| – | | | **P3** | | | | | | **P9** | | |
| | | | D | | | | | | Y | | HLA-A1 |
| | | | E | | | | | | | | |

500 different peptides have been estimated to constitute the self-peptide repertoire of HLA-A2 expressed on the surface of a B cell; however, none of these peptides binds to HLA-A1 and vice versa. The reason is that HLA-A1 and HLA-A2 differ strongly in the anchor motif of their peptide ligands.

Today, more than 220 distinct HLA-A and more than 450 distinct HLA-B type alleles are known. Not all of the alleles have distinct anchor motifs, however, the high diversity of allelic MHC products observed in the human population allows a high diversity of self- and foreign peptides to be bound. This strategy reduces the risk that some pathogens may escape recognition by the immune system due to holes in the binding capacity of MHC molecules.

## 5.2 MHC Class II Molecules

The crystallographic structure of MHC class II molecules shows that they are folded very much like MHC class I molecules, although class II dimers consist of a noncovalent complex of an $\alpha$- and a $\beta$-chain, which are both encoded within the MHC. Again, highly polymorphic regions are located in the peptide-binding cleft, consisting of 8$\beta$-strands and 2$\alpha$-helical segments that serve to sandwich peptide ligands (Fig. 4).

The only and decisive structural difference is that the class II binding cleft is open at both ends, whereas the class I cleft is closed. A critical consequence is that naturally processed peptides binding to class II MHC molecules are at least 12 to 14 amino acids long and can be much longer. The majority of class II–associated peptides are 15- to 17-mers. The binding forces of the class II peptide-binding cleft are distributed to 3 to 4 specificity pockets, most often localized at relative positions P1, P4, P6, and P9 (Table 4). Moreover, about a dozen hydrogen bonds contacting the whole backbone of a typical 15-mer stabilize the class II–peptide interaction. In further contrast to class I molecules, the specificity pockets of the class II groove are more permissive, often allowing 3 to 5 structurally similar amino acid side chains to fit into a particular pocket (Table 4).

The advantage of this peculiarity is that class II molecules encoded by the same MHC class II allele are able to accommodate a large repertoire of different peptide sequences. This may be viewed as an adaptation to pathogens that constantly change their immunodominant antigens by high mutation rates. The disadvantage is that MHC class II allele-specific peptide ligand motifs are rather complex and class II–restricted immunodominant epitopes are difficult to predict.

## 5.3 CD1 Molecules

CD1 molecules are nonpolymorphic transmembrane glycoproteins encoded by genes outside the MHC. The overall structure of CD1 is similar to that of MHC class I molecules in that CD1 molecules form heterodimers of ~45-kDa heavy chains associated with $\beta_{2m}$. However, unlike MHC class I molecules, CD1 is not retained within the endoplasmic reticulum but is targeted at endocytic compartments where it binds its ligands. CD1 molecules are expressed only on dendritic cells, monocytes, and some thymocytes. Like classical MHC class I and class II molecules, CD1 molecules are recognized by T cells. However, unlike classical MHC molecules, group I CD1 molecules (CD1a, CD1b, CD1c), which are found in humans but not in mice, present mycobacterial lipids, mycolic acids, lipoarabinomannan,

**Tab. 4** Self-peptides eluted from the class II molecule HLA-DR4 of human monocyte-derived dendritic cells, stimulated with TNF-$\alpha$.

| ***Antigen*** | | | | | | | | | ***Sequence*** | | | | | | | | | | | |
|---|---|---|---|---|---|---|---|---|---|---|---|---|---|---|---|---|---|---|---|---|
| | | | | | | | **P1** | | | **P4** | | **P6** | | | | | | | | |
| HLA-B60 | D | T | Q | F | V | R | **F** | D | S | **D** | A | **A** | S | Q | R | M | | | | |
| HLA-B60 | D | T | Q | F | V | R | **F** | D | S | **D** | A | **A** | S | Q | R | | | | | |
| HLA-B60 | | T | Q | F | V | R | **F** | D | S | **D** | A | **A** | S | Q | R | | | | | |
| HLA-B60 | | | Q | F | V | R | **F** | D | S | **D** | A | **A** | S | Q | R | | | | | |
| HLA-B60 | | | | F | V | R | **F** | D | S | **D** | A | **A** | S | Q | R | | | | | |
| HLA-B60 | | | | F | V | R | **F** | D | S | **D** | A | **A** | S | Q | R | M | E | P | | |
| $\beta_2$m | | | | Y | L | L | **Y** | Y | T | **E** | F | **T** | P | T | E | K | D | E | | |
| $\beta_2$m | | | | | L | L | **Y** | Y | T | **E** | F | **T** | P | T | E | K | D | E | | |
| Mannose receptor | | | F | E | N | K | **W** | Y | A | **D** | C | **T** | S | A | G | R | S | D | G | |
| Calreticulin | | | D | N | P | E | **Y** | S | P | **D** | P | **S** | I | Y | A | Y | D | N | | |
| Transferrin receptor | | | T | G | Q | F | **L** | Y | Q | **D** | S | **N** | W | A | S | K | Y | | | |
| Apolipoprotein d | | V | L | N | Q | E | **L** | R | A | **D** | G | **T** | V | N | Q | I | E | G | | |
| Apolipoprotein d | | | L | N | Q | E | **L** | R | A | **D** | G | **T** | V | N | Q | I | E | G | | |
| Apolipoprotein d | | | | | Q | E | **L** | R | A | **D** | G | **T** | V | N | Q | I | E | G | | |
| IL-10 receptor | | | D | K | L | S | **V** | I | A | **E** | D | **S** | E | S | G | K | Q | N | P | G |
| Rab-7 | | | F | P | E | P | **I** | K | L | **D** | K | **N** | D | R | A | K | A | S | A | |
| Rab-7 | | | F | P | E | P | **I** | K | L | **D** | K | **N** | D | R | A | K | A | S | | |
| Rab-7 | | | F | P | E | P | **I** | K | L | **D** | K | **N** | D | R | A | K | A | | | |
| | | | | | | | | | Anchor motif | | | | | | | | | | | |
| | | | | | | | F | | | D | | A | | | | | | | | |
| | | | | | | | Y | | | E | | T | | | | | | | | |
| | | | | | | | W | | | | | S | | | | | | | | |
| | | | | | | | I | | | | | N | | | | | | | | |
| | | | | | | | L | | | | | | | | | | | | | |
| | | | | | | | V | | | | | | | | | | | | | |

phosphatidylinositol mannosides, and hexosyl-1-phosphoisoprenoids to cytotoxic T cells. These ligands are derived either from internalized mycobacteria or from the uptake of lipoarabinomannans by the mannose receptor expressed on the surface of monocytes and dendritic cells. No naturally processed bacterial antigens have been identified in the context of the group II molecule CD1d molecule, which is expressed in both humans and mice. CD1d molecules present bacterial $\alpha$-galactosylceramide to NK T cells. The relationship between the peptide- and lipid-binding capacities of CD1 is not clear, as yet. Structural studies show that CD1 molecules bear a deep and hydrophobic ligand-binding cleft in which glycolipids bind – it is open whether peptides associate to the same site (Fig. 4).

Compared to MHC class I and class II molecules, CD1 proteins appear to represent a separate lineage of antigen-presenting molecules specialized in presenting microbial lipids, glycolipids, and a subset of peptide antigens to T cells.

## 6
## Antigen Processing by APCs

Virus-infected cells can be destroyed by $CD8^+$ cytotoxic T cells provided that the respective cells present viral peptide antigens bound to MHC class I molecules on their cell surface. Likewise, exogenous foreign antigens that are pino- or phagocytosed by APCs at sites of inflammation lead to activation of CD4$^+$ helper T cells when APCs present antigen-derived peptides in the context of MHC class II molecules. The concept that foreign antigenic peptides stimulate T cells only in the context of MHC molecules has been introduced by Zinkernagl and Doherty as *MHC restriction*. The conversion of protein antigens derived from the extracellular space or the cytosol into peptides and the conservation of this antigenic information through loading onto MHC class II or class I molecules is accomplished by pathways of antigen processing.

The different fates of vesicular and cytosolic antigens are mainly due to the segregated pathways of biosynthesis and assembly of class I and class II molecules. Both pathways, however, have evolved as adaptations of basic cellular functions, such as endocytosis, transmembrane transport, and protein degradation, which are not exclusively used for antigen presentation.

## 6.1 MHC Class I Processing Pathways

The peptide-receptive binding cleft of class I molecules is localized on the luminal side of the ER. Class I–restricted antigenic peptides, however, are derived from cytosolic or nucleic protein antigens. Therefore, accessory molecules are necessary for antigen processing. IFN-$\gamma$ increases the expression of these accessory molecules and MHC class I molecules (Fig. 5).

### 6.1.1 Generation of Peptide Ligands

The majority of antigenic peptides presented by MHC class I molecules appears to be generated by the multicatalytic proteinase complex, named the proteasome. The proteasome is localized in the cytosol and nucleus and consists of 28 subunits that form a central channel where unfolded polypeptides are cleaved into short peptides (Fig. 5). In the presence of IFN-$\gamma$, the catalytic activity of the proteasome changes due to the replacement of 3 constitutive subunits for the new subunits LMP2, LMP7, and MECL-1. The proteasome containing these new subunits is called the immunoproteasome and preferably generates peptides with a C-terminal hydrophobic or basic amino acid. Peptides with these structural features are superior to other peptides with regard to binding to MHC class I molecules.

Apart from the proteasome, other proteases and peptidases are involved in processing of antigens in the MHC class I pathway: cytosolic calpains and aminopeptidases are sometimes required, the actual involvement being strongly influenced by the type of antigen to be cleaved. Moreover, the ER-resident aminopeptidase ERAP-1 has recently been described to be mainly responsible for N-terminal trimming of precursor peptides, thereby giving rise to typical 8-mer, 9-mer, and 10-mer epitopes.

### 6.1.2 TAP

The transporter associated with antigen processing (TAP) belongs to the ABC family of heterodimeric transmembrane transporters, which are fueled by ATP hydrolysis. TAP is localized in the membrane of the ER and shuttles peptides of variable length ($n = 6$ to 30 amino acids) from the cytosol into the ER lumen (Fig. 5). This is a prerequisite

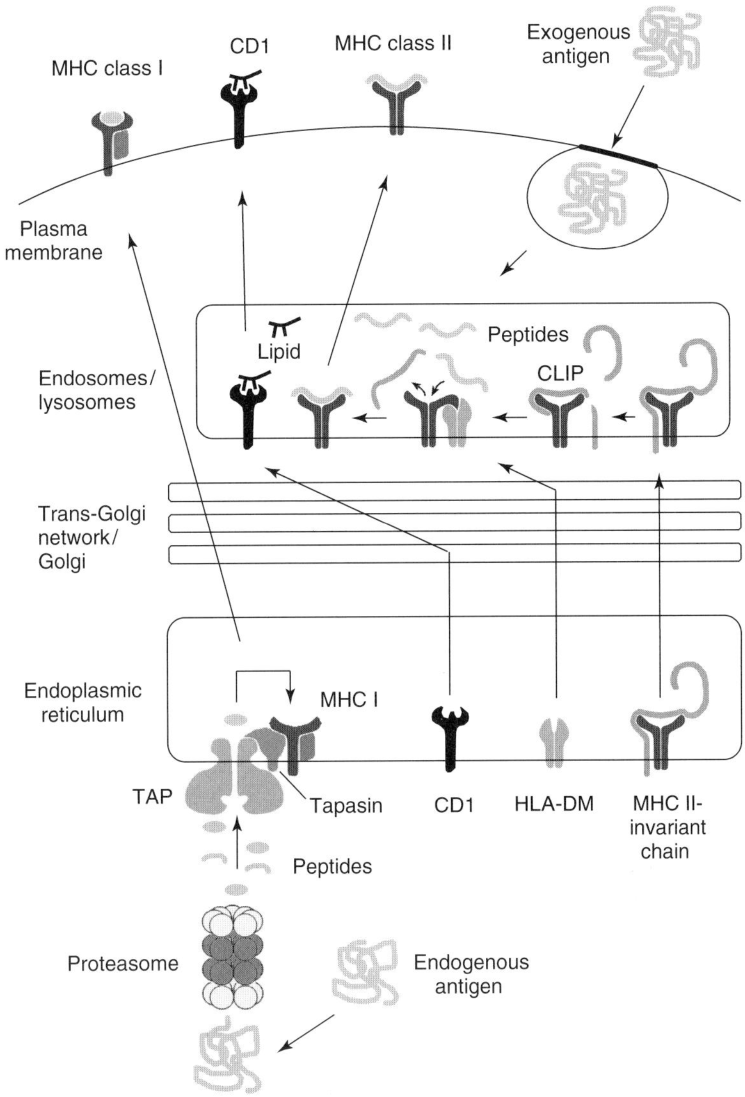


**Fig. 5** Antigen processing pathways leading to the generation of MHC class I-peptide, MHC class II–peptide, and CD1-lipid complexes.

for the regular supply of MHC class I molecules with peptide antigen. Moreover, TAP is an essential constituent of a multimolecular peptide–loading complex in the ER that regulates loading of MHC class I molecules with cognate antigenic peptides.

### 6.1.3 The MHC Class I Loading Complex

Newly synthesized MHC class I heavy chains ($\alpha$) are stabilized by binding to the chaperone calnexin until the heavy chain forms dimers with $\beta_2$-microglobulin ($\beta_{2m}$). After dissociation of calnexin, the chaperone calreticulin allows $\alpha$:$\beta_{2m}$

complexes to enter a multimolecular loading complex, which is responsible for the formation of cognate class I-peptide complexes. Constituents of this loading complex are $\alpha$:$\beta_{2m}$, calreticulin, TAP, the protein-disulfide-isomerase ERp57, and the TAP-associated chaperone tapasin (Fig. 5). During peptide loading, ERp57 gives allowance for the regulated opening and closing of a disulfide bond in the binding cleft domain of the class I heavy chain. Tapasin bridges TAP and $\alpha$:$\beta_{2m}$, and retains peptide-free $\alpha$:$\beta_{2m}$ dimers in the ER until a stably binding peptide occupies the groove. Consequently, preferentially long-lived $\alpha$:$\beta_{2m}$-peptide complexes leave the ER and make it to the cell surface. Thus, tapasin is thought to function as a peptide editor, similar to HLA-DM in the MHC class II processing pathway (see below).

## 6.2 MHC Class II Processing Pathways

MHC class II-associated peptides originate from two sources: endocytosed exogenous antigens or endogenous self-antigens, the majority of them being membrane proteins derived from endosomes or the nucleus (Table 4).

### 6.2.1 Endosomal Generation of Peptides

Professional APCs are equipped to internalize whole pathogens, for example, bacteria or viruses, small vesicles, multimolecular complexes, proteins or peptides via phagocytosis, pinocytosis or in a receptor-mediated fashion. In all these cases, endocytosed antigens end up in endosomal compartments, known for their low pH and the high redox potential (Fig. 5). These conditions favor denaturation of protein antigens and proteolytic generation of peptides via hydrolases, amino- and carboxypeptidases, and endopeptidases. Endosomal/lysosomal compartments of APCs, which are enriched in MHC class II molecules, as determined by electron microscopy or immunohistochemically, have been termed *MIICs* (MHC class II compartments).

### 6.2.2 Invariant Chain and CLIP

The invariant chain (Ii) is a chaperone dedicated primarily to MHC class II molecules. In DCs, Ii also binds to CD1 molecules. In general, Ii trimers bind to newly synthesized class II $\alpha\beta$ dimers, allowing them to obtain their native conformation. Since Ii binding blocks the sole peptide-binding cleft, Ii-associated class II molecules cannot bind antigens. Owing to a targeting signal in the cytosolic tail of Ii, Ii class II complexes are sorted from the trans-Golgi-network directly to endosomal/lysosomal MIICs (Fig. 5).

In MIICs, Ii is proteolytically degraded whereas $\alpha\beta$ dimers remain intact. The proteases mainly responsible for Ii degradation are cathepsin S in DCs and B cells and cathepsin L in cortical epithelial cells. The terminal Ii fragment that still occupies the peptide-binding cleft of class II molecules is termed *CLIP* (class II MHC-associated Ii peptide). Under steady state conditions, CLIP is part of the self-peptide repertoire of class II molecules. Since it dissociates from the majority of class II allelic proteins rather slowly, CLIP has to be actively removed from the class II binding cleft.

### 6.2.3 HLA-DM

HLA-DM (termed H2-M in mice) is a non-classical MHC class II protein, displaying 20 to 25% sequence homology to both classical class I and class II molecules. X-ray analysis revealed that HLA-DM lacks a peptide-binding groove and, thus, is unable to function as an antigen-presenting molecule. It bears a targeting signal for MIICs where it accumulates and transiently engages in complexes with classical class II MHC molecules. HLA-DM has three important functions (Fig. 5): (1) it catalytically removes the Ii-derived precursor peptide CLIP, thereby promoting peptide loading; (2) it remains bound to empty class II molecules and prevents their unfolding as a molecular chaperone; and (3) it functions as a peptide editor, as it removes peptides that bind with low kinetic stability to class II dimers. Consequently, preferentially high-stability class II peptide complexes are displayed on the surface of APCs, as long as these cells bear sufficient HLA-DM. This principle allows APCs, once loaded with cognate antigen, to activate $CD4^+$ T cells for prolonged periods of time, even a couple of days after encounter of antigenic material.

#### 6.2.4 Tetraspan Network

Professional or nonprofessional APCs and even T cells express proteins, such as CD9, CD37, CD53, CD63, CD81, CD82, and CD151, belonging to a family termed *tetraspan proteins* or *tetraspanins*. The nomenclature relates to the fact that all of them traverse the membrane four times. It is well established that they are able to form homo- and heterodimers and thus form two-dimensional networks in the membrane. In APCs, tetraspanins form clusters with various integrins and MHC molecules. In MIICs, CD82 and CD63 associate with HLA-DM and classical class II molecules. On the cell surface, CD81, CD9, and CD53 form microdomains together with MHC class II and class I molecules. The function of tetraspan microdomains may be to increase the local density of MHC class II peptide complexes in loading compartments and on the surface, so that the avidity of an APC–T cell encounter is increased, thereby raising the efficacy of T-cell activation.

## 7 Activation of T Cells by APCs

After leaving the thymus, mature T cells recirculate between blood and peripheral lymphoid tissue until they recognize an MHC-peptide complex on the surface of an APC. T cells that have not yet been activated via encounter of an antigen are termed *naive T cells*. In contrast, T cells that have already contacted MHC molecules carrying foreign peptide antigens start to proliferate and differentiate into cells capable of contributing to the removal of the antigen. They are termed *effector T cells*.

Effector T cells can be subdivided into three classes, according to the type of pathogen that may skew the efficiency of peptide entry into different processing pathways. Peptides from pathogens that multiply in the cytosol, such as influenza virus, vaccinia, and *Listeria monocytogenes*, are presented by MHC class I molecules and activate $CD8^+$ cytotoxic T cells that kill infected target cells. Pathogens that accumulate in endosomal vesicles of macrophages, such as mycobacteria or *Leishmania*, facilitate the differentiation of $T_{H1}$ helper cells that express Fas ligands enabling them to kill infected macrophages. Extracellular antigen derived from viruses tend to stimulate the production of $T_{H2}$ helper cells that initiate a humoral immune response by activating naive B cells to produce neutralizing IgM antibodies.

The activation of naive T cells following engagement with MHC–peptide complexes in APCs constitutes a primary immune response. In the case of such a primary response, a subset of long-lived T cells is generated that gives an accelerated response upon seeing the same antigen for the second time. These T cells are called *memory T cells.* Memory T cells differ in several instances from naive T cells, but like naive T cells they require activation by professional APCs in order to become effector T cells.

Finally, it has been shown that T cells need to contact APCs continuously in order to stay alive. This is accomplished by the recognition of MHC-self-peptide or MHC-foreign peptide complexes on the surface of APCs.

## 7.1
## Cell Adhesion Molecules

The migration of naive T cells through lymph nodes and their initial contacts with APCs depend on interactions that are not antigen-specific, hence are independent from MHC-peptide complexes. The initial cell–cell contact between APCs and T cells is controlled by an array of adhesion molecules on the surface of T cells that recognize a complementary array of adhesion molecules on the surface of an APC. The main classes of adhesion molecules are the selectins, the integrins, members of the Ig superfamily, and mucin-like molecules.

Selectins are particularly important for the T cell homing into particular tissues. For example, L-selectins on the T-cell surface binds to addressins, such as CD34 or MAdCAM-1, on vascular endothelial cells.

To guide naive T cells to DCs, chemokines play a role: the chemokine MIP-3$\beta$ directs naive T cells and mature DCs into lymphoid tissues. Moreover, DCs in lymphoid tissues express the chemokine SLC (secondary lymphoid tissue chemokine) that binds to the chemokine receptor CCR7 expressed on naive T cells. This event obviously reinforces the strength of successive T cell–DC interactions mediated via integrins and proteins of the Ig superfamily.

Integrins are heterodimeric cell surface proteins, consisting of a large $\alpha$-subunit and a smaller $\beta$-subunit. There are several subfamilies of integrins broadly defined by their common $\beta$-chains. An important integrin expressed mainly by T cells, but also by macrophages, is LFA-1 (lymphocyte function-associated antigen-1), an $\alpha_L\beta_2$ integrin. LFA-1 mediates migration of both naive and effector T cells out of the blood and the initial contact with DCs. Another integrin VLA-4 (very late activation antigen-4), which is particularly strongly expressed by activated effector T cells, and the Ig family member CD2 function as substitutes for LFA-1. The majority of cell surface adhesion molecules being involved in T cell–APC interactions are members of the Ig superfamily. Three very similar intercellular adhesion molecules (ICAM-1, ICAM-2, ICAM-3) bind to the integrin LFA-1. ICAM-1 (CD54) and ICAM-2 (CD102) are expressed on APCs and on epithelia, whereas ICAM-3 (CD50) is only expressed on T cells and B cells. In addition to binding to LFA-1, ICAM-3 has a high affinity for the lectin DC-SIGN (DC-specific ICAM-3-grabbing nonintegrin; CD209), which is only found on DCs. The ICAM-3/DC-SIGN interaction appears to be exclusive to the contact between naive T cells and DCs, whereas CD2 binding to LFA-3 synergizes with LFA-1 binding to ICAM-1 and ICAM-2 on all types of APCs.

Importantly, it turned out that the strength of the LFA-1/ICAM interaction of a naive T cell and an APC depends on the type of MHC–peptide complexes being presented by the APC; in those cases in which the T cell recognizes a cognate MHC-peptide ligand, a conformational change in LFA-1 increases its affinity for ICAM-1 or ICAM-2 of the APC. This phenomenon allows that the T cell–APC contact can persist for several days, so that a naive T cell can proliferate and become an effector T cell. In the majority of encounters, the T cell will not recognize an appropriate peptide antigen and separates from APCs, keeping on migrating through the lymph node. Thus, the transient nature of binding of naive T cells to APCs via adhesive interactions is crucial in the sampling of large numbers of MHC molecules and APCs.

## 7.2 Costimulatory Molecules

MHC–peptide complexes expressed on APCs are recognized by ligation of a corresponding T-cell receptor on the T-cell side. This interaction can be viewed as a first signal and it is critical for the antigen-specificity of a cellular immune response. However, in order to trigger clonal expansion and differentiation of a naive T cell, a second or costimulatory signal delivered by the same APC is required.

The best-studied costimulatory molecules solely expressed by APCs are B7.1 (CD80) and B7.2 (CD86). Both are homodimeric members of the Ig superfamily. Both B7 molecules share a common receptor on T cells, CD28. Ligation of CD80 or CD86 by CD28 costimulates the clonal expansion of naive T cells. Lack of B7 molecules abrogates T-cell proliferation.

Other costimulatory molecules on APCs are the TNF family member CD40 and 4-1BBL, which bind to CD40 ligand and 4-1BB (CD137) on the T-cell side, respectively. Both pairs of molecules are induced upon B7/CD28-mediated activation and function in sustaining the development of a full T-cell response. Notably, not only does the T cell receive a signal via the CD40/CD40L engagement but the APC also receives a signal. The best-characterized case is the conditioning of developing DCs via the engagement of activated $CD4^+$ helper T cells, so that the DC gains the capacity to fully activate $CD8^+$ cytotoxic T cells ("license-to-kill model"; cf Sect. 10.5.1.).

Activated T cells express an additional receptor for B7 molecules, denoted as CTLA-4, a homolog of CD28. However, CTLA-4 has a manyfold higher affinity to B7 than CD28 and transmits an inhibitory signal to T cells. Thus, CTLA-4 upregulation on the surface of T cells serves to downregulate proliferation and cytokine secretion of activated T cells.

A further CD28-like costimulator is ICOS, which recognizes LICOS produced on activated B cells, monocytes, and DCs. ICOS is poorly investigated so far – it is only known to induce IL-10 secretion of activated T cells.

The physiological relevance of the involvement of costimulatory molecules is to prevent destructive immune responses in the absence of signals mediated by pathogens, for example, responses against self-tissues. This is accomplished through the limitation that APCs express costimulatory B7 molecules only upon receiving danger signals from bacterial or viral products or equivalent proinflammatory signals. Consequently, naive T cells become effector T cells only when they receive signal 1 and signal 2 from the same APC. When signal 2 is lacking, the T cell is rendered anergic. Anergy means that the T cell becomes refractory to signals, even when signal 1 has been sent by professional APCs.

## 7.3 The Immunological Synapse

High-resolution confocal microscopy revealed that the key molecules involved in the formation of heterotypic junctions between APCs and T cells are reorganized in a highly characteristic manner. This specialized junction that results from cytoskeleton-driven clustering of MHC-peptide complexes, T-cell receptor molecules, adhesion molecules, and costimulators has been termed the *immunological synapse*. It is the close apposition of two membranes and the peculiar shape of the T cell that is reminiscent of neurological synapses (Fig. 6).

The immunological synapse consists of two concentric rings. In the center, proteins of the so-called central supramolecular activation cluster (c-SMAC), such as the T-cell receptor (TCR), CD2, CD3, the costimulator CD28, the protein kinases lck, fyn, and PKC-$\theta$ cocluster on the T-cell surface. Complementarily, MHC-peptide complexes and the costimulator CD80 have been described to cluster in the c-SMAC zone of APCs. The c-SMAC is surrounded by a second zone, the peripheral supramolecular activation cluster (p-SMAC). On T cells, the p-SMAC contains the integrin LFA-1 and the cytoskeletal protein talin. Accordingly, the adhesion molecule ICAM-1, known to be a ligand of LFA-1, coalesces in the p-SMAC of APCs.

The p-SMAC provides adhesive anchoring of the T cell to the APC, whereas the c-SMAC represents a protected zone for sustained signaling via the TCR. Immunological synapses are maintained for at least 1 h. Outside the p-SMAC is a third area containing proteins that are excluded from the synapse, such as the mucin CD43 or the phosphatase CD45. Exclusion of both types of proteins from the T cell–APC contact area is essential, as both molecules are very large and would therefore prevent

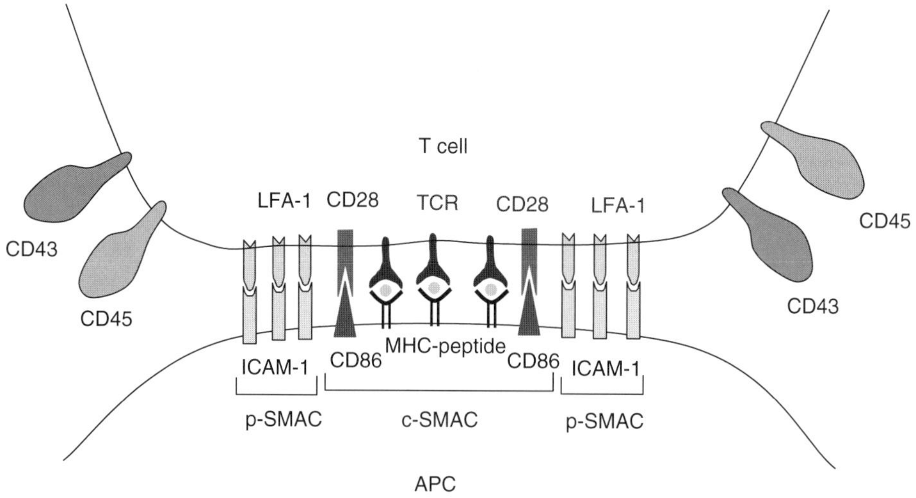


**Fig. 6** The immunological synapse between a T cell and an APC. In the central zone (c-SMAC), MHC-peptide complexes, costimulators, for example, CD86, TCRs, and CD28 accumulate. In the surrounding ring zone, termed p-SMAC, the adhesion molecules ICAM-1 and LFA-1 are found. Excluded are large molecules, for example, CD43 and CD45.

small molecules, such as the TCR and MHC molecules, from interacting with each other.

Exploratory adhesive interactions of T cells and APCs precede formation of the synapse. The adhesion molecule ICAM-3 has been reported to initiate antigen-independent scanning of the APC surface by T cells followed by early signaling events and rearrangements of the T cellular cytoskeleton. The synapse itself appears to be a highly dynamic structure. At the beginning, cognate TCR ligands localize to an outermost ring of the nascent synapse, whereas LFA-1 and ICAM-1 constitute the c-SMAC. After 5 to 10 min, ICAM-1 moves to the p-SMAC and MHC-peptide clusters concentrate at the heart of the synapse. Finally, the coreceptor CD4 progressively migrates out of the c-SMAC in the course of synapse maturation. Strikingly, synapses between naive $CD4^+$ or $CD8^+$ T cells and DCs can form in the absence of antigen or MHC molecules. This finding underscores the fact that the synapse is a flexible rather than a static entity. In conclusion, the immunological synapse can be viewed as a platform that favors interactions between costimulatory molecules, such as CD28, thereby facilitating priming of naive T cells.

## 8 Macrophages

Macrophages belong to the first line of innate defense to protect the vertebrate body from invasive microorganisms. When a microorganism crosses an epithelial barrier and replicates in a host tissue, it is rapidly recognized by mononuclear phagocytes, or macrophages. Macrophages differentiate from monocytes that leave the blood circulation to migrate into tissues throughout the body. Together with neutrophils, macrophages are key players of the innate immune response because they are equipped to recognize, ingest, and destroy many pathogens without the aid of an adaptive immune response. It is noteworthy that macrophages endocytose the equivalent of their entire cell volume in about 30 min and, on the other hand, secrete more distinct products than even hepatocytes, which secrete numerous serum proteins.

As macrophages react very rapidly on encountering an infecting microorganism, and since the immune system of invertebrates relies entirely on macrophage-driven innate responses, in the early 1960s, Elie Metchnikoff believed that the same would be true for vertebrates. In the meantime, we know that macrophages are of superior importance in the nonadaptive branch of our immune system, but play a less dominant role in mounting an adaptive immune response, since they express comparably low amounts of MHC and costimulatory molecules. Nevertheless, they do cross talk extensively with B and T cells and may be very important for targeted effector T-cell functions in the context of infected cells.

### 8.1 Evolutionary and Cellular Origin

Macrophages in culture behave very similar to amoeba feeding on microorganisms. From this and other similarities, it was concluded that the amoeba is one of the earliest or the earliest form of a phagocytic cell or macrophage. The evolutionary pathway from the ancient amoeba to the modern macrophage is unknown. However, the urgent need for an amoeba to distinguish between other amoeba that should be left untouched and nutrient microorganisms that should be engulfed and degraded, is obvious. Hence, any surface receptor that allowed the amoeba to specifically recognize food would have been a great evolutionary progress. Such a receptor could have been the ancestor of a pattern-recognition receptor (PRRs). Such PRRs have originally been proposed by Janeway. They could serve in distinguishing bacteria or viruses, which carry regularly organized membrane or coat proteins named *pathogen-associated molecular patterns* (PAMPs) from self-proteins, which are not organized in such patterns. In the meantime, equivalents of PRRs have been found: the family of Toll-like receptors (TLRs), which recognize various types of viral or bacterial substances, such as lipopolysaccharide (LPS) or dsRNA. And indeed, TLRs are expressed on macrophages and on DCs.

All vertebrates and many invertebrates have a population of phagocytic cells that patrol their bodies. Macrophages are found in higher vertebrates in especially large numbers in connective tissue, in association with the gastrointestinal tract, in the lung, where they are found in both the interstitium and the alveoli, along blood vessels in the liver (termed *Kupffer cells*), throughout the spleen, where they remove senescent blood cells, and in the thymus, where they aid in thymocyte education.

Beyond that, macrophages are also found at several strategically important locations in the lymph nodes. They are particularly enriched in the marginal sinus zones where the afferent lymph enters the lymphoid tissue, and in the medulla, where the efferent lymph collects before having access to the blood. Here, they obviously prevent antigens from entering the blood stream, a critical event, as otherwise sepsis would occur.

Irrespective of their tissue localization, they differentiate from bone marrow–derived monocytes, which are considerably smaller and not phagocytically active.

For targeting macrophages into sites of infection, macrophages bear low levels of chemotactic receptors, such as the f-Met-Leu-Phe or 7-transmembrane-$\alpha$-helical receptor. It binds to N-formylated peptides produced by bacteria, and rarely by mitochondria.

## 8.2 Antigen Recognition Receptors

Macrophages express several receptors on their surface, which are able to recognize a large variety of bacterial or viral constituents (Table 5). The peculiarity of some of these surface receptors is that they can bind pathogen surfaces directly. These receptors are also known as pattern-recognition molecules, as discussed above. Receptors of the innate immune system mediate different functions. Several of the

**Tab. 5** Antigen recognition receptors on macrophages.

| *Receptor* | *Ligand(s)* |
|---|---|
| Mannose receptor (CD206) | Terminal mannose or fucose of glycoprotiens or lipids |
| Scavenger receptor (CD36) | Anionic polymers on microbial surfaces, apoptotic bodies |
| f-Met-Leu-Phe receptor | N-formylated peptides |
| TLR2 | Peptidoglycans<br>Lipopeptides |
| TLR3 | dsRNA |
| TLR4/CD14 | Endotoxin (LPS) |
| TLR4/CD91 | Hsps |
| TLR5 | Bacterial flagellin |
| TLR9 | CpG oligonucleotides |

known receptors are phagocytic receptors that stimulate ingestion of pathogens. One of these receptors is the mannose receptor (MR). The MR is also expressed on DCs, but not on monocytes or neutrophils. The MR is a C-type lectin that binds specific arrays of terminal mannose or fucose residues of glycoproteins and glycolipids, as they are found on the surface of many bacteria and some viruses, including the human immunodeficiency virus (HIV).

Mammalian equivalents contain terminal sialic acid or N-acetylgalactosamine, which is why the MR recognizes microbes and not host cells. Bacterial carbohydrates are also bound by the glucan receptor. Another receptor is the scavenger receptor (CD36). Originally, this receptor was found to bind to aberrant low-density lipoproteins (LDLs) that are unable to be taken up via the LDL receptor. Today it is known that CD36 is involved in removal of old red blood cells and in recognition and removal of pathogens.

Moreover, macrophages express several types of Fc$\gamma$ receptors suitable for binding the Fc portion of IgG antibodies. During a humoral response, soluble IgG molecules coat the microbes and thereby promote phagocytosis by macrophages. Host proteins such as IgG that promote phagocytosis of microorganisms, such as bacteria, are called *opsonins*. Further opsonins are fragments of the complement factor C3 and plasma proteins, such as fibronectin, fibrinogen, C-reactive protein, or the mannose-binding lectin. They coat microbes early in the course of an infection before specific antibodies are available and bind to receptors, such as the integrins $\alpha_V\beta_3$ and Mac-1 (CD11b/CD18).

A third class of receptors that binds microbial components induces effector molecules that mediate initiation of an innate immune response. The best-defined activation pathway of this type is triggered through the family of Toll-like receptors. The name for these receptors reflects the fact that their extracellular regions are homologous to the protein Toll of the fruit fly Drosophila. In this invertebrate, Toll triggers the production of antifungal peptides in response to fungal infection. In mammals, a Toll-family protein called *Toll-like receptor 4*, or TLR-4, signals the presence of the gram-negative cell-wall component LPS, also named *endotoxin*. The polysaccharide moieties of LPS vary strongly between bacterial strains and are major antigens in inducing an adaptive immune response. The lipid moiety, by contrast, is conserved and is a good example of a molecular pattern recognized by the innate immune system.

The LPS recognition system of macrophages consists of 3 components: (1) a plasma protein, termed *LPS–binding protein* (LBP); (2) CD14, which binds LPS bound to LBP; and (3) TLR-4, the signal-transducing receptor. Circulating LPS is captured by LBP, whereupon the LPS-LBP protein binds to CD14. Upon dissociation of LBP, LPS-CD14 engages TLR-4 leading to activation of the macrophage. This is accomplished via a signal transduction pathway sharing structural homology with the type-I IL-1 receptor pathway. Both IL-1 and LPS stimulate activation of the transcription factor NF-$\kappa$B, leading to cytokine secretion and activation of inflammation.

Another mammalian Toll-like receptor, TLR-2, signals the presence of another set of bacterial components that mainly include proteoglycans of gram-positive bacteria. TLR-2 and TLR-4 induce similar but distinct signals. Macrophages are also able to sense the intracellular presence of bacterial DNA. Bacterial DNA contains unmethylated cytidine–guanidine sequences, the so-called CpG nucleotides.

Unmethylated CpG nucleotides, organized in palindromic stretches are rather rare in mammalian DNA. If enough of these CpG sequences accumulate intracellularly, macrophages start secreting IFN-$\gamma$ and IL-12. According to recent results, this response is mediated via TLR-9.

Two other receptors of the TLR family, which are functionally investigated are TLR-5 and TLR-3. TLR-5 was found to bind the bacterial motor protein flagellin, whereas TLR-3 is discussed to bind viral dsRNA. The functions of other proteins of the TLR family, which comprises 10 members, are currently being explored.

### 8.3 Macrophage Activation

Unactivated macrophages stand out for their high rate of phagocytosis and chemotactic movements and they proliferate. This is sufficient to cope with low numbers of pathogens in an antigen-independent way. However, in order to function as an APC and to gain high antimicrobial effectiveness, macrophages have to be activated. Such an activated macrophage can damage a broad spectrum of microbes and even certain tumor cells, but also healthy self-tissue. Therefore, macrophage activation must be tightly regulated. This regulation is achieved through the requirement of two coinciding signals. Signal 1 is provided by IFN-$\gamma$, signal 2 can be provided by different means, for example, LPS, engagement of CD40L, or cognate TCR (Fig. 7).

Activated CD4$^+$ $T_{H1}$ cells can deliver both signals very efficiently: they secrete high doses of IFN-$\gamma$ and express CD40L. Activated CD8$^+$ CTLs also secrete IFN-$\gamma$; however, recognition of MHC-antigenic peptide complexes by macrophages is not always sufficient to attain full activation of macrophages – the presence of low amounts of LPS or membrane TNF-$\alpha$ can compensate for the lack of CD40L on CD8$^+$ T cells.

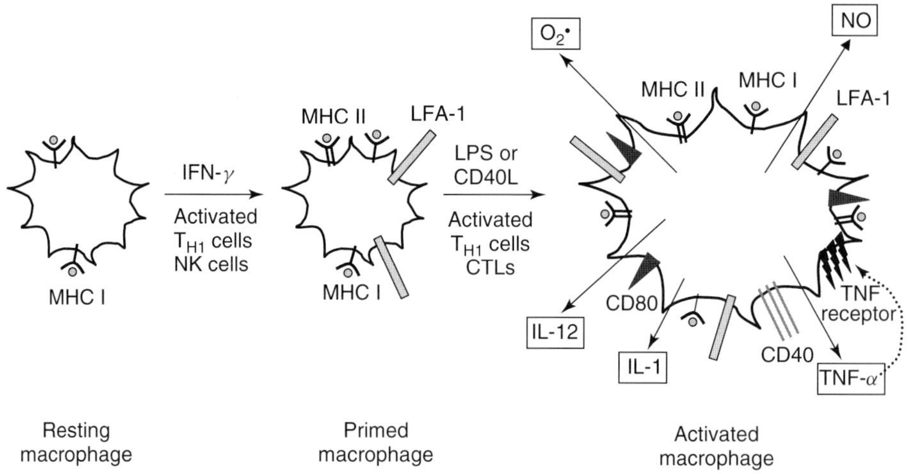


**Fig. 7** The two-step model of macrophage activation. In the first step, resting macrophages are primed by IFN-$\gamma$ secreted by activated $T_{H1}$ or NK cells. Fully activated macrophages are generated in a second step upon encounter of CD40L or LPS. Activated macrophages express MHC class II molecules, costimulators, for example, CD80, CD40, and the TNF receptor. They secrete the cytokines IL-12, IL-1, and TNF-$\alpha$ and the radicals NO and superoxide ($O_2$.).

Another source of IFN-$\gamma$, which is rapidly available at the onset of an infection, is NK cells, thus supporting $T_{H1}$-mediated macrophage activation. $T_{H2}$ cells counteract macrophage activation in that they secrete IL-10 but not IFN-$\gamma$. Macrophage activation by $T_{H1}$ cells expressing CD40L and secreting IFN-$\gamma$ is central to the host response to pathogens that proliferate in macrophage phagosomes, such as *Mycobacteria tuberculosis*, *Mycobacteria leprae* or *Leishmania* species. The same is true for vaccinia virus. Activated macrophages can clear internal microbes by overcoming the block in fusing phagosomes with lysosomes, so that lysis and degradation in phagolysosomes can occur. More importantly, activated macrophages produce the bactericidal nitrogen metabolite NO (see below).

By the late 1960s, the basis of acquired cellular immunity to facultative and obligate intracellular parasites was ascribed to activated macrophages. E. Metchnikoff wrote "The acquisition of immunity against microorganisms is therefore due not only to the change from negative to positive chemotaxis but also to the perfection of the phagocytic and digestive powers of the leukocytes – a general superactivity and adaptation of the phagocytic reaction is produced."

## 8.4 Immunological and Nonimmunological Effector Functions

### 8.4.1 Microbial Killing

$T_{H1}$ cells activate infected macrophages through their TCR engaging MHC class II–antigenic complexes on the macrophage surface and the focal secretion of IFN-$\gamma$. The consequence is a series of events that converts the macrophage into a potent antimicrobial effector cell. An important intracellular event, which can be nicely observed by modern microscopy technologies, is the induced fusion of phagosomes with lysosomes, thereby exposing intracellular or recently ingested microbes to microbicidal lysosomal hydrolases leading to their destruction.

Other changes render macrophages more potent APCs and thereby help amplify an adaptive immune response: surface MHC class II, CD80, CD86, LFA-1, CD40, and TNF receptor are upregulated, thereby increasing the efficacy of macrophages in presenting antigen to resting $CD4^+$ T cells and the responsiveness to CD40L and TNF-$\alpha$.

TNF-$\alpha$ synergizes with IFN-$\gamma$ in inducing the enzyme inducible NO synthetase (iNOS), which produces the reactive nitrogen metabolite NO. NO together with oxygen radicals are secreted by activated macrophages leading to cell damage in the close neighborhood. Together with secreted proteases, NO and $O_2$ radicals are even able to attack large extracellular pathogens such as parasitic worms, which cannot be ingested, or intracellular pathogens that resist or survive phagolysosome fusion. The price to pay is that host tissue is destroyed as well by radicals and other toxic mediators.

Fully activated macrophages also secrete various types of cytokines, for example, TNF-$\alpha$, IL-1, IL-6, and IL-12. IL-1 increases the access of effector T cell, TNF-$\alpha$ is an autocrine stimulus and increases the vascular permeability, IL-6 favors lymphocyte activation and IL-12 favors differentiation of $CD4^+$ T cells into $T_{H1}$ cells and activates NK cells.

Besides priming macrophages through the focal secretion of IFN-$\gamma$, $T_{H1}$ cells are very important in the recruitment of macrophages to sites of infection. This is accomplished in three ways: (1) $T_{H1}$ cells secrete the hematopoietic growth factors IL-3 and GM-CSF (granulocytes-macrophage-colony-stimulating factor), which stimulate generation of monocytes and neutrophils in the bone marrow; (2) $T_{H1}$ cells at sites of infection secrete TNF-$\alpha$ and TNF-$\beta$ which promote diapedesis of monocytes in endothelia; (3) $T_{H1}$ cells in inflamed tissues secrete the chemokine MCP-1 (macrophage chemotactic protein), which attracts macrophages.

### 8.4.2 Tissue Remodeling

Macrophages, in particular activated ones, can secrete a number of proteins not directly related to their functions in innate or acquired immune responses but leading to local changes in the architecture of the tissue where macrophages are residing. Macrophages are the principal source of angiogenic factors, for example, vascular-endothelial growth factor (VEGF), factors that stimulate fibroblast proliferation, for example, platelet-derived growth factor (PDGF), and factors that regulate connective tissue biosynthesis, for example, transforming growth factor-$\beta$ (TGF-$\beta$). Beyond that, unlike neutrophils, macrophages secrete proteases belonging to the matrix-metalloproteinases that degrade extracellular matrix proteins. At the same time, they activate fibroblasts to synthesize new matrix proteins. In settings of prolonged activation, macrophages even mediate tissue fibrosis.

## 8.5 Deactivated and Alternatively Activated Macrophages

In order to minimize local tissue damage and energy consumption, macrophage activation needs to be downregulated at a certain point of time. One way of achieving this is by $T_{H1}$ cells regulating the half-life of their mRNA encoding IFN-$\gamma$. Activation of $T_{H1}$ cells through engaging MHC class II–peptide complexes, CD80/CD86, and CD40 on the macrophage surface induces a new protein that promotes cytokine mRNA degradation, including IFN-$\gamma$ mRNA. Another regulatory mechanism is the production of deactivating glucocorticoids and cytokines, such as TGF-$\beta$, IL-4, IL-10, and IL-13, especially by $T_{H2}$ cells. Thus, the induction of differentiation of $T_{H2}$ cells is a critical pathway for controlling the effector functions of activated macrophages.

More recently, however, IL-4 and glucocorticoids were found to induce increased expression of the MR and to enhance the capacity for endocytosis and antigen presentation of macrophages. This argues against mere deactivation and gave rise to the concept of alternative activation of macrophages. Alveolar and placental macrophages are typical examples. They express certain molecules selectively, such as the chemokine AMAC-1 (alternative macrophage activation-associated chemokine-1). AMAC-1 is related to MIP-1$\alpha$, induced by classical macrophage activators, such as LPS, and inhibited by IL-4. Conversely, AMAC-1 is specifically induced by IL-4 and IL-10 and inhibited by IFN-$\gamma$. AMAC-1 is supposed to facilitate a downregulatory $T_{H2}$ circuit in inflammatory reactions. Moreover, tumors secreting IL-10 and TGF-$\beta$ may systemically and locally induce alternative macrophage activation. These macrophages may facilitate immune escape mechanisms, observed in malignant tumors, by secretion of growth and angiogenic factors supporting enhanced vascularization and nutrition of the tumor. More insight into the role of alternatively activated APCs is required.

## 9
## B lymphocytes

B lymphocytes (short: B cells) can also serve as APCs, in particular, in the phase after a T-cell response has been initiated. In contrast to macrophages, B cells are known for their constitutive expression of MHC class II molecules. Furthermore, only B cells and some medullary thymic epithelial cells, but not other APCs, express the nonclassical MHC class II allele HLA-DO which is thought to regulate antigen processing in a B cell–specific manner. Similar to macrophages, B cells need an activation signal to express costimulatory molecules, thus, avoiding interaction with self-reactive T cells.

It has been known for a while that B cells are the key players in a humoral immune response, as they are the only cells of the immune system which can produce and secrete antibodies recognizing antigenic molecules. We will focus here on the capacity of B cells to function as APCs. Interestingly, it is still open as to how important B cells are in priming naive T cells in natural immune responses. This relates to the fact that B cells are uniquely adapted to bind and internalize mainly soluble molecules through their B-cell receptor (BCR), which is surface immunoglobulin (cf Fig. 2). However, during a bacterial or viral infection, particulate antigen is abundant but soluble antigen is rare. Therefore, it is not very likely that the limited number of B cells will detect soluble antigen that is only present at very low concentrations and that such an event is needed to prime a helper T cell. It is by far more likely that the main task of B cells is to present peptide antigens in the context of MHC class II molecules in order to receive a signal from activated T cells leading to their differentiation into antibody-secreting plasma cells.

### 9.1
### Antigen-induced B-Cell Activation

The activation of B cells in an antigen-specific manner is initiated by the binding of antigen to the BCR. The BCR serves two key roles in B-cell activation: (1) antigen-mediated clustering of the BCR–antigen complexes initiates a signaling cascade; (2) BCR mediates endocytosis of BCR–antigen complexes into MIIC compartments where MHC class II molecules capture antigenic peptides to be presented to $CD4^+$ helper T cells.

Delivery of signals by the BCR starts upon clustering of at least two BCR complexes. *In vivo*, clustering is thought to occur by multivalent antigen. In resting B cells as well as in B cells that have undergone isotype switching, membrane molecules are associated with two other molecules, termed Ig$\alpha$ and Ig$\beta$. Both of them are disulphide-linked heterodimers. The cytoplasmic domains of Ig$\alpha$ and Ig$\beta$ contain immunoreceptor tyrosine-based activation motifs (ITAMs), which are tyrosine enriched. Cross-linking of IgM or IgD molecules brings several ITAMs into closer proximity, thereby triggering subsequent signaling events. Initiation of signaling is accomplished via src family protein tyrosine kinases, such as Lyn, Fyn, or Blk, which are associated with the BCR and trans-phosphorylate ITAMs. Downstream signaling events ultimately activate transcription factors such as Fos, JunB, or NF-$\kappa$B that induce the expression of genes whose products facilitate B-cell activation.

B-cell activation requires, in addition to antigen, second signals, which may be provided by complement proteins; for example, C3d binds to the type 2 complement–receptor CR2 on B cells (Fig. 8). Since C3d binds covalently to microbial antigens in the course of complement activation, antigen-C3d complexes can bridge the BCR and CR2. Binding of C3d to the B-cell complement receptor recruits CD19 and the tetraspanin CD81 into the complex. CR2-CD19-CD81 is often called the B cell coreceptor complex. Phosphorylation of the cytoplasmic tail of CD19 by the BCR-associated kinases leads to augmentation of the humoral immune response by 100- to 1000-fold. In summary, a proteolytic fragment of complement, which is induced by microbes, provides the second signal for B-cell activation. This is reminiscent of T-cell activation depending not only on signal 1 (TCR-MHC-peptide interaction) but also on costimulation (CD80/CD86-CD28 interaction).

## 9.2 Antigen Presentation Outside the Follicular Center

Within 1 to 2 days after antigen administration, B cells recognize antigen in the follicles of peripheral lymphoid organs, are activated, and begin to migrate out of the follicles toward the T-cell zones. The initial encounters between antigen-stimulated B and T cells occur at the interphase of the follicles and the T-cell zones (Fig. 9). In striking contrast to other APCs, B cells do not migrate into tissues where pathogens enter the body, but the antigens have to be transported into lymph nodes or to the spleen via the lymphatic vessels or blood stream to meet B cells.

### 9.2.1 B-cell Epitope versus T-cell Epitope

The three-dimensional molecular structure of an antigen that is recognized by the BCR of a B cell is termed the *B-cell epitope*, which is a conformational epitope.

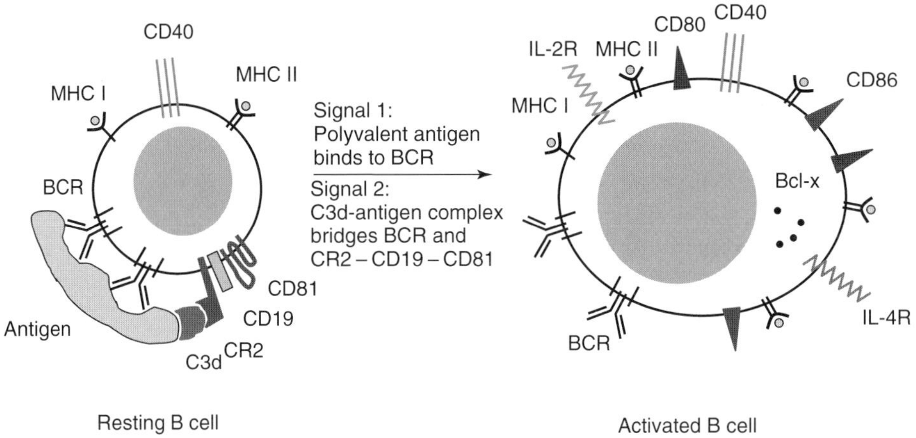


**Fig. 8** Activation of B cells: resting B cells bind polyvalent antigen via the B-cell receptor (BCR) thereby generating signal1. The complement factor C3d bridges the antigen with another signaling complex consisting of the complement receptor CR2, CD19, and CD81, thereby generating signal 2. Activated B cells express costimulatory molecules, for example, CD80 and CD86, the IL-2 receptor (IL-2R), and the IL-4 receptor (IL-4R).

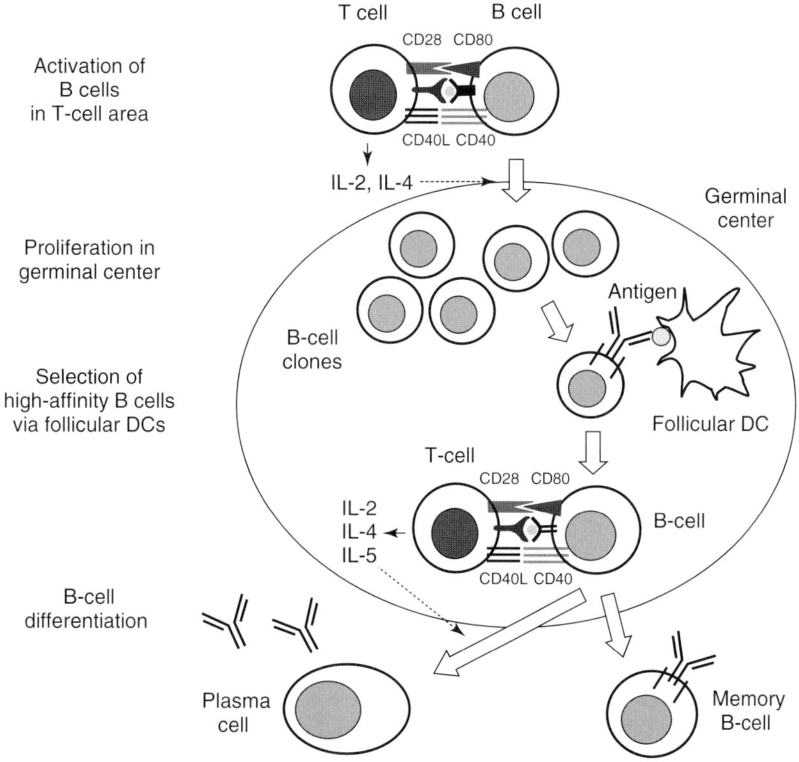


**Fig. 9** T cell–B cell interactions outside and inside germinal centers: B cells encountering activated T cells in the T-cell areas of a lymph node are stimulated via cytokines to proliferate and form the germinal center. Binding to follicular DCs, which present native antigen, leads to selection of high-affinity B cells. The B cells that have survived reencounter activated T cells thereby receiving a differentiation signal. This gives rise to antibody-secreting plasma cells and memory B cells.

The peptide that is generated by proteolytic digestion of the antigen, bound and presented by MHC class II molecules, and recognized by a $CD4^+$ T cell is termed "*T-cell epitope.*" It is a linear epitope.

In the majority of cases, B- and T-cell epitopes are distinct entities with different functions. This can be nicely illustrated with hapten–protein conjugates: hapten-specific B cells bind the hapten via the BCR, which mediates efficient internalization. This means that the hapten contains the B cell epitope. The hapten, however, cannot be bound by MHC class II molecules. This is why a carrier protein is necessary: the T-cell epitope to be presented to $CD4^+$ T cells is a peptide derived from the carrier protein. The requirement for MHC-associated presentation of the T-cell epitope for T-cell activation accounts for the MHC restriction of B cell–T cell interactions.

Which mechanisms prevent B cells that do not possess an antigen-specific BCR but are localized close to an antigen-specific B cell, from so-called *bystander activation*? There are several characteristics of the B cell–T cell interactions that counteract activation of bystander B cells: (1) the first activation signal is dependent on an antigen-specific BCR; (2) binding of an antigen, endocytosis, processing and loading onto MHC class II molecules occurs at $10^4$ to $10^6$-fold lower antigen dose in B cells expressing an antigen-specific BCR than in other cells; (3) only B cells that remain in prolonged and tight contact with a T cell have access to cytokines secreted by T cells in a polarized fashion upon activation. This is accomplished by formation of the immunological synapse (cf. Sect. 7.3.).

### 9.2.2 CD40–CD40L Engagement

CD40 is a member of the TNF receptor family and constitutively expressed on the surface of B cells. Its ligand, CD40L, is only found on the surface of T cells upon seeing cognate antigenic peptide in the context of MHC class II molecules and costimulatory molecules CD80 and/or CD86. When these activated helper T cells bind antigen-presenting B cells, CD40–CD40L interactions lead to CD40 oligomerization. This oligomerization initiates enzyme cascades finally leading to the activation and nuclear translocation of transcription factors, including NF-κB and AP-1. This pathway has already been discussed in the context of T cell–mediated macrophage activation; hence, it is a general mechanism for the stimulation of target cells by helper T cells.

The importance of the CD40–CD40L interaction in humoral immunity is underscored by the observations that CD40 or CD40L gene knockout mice and humans with mutations in the CD40L gene exhibit profound defects in affinity maturation of immunoglobulin and memory B-cell generation. Interestingly, a transforming protein of Epstein–Barr Virus (EBV), which infects human B cells, utilizes the same signaling pathway as CD40, with the consequence that EBV-transformed B cells proliferate rather rapidly, leading to lymphomas.

Three helper T-cell derived cytokines, IL-2, IL-4, and IL-5, contribute to proliferation of activated B cells (Fig. 9). Besides CD40-mediated signals, it is again cytokines, mainly IL-2, IL-4, and IL-6, that stimulate antibody synthesis, the production of secreted Ig and switching from IgM to the IgG, IgA, or IgE isotype. There is a clear preference of some cytokines to induce secretion of particular Ig isotypes, for example, IFN-$\gamma$ promotes secretion of IgG2a, IL-4 favors IgG1, and IgE ad IL-5/TGF-$\beta$ favor IgA production.

Ig-secreting B cells are found in the red pulp of the spleen, the medulla of the lymph nodes, and the bone marrow. Many of the antibody-secreting B cells change into plasma cells, which are specialized in producing and secreting soluble antibody.

## 9.3 Antigen Presentation in the Germinal Center

Within 4 to 7 days after antigen exposure, some of the activated B cells migrate back toward the center of the follicle and begin to proliferate rapidly, thereby forming the germinal center. The B cells in the germinal center are named *centroblasts* and *centrocytes*. They have a doubling time of 6 to 12 h, so that 1 B cell gives rise to a clone of about 5000 centroblasts in 5 days. It was estimated that on average, three B-cell clones colonize each follicle after a single immunization. In particular, CD40–CD40L interactions and presentation of cognate peptide antigens on MHC class II molecules are required to maintain proliferation of centroblasts and to allow further differentiation.

### 9.3.1 B Cell/Follicular DC Contacts

Some of the mutations generated during somatic mutations in immunoglobulin genes are likely to be useful because antibodies with higher affinity are generated. However, the majority of the mutations may lead to a decline or even loss of antigen binding or to recognition of self-antigen. Therefore, the next step in the process of affinity maturation is the selection of the useful, high-affinity B cells that do not react with self.

To this end, follicular dendritic cells (FDCs) are required to engage hypermutated B cells. FDCs have long cytoplasmic processes that form a network around which germinal centers are formed (Fig. 9). FDCs do not express MHC class II molecules, and hence, are not related to classical DCs, but they express several complement receptors, Fc receptors, and CD40L. These receptors bind and display intact antigen that are decorated with Ig or complement factors. Signals generated by antigen binding – mediated by FDCs – to surface Ig on the B cell block a default cell death pathway by inducing antiapoptotic protein of the Bcl family.

With increasing numbers of antigen-specific antibodies being available, more and more of the antigen is captured, and thus, the concentration of available antigen in the germinal center decreases. Therefore, these B cells that are still able to capture antigen have to have a high affinity for the antigen – and only they will survive. The survival of the respective B cells involves uptake and processing of antigen hold on FDCs, and its presentation to T cells via MHC class II molecules (Fig. 9). This event will lead to T-cell activation – the CD40–CD40L interaction conveys long-term survival to the respective centrocyte. The net result of this selection process is a population of B cells producing antibodies with significantly higher affinities for antigen than the antibodies produced by the same clones of B cells earlier in the immune response.

Some antigen-activated B cells acquire the ability to survive for long periods of time. They are termed *memory B cells* capable of mounting rapid responses to subsequent infections. It is possible that memory cells are continually generated and maintained by antigenic stimulation, facilitated by the help of FDCs. Memory B cells typically bear high-affinity Ig. The production of large quantities of high-affinity antibodies is strongly accelerated after secondary exposures to antigen. This can be attributed to the activation of B cells in germinal centers and the rapid formation of immune complexes presented by FDCs and processed by memory B cells in order to contact cognate T helper cells.

### 9.3.2 HLA-DO

HLA-DO is an MHC class II molecule that is unable to bind peptide but serves as a regulator of the peptide editor and chaperone HLA-DM. HLA-DO is expressed in B cells, but not in other APCs. In further contrast to the classical MHC class II molecules and to HLA-DM, HLA-DO expression is initiated only after B-cell development is complete, that is, when B cells start becoming capable of interacting successfully with T cells.

HLA-DO is tightly bound to HLA-DM and is rapidly degraded in the absence of HLA-DM. HLA-DM–HLA-DO complexes are poorly active in removal of the CLIP peptide, especially at moderate pH found in early and late endosomes, suggesting that HLA-DO is a negative regulator of antigen presentation. However, at the same time, HLA-DO acts as a cochaperone of HLA-DM, which is particularly relevant at the low pH found in lysosomal MIICs of B cells. Thus, by limiting the catalytic function of HLA-DM in all but the most acidic MIICs, HLA-DO appears to selectively promote loading and presentation of antigenic peptides derived from proteins internalized through the BCR. The rationale is that very low pH, a high redox potential, and a high abundance of proteases, typical characteristics of lysosomal MIICs, are required for efficient processing of antigens tightly bound to BCRs.

Strikingly, the capacity of B cells to present antigen to helper T cells is controlled via differential expression of HLA-DO in quiescent B cells as compared to activated B cells in germinal centers. In peripheral blood B cells, 50 to 100% of HLA-DM is associated to HLA-DO. This leads to inefficient CLIP removal, and hence to high surface levels of MHC class II-CLIP complexes and to a reduced capacity to process self- or foreign antigen. The conclusion is that due to the comparably high abundance of HLA-DO, peripheral B cells are reduced in their potential to present foreign antigen. The role of class II-CLIP complexes, as typical representatives of self, in peripheral tolerance induction with regard to self-reactive T cells has to be elucidated.

In germinal centers, however, HLA-DO is markedly downregulated so that CLIP is efficiently released from MHC class II molecules and antigenic peptides derived from internalized BCR–antigen complexes have access to the binding cleft of class II molecules. As 20 to 30% of HLA-DM is still associated to HLA-DO, peptide receptive "empty" class II molecules are chaperoned by HLA-DM–HLA-DO complexes until BCR-mediated uptake of foreign antigen occurs. In summary, regulated HLA-DO expression allows optimization of antigen processing and presentation to helper T cells in germinal centers during the development of humoral immune responses.

## 10
## Dendritic Cells (DCs)

Dendritic cells (DCs) represent a highly heterogeneous population of APCs, at the same time being the most potent APCs. They are bone marrow–derived and reside in most peripheral tissues as sentinels of the adaptive immune system. As DCs, similar to cells of the innate immune system, express various types of TLRs that mediate recognition of microbial danger motifs, DCs are also most critical in bridging innate and adaptive immunity.

DCs are specialized for the uptake, transport, processing, and presentation of antigens to T cells. Any encounter with microbial products or tissue damage initiates the migration of DCs out of peripheral tissues to lymph nodes and their differentiation into the mature state. Mature DCs express strongly upregulated levels of MHC and costimulatory molecules so that, in lymph nodes, DCs can trigger an immune response by any T cell with a receptor that is specific for MHC-foreign peptide complexes on the DC surface. Since subsets of DCs are constantly carrying self-antigens into lymph nodes without receiving a maturation stimulus, DCs are thought to play a major role in the maintenance of peripheral tolerance against self as well.

## 10.1 DC Subsets

Phagocytic cells carrying numerous dendrites and expressing considerable amounts of MHC and costimulatory molecules in their unactivated state are found in distinct microenvironments of our body, for example, in peripheral tissues, in lymph nodes, and in the thymus. In addition, DCs do not only interact with T cells, but also with B cells and with NK cells. These aspects render it most likely that the various and often opposing functions ascribed to DCs can only be performed by different sets of DCs. According to the functional plasticity model, only a single DC lineage (hematopoietic precursor, DC precursor, immature DC, mature DC) exists and local environmental factors cause functional diversity (Fig. 10). According to the specialized lineage model, however, specialized DC subtypes are the product of entirely separate developmental lineages, with DC precursors being already functionally committed. It appears that in reality, we are facing a complicated mixture of both models.

### 10.1.1 Myeloid Versus Lymphoid DCs

Distinct DC subtypes were initially more evident among mouse DCs than among human DCs because of the availability of

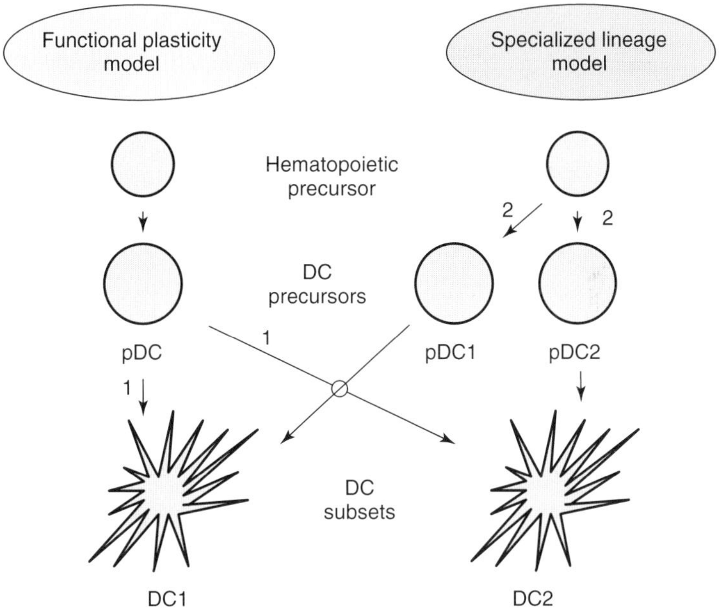


**Fig. 10** Generation of functionally distinct DC subsets. According to the functional plasticity model, all DCs belong to a single hematopoietic lineage, the subtypes DC1 and DC2 being generated by local environmental factors in the periphery (1). The specialized lineage model proposes that DC subsets originate from early divergences in the development, triggered by the microenvironment of the bone marrow (2), giving rise to distinct precursors, pDC1 and pDC2. Subsequently, pDC1 and pDC2 function as the precursors of the DC1 and DC2 subsets respectively.

different murine lymphoid tissues and the expression of distinct markers not present on human DCs. Hence, there was evidence that the $CD8^+$ DCs of the thymus shared early steps of development with T cells. They were thought to derive from lymphoid-restricted precursors and were named *lymphoid DCs*. All other DCs being $CD8^-$ were thought to originate from myeloid-restricted precursors; they were called *myeloid DCs*. In the meantime, when myeloid- and lymphoid-restricted precursors were isolated from bone marrow, it became evident that each precursor could produce all the mature splenic and thymic DC subtypes. Besides CD8 and CD4, the integrin $\alpha_M$ chain (CD11b), the lectin DEC205 (CD205) and the integrin $\alpha_X$ chain (CD11c) are used as surface markers to distinguish murine DC subtypes. Currently, the terms *myeloid* and *plasmacytoid* (Table 6) are used to distinguish DC lineages in both mouse and man.

### 10.1.2 Human DC Subtypes

In contrast to the situation in mice, only blood and bone marrow can be used as an *ex vivo* source of human DCs. Human blood DCs are highly heterogeneous in their marker expression, with many markers reflecting differences in the maturation or activation state rather than separate lineages. A comparison to mouse subtypes is difficult, since human DCs lack expression of CD8. Only human thymic DCs make a strong case for subset segregation, similar to that in mouse: most human thymic DCs are $CD11c^+$ $CD11b^-$ and lack myeloid markers, and so resemble murine thymic $CD8^+$ DCs.

As yet, most insight into DC subsets has come from studies of their development in culture from precursors. Essentially three distinct precursor types have been used: $CD34^+$ bone marrow–derived precursors, blood monocytes, and plasmacytoid cells (Table 7).

**$CD34^+$ precursor pathway** The earliest precursors used are $CD34^+$ and derived from bone marrow or umbilical-cord blood. Liquid culture with granulocyte-macrophage colony-stimulating factor (GM-CSF) and TNF-$\alpha$ leads to two

**Tab. 6** Tissue distribution of murine DC subtypes.

| *Tissue* | *Percentage of total DCs*[a] | | | |
|---|---|---|---|---|
| | *Myeloid DCs* | | *Plasmacytoid DCs* | |
| | *$CD4^+$ $CD8^-$ $CD205^-$ $CD11b^+$* | *$CD4^-$ $CD8^-$ $CD11b^+$* | *$CD4^-$ $CD8^+$ $CD205^+$ $CD11b^-$ lymphoid* | *$CD4^-$ $CD8^+$ $CD205^+$ $CD11b^+$ Langerhans DCs* |
| Spleen | 56 | 20 | 23 | <1 |
| Thymus | – | – | 60 | – |
| Mesenteric lymph nodes | 4 | 63 | 19 | <4 |
| Skin-draining lymph nodes | 4 | 37 | 17 | 33 |

[a] Adapted from Shortman, K., Lin, Y.J. (2002) *Nature Reviews* **2**, 151–161.

**Tab. 7** Different subsets of human dendritic cells (DCs).

| *Name* | *Monocyte-derived DC* | *Langerhans DC* | *IL-3 receptor$^+$ DC* |
|---|---|---|---|
| Origin | Blood | Bone marrow cord blood | Blood |
| Precursor | CLA$^-$/CD14$^+$ | CD34$^+$/CLA$^+$ | pDC2 |
| Lineage | Myeloid | Plasmacytoid | Plasmacytoid |
| Phenotype (immature): | | | |
| MHC class II | ++ | ++ | − |
| CD207 (Langerin) | − | +++ | − |
| CD11c | + | + | − |
| IL-3 receptor | − | − | ++ |
| CD4 | + | + | +++ |

apparently separate pathways of DC development, which can be distinguished by the expression of a skin-homing factor, known as *cutaneous lymphocyte-associated antigen* (CLA).

One pathway (via CD34$^+$/CLA$^+$ intermediates), depending on TGF-$\beta$, leads to DCs resembling Langerhans cells: they express the Langerhans cell–associated antigens Langerin and E-cadherin, CD11c, CD1a, and have Birbeck granules, characteristic structures known from electron microscopy studies of Langerhans epidermal DCs (Table 7).

The second pathway (via CD34$^+$/CLA$^-$, CD14$^+$ intermediates, which resemble blood monocytes) produces DCs reminiscent of intestinal DCs (Table 7). They lack Langerin and E-cadherin, but express the tetraspanin CD9, the coagulation factor XIIIa and CD68.

**CD14$^+$ monocyte pathway** Peripheral blood monocytes are the most commonly used precursors for generating human DCs in culture. CD14$^+$ monocytes differentiate into DCs (termed DC1) in the presence of GM-CSF and IL-4. In the presence of M-CSF, macrophages are generated. The monocyte-to-DC differentiation can also be accomplished by seeding monocytes onto a layer of endothelial cells over a collagen matrix, mimicking the entry of monocytes into tissues. Most of the monocytes that have migrated into the collagen matrix remain there and become macrophages. A subset of them, however, migrate back through the endothelial barrier and become DCs (Table 7).

**Plasmacytoid cell pathway** Human plasmacytoid cells, termed DC2, were found by their plasma cell–like morphology and their unique surface phenotype: they are strongly positive for the IL-3 receptor, but negative for CD11c and for myeloid mineage markers (Table 7). Strikingly, they produce mRNA for germ-line Ig$\kappa$ and pre-T-cell receptor $\alpha$. Plasmacytoid cells are found in blood and many lymphoid tissues. They respond to viral and microbial stimuli by producing IFN-$\alpha$ and IFN-$\beta$.

## 10.2 Antigen Uptake Mechanisms

Tissue DCs capture pathogens, infected cells, dead cells, or their derived products, and also soluble material to use for antigen processing and presentation. Importantly, binding and uptake of certain pathogen-derived components are in many cases linked to DC maturation. During DC maturation, they downregulate their endocytic capacity. The consequence is that antigen uptake is no longer possible during migration to and in the lymph nodes, but only in peripheral organs. Down-modulation of endocytosis is achieved in at least two ways: (1) a decrease in surface expression of several antigen-binding receptors; (2) inactivation of the GTPases cdc42 and rac1 widely abolishes macropinocytosis and phagocytosis.

### 10.2.1 Macropinocytosis Versus Phagocytosis

Macropinocytosis is a cytoskeleton-dependent type of fluid-phase endocytosis that certain growth factors can induce in macrophages. In immature DCs, however, macropinocytosis is a constitutive phenomenon. It allows DCs to sample large amounts of extracellular fluid in short periods of time. This is the nonspecific way of DCs to search for antigenic material. Antigens that have been captured by macropinocytosis have access to MIICs and hence can be loaded onto MHC class II or CD1 molecules.

Phagocytosis is initiated by the engagement of specific surface receptors; hence, it is to a certain extent antigen-specific. Immature DCs phagocytose whole bacteria, irrespective of whether they are gram-positive or gram-negative, but also yeast cells and hyphae. As part of the GALT, interstitial DCs penetrate tight junctions so that their dendrites have access to bacteria in the gut lumen.

Immature DCs also internalize cell debris from cells that have undergone cell death by apoptosis or necrosis. Human monocyte-derived DCs take up apoptotic and necrotic bodies derived from B or T cells, virus-infected apoptotic monocytes, or tumor cells, including melanoma cells or various carcinoma cells. Moreover, DCs have been shown to transport apoptotic intestinal epithelial cells to T-cell areas of mesenteric lymph nodes in rats. Phagocytosis of apoptotic bodies mainly occurs through a complex including the scavenger receptor CD36 and the integrins $\alpha_V\beta_5$ and $\alpha_V\beta_3$. The receptors needed for phagocytosis of necrotic cells are less well defined.

### 10.2.2 Receptors for Endocytosis

Receptors for the $F_C$ portion of immunoglobulin belong to the prominent molecules on the DC surface: human monocyte-derived DCs express mainly Fc$\gamma$RII and Fc$\alpha$R, LCs express Fc$\gamma$RI and Fc$\varepsilon$RI, whereas blood DCs are positive for Fc$\gamma$RII and Fc$\gamma$RI. Recently, a novel member of the Ig superfamily, named Ig-like transcript (ILT)-3 was found to facilitate antigen processing in DCs, but also in monocytes and macrophages.

In contrast to macrophages, immature DCs express only the complement receptors CR3 and CR4, but not CR1 or CR2. Similar to macrophages, immature monocyte-derived DCs, blood DCs, and interstitial DCs of the dermis display high levels of the MR. The MR not only allows DCs to endocytose a large variety of bacterial and yeast antigens, but also desialylated immunoglobulin. Strikingly, LCs do not express the MR, but Langerin (CD207), which is also a lectin with mannose specificity. Unlike the MR, CD207 induces the formation of Birbeck granules. Finally, DCs are also very potent in binding and taking up Hsps, such as hsp70 and gp96. Recently, CD91, known as the $\alpha$2–macroglobulin receptor, and CD40 were described to mediate endocytosis of Hsps of the Hsp70 and Hsp90 protein family. These findings are intriguing, as endogenous as well as foreign Hsps are discussed to mediate maturation of DCs. In particular, the activatory capacity of necrotic cells has been ascribed to Hsps

#### 10.2.3 The Adhesion Receptor DC-SIGN

DC-SIGN (DC-specific ICAM-3-Grabbing Nonintegrin), also termed CD209, is a mannose-binding C-type lectin that is strongly and selectively expressed on DCs, irrespective of whether they are immature or mature.

DC-SIGN recognizes with high affinity the adhesion molecule ICAM-2 on vascular and lymphoid endothelia. ICAM-2 serves as a rolling counterreceptor, thereby enabling DC-SIGN$^+$ DCs to tether to and roll along endothelia. This is a prerequisite for trans-endothelial migration into peripheral tissues and secondary lymphoid organs. Thus, one function of DC-SIGN is to function as a DC-specific rolling receptor.

Upon reaching T cell–rich areas in secondary lymphoid organs, DC-SIGN mediates initiation of T cell–dependent immune responses: it enables transient DC–T-cell interactions through binding to ICAM-3, which is expressed on the surface of naive T cells. In contrast, DC-SIGN does not bind to the principal ligand of LFA-1, ICAM-1.

Finally, DC-SIGN functions as an HIV-1 trans-receptor important in the dissemination of HIV-1. HIV-1 is captured on DCs present in the periphery by DC-SIGN, which binds to the HIV-1 coat protein gp120. Thus, HIV-1 is transported by DCs that migrate into lymphoid tissues where DC-SIGN-associated HIV-1 efficiently infects target CD4$^+$ T cells.

## 10.3
## Maturation of DCs

The various subtypes of human DCs described above vary considerably with regard to the expression of marker proteins and their behavior in the immature state. However, as soon as they are subjected to inflammatory stimuli, they share a common program, termed maturation, which transforms immature into mature DCs that are highly efficient in priming and activating T cells. They are by far more potent than activated macrophages or B cells in initiating a cellular immune response. This is mainly due to the fact that key proteins, for example, MHC molecules, costimulators, and accessory molecules, are much more abundant on mature DCs than on stimulated macrophages or B cells.

#### 10.3.1 Maturation Stimuli

Immature DCs respond to two types of maturation stimuli: (1) direct recognition of pathogens via TLRs and other PPRs; and (2) inflammatory cytokines and other exogenous or endogenous mediators that are secreted during an infection or in other acute alarm situations (Fig. 11). In detail, DCs can recognize bacterial peptidoglycans, lipopeptides, and mycoplasma lipoproteins via TLR2, LPS via TLR4, viral dsRNA via TLR3 and certain DNA oligonucleotides via TLR9. It should be emphasized that TLR2 is not present on the cell surface, but specifically recruited to yeast-containing phagosomes. Likewise, TLR4 is not compulsory for internalization of LPS-carrying bacteria. This suggests that TLRs are involved in discrimination of microbes and not necessarily in their uptake. Apart from that, DCs have receptors in order to sense infections and

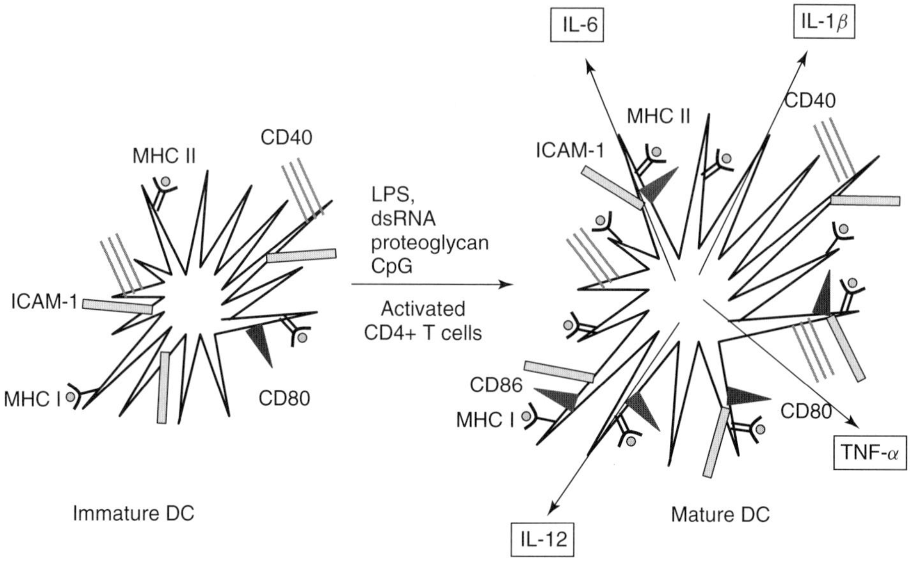


**Fig. 11** Maturation of dendritic cells (DCs). Several microbial signals, including LPS, dsRNA, proteoglycan, and CpG, or the encounter of activated T helper cells induces the mature phenotype. Mature DCs strongly upregulate MHC, costimulatory and adhesion molecules and secrete the proinflammatory cytokines IL-12, IL-6, IL-1$\beta$, and TNF-$\alpha$.

danger indirectly, via binding of TNF-$\alpha$, IL-1$\beta$, or $PGE_2$, whose secretion is triggered by pathogens.

Moreover, whole viruses or bacteria decorated with immunoglobulins induce DC maturation via binding to FcR, including Fc$\gamma$RI, Fc$\gamma$RIII, and Fc$\gamma$R. Even more importantly, activated CD4$^+$ T cells provide a strong stimulus via CD40L that triggers CD40 of DCs. This signal appears to be essential for conditioning DCs in such a way that they are able to fully activate certain types of CD8$^+$ CTLs.

There is accumulating evidence that cell death gives rise to signals that render DC mature, as well. In particular, necrosis induces DC maturation, however, the mediators of this activation are still unknown. Apoptotic cells, at least after heat stress, are also capable of stimulating immature DCs. Nucleotides, such as ATP and UTP, and Hsps, including gp96, Hsp90, and Hsc70 are liberated extensively in the course of necrosis and have been proposed to function as danger signals being able to trigger DC differentiation. However, contaminations with LPS have been implicated in this context, so that it is doubtful, as yet, whether Hsps per se or Hsp-LPS complexes have induced DC maturation. Moreover, Hsp60 and Hsc70 have been described to activate macrophages through a CD14/TLR4 complex, but receptors on DCs are not characterized, yet.

### 10.3.2 **The Danger Hypothesis**

The model of self–nonself discrimination foresees that the baseline state of the immune system is "OFF." To initiate a cellular immune response, the key players, namely, APCs and T cells have to be activated. T cells get activated by the encounter of mature APCs. Such mature

APCs provide signal 1 via MHC-peptide complexes engaging TCRs. Since APCs appear to be unable to distinguish self-peptide from foreign peptide and since self-reactive T cells do exist in our body, the safe guard of the self–nonself discrimination model is signal 2. Signal 2 is only provided via costimulators of APCs that have recognized patterns of nonself molecules found only on evolutionary distinct organisms, such as bacteria or viruses. This is thought to be accomplished via PRRs, for example, TLRs. However, even Janeway, who pioneered the idea of pattern recognition serving for costimulation, pointed out that PRRs cannot explain autoimmunity or immune responses to tumors or transplants. Moreover, it is still open how the immune system deals with a changing self. The self–nonself discrimination model, for example, fails to give us a clue why vertebrates do not attack themselves at puberty and why females do not reject their own newly lactating breasts when they begin to produce milk proteins that were not part of self until that time.

The danger model pioneered by Matzinger gives an answer to these questions. This model is based on the hypothesis that the ultimate controlling signals are endogenous, not exogenous. They are the alarm signals that emanate from stressed or injured tissues (Fig. 12). Hence, cells of normal bodily tissue, when distressed, send signals that are called *danger signals*, which serve to activate local APCs. Healthy cells, in fact, may send calming signals to local APCs, whereas cells that are damaged or die by necrosis should favor maturation of APCs. In contrast, cells that die by apoptosis, a process of normal programmed cell death, should merely send signals that may favor phagocytosis but should not induce the expression of costimulatory molecules. This view is consistent with findings where necrotic cells do induce maturation of DCs, but not apoptotic cells. This correlates with the liberation of Hsps from necrotic but not apoptotic cells. Strikingly, apoptotic cells that have undergone heat stress were also able to activate DCs. As yet, Hsps are the only candidate for endogenous mediators of danger.

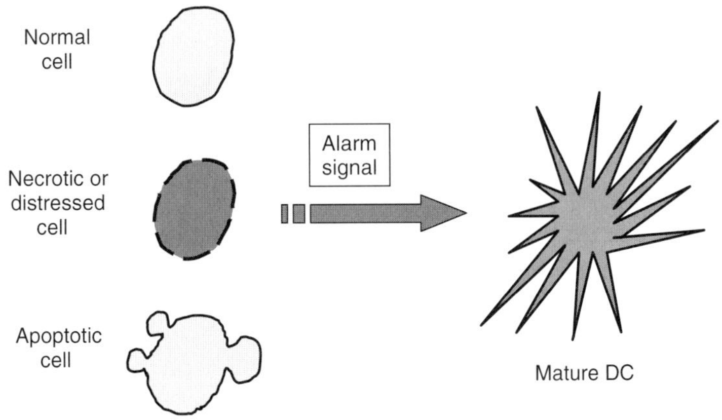


**Fig. 12** The danger model DCs receive an alarm signal from injured, distressed, or necrotic cells leading to their maturation. This is not the case with normal cells or in cells dying by apoptosis.

## 10.4 Antigen Processing and Presentation

### 10.4.1 Classical and Cross-presentation via MHC Class I

Like other APCs and other nucleated cells, DCs present self- or virus-derived endogenous antigens to $CD8^+$ CTLs in an MHC class I–restricted manner. In contrast to other cells, DCs express low levels of the immunoproteasome already in the immature state and upregulate it upon maturation. Moreover, the abundance of MHC class I–peptide complexes increases during maturation. However, in contrast to the MHC class II pathway, mature DCs still synthesize and replace surface MHC class I molecules in their mature state. This finding raises the question, whether DCs that have captured a class I–restricted antigen in peripheral tissues, are the same DCs that will present this antigen to $CD8^+$ T cells in secondary lymphoid organs. Alternatively, DCs may capture antigens in the lymph node. In this case, DCs should be able to take up exogenous antigen and transfer it onto MHC class I molecules. This nonclassical mechanism had already been proposed by Bevan and colleagues: they showed that priming of CTL responses *in vivo* can occur after presentation of exogenous antigens by class I MHC molecules. These phenomena were referred to as *cross-presentation* of antigen and *cross-priming* of $CD8^+$ T cells.

In the meantime, cross-presentation has been shown to be a major pathway utilized by DCs in the context of various sources of exogenous antigens. Physiologically, phagocytosis appears to be a major route for antigen uptake and cross-presentation, as FcR-mediated uptake of immune complexes or opsonized dead cells promotes cross-priming. Furthermore, phagocytosis of bacteria, bacteria-derived antigens, and apoptotic cells results in cross-presentation of bacterial antigens and the same has been observed for viral and tumor antigens. Likewise, peptides bound to Hsps, including gp96, Hsp90, Hsp60, and Hsc70, are cross-presented by mature DCs.

Two pathways for cross-presentation were reported: the first foresees peptide loading in the endocytic system, and hence is TAP- and proteasome-independent. Recycling MHC class I molecules are, indeed, found in endosomes of immature DCs, where low pH is supposed to favor exchange of prebound peptide for exogenous peptide. The second pathway is TAP-dependent and was reported in DCs in several experimental settings. This pathway foresees the existence of a membrane transport system linking the lumen of endosomes and the cytosol. Other cells, for example, thymic epithelia cells or liver sinusoidal epithelial cells, are also able to present soluble exogenous antigen in a TAP-dependent fashion, albeit less efficiently than DCs.

As an alternative to explain cross-presentation, one could envisage that ER compartments fuse with nascent endosomal membranes before endosomes separate from the plasma membrane, as most recently shown for macrophages. This would allow loading of exogenous antigens onto MHC class I molecules in the ER without the need for a specialized transporter in the endosomal membrane. Physiologically, this mechanism could serve to feed the plasma membrane with phospholipids from the ER to compensate the loss of membranes due to endocytosis.

### 10.4.2 MHC Class II–restricted Antigen Presentation

The cell surface display of MHC class II–peptide complexes is tightly regulated during the life cycle of DCs. An immature DC exposes around $10^6$ MHC class II molecules on the surface. This is already more than activated B cells or macrophages can ever express, however, 5 to 10 times less than in mature DCs. Several reasons account for that: the efficiency of major proteases involved in antigen degradation in immature DCs, for example, cathepsins B and S, is low. This may be due to comparably high levels of endogenous protease inhibitors such as cystatin C and the Ii splice variant Iip41. Another aspect is that not all endosomes or phagosomes may have access to MIICs, where MHC class II molecules are loaded with peptide. Finally, MHC class II–peptide complexes have a short half-life ($\leq$10 h) in immature DCs, since they are constantly internalized and transported to recycling endosomes, where peptide exchange occurs, or is degraded in lysosomes. Nevertheless, it should be pointed out that immature DCs are able to present self- and foreign antigen peptides to $CD4^+$ T cells. According to current concepts, this process serves mainly to tolerize T cells against self-antigens.

Inflammatory signals activate a program leading to profound changes in antigen processing for class II molecules. MHC class II biosynthesis is transiently upregulated, the activity of processing proteases increases accompanied by their relocalization into MIICs and recycling of class II molecules ceases widely. Following peptide loading in MIICs, subsets of class II molecules associate with the costimulator CD86 and the tetraspanins CD82 and CD63, thereby becoming a part of multimeric clusters that are transported to the cell surface. At the cell surface, class II peptide complexes remain organized in a multimerized fashion, mainly associated to tetraspanins. The tetraspan microdomains, which are strongly upregulated in mature DCs, not only concentrate MHC-peptides complexes but also enrich MHC class II molecules loaded with particular peptide. This increases the local avidity to T cells and, hence, increases the efficacy of T-cell activation. This is further supported by the fact that the mature DCs no longer endocytose class II molecules so that the latter become rather long-lived. After 10 to 20 h of maturation, no further class II molecules are synthesized and the activity of HLA-DM is strongly downregulated so that peptide loading ceases. Thus, mature DCs that have reached the T-cell areas of lymph nodes mainly present those peptide antigens to $CD4^+$ T cells that they have captured before or during induction of maturation in peripheral tissues.

## 10.5 T Cell Priming and Polarization

### 10.5.1 Priming of $CD8^+$ and $CD4^+$ T Cells

Several lines of evidence support the concept that DCs are the principal APCs for T-cell priming. This is based on *in vitro* observations, where DCs were compared to other APCs for priming alloreactive, naive T cells or the expansion of naive precursors from polyclonal populations. Furthermore, injection of antigen-loaded DCs induces potent $CD4^+$ and $CD8^+$ T-cell primary responses. By immunofluorescence on lymph node sections, interactions of DCs and T cells could be directly visualized. Beyond that, DCs break peripheral tolerance against transplantation antigens and tumor antigens. Moreover, the exclusive expression of MHC class II molecules in DCs was sufficient for the establishment of central tolerance through V$\beta$-specific deletion of T cells by a superantigen presented by DCs in the thymus medulla.

The outstanding T-cell priming capacity of mature DCs is mainly due to the enhanced surface levels of costimulatory CD80 and CD86 and adhesion molecules, but also connected to a rapid relocalization of antigen-bearing DCs to the T-cell zones of secondary lymphoid organs. The *in vivo* relevance of DC maturation in T-cell priming is particularly evident in the case of $CD8^+$ T cells. Although there are examples of direct priming of CTLs by DCs without the involvement of $CD4^+$ T cells, numerous cases have been described, where $CD8^+$ naive T cells are dependent on $CD4^+$ T-cell help. It was found that CD40 ligation on DCs by CD40L on $CD4^+$ T cells was necessary and sufficient to confer to DCs the ability to prime CTLs. At the same time, DCs are rendered mature (Fig. 13).

This mechanism allows a temporal dissociation of interactions between DCs, $CD4^+$ helper T cells, and $CD8^+$ CTLs: once licensed by $CD4^+$ T cells, mature DCs are competent to prime $CD8^+$ T cells. This model, also known as the *license-to-kill model* bypasses the requirement of a tricellular interaction of DCs, activated helper T cells and naive CTLs in lymph nodes, which is rather unlikely to occur. Apart from helper T cells, pathogen-derived products, such as LPS, viruses or apoptotic bodies carrying CD40L have been shown to deliver licensing signals to DCs.

### 10.5.2 T Helper Cell Polarization

In the most recent literature, the issue of T-helper cell polarization by DCs has received considerable attention. Various pathways of regulation are operative. Immature DCs *per se* favor $T_{H2}$ polarization, but they can be reinstructed by the characteristics of the antigen or its accompanying adjuvant to induce a $T_{H1}$, $T_{H2}$, or an undifferentiated $T_{H0}$ response (Table 8).

Microbe-derived pattern molecules, such as LPS or CpG oligonucleotides, drive $T_{H1}$ polarization by induction of IL-12, IL-6, and IL-1$\beta$ secretion through binding to TLRs. However, microbial products often need additional host-derived microenvironmental instruction to

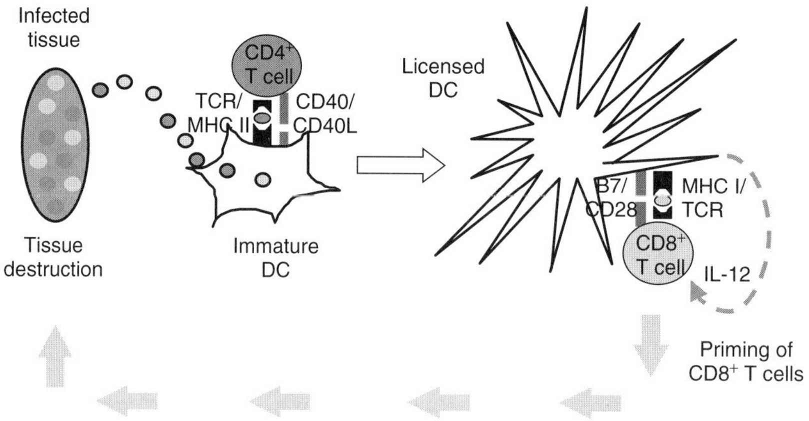


**Fig. 13** The license-to-kill model of DCs. Immature DCs that encounter activated $CD4^+$ Thelper cells receive a signal through CD40/CD40L engagement. DCs that have received this signal are able to render cytotoxic $CD8^+$ T cells highly potent in killing infected target tissues. Thus, DCs conditioned by Thelper cells bear the license to kill.

**Tab. 8** T helper cell polarization mediated by DCs.

| | $T_{H1}$ *bias* | $T_{H2}$ *bias* |
|---|---|---|
| Antigen dose | High | Low |
| Type of antigen | *Candida albicans yeast* Toxoplasma antigen | *Candida albicans hyphae* Allergens |
| Adjuvants | CpG oligonucleotides<br>LPS | $PGE_2$<br>Cholera toxin |
| MHC-TcR interaction | High affinity<br>Short interaction | Low affinity<br>Sustained interaction |
| Stimulator/responser ratio | High | Low |
| Costimulators | ICAM-1<br>CD40 | OX40-L<br>ICOS-L |
| DC state | Mature | Immature |
| DC type | $CD1a^{+}$ (myeloid) | $CD1a^{-}$ (myeloid) |
| DC localization | | Mucosal DC |
| Cytokine status | High IL-12<br>Low IL-10 | Low IL-12<br>High IL-10<br>High IL-6 |

induce stable IL-12 production by DCs, for example, high doses of antigen, IFN-$\gamma$, or IL-4. Interestingly, monocyte-derived DCs prime strong $T_{H1}$ responses early after onset of maturation, whereas at later time points, the same cells prime $T_{H2}$ or $T_{H0}$ responses. Likewise, when DCs stimulate naive T cells at low stimulator/responder ratios, $T_{H2}$ polarization is favored despite the presence of IL-12. Recently, it was proposed that different developmental lineages of DCs could induce either $T_{H1}$ or $T_{H2}$ responses, hence the terminology DC1 or DC2 (Fig. 10) was introduced. Monocyte-derived or myeloid DCs were thought to prime T cells towards the $T_{H1}$ phenotype, whereas plasmacytoid DCs were believed to exclusively prime towards the $T_{H2}$ phenotype. However, this is not the case, as evidenced by at least two recent observations: CD1a-negative subpopulations of monocyte-derived DCs do not secrete IL-12, even after stimulation with LPS, IFN-$\gamma$, or CD40L. Likewise, plasmacytoid DCs can be very potent in priming naive T cells towards $T_{H1}$ differentiation, especially after viral infection.

In conclusion, T helper cell priming by DCs is very flexible, allowing the simultaneous generation of $T_{H1}$, $T_{H2}$, and $T_{H0}$ responses.

## 10.6
## Immune Evasion and Plasticity of DCs

In general, DCs, after having met viruses or other pathogens, will mature and secrete IL-12, thereby skewing $T_{H1}$ differentiation. However, infection of immature human DCs with herpes simplex virus type 1 (HSV-1) or vaccinia virus has been shown to inhibit DC maturation. Likewise, the bacterial metabolite n-butyrate, which is found in high concentrations in the gastrointestinal tract, can interfere with LPS-induced IL-12 release, but not of IL-10. IL-10 released during infection with *Bordetella pertussis* not only promotes the differentiation of monocytes into macrophages but also the differentiation of T regulatory cells, which suppress protective $T_{H1}$

responses (Fig. 14). In response to viruses and *Streptococcus pneumoniae,* DCs can activate B cells via secretion of proinflammatory type 1 interferons, thereby promoting immunoglobulin isotype switching. Most recently, it has been observed that there are reciprocal activatory interactions between NK cells and DCs. The cross talk between both cell types appears to be relevant in innate antitumor responses.

### 10.7 DCs in Immunotherapy

The superior role of DCs in triggering antigen-specific T-cell responses has provided the rationale for developing new vaccination strategies based on the injection of DCs loaded with antigen *in vitro.* In most immunotherapy protocols, autologous $CD34^+$ precursor-derived DCs or monocyte-derived DCs from healthy volunteers or patients are being used.

In vaccination against tumors, the ultimate goal is to prime $CD4^+$ and $CD8^+$ T cells with antitumor reactivity. To achieve this, efficient and stable loading of MHC molecules of DCs with cognate tumor antigens is necessary. Two strategies are employed to provide antigens: an endogenous and an exogenous way. Endogenous antigen expression was obtained by using bulk RNA prepared from tumors or retroviral, adenoviral, or poxviral vectors to transduce DCs with genes coding for the respective tumor antigens. Another

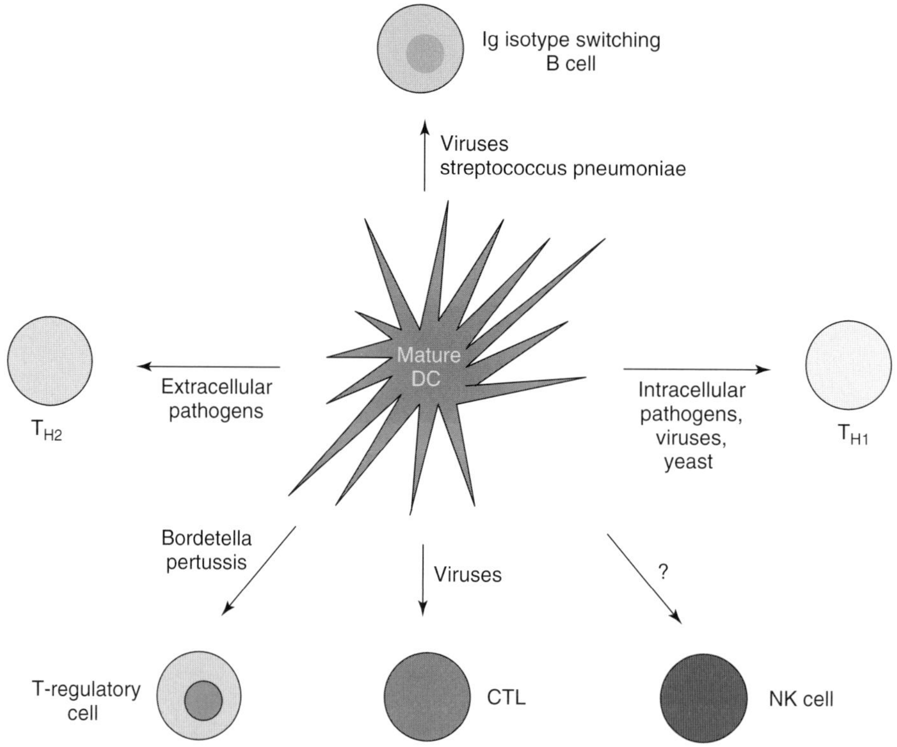


**Fig. 14** The plasticity of DCs. Mature DCs in the draining lymph node are capable of delivering different signals according to the microbes they encounter in the periphery.

approach to obtain endogenous antigen synthesis in DCs has been to fuse DCs to tumor cells. The immunogenicity of such vaccines was confirmed in several experimental systems, although the mechanisms leading to the clinical success of allogeneic approaches remained open.

The alternative strategy foresees to provide tumor antigens exogenously. To this end, tumor cell lysates and defined tumor antigens, applied as a protein or a peptide, were coincubated with autologous DCs so that MHC class II–restricted peptide antigens become cross-presented. An alternative to this latter strategy is to prepare exosomes from the respective tumor cells. Exosomes are the internal vesicles of multivesicular endosomes. They were shown to contain tumor rejection antigens and to be able to transfer them to DCs, so that tumor-specific CTL responses are generated in mice. Although several clinical trials – where DC-based vaccination was employed in the context of malignant melanoma, lymphoma, myeloma, prostate, and renal cancer – were successful to a certain extent, there are still several potential caveats that have to be taken into account. Some tumor antigens are not tumor-specific but tissue-specific and hence widely expressed in normal tissues. In case of vaccination with melanocyte/melanoma antigens, this leads to vitiligo due to autoimmunity at the vaccination site. Conversely, the immunoproteasome of DCs may process tumor antigens in a way that is different from processing by the normal proteasome being active in tumor cells. Moreover, feeding DCs with whole tumor cell lysates is risky since T cells that are not directed against tumor antigens may be primed as well. Therefore, precise knowledge of MHC class I–and MHC class II–associated tumor-specific peptide antigens is necessary. Unfortunately, only a few class II–restricted helper epitopes are known so far. Therefore, further effort is needed to obtain a more comprehensive set of tumor-specific peptide epitopes in order to optimize DC-based vaccination against major malignancies.

## Bibliography

### Books and Reviews


Banchereau, J., Steinman, R.M. (1998) Dendritic cells and the control of immunity, *Nature* **392**, 245–252.

Beck, S., Trowsdale, J. (1999) Sequence organisation of the class II region of the human MHC, *Immunol. Reviews* **167**, 201–210.

Fearon, D.T., Carter, R.H. (1995) The CD19/CR2/TAPA-1 complex of B lymphocytes: linking natural to acquired immunity, *Ann. Rev. Immunol.* **13**, 127–149.

Fehder, W.P., Ho, W.Z., Douglas, S.D. (2000) Macrophages, in: Fink, G. (Ed.) *Encyclopedia of Stress*, Academic Press, San Diego pp. 382–398.

Greenfield, E.A., Nguyen, K.A., Kuchroo, V.K. (1998) CD28/B7 co-stimulation: a review, *Crit. Rev. Immunol.* **18**, 389–418.

Humphreys, R.E., Pierce, S.K. (Eds.) (1994) *Antigen Processing and Presentation*, Academic Press, New York.

Kronenberg, M., Brossay, L., Kurepa, Z., Forman, J. (1999) Conserved lipid and peptide presentation functions of nonclassical class I molecules, *Immunol. Today* **20**, 515–521.

Kropshofer, H., Hammerling, G.J., Vogt, A.B. (1999) The impact of the non-classical MHC proteins HLA-DM and HLA-DO on loading of MHC class II molecules, *Immunol. Reviews* **172**, 267–278.

Lotze, M., Thompson, A. (Eds.) (1998) *Dendritic cells: Biology and Clinical Application*, Academic Press, New York.

Ma, X., Trinchieri, G. (2001) Regulation of interleukin-12 production in antigen presenting cells, *Adv. Immunol.* **79**, 55–92.

Matzinger, P. (2001) The danger model in its historical context, *Scand. J. Immunol.* **54**, 4–9.

Parker, D.C. (1993) T cell-dependent B cell activation, *Ann. Rev. Immunol.* **11**, 331–360.

Paulnock, D.M. (1992) Macrophage activation by T cells, *Curr. Opin. Immunol.* **4**, 344–349.

Srivastava, P.K., Menoret, A., Basu, S., Binder, R.J., McQuade, K.L. (1998) Heat shock proteinscome of age: primitive functions acquire new roles in an adaptive world, *Immunity* **8**, 657–665.

Stoy, N. (2001) Macrophage biology and pathobiology in the evolution of immune responses: a functional analysis, *Pathobiology* **69**, 179–211.

Vogt, A.B., Spindeldreher, S., Kropshofer, H. (2002) Clustering of MHC peptide complexes prior to their engagement in the immunological synapse: lipid raft and tetraspan microdomains, *Immunolog. Reviews* **189**, 136–151.

Yewdell, J.W., Norbury, C.C., Bennink, J.R. (1999) Mechanisms of exogenous antigen presentation by MHC class I molecules in vitro and in vivo: implications for generating $CD8^+$ T cell responses to infectious agents, tumors, transplants, and vaccines, *Adv. Immunol.* **73**, 1–77.

Zinkernagel, R.M. (2000) Localization, dose and time of antigens determine immune reactivity, *Semin. Immunol.* **12**, 215–219.

## Primary Literature

Aderem, A., Ulevitch, R.J. (1998) Toll-like receptors in the induction of the innate immune response, *Nature* **406**, 782–787.

Bachmann, M.F., McKall-Faienza, K., Schmidts, R., Bouchard, D., Beach, J., Speiser, D.E., Mak, T.W., Ohashi, P.S. (1997) Distinct roles for LFA-1 and CD28 during activation of naïve T cells: adhesion versus co-stimulation, *Immunity* **7**, 549–557.

Basu, S., Binder, R.J., Ramalingam, T., Srivastava, P.K. (2001) CD91 is a common receptor for heat shock proteins gp96, hsp90, hsp70, and calreticulin, *Immunity* **14**, 303–313.

Cella, M., Engering, A., Pinet, V., Pieters, J., Lanzavecchia, A. (1997) Inflammatory stimuli induce accumulation of MHC class II complexes on dendritic cells, *Nature* **388**, 782–787.

Cella, M., Facchetti, F., Lanzavecchia, A., Colonna, M. (2000) Plasmacytoid dendritic cells activated by influenza virus and CD40L drive a potent TH1 polarization, *Nat. Immunol.* **1**, 305–310.

Chen, X., Laur, O., Kambayashi, T., Li, S., Bray, R.A., Weber, D.A., Karlsson, L., Jensen, P.E. (2002) Regulated expression of human histocompatibility leucocyte antigen (HLA)-DO during antigen-dependent and antigen-independent phases of B cell development, *J. Exp. Med.* **195**, 1053–1062.

Gadola, S.D., Zaccai, N.R., Harlos, K., Shepherd, D., Castro-Palomino, J.C., Ritter, G., Schmidt, R.R., Jones, E.Y., Cerundola, V. (2002) Structure of human CD1b with bound ligands at 2.3 Å, a maze for alkyl chains, *Nat. Immunol.* **3**, 721–726.

Garside, P., Ingulli, E., Merica, R.R., Johnson, J.G., Noelle, R.J., Jenkins, M.K. (1998) Visualization of specific B and T lymphocyte interactions in the lymph node, *Science* **281**, 96–99.

Goerdt, S., Politz, O., Schledzewski, K., Birk, R., Gratchev, A., Guillot, P., Hakiy, N., Klemke, C.D., Dippel, E., Kodelja, V., Orfanos, C.E. (1999) Alternative versus classical activation of macrophages, *Pathobiology* **67**, 222–226.

Gorak, P.M., Engwerda, C.R., Kaye, P.M. (1998) Dendritic cells but not macrophages, produce IL-12 immediately following *Leishmania donovani* infection, *Eur. J. Immunol.* **28**, 687–695.

Grakoui, A. Bromley, S.K., Sumen, C., Davis, M.M., Shaw, A.S., Allen, P.M., Dustin, M.L. (1999) The immunological synapse: a molecular machine controlling T cell activation, *Science* **285**, 221–227.

Gunzer, M., Schafer, A., Borgmann, S., Grabbe, S., Zanker, K.S., Brocker, E.B., Kampgen, E., Friedl, P. (2000) Antigen presentation in extracellular matrix: interactions of T cells with dendritic cells are dynamic, short-lived, and sequential, *Immunity* **13**, 323–332.

Humphrey, J.H., Grennan, D., Sundaram, V. (1984) The origin of follicular dendritic cells in the mouse and the mechanism of trapping of immune complexes on them, *Eur. J. Immunol.* **14**, 1859–1867.

Kropshofer, H., Arndt, S.O., Moldenhauer, G., Hammerling, G.J., Vogt, A.B. (1997) HLA-DM acts as a molecular chaperone and rescues empty HLA-DR molecules at lysososmal pH, *Immunity* **6**, 293–302.

Kropshofer, H., Spindeldreher, S., Roehn, T.A., Platania, N., Grygar, C., Daniel, N., Woelpl, A., Langen, H., Horejsi, V., Vogt, A.B. (2002)

Tetraspan microdomains distinct from lipid rafts enrich select peptide-MHC class II complexes, *Nat. Immunol.* **3**, 61–68.

Kupfer, H., Monks, C.R., Kupfer, A. (1994) Small splenic B cells that bind to antigen-specific T helper (Th) cells and face the site of cytokine production in the Th cells selectively proliferate: immunofluorescence microscopic studies of Th-B antigen-presenting cell interactions, *J. Exp. Med.* **179**, 1507–1515.

Langenkamp, A., Messi, M., Lanzavecchia, A., Sallusto, F. (2000) Kinetics of dendritic cell activation: impact on priming of TH1, TH2 and nonpolarized T cells, *Nat. Immunol.* **1**, 311–316.

Lohoff, M., Koch, A., Rollinghoff, M. (1992) Two signals are involved in polyclonal B cell stimulation by T helper type 2 cells: a role for LFA-1 molecules and interleukin 4, *Eur. J. Immunol.* **22**, 599–602.

Ludewig, B., Odermatt, B., Landmann, S., Hengartner, H., Zinkernagel, R.M. (1998) Dendritic cells induce autoimmune diabetes and maintain disease via de novo formation of local lymphoid tissue, *J. Exp. Med.* **188**, 1493–1501.

Mitchell, D.A., Nair, S.K., Gilboa, E. (1998) Dendritic cell/macrophage precursors capture endogenous antigen for MHC class I presentation by dendritic cells, *Eur. J. Immunol.* **28**, 1923–1933.

Munoz-Fernandez, M.A., Fernandez, M.A., Fresno, M. (1992) Synergism between tumor necrosis factor-$\alpha$ and interferon-$\gamma$ on macrophage activation for the killing of intracellular *Trypanosoma* crusi through a nitric oxide-dependent mechanism, *Eur. J. Immunol.* **22**, 301–307.

Nestle, F.O., Alijagic, S., Gilliet, M., Sun, Y., Grabbe, S., Dummer, R., Burg, G., Schadendorf, D. (1998) Vaccination of melanoma patients with peptide- or tumor lysate-pulsed dendritic cells, *Nat. Med.* **4**, 328–332.

Nestle, F.O., Banchereau, J., Hart, D. (2001) Dendritic cells: on the move from bench to bedside, *Nat. Med.* **7**, 761–765.

O'Rourke, A.M., Mescher, M.F. (1992) Cytotoxic T lymphocyte activation involves a cascade of signaling and adhesion events, *Nature* **358**, 253–255.

Ridge, J.P., Di Rosa, F., Matzinger, P. (1998) A conditioned dendritic cell can be a temporal bridge between a $CD4^+$ T-helper and a T-killer cell, *Nature* **393**, 474–478.

Roncarolo, M.G., Levings, M.K., Traversari, C. (2001) Differentiation of T regulatory T cells by immature dendritic cells, *J. Exp. Med.* **153**, 5–9.

Sauter, B., Albert, M.L., Francisco, L., Larsson, M., Somersan, S., Bhardwaj, N. (2000) Consequences of cell-death: exposure to necrotic tumor cells, but not primary tissue cells or apoptotic cells, induces the maturation of immunostimulatory dendritic cells, *J. Exp. Med.* **191**, 423–434.

Sallusto, F., Lanzavecchia, A. (1994) Efficient presentation of soluble antigen by cultured human dendritic cells is maintained by granulocyte/ macrophage colony-stimulating factor plus interleukin 4 and down-regulated by tumor necrosis factor alpha, *J. Exp. Med.* **179**, 1109–1118.

Shi, G.P., Villadangos, J.A., Dranoff, G., Small, C., Gu, L., Haley, K.J., Riese, R., Ploegh, H., Chapman, H.A. (1999) Cathepsin S required for normal MHC class II peptide loading and germinal center development, *Immunity* **10**, 197–206.

Stout, R., Bottomly, K. (1989) Antigen-specific activation of effector macrophages by interferon-$\gamma$-producing (TH1) T-cell clones: failure of IL-4 producing (TH2) T-cell clones to activate effector functions in macrophages, *J. Immunol.* **142**, 760–768.

Sugita, M., Grant, E.P., van Donselaar, E., Hsu, V.W., Rogers, R.A., Peters, P.J., Brenner, M.B. (2000) Separate pathways for antigen presentation by CD1 molecules, *Immunity* **11**, 743–752.

Underhill, D.M., Bassetti, M., Rudensky, A., Aderem, A. (1999) Dynamic interactions of macrophages with T cells during antigen presentation, *J. Exp. Med.* **190**, 1909–1914.

# 4
# Cell Mediated Immune Defence

*Martin F. Bachmann[1] and Thomas M. Kundig[2]*
*[1] Cytos Biotechnology AG, Zürich-Schlieren, Switzerland*
*[2] University of Zürich, Zürich, Switzerland*

**1 Introduction 95**

**2 Innate Immunity 97**
2.1 Epithelial Cells 97
2.2 Neutrophils 98
2.3 Monocytes and Macrophages 99
2.4 Natural Killer Cells 99
2.5 Eosinophil Granulocytes 100
2.6 Basophils and Mast Cells 101
2.7 Tissue Cells 101
2.8 Platelets 102

**3 Adaptive Immunity 102**
3.1 Cell Mediated versus Antibody Mediated Immunity 102
3.2 Induction of Cell-mediated Immunity: T-Cell Activation 103
3.2.1 Costimulation 103
3.2.2 Duration of Antigen Presentation 104
3.2.3 DC Maturation 104
3.3 Extravasation 105
3.4 Delayed Type Hypersensitivity Reaction 105
3.5 Lymphoid Organ-like Structures in Chronic Inflammation 106

**4 Mechanisms of T-Cell Mediated Immunity 107**
4.1 Th Cells (CD4+) 107
4.2 Cytotoxic T Cells (CD8+) 108

**5 Silencing of the Immune Response: What Goes Up Must Come Down 108**

*Immunology. From Cell Biology to Disease.* Edited by Robert A. Meyers.

ISBN: 978-3-527-31770-7

**6 T-Cell Memory 109**

**Bibliography 111**
Books and Reviews 111
Primary Literature 112

## Keywords

**B cells**
Lymphocytes generated in the bone marrow that secrete antibodies after activation.

**T cells**
Lymphocytes generated in the bone marrow and educated in the thymus to recognize peptides presented by major histocompatibility antigens.

**T helper cells (Th cells)**
A subset of T cells orchestrating the immune response and helping B cells to produce antibodies.

**Th1 cells**
A subset of Th cells, characterized by the production of IFN$\gamma$ and driving effector responses against intracellular pathogens, bacteria, and fungi.

**Th2 cells**
A subset of Th cells, characterized by the production of IL-4, IL-5, and IL-13, driving effector responses against multicellular parasites.

**Cytotoxic T cells (CTLs)**
A subset of T cells killing infected host cells and tumor cells.

**Dendritic cells (DCs)**
The most potent antigen-presenting cells. Essential for the induction of T-cell responses.

Many cell types make up the arsenal of the immune system and are directly involved in the destruction of invading pathogens. Broadly, two types of cells can be distinguished: firstly, some cells recognize signals that are associated with pathogens, such as tissue damage or patterns found on pathogens but not in the host. These cells belong to the innate immune system and appeared first during evolution. The second type of cells, the lymphocyte, has developed a complex machinery that allows each lymphocyte to express a single and unique receptor with random specificity, the B- and T-cell receptors. Once these receptors recognize a pathogen, the lymphocyte encoding the receptor undergoes clonal expansion and the lymphocyte population differentiates into effector cells, ridding the pathogen. Lymphocytes therefore make up the specific or adaptive immune system. This article discusses how the various cell types protect from infection with pathogens and elucidates some of the complex interactions between the two systems.

## 1
## Introduction

Metazoan organisms have evolved defense systems to a level of near perfection, such that severe or sustained infections are rare. The immune system is divided into an innate system and an adaptive system. The innate immune system acts rapidly and is the most important part of immune defense, as most organisms are able to survive through innate immune mechanisms alone. Only vertebrates have evolved additional mechanisms for pathogen recognition and elimination, so-called *adaptive immunity*.

The innate immune system involves a broad array of different cell types, as many cells and organs of the host have evolved to defend themselves against microbes. For decades, the innate immune system has been regarded as somewhat "unspecific" or even "primitive." However, in recent years it has become more and more apparent that the strategy of the innate immune system is to detect molecules that are common and conserved in a broad range of microorganisms and that are different from molecules of the host itself. It has been discovered that the innate immune system recognizes certain molecular patterns that are present in microbes but not in the host, so-called *pathogen-associated microbial patterns* (PAMPs). The receptors recognizing those microbial patterns are called *pattern recognition receptors* (PRRs). Many cell types of the innate immune system express PRRs, so that the innate immune system can react very rapidly.

In contrast to the innate immune system that has certain specificity for microbes, late evolution in vertebrates has developed an even more specific and more complex system of so-called *adaptive immunity*. The adaptive immune system is able to specifically recognize any protein (or even sequences thereof) that is not normally present in the host, and thus represents an invader. The key players in mounting an adaptive immune response are dendritic cells that induce T- and B-cell responses. The adaptive immune system's high specificity has the disadvantage that early in an infection, only few T and B cells that can recognize the microbe are present and these few specific T and B cells first have to proliferate to an "army of relevant size." This proliferation takes several days during which the innate immune system is the one fighting the microbes. Also, the innate immune system produces an array of different signals in the form of cytokines and chemokines as well as cognate interactions to support the adaptive immune system. Thus, the battle of the innate immune system against a microbe may be interpreted as a "wake-up call" to the adaptive immune system. Once an adaptive immune response has been generated, which includes a microbe specific antibody- and T-cell response, the adaptive immune system also supports the innate immune system, for example, by coating microbes with antibodies so that they can be better phagocytosed (so-called *opsonization*), or by providing T-cell dependent cytokines that enhance the phagocytic activity of the innate immune cells.

Only very few of the microorganisms present in the environment pass all the lines of defense in a vertebrate host. Already in order to pass the first barrier, that is, chemical and physical barriers such as mucus, saliva, cilia movements, and epithelia with tight junctions, specialized pathogen receptors are required for pathogen adherence and penetration. Even fewer microorganisms will pass the innate immune system that has cells positioned in close association with the skin and mucosal surfaces, which are the usual sites of microbe entry. Microbes that have been eliminated either by physical or chemical barriers or by the innate immune system do not usually cause clinically manifest disease. Only those microbes that have passed all these lines of defense are dangerous for the host and will elicit a specific immune response and should actually be called *pathogens*. The adaptive immune response, in concert with an enhanced innate immune response, finally eliminates the pathogen. The immune

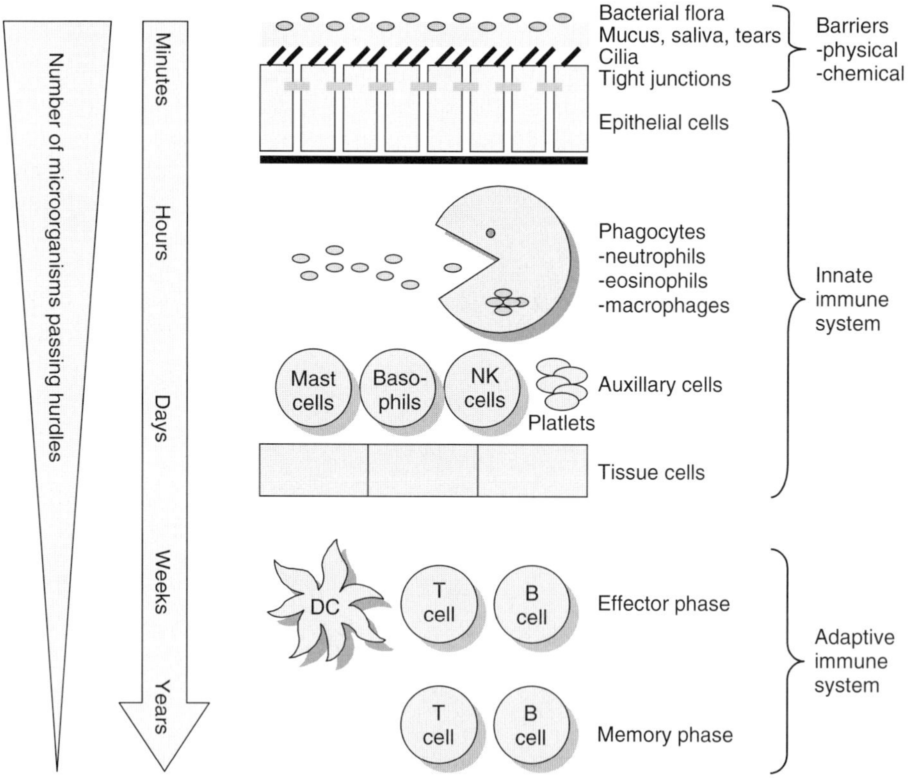


**Fig. 1** Cells of the innate and adaptive immune response. Number of microorganisms that pass lines of defense and induce a response, as well as approximate time for response.

system thereafter tries to keep these dangerous microbes in mind by generating an immunological memory. Even here there are different levels of immunological memory that again depend on the virulence and persistence of the pathogen. Against persisting pathogens, for example, tuberculosis or other pathogens inducing granulomas, immunological memory is strong with pathogen-specific T cells constantly activated and alert. In addition, B cells are repeatedly restimulated to produce high and persistent levels of antibody. In contrast, if the pathogen does not persist, the immune system keeps the pathogen-specific B- and, in particular, T cells at a functionally lower level.

Figure 1 summarizes the three major lines of defense, that is, physical and chemical barriers, followed by the innate immune system consisting of phagocytes, auxiliary cells, and tissue cells. Only very few pathogens pass these first lines of defense so that an adaptive immune response is generated. An arrow indicates the approximate time frames it takes for the immune system to react.

## 2 Innate Immunity

### 2.1 Epithelial Cells

Epithelial cells have a variety of immunological functions. Most microorganisms that an individual encounters do not penetrate the body surface, but are prevented from entering by external or internal epithelial surfaces. Epithelial cells are joined by tight junctions that form an effective seal against the outside environment. Moreover, the epithelial lining of the respiratory tract is covered by sticky mucus. Possible invaders are trapped in the mucus and are then transported outward by moving cilia on the respiratory epithelium.

In addition to these mechanical defense mechanisms, the skin and mucosa also produce antimicrobial chemical substances that cover their surfaces. Tears and saliva contain the antimicrobial enzyme lysozyme. Epithelial cells also release low-molecular-weight antimicrobial peptides. The best characterized peptides belong to the family of so-called *defensins*, which have molecular masses between 3 to 6 kD and have the physicochemical properties to insert into membranes and disrupt their structure. Defensins are subdivided into two structural classes, alpha ($\alpha$) and beta ($\beta$). So far, six $\alpha$-defensins and three $\beta$-defensins have been described in humans. The $\alpha$-defensins are produced primarily by intestinal Paneth cells, whereas airway epithelial cells produce mostly $\beta$-defensins. Defensins have a broad-spectrum antimicrobial activity ranging from gram-negative to gram-positive bacteria, as well as some viruses and fungi. The digestive tract produces antimicrobially active acid, digestive enzymes, and $\alpha$-defensins. The lungs produce surfactant proteins A and D that coat the pathogen surface and enhance phagocytosis by macrophages. Last but not least, most epithelia are covered with a flora of nonpathogenic bacteria that compete with pathogenic bacteria for nutrients and can also produce additional antibacterial peptides and other substances.

Apart from the above-mentioned direct antimicrobial activity, epithelial cells are important regulators of the immune response. Epithelial cells produce numerous cytokines, cytokine antagonists, chemokines, colony-stimulating factors, and growth factors that all have the capacity to recruit multiple inflammatory cells.

Epithelial cells of mucosal surfaces also contribute to mucosal immunity through secretion of IgA. The basal epithelial cell surface expresses the secretory component, acting as a specific receptor for IgA dimers. This complex is then transported and secreted through the apical surface. Such secretory IgA contributes to antiviral and antibacterial immunity.

Epithelial cells can further internalize antigen and express major histocompatibility complex (MHC) class I and class II molecules. Epithelial cells also express costimulatory molecules, such as CD80 and CD86, which are essential for activating T cells. Thus, epithelial cells may be able to recruit inflammatory cells and sustain a local T-cell response through antigen presentation and costimulation.

## 2.2 Neutrophils

Neutrophils constitute the majority of the blood leukocytes. They develop in the bone marrow from the same precursors as monocytes and macrophages. After release from the bone marrow, mature neutrophils have a short transit time of only 6 to 8 h in the circulation. Then, neutrophils are stored in tissue sites through so-called *margination*. This mechanism guarantees that upon infection, a large pool of mature neutrophils is rapidly available. Because neutrophils are present in larger numbers than any other inflammatory cell in the circulation, they are the most important cells in initiating the acute tissue response to injury and infection.

In order to enter the tissue, neutrophils must adhere to the endothelium and migrate through the blood vessel, a process that is regulated by cell adhesion molecules (CAMs). These CAMs are present on neutrophils and endothelial cells where they are upregulated at sites of inflammation. Apart from CAMs, neutrophil migration also requires chemotactic mediators. Neutrophil chemotactic mediators include chemokines, most importantly Interleukin-8 (IL-8), bacterial products, complement split products, as well as leukotrienes, and other lipid mediators.

Classically, neutrophils are regarded as phagocytes. Indeed, neutrophils are highly efficient at taking up and digesting bacteria and other pathogens. Since neutrophils express a variety of receptors for the constant part of antibodies (Fc), they are most potent at eliminating antibody-decorated pathogens. Moreover, neutrophils have recently been reported to function as scavenger cells, which deliver antigens, in particular bacteria, into the spleen for induction of an adaptive immune response. However, neutrophils are far more than that and produce antimicrobial agents such as proteases, oxygen radicals, and defensins upon activation, as well as lipid mediators that lead to further cellular recruitment.

A variety of highly active substances are stored preformed in the granules of granulocytes. The serine proteinases neutrophil elastase, cathepsin G, and proteinase-3 are "ready-to-use" stored in the azurophilic granules of mature neutrophils. From there, they can either be transferred to the phagolysosomes or are released upon stimulation. Once released, neutrophil serine proteinases digest a number of extracellular matrix proteins, including elastin, collagen (types I to IV), fibronectin, laminin, and proteoglycans, which may lead to extensive tissue damage.

Release of oxygen radicals from neutrophils or other inflammatory cells apparently plays an important role in causing tissue damage at sites of inflammation.

Among the lipid mediators released by activated neutrophils are leukotrienes ($LTB_4$) and platelet-activating factor (PAF). Both cause further neutrophil migration and activation, thus amplifying the neutrophil inflammatory response.

Human neutrophil defensins, or human neutrophil peptides (HNP-1 to HNP-4) are small (relative molecular mass 3.54 kD) cationic peptides defined by their antimicrobial activity, but may, however, have further immunoregulatory functions.

## 2.3 Monocytes and Macrophages

Monocytic cells are released from the bone marrow and reach the blood as monocytes. From there, monocytes leave the circulation to reach their resident site in tissues where they differentiate to so-called *macrophages*. Macrophages are recruited to foci of inflammation.

Monocytes and macrophages play a critical role in host defense, not only in innate immunity but also in regulation of the subsequent adaptive immune response. Monocytes and macrophages produce an array of inflammatory mediators and cytokines, such as IL-1$\beta$, TNF-$\alpha$, GM-CSF, IFN-$\gamma$, IL-8, iNOS, RANTES and also lipid mediators such as leukotrienes ($LTC_4$) and prostaglandins ($PGE_2$). Their function is mostly proinflammatory; under certain circumstances, monocytes and macrophages can, however, also suppress an immune response by producing IL-10.

Macrophages and monocytes are also important in adaptive immunity, as they express receptors for immunoglobulins G and E (IgG, IgE). Also, T cells and NK cells activate the phagocytic and microbicidal function of macrophages and monocytes by interferon gamma (IFN-$\gamma$). IFN-$\gamma$ enhances the capacity of monocytes and macrophages to present antigen, including the upregulation of MHC class I and II molecules.

## 2.4 Natural Killer Cells

Natural killer (NK) cells are bone marrow-derived lymphocytes. They comprise the third major lymphocyte population and can be distinguished from other lymphocytes by the absence of B- and T-cell antigen receptors, that is, sIg and T-cell receptor (TCR) respectively. NK cells are capable of lysing certain tumor cells without prior sensitization. Because of this capacity, termed *natural killing*, NK cells were initially thought to serve primarily in defense against tumor cells. Over the years, however, NK cells have been shown to have additional functions. They produce cytokines that regulate the development of acquired, specific immunity, and are involved in the regulation of Th1 versus Th2 development.

NK cells provide early defense against pathogenic organisms during the initial response period against a variety of microorganisms, most importantly against viruses, but also against bacteria and protozoa. NK cells recognize their targets via two general types of receptors. NK cells express inhibitory receptors specific for MHC class I molecules on target cells. This inhibitory receptor prevents the NK cell from activation and killing healthy tissues of the host ("missing-self" hypothesis). NK cells also express activation receptors that recognize target cell ligands and can trigger perforin-dependent natural killing. Many of these activating receptors, such as Ly49D and Ly49H, are structurally related to the inhibitory receptors. The two types of receptors, inhibitory and activating, cross-regulate each other and the balanced signal of the two will determine NK cell activation. Stimulation of NK cells through the activation receptors can lead to production of proinflammatory cytokines such as IFN$\gamma$, TNF-$\alpha$, and granulocyte-macrophage colony-stimulating factor (GM-CSF).

NK cells are also activated by cytokine stimulation. Infected or activated dendritic cells (DCs) and macrophages produce cytokines and chemokines such as IFN$\alpha/\beta$, IL-12, IL-15, and IL-18 that stimulate NK cells to rapidly produce other cytokines (including IFN$\gamma$, TNF$\alpha$, and GM-CSF) and chemokines (such as ATAC/lymphotactin, Mig, and MIP-1$\alpha$), resulting in further activation of immune cells.

## 2.5 Eosinophil Granulocytes

Eosinophils derive from the bone marrow. They are considered to merely pass through the blood circulation to finally reside in the tissues. The ratio of tissue eosinophils to blood is around 100 : 1. The highest density of eosinophils is found in mucosal tissues, a localization pattern that attests to their role in the mucosal immune response.

Eosinophils possess an arsenal of numerous highly basic and cytotoxic granule proteins, as well as enzymes designed to inflict oxidative damage on biologic targets. This weaponry is released on eosinophil activation.

The eosinophil contains an array of cytotoxic granule proteins. Major Basic Protein-2 (MBP-2) is a potent cytotoxin and helminthotoxin. It is also bactericidal, and virucidal. Eosinophil Cationic Protein (ECP) is a highly potent toxin for parasites. Eosinophil-Derived Neurotoxin (EDN) is a potent neurotoxin. Eosinophil peroxidase (EPO) in the presence of $H_2O_2$ kills microorganisms and tumor cells. A variety of other enzymes are also found in the eosinophil, including Arylsulfatase B and lysozyme.

Apart from a direct cytotoxic and antimicrobial function, the eosinophil also acts as a regulator of the immune response. It contains lipid bodies that serve as a storage site for the production of lipid mediators, predominantly leukotriene $C_4$ ($LTC_4$). Eosinophils are also sources of growth factors as well as regulatory or proinflammatory cytokines and chemokines, such as IL-3, IL-4, IL-5, IL-8, IL-16, GM-CSF, TGF-$\alpha$, TGF-$\beta_1$, RANTES, PAF, TNF-$\alpha$, and MIP-1$\alpha$.

Eosinophils express various membrane receptors that regulate their trafficking and activation, including adhesion receptors, receptors for the constant portion of immunoglobulins (Fc), as well as receptors for cytokines and lipid mediators.

The major role of eosinophils appears to lie in defense against large organisms, such as multicellular parasites, which are too large for phagocytosis. Eosinophils may respond to carbohydrate ligands on the parasite surface, such as the Lewis$^X$-related molecules, and cell adhesion molecules that appear similar to selectins. Interestingly, eosinophil action in the lung is under the control of Th cells. Specifically, while eosinophils may be recruited to the lung during an allergic reaction in a Th cell independent fashion, the presence of Th cells in the lung is required for the degranulation of eosinophils. This is yet another example of the complexity of the interaction between cells of the innate and adaptive immune system and how the systems cross-regulate each other in order to focus the responses on potentially dangerous pathogens.

## 2.6 Basophils and Mast Cells

Basophils account for less than 0.2% of all leukocytes. The granules in basophiles contain an abundance of mediators, the most important ones being histamine, heparin, and leukotrienes. Upon cross-linking of the high affinity IgE receptor (Fc$\varepsilon$RI), basophils degranulate and can cause severe systemic symptoms of allergy including anaphylactic shock. However, basophils also play a role in the immune response against parasites by augmenting inflammation.

Mature human mast cells are rounded or spider-like cells of 9 to 12-μm diameter, which contain a striking amount of granules that, when stained, behave metachromatically. Mast cells are present in nearly every organ tissue, but high numbers of mast cells are found in potential entry ports for pathogens, such as the skin, and the mucosal surfaces of the eye, the gastrointestinal tract, and the respiratory tract.

Two types of mast cells, differing in their granule contents and in their patterns of mast cell–specific proteases can be distinguished. Human mast cell nomenclature is based on neutral protease composition. Whereas $MC_T$ cells express tryptase alone, $MC_{TC}$ cells produce tryptase, chymase, mast cell carboxypeptidase, and cathepsin G. It can be said that $MC_T$ cells predominate in the normal lung, particularly in the alveoli, and in the small intestine mucosa. $MC_{TC}$ cells are the predominant type found in normal skin, blood vessels, gastrointestinal submucosa, and synovium.

Mast cells are well-known effectors of allergic disease, as they express the high-affinity receptor for immunoglobulin E (IgE). If cross-linked by an allergen, the mast cell immediately degranulates an array of mediators, one of the most rapidly acting being histamine.

While it is clear that mast cells are responsible for making patients suffer from their allergies against harmless substances such as pollen or food, it is extremely difficult to find a positive role for mast cells. It is not until recently that a new and positive role of the mast cell in protective immunity against parasites and bacteria is emerging. Animal studies with parasites showed a correlation of mast cell numbers with parasite expulsion that can be readily attributed to mast cell mediators such as histamine, proteases, and cytokines. The immune response initiated by mast cell-released mediators also involves eosinophils, and Th2 cells. More recent animal studies using mast cell-deficient mice, however, suggest that it is not the mast cells themselves, but the Th2 type immune response that is the key to antiparasite immunity. Interestingly, mast cell–deficient mice also appear more susceptible to bacterial infections, suggesting a role for the mast cell in innate immunity. Mast cells produce tumor necrosis factor alpha (TNF-$\alpha$), a strong chemo attractant for neutrophils, offering an explanation for the antibacterial activity of mast cells.

## 2.7 Tissue Cells

Although usually not regarded as part of the immune system, tissue cells possess antimicrobial defense mechanisms that are the key to host survival. Here, interferons must be mentioned. Host cells, upon viral infection, start producing IFN-$\alpha$ and IFN-$\beta$. These interferons induce resistance to viral replication in neighboring uninfected cells. Interferons also upregulate class I MHC molecules, thus making virally infected cells more susceptible to CD8+ T cell-mediated killing, so that the virus is more rapidly eliminated. Finally, interferons activate NK cells, which then kill virus-infected cells.

### 2.8 Platelets

In addition to their role in blood clotting, platelets are involved in inflammation. Each day an adult human produces $10^{11}$ platelets that are split off from megakaryocytic cells in the bone marrow. Injury to endothelial cells induces platelet adherence and aggregation, leading to platelet activation and degranulation. Platelet granula contain serotonin so that capillary permeability is increased, as well as fibrinogen that activates complement and attracts leukocytes. Interestingly, platelets express CD40 ligand, a molecule critical for activation of B cells, macrophages, and dendritic cells, suggesting a possible role of platelets in the regulation of innate and adaptive immune responses.

## 3 Adaptive Immunity

### 3.1 Cell Mediated versus Antibody Mediated Immunity

The adaptive immune system is based on two pillars, T cells and B cells. T cells are responsible for cell-mediated immunity while B cells produce antibodies, the basis of humoral immunity. Although the two arms are intimately linked with T cells regulating B-cell responses (in particular, through T help) and vice versa, B cells influencing T-cell responses, the basic functions and tasks of B and T cells are quite distinct (Table 1). Antibodies are important for protection against a variety of pathogens, in particular, many viruses and bacteria. Although antibodies are essential for protection against most secondary infections, they also play a major role during many primary infections and absence of antibodies results in increased susceptibility to many viruses, such as rotaviruses and polio virus. T cells also play a major role during many primary viral infections. Cytotoxic T cells (CTLs) are the key to the elimination of many non- or poorly cytopathic viruses, such as hepatitis B and C. CTLs also play a major role in keeping persistent infections under control. Th cells, on the other hand, play a

**Tab. 1** Effector mechanisms for the elimination of pathogens and tumors.

| *Type of disease* | *Dominant effector mechanism* | *Example* |
|---|---|---|
| Primary infection | Antibody, T cell, innate | Rota virus (Ab), hepatitis B (T cell), Streptococcus (innate) |
| Secondary infection | Antibody | Rota virus, polio, hepatitis B |
| Chronic infection | T cells | HIV, HCV |
| Solid tumor | T cells | Melanoma |
| Single-cell tumors | Antibody (monoclonal) | B-cell lymphoma |

more complex role. Th cells are critical for the B-cell response, which remains abortive in the absence of Th. This not only helps in eliminating pathogens in acute infections but may also be essential in preventing the loss of CTL-memory during chronic infection. In addition, Th cells act more directly on CTLs by facilitating the induction of proliferation competent memory CTLs (see Sect. 4).

The situation for cancer appears different since antibodies play a minor role in control of tumor growth. As antibodies do not efficiently penetrate tissues, their ability to kill tumor cells remains limited. The monoclonal antibody herceptin, which is directed against breast cancer cells, and has proven successful in patients, is probably a rather exceptional case. Another exception is single-cell tumors, such as lymphomas, as they are more accessible to antibodies than solid tumors. This is well documented by the clinical efficacy of a variety of monoclonal antibodies, such as anti-CD20 antibodies (Rituximab), which eliminate B cells in an Fc receptor dependent fashion. Yet, the importance of naturally induced lymphoma specific antibodies remains to be documented. In addition to a possible direct antitumor activity, antibodies may facilitate presentation of tumor-associated antigens by DCs and other antigen-presenting cells (APCs), leading to heightened T-cell responses. It is this T-cell response that appears most apt to reduce tumor burden, since T cells have the ability to enter solid tumor tissues. The key antitumor effector cells are CTLs, which can induce direct lysis of tumor cells. In addition, Th cells locally produce cytokines in the tumor mass, creating a proinflammatory milieu, facilitating elimination of tumor cells by recruitment and activation CTLs and nonspecific cells such as macrophages or eosinophils.

## 3.2 Induction of Cell-mediated Immunity: T-Cell Activation

T-cell activation is a carefully orchestrated process optimally driven by professional antigen-presenting cells, such as DCs, which express high levels of accessory and costimulatory molecules. Several additional parameters regulate T-cell activation, including the kinetics and duration of antigen presentation as well as the activation status of DCs. The following sections will discuss these parameters in more detail.

### 3.2.1 Costimulation

Antigen-specific stimulation of T cells through the TCR alone is not usually sufficient for full T-cell activation. On the contrary, T-cell responses remain abortive in the absence of costimulation by accessory molecules, resulting in T-cell tolerance or anergy, rather than activation. T-cell anergy characterizes nonfunctional T cells that fail to proliferate upon exposure to antigen and therefore are not able to mediate an effective immune response. The most prominent costimulatory molecule on T cells is CD28, a receptor triggered by the B-7 family members CD80 and CD86 expressed on professional antigen-presenting cells, in particular dendritic cells. Indeed, engaging CD28 together with the TCR results in a reduced number of TCRs that need to be triggered for T-cell activation and enhances TCR-mediated signal transduction. Moreover, CD28 delivers an additional, poorly understood, signal into T cells, preventing induction of T-cell anergy. Yet, many viruses and other pathogens are able to induce strong and protective T-cell responses in the absence of CD28.

This may not reflect truly costimulation-independent T-cell activation but may rather be explained by the usage of alternative costimulatory molecules, such as ICOS. Indeed, ICOS has been shown to be responsible for CD28-independent activation of T-cell responses upon infection with vesicular stomatitis virus (VSV). OX40 and 41BB, members of the TNF-receptor family are additional molecules involved in T-cell activation, presumably through costimulation-like processes. In addition to such costimulatory molecules, there are accessory molecules that facilitate T-cell activation through adhesion of T cells to DCs. LFA-1 and CD2 may be good examples, which, in addition to a postulated signaling function, enhance TCR triggering by facilitating T-cell APC interactions. Interestingly, effector function of T cells occurs rather independent of costimulation, ensuring proper activity of effector cells in the periphery, where costimulatory molecules are scarce. Thus, T cells go through an initial phase of costimulation-dependent activation followed by costimulation-independent effector function.

### 3.2.2 Duration of Antigen Presentation

It has been shown that a single brief exposure to antigen allows T cells to undergo a full program of activation, proliferation, and effector cell differentiation. Yet, extended antigen presentation is essential for the generation of protective T-cell responses and also for the induction of CD28-independent T-cell activation as observed after viral infection. In fact, one of the key differences between pathogens and recombinant, nonreplicating antigens is the duration of antigen presentation. While antigens derived from viruses, bacteria, and other pathogens are presented for at least as long as the pathogens replicate, recombinant antigens are presented for much shorter time spans; peptides are, for example, usually only presented for a few hours. This, in part, explains the poor immune response usually obtained with recombinant antigens and also indicates why adjuvants enhance immune responses, since adjuvants, such as aluminum hydroxide or Freund's adjuvants, form antigen depots, leading to a prolonged antigen release. In addition to forming antigen depots, modern vaccines also contain substances that activate the innate immune system, as discussed in the next Section (3.2.3).

### 3.2.3 DC Maturation

DCs are cells that are potent in taking up and processing antigens. In addition, DCs are the most efficient cell type for the induction of protective T-cell responses. Yet, DCs are surprisingly efficient at inducing T-cell tolerance if they are not activated by additional stimuli. Thus, only activated DCs induce protective T-cell responses. When DCs are resting or immature, they induce T-cell tolerance rather than activation. The inability of immature DCs to trigger T-cell activation may in part be explained by reduced levels of costimulatory molecules; indeed, during DC maturation, costimulatory molecules such as CD80/CD86 are upregulated, leading to increased costimulation. Nevertheless, resting DCs do express significant levels of CD86 and it seems unlikely that absence of costimulation alone is responsible for inducing tolerance rather then immunity. Additional factors, such as lack of cytokine secretion by resting DCs or their short life span that leads to a short duration of antigen presentation are probably also contributing to the inability of immature DCs to induce protective T-cell responses.

Two types of stimuli cause activation and maturation of dendritic cells: cytokines and other proteins produced by T cells, macrophages, DCs, and/or other cell types and patterns associated with pathogens (PAMPs) recognized by specific receptors on DCs. Good examples of such patterns are LPS or other cell wall components of bacteria as well as the double-stranded RNA of viruses. The best described receptors for such patterns (PRRs) are toll-like receptors 1 to 11. Interestingly, different toll-like receptors appear to stimulate the immune system in slightly different ways. DCs, for example, express, amongst others, toll-like receptors 2, 4, and 9. For the induction of CTL responses, toll-like receptor 9 appears to be much more important than the other two receptors, while for enhancing antibody responses against allergens, toll-like receptor 2 appears to be most efficient.

Simulation of DCs and other APCs by toll-like receptors is most important for the early immune response, directing the type of immune response that is induced. Specifically, most toll-like receptors direct the immune response toward a Th1 type and CTLs, namely the type of T- cell response that is critical to rid viruses and bacteria. Many multicellular parasites are eliminated by the immune system mounting a Th2 response. In line with this, most parasites are not known to engage toll-like receptors.

Cytokines produced by activated T cells are playing a major role in reinforcement, sustaining and amplifying the type of response that has been initiated by the toll-like receptors. In this way, the innate and adaptive immune system interacts intimately early in the response and the adaptive immune system only becomes independent of the innate immune system once the T-cell response has reached a certain threshold required for a sustained response.

## 3.3 Extravasation

Once activated, T cells need to leave lymphoid organs and emigrate into tissues. A key step in this process is the trans-endothelial migration of lymphocytes. Three steps dominate this migration (Fig. 2). In the first step, lymphocytes attach loosely to the endothelial layer by way of interaction between selectins expressed on T cells and adressins expressed on endothelial cells. In this phase, lymphocytes roll over the endothelium of blood vessels. In the second step, lymphocytes become activated by local chemokines bound to the extracellular matrix on endothelial cells. This leads to cellular activation and increased avidity of LFA-1 to its ligand I-CAM, resulting in arrest of the cells at the local site and trans-endothelial migration (diapedesis). At the site of injury or infection, CTLs may then directly lyse target cells or inhibit pathogen replication by secretion of cytokines. Similarly, Th cells, if activated by local MHC class II expressing APCs, also secrete cytokines, chemokines, and other mediators, leading to increased permeability of the endothelial cell layer and further cell-recruitment. In addition, cytokines and chemokines may trigger local activation of APCs and effector cells, such as granulocytes and CTLs.

## 3.4 Delayed Type Hypersensitivity Reaction

A well-known example of T-cell activation, extravasation, and local release of

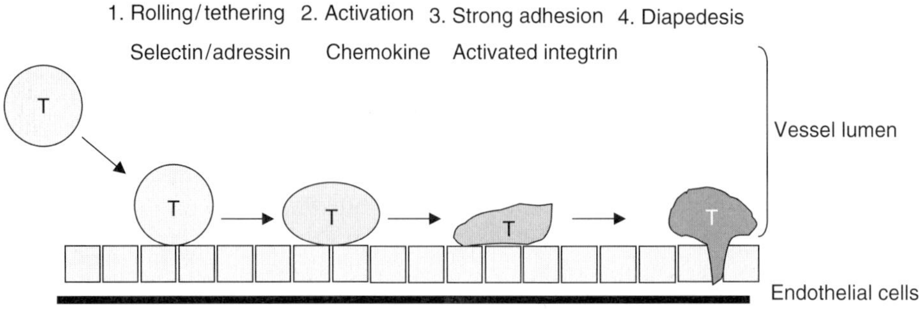


**Fig. 2** Multistep model of lymphocyte activation for extravasation from blood into the tissue.

immunological mediators is the delayed type hypersensitivity reaction. Upon injection of proteins into the skin, antigen drains into T-cell areas of the lymph node, leading to MHC class II-associated antigen presentation. At the same time, Langerhans cells in the skin take up the antigen, process it, and transport it into the lymph node, further increasing antigen presentation. This leads to local activation of T cells. If the organism is exposed for the first time to the antigen, the kinetics of the T-cell response induced is slow and by the time activated T cells leave the lymph node, the antigen in the skin has been eliminated and no local reaction becomes visible. In contrast, in the presence of memory T cells, activated T cells leave the lymph node quickly while the injected antigen is still present in the skin. This results in a local inflammatory response, which is readily visible as a local rash and swelling, due to increased permeability of the vessels. Since no inflammatory response occurs in the absence of preexisting T-cell memory, this delayed type hypersensitivity (DTH) reaction is often used to measure T-cell memory and to probe for persistent infections, as is done with the tuberculin test for diagnosis of latent infection with tuberculosis.

### 3.5
### Lymphoid Organ-like Structures in Chronic Inflammation

Induction and maintenance of T-cell responses usually requires the transport of antigen into lymphoid organs, where T-cell activation occurs. This is true both for the induction of protective T-cell responses against pathogens as well as for the generation of T-cell responses directed against self, potentially causing autoimmunity. For these latter autoimmune responses, however, the reverse may also occur, namely, that lymphocytes are attracted to self-antigens in the periphery, causing sustained T-cell responses and damage in the afflicted organs. An important molecule for this process is the chemokine BLC (CXCL13, BCA-1), which attracts B cells. Under normal conditions, BLC orchestrates the generation and maintenance of B-cell follicles in lymph nodes and spleen, by attracting CXCR5 receptor expressing B cells. B cells that have been stimulated by BLC express lymphotoxin $\beta$(LT-$\beta$), which is essential for the generation of follicular dendritic cells and formation of B-cell follicles. Around these B-cell follicles, T-cell areas are formed leading to the generation of the typical structure of lymphoid organs. Expression of BLC in the periphery is sufficient to drive the generation of such structures within peripheral organs, causing chronic stimulation of self-specific T cells within these ectopic structures. Indeed, in a mouse model of lupus erythematosus, a severe autoimmune disease in humans, BLC expression was increased up to 10 000 fold in organs such as thymus and kidney, that are typically infiltrated by lymphocytes. It is likely that lymphoid organ-like structures observed in patients with Hashimoto's thyroiditis, type I diabetes, or rheumatoid arthritis may have a similar origin and are caused by local expression of BLC or lymphotoxin-$\beta$.

## 4 Mechanisms of T-Cell Mediated Immunity

### 4.1 Th Cells (CD4+)

Th cells have a limited ability to directly interfere with pathogen replication. Through secretion of cytokines, in particular IFN$\gamma$, Th cells can induce an antiviral state in neighboring cells, halting replication of viruses and intracellular bacteria. In addition, members of the TNF superfamily (TNF and Fas-ligand) may have the ability to cause apoptosis in infected target cells, also reducing pathogen replication. The most important function of Th cells is, however, the orchestration of the various antimicrobial effector cells (Fig. 3). Through the secretion of chemokines and cytokines, Th cells are able to recruit innate effector cells, such as macrophages, granulocytes, or eosinophils and activate DCs, enhancing T-cell priming. In addition, Th cells are able to cause isotype-switching in B cells and the generation of protective memory antibody levels. Moreover, Th cells also act directly on CD8$^{+}$ T cells facilitating the induction of CTL responses against weak antigens and enabling development of proliferation competent memory CTLs. Thus, although Th

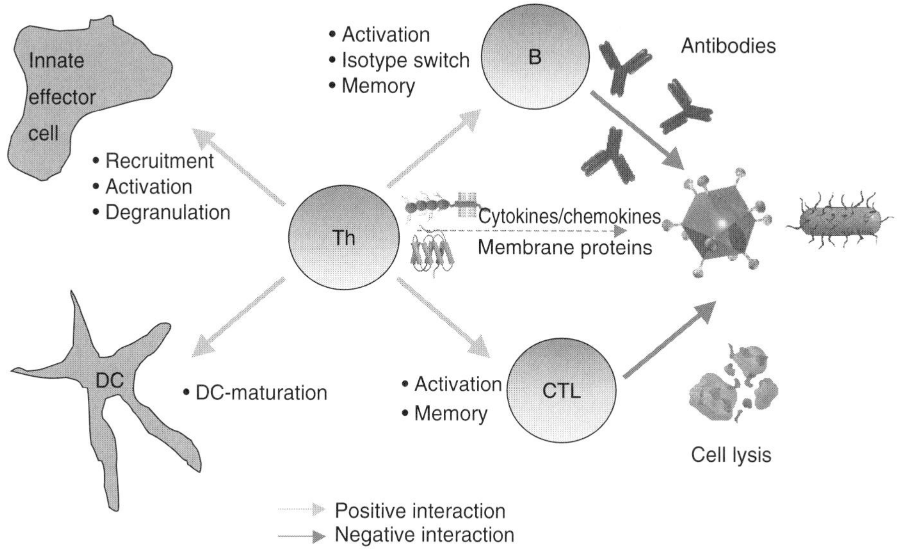


**Fig. 3** Cross-talk between the innate and adaptive immune system and the critical role of Th cells in orchestrating the immune response (see color plate p. xvi).

cells have little immediate effector function, they are key cells for the induction and maintenance of immunity by activating an army of innate effector cells and facilitating the generation of B- and T-cell memory.

### 4.2
### Cytotoxic T Cells (CD8+)

In contrast to Th cells, CTLs are key effector cells of the adaptive immune system. Similar to Th cells, CTLs are secreting a variety of chemokines, cytokines, and IFN$\gamma$ enhancing cellular recruitment and activation as well as causing a local antiviral state. In addition, CTLs harbor lytic granules, which upon stimulation are released in a directed fashion at the immunological synapse onto the recognized target cells. The lytic granules contain a variety of proteins that can cause death of the recognized target cells. The most important proteins are perforin, which acts by "punching" holes into target cells as well as serine proteases belonging to the family of granzymes. Although it was originally thought that perforin mediates cell death by destruction of membrane integrity, it has recently become clear that the main function of perforin is to enable granzymes to enter target cells. This causes death of the target cells, because granzymes activate caspases, leading to apoptosis.

Granzymes are inactive at the low pH while they are stored in the granules and they only become activated once they are released from the granules, that is, in the target cells. The importance of the lytic perforin/granzyme pathway versus the secretion of cytokines by CTLs varies with the pathogen. For lytic viruses, which rapidly kill infected target cells, the perforin/granzymes pathway is of limited importance and cytokines and TNF family members are more important for protection. For nonlytic viruses, which do not kill the infected target cells, the perforin/granzymes pathway is more pivotal and cytokines as well as other soluble mediators play a minor role.

## 5
## Silencing of the Immune Response: What Goes Up Must Come Down

So far, we have mainly been concerned with the buildup of protective T-cell responses. However, the silencing of ongoing immune responses to limit damage in the host caused by the immune system (immunopathology) is equally important. What mechanism(s) are available to switch off T-cell responses? For responses against acute infections, the most efficient mechanism is the elimination of antigen. Once the pathogen has been cleared from the host, the majority of T cells undergo apoptosis and, in the absence of further stimulation, the T cells become quiescent memory T cells (Fig. 4). As an additional mechanism, antigen-specific attenuation of T-cell responses by membrane-bound receptors has been postulated for CTLA-4, PD-1, a CTLA-4 homolog and further homologs of this family of receptors (e.g. BTLA). Indeed, mice deficient in the respective genes have a striking phenotype and rapidly develop symptoms of autoimmune disease. However, the mechanism of this attenuation is still a matter of debate. In fact, rather than downregulating T cells in an antigen-specific manner, receptors such as CTLA-4 may facilitate the induction of regulatory T helper cells that secrete anti-inflammatory cytokines.

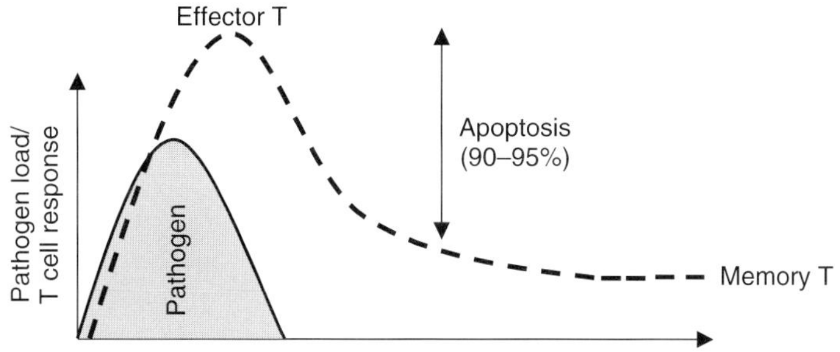


**Fig. 4** Regulation of T-cell responses and establishment of memory.

TGF-$\beta$ and IL-10 are the most potent immunosuppressive cytokines. Production of these anti-inflammatory cytokines has been shown to be of great importance during an infection with parasites, which may otherwise induce overshooting T cells responses resulting in the death of the host. The ambivalent function of immunosuppressive or anti-inflammatory cytokines is well illustrated by IL-10. On the one hand, acute infection of IL-10 deficient mice with Toxoplasma gondii or the malaria parasite *Plasmodium* results in a fulminate inflammatory response, causing death of the infected mice due to immunopathology usually kept in check by IL-10. On the other hand, due to increased inflammatory T-cell responses, IL-10-deficient mice are better protected from primary and secondary systemic infection with *Listeria monocytogenes* and show improved clearance of pathogens such as *Mycobacteria* or *Leishmania major* that usually persist in their hosts.

In case the pathogen cannot be cleared from the host, as is the case for the parasite *L. major*, the member of the TNF-receptor family Fas has been shown to be important by causing apoptosis of highly stimulated T cells, preventing their accumulation. Indeed, Fas-deficient mice amass antigen-specific T cells upon infection with *L. major*, resulting in grossly enlarged draining lymph nodes.

In conclusion, T-cell responses induced by acute infection that can be resolved by the host are usually downregulated by clearance of the antigen. Under condition of excessive antigen load or undue stimulation of innate immunity or when the pathogen cannot be eliminated, additional mechanisms may kick in, such as inhibitory receptors and, in particular, the production of anti-inflammatory cytokines.

## 6
## T-Cell Memory

Upon infection or vaccination, specific T cells proliferate and expand as a population. Peak frequencies are usually reached within one or two weeks. Once the pathogen has been cleared, T-cell frequencies start to drop and more than 90% of the T cells undergo apoptosis. The remaining cells differentiate into memory cells and are maintained at relatively constant frequencies over extended time periods (Fig. 4). If the pathogen is reencountered, these T cells respond with accelerated kinetics, leading to an enhanced secondary

T-cell response. Hence the term "T-cell memory"; the immune systems remember that it has previously responded to the pathogen.

This enhanced T-cell response is not only due to increased numbers of memory T-cell precursor. Memory T cells also show an altered physiology when compared to naïve T cells. Most importantly, memory T cells can secrete cytokines within hours after stimulation while naïve T cells require days for initiation of cytokine secretion. Moreover, memory T cells can be activated in the absence of costimulation, apparently due to a rearranged signal transduction machinery. Thus, if compared to naïve T cells, memory T cells are also more potent at the single-cell level.

Various memory T-cell subsets have been described. Particularly prominent is the concept of central versus effector memory T cells. Originally, it was thought that central memory T cells reside in lymph nodes and the spleen and rapidly respond with proliferation to antigen reexposure. Effector memory T cells, on the other hand, were suggested to reside in peripheral tissues and exhibit immediate effector function. Hence, effector memory T cells were thought to be mainly responsible for immediate protection against reinfection. However, this concept has meanwhile been challenged in many ways and it was found that central memory T cells are often found in peripheral tissues and exhibit similar immediate effector function as effector memory T cells. In addition, it was even reported that central memory T cells are more potent at mediating protection against reinfection than effector memory T cells. Thus, the physiological significance of the two types of memory T cells remains to be established.

The importance of persisting antigen for the maintenance of T-cell memory and for the balance between central and effector memory T cells has been a matter of extensive debate. Nevertheless, it is now emerging that memory T cells do not need persisting antigen for their survival. Yet, presence of persisting antigen keeps the T cells in a more activated state, facilitating maintenance of immediate effector function. Thus, optimal long-term protective T-cell memory is induced by viruses that persist at a low level in the host, such as lymphocytic choriomeningitis virus or cytomegalovirus. In fact, it is the chronically stimulated T cells that keep the virus in check and prevent widespread replication. Indeed, containment of persisting viruses and other pathogens is probably one of the most important tasks of T-cell memory. The importance of chronic antigen exposure for the maintenance of T-cell effector function is also illustrated by the examples described above of lymphoid organ-like structures in peripheral organs of patients with autoimmune diseases. It is the continuous antigenic exposure in an immunostimulatory environment that keeps T cells activated and capable of mediating tissue destruction.

In conclusion, upon elimination of the bulk of the antigen, most of the effector T cells undergo apoptosis and few of the cells differentiate into memory T cells. These memory cells are maintained at increased frequencies and are able to respond to antigenic reexposure with enhanced kinetics. One key function of memory T cells is to keep persistent infections under control and prevent widespread replication of the pathogens.

## Bibliography

### Books and Reviews

Ahmed, R., Gray, D. (1996) Immunological memory and protective immunity:

understanding their relation, *Science* **272**(5258), 54–60.

Akira, S. (2003) Mammalian Toll-like receptors, *Curr. Opin. Immunol.* **15**(1), 5–11.

Bachmann, M.F., Kopf, M. (2002) Balancing protective immunity and immunopathology, *Curr. Opin. Immunol.* **14**(4), 413–419.

Beutler, B. (2002) Toll-like receptors: how they work and what they do, *Curr. Opin. Hematol.* **9**(1), 2–10.

Biron, C.A. (2001) Interferons alpha and beta as immune regulators–a new look, *Immunity* **14**(6), 661–664.

Campbell, D.J., Kim, C.H., et al. (2003) Chemokines in the systemic organization of immunity, *Immunol. Rev.* **195**, 58–71.

Coyle, A.J., Gutierrez-Ramos, J.C. (2001) The expanding B7 superfamily: increasing complexity in costimulatory signals regulating T cell function, *Nat. Immunol.* **2**(3), 203–209.

Delves, P.J., Roitt, I.M. (2000) The immune system. First of two parts. The immune system. Second of two parts, *N. Engl. J. Med.* **343**(1), 37–49; *N. Engl. J. Med.* **343**(2), 108–117.

Dutton, R.W. (1967) In vitro studies of immunological responses of lymphoid cells, *Adv. Immunol.* **6**, 253–336.

Freedman, J.E. (2003) CD40-CD40L and platelet function: beyond hemostasis, *Circ. Res.* **92**(9), 944–946.

Galli, S.J., Maurer, M., Lantz, C.S. (1999) Mast cells as sentinels of innate immunity, *Curr. Opin. Immunol.* **11**(1), 53–59.

Germain, R.N., Stefanova, I. (1999) The dynamics of T cell receptor signaling: complex orchestration and the key roles of tempo and cooperation, *Annu. Rev. Immunol.* **17**, 467–522.

Gleich, G.J., Adolphson, C.R. (1986) The eosinophilic leukocyte: structure and function, *Adv. Immunol.* **39**, 177–253.

Hoebe, K., Janssen, E., Beutler, B. (2004) The interface between innate and adaptive immunity, *Nat. Immunol.* **5**(10), 971–974.

Hoffmann, J.A., Kafatos, F.C., et al. (1999) Phylogenetic perspectives in innate immunity, *Science* **284**(5418), 1313–1318.

Janeway, C.A. Jr., Goodnow, C.C., et al. (1996). Danger – pathogen on the premises! Immunological tolerance, *Curr. Biol.* **6**(5), 519–522.

Kitamura, Y. (1989) Heterogeneity of mast cells and phenotypic change between subpopulations, *Annu. Rev. Immunol.* **7**, 59–76.

Klion, A.D., Nutman, T.B. (2004) The role of eosinophils in host defense against helminth parasites, *J. Allergy Clin. Immunol.* **113**(1), 30–37.

Lee, W.L., Harrison, R.E., Grinstein, S. (2003) Phagocytosis by neutrophils, *Microbes Infect.* **5**(14), 1299–1306.

Lieberman, J. (2003) The ABCs of granule-mediated cytotoxicity: new weapons in the arsenal, *Nat. Rev. Immunol.* **3**(5), 361–370.

Mellman, I., Steinman, R.M. (2001) Dendritic cells: specialized and regulated antigen processing machines, *Cell* **106**(3), 255–258.

Parker, D.C. (1993) T cell-dependent B cell activation, *Annu. Rev. Immunol.* **11**, 331–360.

Raulet, D.H. (2004) Interplay of natural killer cells and their receptors with the adaptive immune response, *Nat. Immunol.* **5**(10), 996–1002.

Sallusto, F., Lanzavecchia, A. (2000) Understanding dendritic cell and T-lymphocyte traffic through the analysis of chemokine receptor expression, *Immunol. Rev.* **177**, 134–140.

Scapini, P., Lapinet-Vera, J.A., Gasperini, S., Calzetti, F., Bazzoni, F., Cassatella, M.A. (2000) The neutrophil as a cellular source of chemokines, *Immunol. Rev.* **177**, 195–203.

Schutte, B.C., McCray, P.B. Jr. (2002) [beta]-defensins in lung host defense, *Annu. Rev. Physiol.* **64**, 709–748.

Sprent, J., Surh, C.D. (2002) T cell memory, *Annu. Rev. Immunol.* **20**, 551–579.

Takeda, K., Kaisho, T., Akira, S. (2003) Toll-like receptors, *Annu. Rev. Immunol.* **21**, 335–376.

Toes, R.E., Schoenberger, S.P., et al. (1998) CD40-CD40Ligand interactions and their role in cytotoxic T lymphocyte priming and anti tumor immunity, *Semin. Immunol.* **10**(6), 443–448.

Unanue, E.R., Askonas, B.A. (1968) The immune response of mice to antigen in macrophages, *Immunology* **15**(2), 287–296.

Zinkernagel, R.M., Bachmann, M.F., et al. (1996) On immunological memory, *Annu. Rev. Immunol.* **14**, 333–367.

Zinkernagel, R.M., Hengartner, H. (2001) Regulation of the immune response by antigen, *Science* **293**(5528), 251–253.

## Primary Literature


Bachmann, M.F., Kohler, G., et al. (1999) Cutting edge: lymphoproliferative disease in the absence of CTLA-4 is not T cell autonomous, *J. Immunol.* **163**(3), 1128–1131.

Bachmann, M.F., McKall-Faienza, K., et al. (1997) Distinct roles for LFA-1 and CD28 during activation of naive T cells: adhesion versus costimulation, *Immunity* **7**(4), 549–557.

Bachmann, M.F., Wong, B.R., et al. (1999) TRANCE, a tumor necrosis factor family member critical for CD40 ligand-independent T helper cell activation, *J. Exp. Med.* **189**(7), 1025–1031.

Bachmann, M.F., Zinkernagel, R.M., et al. (1998) Immune responses in the absence of costimulation: viruses know the trick, *J. Immunol.* **161**(11), 5791–5794.

Bjorkander, J., Bake, B., Oxelius, V.A., Hanson, L.A. (1985) Impaired lung function in patients with IgA deficiency and low levels of IgG2 or IgG3, *N. Engl. J. Med.* **313**(12), 720–724.

DeBenedette, M.A., Shahinian, A., et al. (1997) Costimulation of CD28-T lymphocytes by 4-1BB ligand, *J. Immunol.* **158**(2), 551–559.

Ehrlich, P. (1878). *Beiträge zur Theorie und Praxis der Histologischen Färbung*, Thesis in Leipzig.

Fleming, A. (1922) On a remarkable bacteriolytic element found in tissues and secretions, *Proc. Roy. Soc. Ser. B* **93**, 306–317.

Greenwald, R.J., Boussiotis, V.A., et al. (2001) CTLA-4 regulates induction of anergy in vivo, *Immunity* **14**(2), 145–155.

Huang, S., Hendriks, W., Althage, A., Hemmi, S., Bluethmann, H., Kamijo, R., Vilcek, J., Zinkernagel, R.M., Aguet, M. (1993) Immune response in mice that lack the interferon-gamma receptor, *Science* **259**(5102), 1742–1745.

Ishizaka, T., De Bernardo, R., Tomioka, H., Lichtenstein, L.M., Ishizaka, K. (1972) Identification of basophil granulocytes as a site of allergic histamine release, *J. Immunol.* **108**(4), 1000–1008.

Kazura, J.W., Grove, D.I. (1978) Stage-specific antibody-dependent eosinophil-mediated destruction of Trichinella spiralis, *Nature* **274**(5671), 588–589.

Kopf, M., Coyle, A.J., et al. (2000) Inducible costimulator protein (ICOS) controls T helper cell subset polarization after virus and parasite infection, *J. Exp. Med.* **192**(1), 53–61.

Kopf, M., Ruedl, C., et al. (1999) OX40-deficient mice are defective in Th cell proliferation but are competent in generating B cell and CTL Responses after virus infection, *Immunity* **11**(6), 699–708.

Kundig, T.M., Shahinian, A., et al. (1996) Duration of TCR stimulation determines costimulatory requirement of T cells, *Immunity* **5**(1), 41–52.

Oxenius, A., Campbell, K.A., et al. (1996) CD40-CD40 ligand interactions are critical in T-B cooperation but not for other anti-viral CD4+ T cell functions, *J. Exp. Med.* **183**(5), 2209–2218.

Ruedl, C., Kopf, M., et al. (1999) CD8(+) T cells mediate CD40-independent maturation of dendritic cells in vivo, *J. Exp. Med.* **189**(12), 1875–1884.

Sallusto, F., Lenig, D., et al. (1999) Two subsets of memory T lymphocytes with distinct homing potentials and effector functions, *Nature* **401**(6754), 708–712.

Schoenberger, S.P., Toes, R.E., et al. (1998) T-cell help for cytotoxic T lymphocytes is mediated by CD40-CD40L interactions, *Nature* **393**(6684), 480–483.

Thomsen, A.R., Nansen, A., et al. (1998) CD40 ligand is pivotal to efficient control of virus replication in mice infected with lymphocytic choriomeningitis virus, *J. Immunol.* **161**(9), 4583–4590.

Unsoeld, H., Krautwald, S., et al. (2002) Cutting edge: CCR7+ and CCR7− memory T cells do not differ in immediate effector cell function, *J. Immunol.* **169**(2), 638–641.

Viola, A., Lanzavecchia, A. (1996) T cell activation determined by T cell receptor number and tunable thresholds, *Science* **273**(5271), 104–106.

# 5
# Immunologic Memory

*Alexander Ploss[1,2] and Eric G. Pamer[1]*
[1] *Sloan-Kettering Institute, New York, USA*
[2] *Weill Graduate School of Medical Sciences of Cornell University, New York, USA*

**1 Introduction 116**

**2 Characteristics of Memory Cells 118**

**3 CD8+ T-Cell Memory 118**
3.1 Phenotyping Memory CD8+ T Cells 118
3.2 Enhanced Responsiveness of Memory CD8+ T Cells: Potential Mechanisms 119
3.3 Generation of Memory CD8+ T Cells 120
3.4 Maintaining CD8+ T-Cell Memory 122
3.5 Models of CD8+ T-Cell Memory Generation 123

**4 CD4+ T-Cell Memory 126**
4.1 Differentiation of Effector and Memory CD4+ T Cells 126
4.2 Phenotype of Memory CD4+ T Cells 126
4.3 Memory Generation and Maintenance 127
4.4 Trafficking of Memory CD4+ T Cells 127

**5 B-Cell Memory 128**
5.1 Generation of B-Cell Memory 128
5.2 Maintenance of B-Cell Memory 129

**6 Conclusions 129**

**Bibliography 129**
Books and Reviews 129
Primary Literature 130

*Immunology. From Cell Biology to Disease.* Edited by Robert A. Meyers.

ISBN: 978-3-527-31770-7

## Keywords

**Adaptive Immune Response**
In contrast to the innate or inborn immune response, the adaptive or acquired immune response is the response of antigen-specific lymphocytes to antigen and is generated by clonal expansion of lymphocytes.

**Affinity Maturation**
A process during the humoral immune response, which selects for survival of B cells with high affinity for their antigen, which becomes particularly apparent during secondary and subsequent immunizations.

**Antigen**
Any substance that stimulates the production of antibodies, hence the name is derived from their ability to *gen*erate *anti*bodies. Here, the term is also used for molecules that are recognized by T cells.

**Apoptosis**
In contrast to necrosis, which is induced by poisoning or anoxia, apoptosis is a programmed cell death in which the cell activates a specialized cellular process to destroy itself. This cell suicide mechanism enables metazoans to control cell numbers and is characterized by nuclear condensation, nuclear degeneration, and DNA degradation.

**Chemokine**
Proteins secreted by cells that stimulate the activity of other cells. Chemokines have chemoattractant properties, act as messengers between cells and have a central role in inflammatory processes.

**Clusters of Differentiation (CD)**
A term that was originally coined to define cell-surface molecules that are recognized by a given set of monoclonal antibodies.

**Cytokine**
Any of various proteins secreted by cells of the immune system that serve to affect the behavior of other cells.

**Cytotoxic T cell (CTL)**
CTLs are lymphocytes that can kill other cells and are particularly important for the host defense against intracellular pathogens. Most CTLs are MHC class I restricted CD8+ T cells but MHC class II restricted CD4+ T cells can also kill in some cases.

**Dendritic Cells (DCs)**
Their name originates from their branched or dendritic morphology. The DCs found in lymphoid tissues are the most potent stimulators of T-cell responses. Follicular DCs present antigen to B cells.

**Helper CD4+ T Cell (Th Cells)**
Th cells are CD4+ T lymphocytes that can help B cells to make antibodies and have been implicated in the generation of functional effector and memory CD8+ T cells. Depending on their cytokine expression profile helper T cells can be divided into Th1 cells, which produce IFN-$\gamma$ and Th2 cells, which characteristically produce IL-4 and IL-5.

**Homeostasis**
A state of sustained equilibrium in which cells maintain stable numbers. In the case of lymphocytes, it refers to maintenance of normal numbers of lymphocytes in a noninfected individual.

**Immunity**
Ability to resist infections.

**Immunoglobulin**
General term for proteins that function as antibodies.

**Interleukin (IL)**
Generic term for cytokines produced by lymphocytes.

**Lymphocyte**
A lymphocyte is any of a group of white blood cells of crucial importance to the adaptive immune system. The two major subsets of lymphocytes are *thymus-dependent lymphocytes* or *T cells* and *B cells*, which are generated in the bone marrow.

**Lymphatic System**
A system of lymphoid channels and tissues that drains extracellular fluid from the periphery via the thoracic duct to the blood. It includes lymph nodes, Peyer's patches, and other organized lymphoid organs apart from the spleen, which communicates directly with the blood.

**Major Histocompatibility Complex (MHC)**
It is the most polymorphic gene clusters in the human genome and is encoded on human chromosome 6 (mouse chromosome 17). MHC molecules are membrane glycoproteins that present peptides to T cells. Other gene products of this region included proteins involved in antigen processing and other aspects of host defense.

**Pathogen**
Any disease-causing organism, such as a virus, bacterium or fungus.

**Plasma Cells**
Terminally differentiated B cells that are the main antibody-secreting cells of the body. They are found in the medulla of the lymph nodes, in splenic red pulp, and in the bone marrow.

**T-Cell Receptor (TCR)**
One of the main characteristics of T cells that distinguishes them from all other lymphocytes. It consists of a disulfide-linked heterodimer of the highly variable $\alpha$- and $\beta$-chains, which form a stable complex with the invariant CD3 chains on the cell surface. T lymphocytes expressing this composition of receptor are referred to as $\alpha{:}\beta$ T cells, distinguishing them from a smaller subset of T cells that express an alternative receptor on the cell surface consisting of variable $\gamma$- and $\delta$-chains in complex with CD3.

**Telomere**
The end of a chromosome. This specialized structure is involved in the replication and stability of linear DNA molecules.

**Transgene**
A gene that is integrated into the germline that is expressed *in vivo*.

**Vaccination**
Process of inducing an adaptive immune response against a given pathogen by stimulation with a dead or attenuated form of the pathogen.

Immunologic memory provides long-term protection against infectious pathogens and is the basis for all vaccinations. Long-term protection is mediated by specialized antigen-specific cells of the adaptive immune system, memory T- and B cell, as well as plasma cells. In response to an invading pathogen, naïve T cells specific for foreign antigen are primed by dendritic cells, expand and differentiate into effector cells that contribute to the clearance of the infection. After the initial immune response, most expanded T cells die and a small population of long-lived memory cells remains. Memory cells are qualitatively distinct from their antigen-inexperienced precursors since they proliferate and acquire their effector functions more rapidly upon repeat exposure to antigen.

## 1 Introduction

Immunologic memory, manifested as specific resistance to a second encounter with a pathogen, has been appreciated for more than 2000 years. In his book *The history of the Peleponnesian war*, Thucydides describes how a plague – the true pathogen is unknown – swept through Athens in 431 BCE. Thucydides made several observations on the plague including the fact

on memory versus naïve T cells is complex and can change with nonspecific activation and cytokine treatment, CD45 is a relatively unreliable marker for memory T-cell definition.

The IL-2 receptor $\alpha$-chain (CD25) and CD69 are rapidly expressed by T cells upon encountering cognate antigen, but their expression is not sustained on memory T cells. Similarly, O-glycosylation on mucin-type surface glycoproteins such as CD43 can distinguish effector from memory cells. Some memory T cells, however, express the IL-2 receptor $\beta$-chain (CD122), a component common to both IL-2 and IL-15 receptors, the latter cytokine being essential for memory CD8+ T-cell maintenance. Ly-6C has also been used as a marker to distinguish effector from memory T cells. Low and intermediate expression of this marker is characteristic of effector T cells, while it is highly expressed on memory T cells.

More recently, classification of human virus-specific T cells into naïve, effector, and memory cells on the basis of CD27 and CD28 expression was described. Human immunodeficiency virus (HIV), Epstein–Barr virus (EBV), cytomegalovirus (CMV) and hepatitis C virus (HCV) specific T cells, while similar during primary infection, give rise to distinct memory T-cell populations during chronic infection, which are characterized by differential expression of various surface antigens and effector molecules. Naïve CD8+ T cells are CD27+ CD28+ and express CCR7. Early and intermediate effector T cells downregulate CD28, CCR7, and CD45RA and express cytotoxic factors such as perforin, granzyme A, and GMP-17 (a marker for cytotoxic granules and lysosomes). Memory cells appear to lose the CD28, CD27, and CCR7 markers and maintain CD45 expression at low levels. They exhibit a greater cytotoxic potential than nonprimed cells and their telomeres are shortened. In this study, T-cell populations responding to persistent viral infections were more diverse than expected. EBV and HCV promoted the development of early effector memory T cells while HIV induced intermediate and CMV induced terminally differentiated memory T cells. In aggregate, recent studies demonstrate the remarkable diversity within memory T-cell populations and illustrate the importance of integrating phenotypic and functional characteristics when assessing memory T cells.

## 3.2
## Enhanced Responsiveness of Memory CD8+ T Cells: Potential Mechanisms

Enhanced sensitivity of memory T cells for cognate antigen likely results at least in part from more rapid signal transduction. Priming of naïve T cells leads to association of the tyrosine kinase Lck with the CD8 coreceptor, thereby enhancing TCR signaling. The association between Lck and CD8 is maintained in memory cells, which contributes to the lower stringency for activation of effector and memory cells. PEST domain-enriched tyrosine phosphatase (PEP) has recently been implicated as a negative regulator for specific aspects of T-cell development and function. Mice deficient in PEP demonstrate enhanced activation of Lck, which results in increased expansion and function of the effector and memory T-cell pool. Additional qualitative differences have been reported further downstream in the signaling cascade, involving the mitogen-activated protein (MAP) kinases ERK1 and 2 (extracellular signal-regulated kinase), which activate transcription factors essential for T-cell activation. In contrast to naïve T cells, memory cells show an increased ability to phosphorylate ERK1 and ERK2, thereby accelerating signal propagation.

In addition to enhancing the rate of T-cell activation, regulation of signaling molecules can also influence activation-induced differentiation of memory cells. For example, transgenic mice expressing a partially calcium-independent mutant of the calcium/calmodulin kinase II (CaMKII) $\gamma$B show an increase in the number of T cells in secondary lymphoid organs with a memory phenotype suggesting that this kinase is important for the development of memory.

The stage of the cell cycle that most naïve and antigen experienced T cells occupy also differs. Naïve T cells are arrested in the G0/G1 stage, with high expression of the cyclin dependent kinase (CDK) inhibitor p27Kip1 and low activity of the CDK6 and CDK2. In contrast, memory CD8+ T cells are also in G0/G1 but have low expression of p27Kip1 and high CDK6 kinase activity. This results in a preactivated cell cycle state of memory T cells that favors rapid division after antigen stimulation.

Lymphocyte differentiation is associated with programmed alterations in gene expression, which is largely regulated by structural changes in chromatin. Distinct gene expression profiles have been reported for memory cells. For example, mRNA levels for several cytokines including RANTES, interferon-$\gamma$ (IFN-$\gamma$), and cytotoxic molecules such as perforin and granzyme B are elevated in memory T cells. Transcripts for these genes are rapidly translated in memory T cells following TCR stimulation, circumventing the time required for transcriptional activation.

## 3.3 Generation of Memory CD8+ T Cells

At the end of an immune response to an invading pathogen, most of the clonally expanded T cells die, leaving a small population of memory T cells, which is maintained for long periods of time. How do memory T cell–precursors survive during the contraction phase of the immune response and which cells become long-lived memory cells? Recent experiments revealed that the homotypic form of CD8, CD8$\alpha\alpha$, is selectively expressed by CD8+ memory T cell–precursors and is required for their survival. Mice that are deficient in the CD8 enhancer ($E8_I$) express the usual heterotypic CD8$\alpha\beta$ molecule but not the homotypic form CD8$\alpha\alpha$. $E8_I$ knock-out mice mount a normal primary immune response but show diminished memory expansion. CD8$\alpha\alpha$ interacts with the non-classical MHC class I like molecule, thymic leukemia (TL) antigen. This interaction has been shown to modulate T-cell responses and leads to enhanced expression of IL-2/IL-15R$\beta$ and IL-7R$\alpha$. CD8+ memory T-cell precursors express IL-7R$\alpha$ chain, which renders them responsive to cytokine mediated survival signals. IL-7 is a potent survival factor for both naïve and memory CD8+ T cells due to the induction of antiapoptotic factors such as Bcl-2 and Bcl-$X_L$.

Studies of CD8+ T-cell memory have been greatly facilitated by the increasing spectrum of mice with targeted genetic deletions. For example, mice deficient in CD27, a member of the TRAF-linked tumor necrosis factor (TNF) receptor family that plays a costimulatory role during T-cell priming, have a markedly diminished virus-specific memory CD8+ T-cell response. On the other hand, mice lacking the CD28 costimulatory molecule, while deprived of CD86-mediated helper signals, mount nearly normal memory CD8+ T-cell responses in response to lymphocytic choriomeningitis virus (LCMV) or *Listeria monocytogenes* infection, supporting the notion that the costimulatory networks contain built-in redundancies that can compensate for the absence of CD28. However, costimulatory requirements for memory T-cell generation can vary significantly, since different pathogens create distinct inflammatory milieus. For example, induction of long-term immunity to influenza virus infection in contrast to LCMV and *L. monocytogenes* infection, requires CD28 signaling. 4-1BB, a TNF receptor family member, is another costimulatory molecule that has an impact on memory T-cell formation. Mice lacking 4-1BB ligand have diminished late expansion of virus-specific effector and memory CD8+ T cells, suggesting that 4-1BB signals play a role in sustaining T-cell activation and memory generation.

TNFR-associated factor 2 (TRAF2) is an adapter protein, which associates with the cytoplasmic tail of several TNF receptor family members including CD27, CD40, OX40, and 4-1BB and thereby links signals to downstream pathways. Hence, the absence of TRAF2 is likely to abrogate signaling through multiple costimulatory pathways. However, studies using dominant negative forms of TRAF2 demonstrated that in response to influenza virus infection, only secondary CD4+ and CD8+ T-cell expansion is decreased, while the primary expansion is not affected.

Cytokines are also implicated in the generation of antigen-specific memory T cells. Infection of IL-15- or IL-15R$\alpha$-deficient mice with LCMV generates a potent primary response but the memory pool decreases gradually over time, leading to impaired secondary CD8+ T-cell expansion. This demonstrates differential IL-15 requirements for naïve and memory CD8+ T cells. In the absence of IL-15 signaling, the memory response is also diminished following vesicular stomatitis virus (VSV) infection; however, the primary CD8+ T-cell expansion also appeared to be impaired. These studies suggest that in some circumstances, IL-15 may play a role in the generation of memory CD8 T cells.

CD4+ T cells have also been implicated in the generation of CD8+ T-cell memory; however, their importance varies with the type of infection. In the setting of LCMV infection, CD4+ T cells are not necessary for primary CD8+ T-cell responses, but they are indispensable for long-term CD8+ T-cell memory. Remarkably, CD4+ T cells contribute to CD8+ T-cell memory at the time of initial priming, since activation of memory CD8+ T cells occurs independently of CD4+ T-cell help. In contrast to viral infections, generation of memory CD8+ T cells during bacterial infections appeared to be less dependent on CD4+ T cells. For example, although mice lacking the class II transactivator (CIITA) have diminished numbers of CD4+ T cells, they generate memory CD8+ T cells following infection with the *L. monocytogenes*. Consistent with this result, CD4+ T cell-depleted, CD4−/−, and MHC class II−/− mice immunized with *L. monocytogenes* 3 to 4 weeks prior to rechallenge, mount protective recall responses. However, the consequences of the lack of CD4+ T cells help in generating functional CD8+ T-cell memory become more apparent many months following the initial antigen encounter. Recent data demonstrated that memory CD8+ T cells generated without CD4 help are defective in their ability to respond to secondary encounters with antigen. Memory CD8+

T-cell generation seems to be dependent on the presence of Th cells during, but not after, priming, suggesting that T-cell help is "programmed" into CD8+ T cells during priming. However, this concept was challenged by studies in which memory CD8+ T cells were adoptively transferred into wild type or CD4+ T cell-deficient recipients. The presence of CD4+ T cells was only important after, but not during, the early CD8+ T-cell programming phase, suggesting that CD4+ T cells are required only during the maintenance phase of long-live memory CD8+ T cells.

An important mechanism by which CD4+ T cells provide help to CD8+ T cell is through CD40-CD154 interactions. Antigen-presenting cells (APCs) were originally thought to be the mediators for these signals. Recent data demonstrates, however, that activated CD8+ T cells that express CD40 may receive direct help for CD4+ T cells expressing CD154. Administration of agonistic anti-CD40 monoclonal antibodies greatly enhances the protective potential of adoptively transferred, *L. monocytogenes*-specific CD8 T cells, providing further support for the role of CD40 in memory T-cell generation and/or maintenance.

Signals transmitted from CD4+ T cells are not uniformly stimulatory. In recent years, CD4+ CD25+ T regulatory cells have been implicated in the control of autoimmunity. During infection with *L. monocytogenes*, CD4+ CD25+ T regulatory cells can suppress memory CD8+ T-cell expansion. Similar observations were reported for memory CD8+ T-cells responses to infection with herpes simplex virus (HSV). Thus, CD4+ T cells play both positive and negative roles in the control of CD8+ T-cell memory.

Primary infection with *L. monocytogenes* induces memory T cells specific for multiple distinct peptides. CD8+ T cells restricted by the nonclassical MHC class Ib molecule H2-M3 and specific for *N*-formyl methionine peptides expand more rapidly than MHC class Ia-restricted CD8+ T cells during a primary infection. However, the memory responses of H2-M3-versus MHC class Ia-restricted T cells are dramatically different. H2-M3-restricted memory T cells express activation markers upon reencountering antigen, but do not proliferate, while MHC class Ia-restricted T cells undergo explosive expansion. It was initially suggested that this disparity may reflect differences in thymic selection of these two CD8+ T-cell types. However, more recent data show that MHC class Ia-restricted memory T cells inhibit the H2-M3 memory response by rapidly terminating *in vivo* antigen presentation of H2-M3 antigen following secondary infection.

### 3.4
### Maintaining CD8+ T-Cell Memory

Most infections are eliminated by the mammalian host, yet memory T cells persist afterwards for many years. Determining the mechanisms that enable long-term memory T-cell persistence in the absence of antigen has been a great challenge. Pioneering studies with adoptively transferred, antigen-specific memory T cells demonstrated the ability of these cells to survive in MHC-deficient mice. These observations provide strong evidence that once the differentiation program is initiated, memory CD8+ T cells can persist in the absence of antigen. However, immunological T-cell memory that is maintained without MHC contact may become functionally impaired. For example, memory CD4+ T cells specific for the male (HY) antigen show defective responses upon rechallenge and also alter their surface phenotype expression if they are kept in MHC-deficient hosts.

If maintenance of memory T cells is MHC and antigen independent, what regulates their persistence? Recent studies have demonstrated that IL-15 is crucial for *in vivo* maintenance of memory CD8+ T cells. Both primary and memory CD8+ T-cell expansion is diminished in IL-15- or IL-15-receptor-deficient mice, and maintenance of memory T-cell populations following immunization is markedly attenuated in the absence of IL-15 signaling. The gradual decline in memory CD8+ T-cell numbers in the absence of IL-15 results from decreased proliferation by memory cells. Interestingly, IL-7 can compensate for the missing survival signals in the absence of IL-15. When IL-7 is overexpressed in IL-15−/− mice, memory CD8+ T cells are generated, while removal of IL-7 and IL-15 completely inhibits the homeostatic proliferation of CD8+ memory T cells. These findings emphasize the joint contribution of these two cytokines for promoting CD8+ memory T-cell proliferation. Surprisingly, IL-2, a factor best known for its ability to stimulate T-cell proliferation, appears to play a negative role in memory T-cell survival.

In comparison to effector CD8+ T cells, memory CD8+ T cells express high levels of bcl-2 mRNA and protein, explaining their relative resistance to apoptosis. One effect of IL-15 is to increase bcl-2 levels in memory T cells, providing a second IL-15-mediated mechanism for memory CD8+ T-cell maintenance.

Cap structures that protect telomeres from degradation and terminal fusion play an essential role in stabilizing chromosome ends. Telomeres tend to progressively shorten when cells proliferate, eventually resulting in death of the proliferating cell. Telomere length, however, can be maintained in cells with activated telomerase, which adds DNA to telomeres with each cell division. Naïve as well as memory T cells undergo massive proliferation following antigen encounter. In order to compensate for telomere shortening, telomerase is more active in memory than naïve or effector T cells, potentially contributing to their long-term survival.

## 3.5 Models of CD8+ T-Cell Memory Generation

While some aspects of memory T-cell maintenance have been characterized, the mechanisms that generate memory CD8+ T cells are less clear. Several models for memory generation have been proposed and are outlined in this section. The recent introduction of methods that precisely quantify antigen-specific T cells during immune responses has provided immunologists with a framework upon which to build these models.

The *linear differentiation model* of memory T-cell formation proposes progressive differentiation of T cells from naïve to effector to memory cells (see Fig. 1). Following initial stimulation, an activation program drives naïve T-cell differentiation into effector cells, followed by further differentiation of a subset of effector T cells into memory T cells. Early support for this model came from TCR repertoire analyses of effector and memory T cells. The TCR repertoire of primary CD8+ T-cell responses is similar to that of memory CD8+ T cells, suggesting that memory cells are stochastically chosen from the population of effector CD8+ T cells activated during the primary infection. While linear differentiation of memory T cells was supported by adoptive transfer studies with T-cell receptor (TCR), transgenic (tg) CD8+ T cells

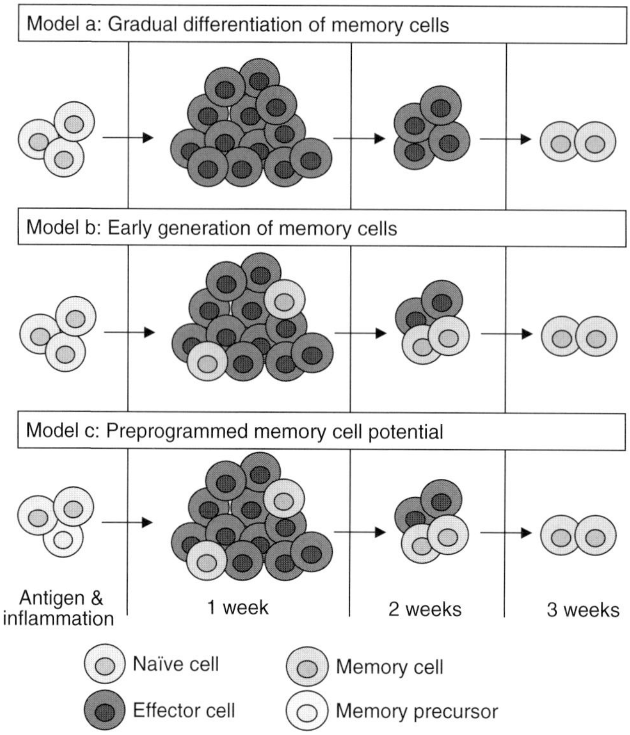


**Fig. 1** Models of programmed memory T-cell generation (see color plate p. xvii).

specific for the male H-Y antigen, the most direct evidence in support of this model came from studies using a clever system for genetically marking T cells that had acquired effector function. In this study, memory T cells were derived from effector T cells. It also remains controversial when memory cells form during the course of an immune response (see Fig. 1). One recent study demonstrated that T cells require several weeks from the time of antigen encounter to differentiate into memory T cells. Other studies showed that rechallenge of mice with a high inoculum of antigen 5 days after primary immunization resulted in a memory-like CD8+ T-cell response, suggesting that memory T cells may already be present during the primary immune response.

A variation of the linear differentiation model proposes that different memory T-cell subsets develop at different times following T-cell priming. According to this model, primed effector T cells can, in the absence of further stimuli, develop into CCR7+ CD62L+ central memory cells, which home to lymph nodes and lack effector functions. In contrast, effector memory cells, which are CCR7− CD62L−, derive from effector cells that have undergone more prolonged stimulation with antigen. This subset of memory cells produces high levels of effector cytokines, maintains cytolytic activity, and trafficks to peripheral tissues. Some evidence suggests that effector memory cells can convert to central memory cells upon antigen clearance. Evidence in support of this model comes from the analysis of T-cell priming by live versus killed bacterial vaccines. Live bacteria, which induce substantial inflammation, result in large effector memory T-cell populations, while killed bacteria, which induce an attenuated inflammatory response, give rise to T cells with a central memory phenotype.

Another model for memory T-cell development is the *decreasing-potential model,* which is based on the progressive loss of proliferative capacity. The underlying idea of this model is that successful memory T-cell generation requires a short duration of antigenic exposure during T-cell priming. According to this model, larger numbers of functional memory T cells should be produced when effector T cells are briefly exposed to antigen, thereby avoiding antigen driven apoptosis (AICD). Support for this model comes from situations of overwhelming antigen dose or during chronic infections, such as HIV or hepatitis C, where antigen-specific T cells either disappear or become dysfunctional.

A variant of the decreasing-potential model suggests that the timing of T-cell priming during an immune response determines which cells enter the memory lineage. The circulation of lymphocytes between the blood and lymphatic system is a rather slow process with turnover times of 12 to 24 h. Thus, the time that it takes for individual lymphocytes to encounter antigen presented in secondary lymphoid organs can vary significantly. Since antigen presentation is transient, cells recruited later will encounter antigen for a shorter time period than cells that are initially recruited. Brief exposure to antigen would be sufficient to induce their proliferation and differentiation into effector cells; however, the signal would be inadequate to trigger death pathways. Further studies are required to provide experimental support for this model.

The third model for memory T-cell generation is the *instructive model,* which posits that effector and memory T-cell differentiation diverges during T-cell priming. According to this model, memory T cells directly differentiate from naïve T cells without passing through the effector stage. Although experimental support for this model is sparse, the notion that naïve T cells might take their cues from the inflammatory and costimulatory context and undergo differentiation into either effector or memory T cells remains plausible.

T-cell programming is a relatively new concept that should be incorporated into models of memory T-cell generation (see Fig. 2). Early studies demonstrated that CD8+ T cells specific for different epitopes undergo coordinate expansion and contraction in response to infection,

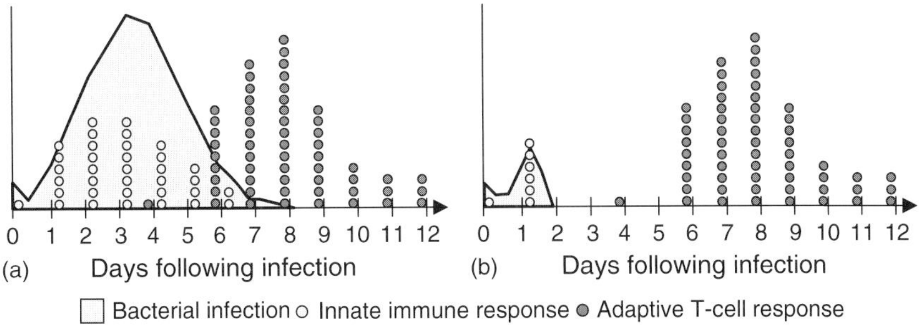


**Fig. 2** Antigen-independent proliferation and memory generation (see color plate p. xvii).

despite substantial differences in antigen expression levels and stability of peptide-MHC complexes. One explanation for this finding is the relatively short duration of effective *in vivo* antigen presentation, which results from downregulation by the nascent T-cell response. While feedback inhibition limits T-cell priming to the very early phase of the infection, the brief period of priming is sufficient to initiate a program of antigen-independent T-cell expansion. Remarkably, the process of T-cell expansion, contraction, and memory formation occur in the absence of both antigen and inflammation.

## 4
## CD4+ T-Cell Memory

While CD8+ T-cell memory has been extensively investigated, less is known about the generation and maintenance of memory CD4+ T cells. One explanation for this disparity is the difficulty to directly identify antigen-specific CD4+ T cells during the course of an immune response. One recent study, however, has compared CD4+ and CD8+ memory T-cell responses to LCMV and found that virus-specific memory CD4+ T cells decline in frequency over time while memory CD8+ T cells are maintained at high frequencies. In this section, we will concentrate on recent studies focusing on memory CD4+ T cells.

### 4.1
### Differentiation of Effector and Memory CD4+ T Cells

Naïve CD4+ T cells, upon *in vivo* priming, differentiate into either Th1 or Th2 effector cells. While the mechanisms leading to Th1 or Th2 differentiation remain incompletely understood, the innate inflammatory response elicited by the invading pathogen plays a major role in this process. Many viruses and bacteria induce IL-12 and interferon-$\gamma$ (IFN-$\gamma$) secretion, driving the differentiation of naïve CD4+ T cells into Th1 cells. In contrast, Th2 immunity, which is induced by intestinal helminth infection, is associated with innate inflammatory responses that produce IL-4, driving naïve CD4+ T-cell differentiation into Th2 cells. While the Th1/Th2 lineage decision has important consequences for the primary immune response, long-term memory T cells maintain the phenotype induced at the time of priming. Thus, while memory CD4 T cells do not produce cytokines in the absence of stimulation, upon reencountering antigen they rapidly reexpress the cytokines elicited during their primary activation. Several studies have demonstrated that epigenetic chromatin remodeling during the primary T-cell response accounts for specific cytokine expression profiles of reactivated memory T cells.

### 4.2
### Phenotype of Memory CD4+ T Cells

Memory CD4+ T cells have a lower threshold for activation than naïve T cells. Indeed, memory CD4+ T cells respond more rapidly to antigen, with enhanced cytokine secretion and greater proliferation in response to low antigen doses. In addition to greater responsiveness to antigen, memory CD4+ T cells are also present at higher frequencies in immune animals. While cytokine production by naïve CD4+ T cells is restricted to IL-2 and IL-3, memory CD4+ T cells secrete a greater variety of cytokines. Adoptive transfer of *in vitro* generated antigen-specific Th1 and Th2 cells leads to long-lived memory cells, which produce a pattern of cytokines closely related to that of the originally transferred cells. As mentioned in the previous section, epigenetic imprinting accounts for the fidelity of memory T-cell cytokine production.

### 4.3 Memory Generation and Maintenance

Mechanisms for CD4+ memory T-cell generation have not been defined. While models similar to those outlined for CD8+ T-cell memory formation may be applicable to CD4+ T-cell memory generation, existing data most strongly support linear differentiation from naïve to effector to memory CD4+ T cells. *In vitro* studies, for example, demonstrated that effector CD4+ T cells can differentiate into memory cells without the requirement for cell division. However, the factors involved in the transition from effector to memory T cells remain unclear. Remarkably, CD4+ T-cell memory is efficiently generated in the absence of IL-2, IL-4, or IFN-$\gamma$. As with CD8+ T cells, initial antigen stimulation is sufficient to initiate programmed expansion in CD4+ T cells, and extended antigenic stimulation is not required for differentiation into memory cells or for long-term persistence. Indeed, memory CD4+ T cells persist in the absence of both antigen and MHC class II molecules.

The cytokine requirements for long-term memory CD4+ T-cell survival are distinct from those described for memory CD8+ T cells. For example, while memory CD8+ T cells decrease in frequency in the absence of IL-15, memory CD4+ T cells are maintained independently of IL-15. Consistent with this finding, CD122 is expressed at low levels on memory CD4+ T cells. In addition, memory CD4+ T cells undergo homeostatic proliferation in the absence of both IL-7 and IL-15, and CD4+ T cells from mice deficient in the common cytokine receptor $\gamma$–chain, which is shared by receptors for IL-2, -4, -7, -9, and 15, can generate long-lived CD4+ memory T cells. IL-7 might not be essential for the homeostasis of CD4+ memory T cells, but contributes to the maintenance of all naïve and memory T-cell subsets, and thereby influences the control of the overall size of the T-cell pool. Differences in CD8+ and CD4+ T cell–memory maintenance have been detected following viral infection: virus-specific CD4+ memory T cells gradually decline, while CD8+ memory T cells remain stable. While many factors have been excluded as essential to CD4+ T cell–memory formation, the underlying mechanisms that drive differentiation of effector into memory T cells remains mysterious.

### 4.4 Trafficking of Memory CD4+ T Cells

CD8+ as well as CD4+ T cells differentiate in memory subsets with distinct effector functions and migration capacities. The differences in homing and trafficking of CD4+ T cells can be attributed to their expression of the selectin CD62L and the chemokine receptor CCR7. Central memory (CCR7+, CD62L+) CD4+ T cells are preferentially home to lymphoid tissues, whereas effector memory (CCR7–, CD62L–) recirculate in peripheral tissues long after immunization. The localization of central and effector memory CD4+ T cells in different organs has been directly visualized by immunohistology on whole body sections.

## 5
## B-Cell Memory

### 5.1
### Generation of B-Cell Memory

Similar to memory-T cells, memory-B cells respond rapidly to secondary exposure to antigen, proliferating and producing high affinity antibodies. B cells require accessory signals provided by helper T cells for full differentiation and are also activated by certain microbial stimuli. There are two distinct developmental pathways for antigen-specific B cells. B-1 and marginal-zone B cells proliferate and differentiate into antibody-secreting plasma cells, thereby providing the most rapid humoral response to antigen. While many plasma cells have a life span limited to a few days or weeks, some develop into long-lived cells that mediate long-term humoral immunity. Some antigen activated B cells migrate into follicles where they proliferate and form germinal centers (GCs), an important site of antibody secretion and affinity maturation. B cells expressing high affinity immunoglobulins receive positive signals from T-helper cells, leading to B-cell proliferation, survival, and maturation.

Lymphotoxin (LT) is important for development of peripheral lymphoid organs since mice lacking lymphotoxin-$\alpha$ (LT-$\alpha$) receptor have no lymph nodes or Peyer's patches and fail to form germinal centers. These structural deficits are linked to defective isotype-switching following primary or secondary immunization with T cell–dependent antigens. The important role of germinal centers in memory B-cell formation was demonstrated in adoptive transfer experiments. B cells from LT-$\alpha$−/− donors were capable of giving rise to a memory population, whereas a memory IgG B-cell responses could not be elicited in LT-$\alpha$−/− hosts. These studies clearly support previous observations, which point out the important role of germinal center structures for memory B-cell development.

Some of the factors determining the lineage decision between plasma cells and memory B cells have been identified, and include cell-surface receptors, cytokines, and transcriptional regulators. Stimulation through CD40-CD154, IL-4, expression of PAX5 and BCL-6 inhibit plasma cell differentiation. On the other hand, absence of CD40, IL-3, IL-6, and IL-10 and the commensurate degradation of BCL-6 and induction of Blimp-1 (B-lymphocyte-induced maturation protein 1) promote plasma cell differentiation.

CD4+ T cells are crucial for the generation of long-term humoral immunity. Direct antigen-specific T-B-cell interactions have been visualized with adoptive transfer systems, and the molecular requirements for CD4+ T-cell help have been identified recently. One important participant in this process is the signaling lymphocyte activation molecule (SLAM), SLAM-associated protein (SAP), which was originally identified as altered in X-linked lymphoproliferative syndrome (XLP). SAP controls signaling via the family of SLAM surface receptors, including CD84, CD150 (SLAM), CD229, and CD244 and hence plays a fundamental role in T-cell function. Mice deficient in the SAP gene (SAP−/−) mount strong antibody responses during an acute viral infection, but they fail to generate long-lived virus-specific plasma and memory-B cells. Adoptive transfer experiments demonstrated that SAP expression by CD4+ T cells, but not B cells, is essential for generation of long-term humoral immunity.

### 5.2 Maintenance of B-Cell Memory

It has been proposed that long-term humoral immunity is maintained by memory B cells, which continuously differentiate into short-lived plasma cells. Recently, however, it was shown that long-lived plasma cells survive for prolonged periods of time and maintain humoral immunity in the absence of memory B cells. Maintenance of humoral immunity appears to be antigen independent since memory B cells persist in the absence of antigen. Some experiments support the hypothesis that survival of long-lived plasma cells is determined by microenvironment rather than by intrinsic factors. *In vitro* experiments, for example, show that plasma cells isolated from tonsils undergo apoptosis unless rescued by stromal cells. Further support for this concept comes from the observation that plasma cell maintenance is restricted by the capacity of the splenic red pulp to provide "space," suggesting that plasma cell homeostasis is regulated by competition for specific survival niches.

Nerve growth factor (NGF) regulates neuronal cell development and survival. The finding that patients with autoimmune disorders often have elevated NGF plasma levels, and that NGF receptor shares structural similarities with some cytokine receptors suggested a role for this cytokine in the immune system. B cells can produce NGF under basal conditions and constitutively express NGF-receptors. It turns out that NGF acts as an autocrine survival factor for memory B cells and its neutralization completely inhibits humoral memory responses.

## 6 Conclusions

The journey to decipher memory T- and B-cell development continues to be a great scientific adventure, and there remain many unanswered questions that will require additional clever models and experiments. The motivation for moving forward, however, is substantial. Delineating basic mechanisms is important and applying the knowledge gained from an enhanced understanding of immunologic memory to the development of vaccines, against both infectious diseases and malignancy, will be especially satisfying.

## Bibliography

### Books and Reviews


Ahmed, R., Gray, D. (1996) Immunological memory and protective immunity: understanding their relation, *Science* **272**, 54–60.

Bevan, M.J. (2004) Helping the CD8(+) T-cell response, *Nat. Rev. Immunol* **4**, 595–602.

Calame, K.L., Lin, K.I., Tunyaplin, C. (2003) Regulatory mechanisms that determine the development and function of plasma cells, *Annu. Rev. Immunol.* **21**, 205–230.

Janeway, C.A., Travers, P., Walport, M., Shlomchik, M. (2004) *Immunobiology*, 6th edition, New York and London.

Kaech, S.M., Wherry, E.J., Ahmed, R. (2002) Effector and memory T-cell differentiation: implications for vaccine development, *Nat. Rev. Immunol.* **2**, 251–262.

Ploss, A., Pamer, E.G. (2004) Memory, in: Kaufmann S.H.E. (Ed.) *Novel Vaccination Strategies*, 1st edition. WILEY-VCH, Weinheim, Germany, pp. 73–88.

Sallusto, F., Geginat, J., Lanzavecchia, A. (2004) Central memory and effector memory T cell subsets: function, generation, and maintenance, *Annu. Rev. Immunol.* **22**, 765–787.

Schluns, K.S., Lefrancois, L. (2003) Cytokine control of memory T-cell development and survival, *Nat. Rev. Immunol.* **3**, 269–279.

Silverstein, A.M. (1989) *A History of Immunology*, 1st edition, Academic Press, Inc., San Diego, CA.

## Primary Literature

Agarwal, S., Rao, A. (1998) Modulation of chromatin structure regulates cytokine gene expression during T cell differentiation, *Immunity* **9**, 765–775.

Appay, V., Dunbar, P.R., Callan, M., Klenerman, P., Gillespie, G.M., Papagno, L., Ogg, G.S., King, A., Lechner, F., Spina, C.A., et al. (2002) Memory CD8+ T cells vary in differentiation phenotype in different persistent virus infections, *Nat. Med.* **8**, 379–385.

Bachmann, M.F., Gallimore, A., Linkert, S., Cerundolo, V., Lanzavecchia, A., Kopf, M., Viola, A. (1999) Developmental regulation of Lck targeting to the CD8 coreceptor controls signaling in naive and memory T cells, *J. Exp. Med.* **189**, 1521–1530.

Badovinac, V.P., Porter, B.B., Harty, J.T. (2002) Programmed contraction of CD8(+) T cells after infection, *Nat. Immunol.* **3**, 619–626.

Becker, T.C., Wherry, E.J., Boone, D., Murali-Krishna, K., Antia, R., Ma, A., Ahmed, R. (2002) Interleukin 15 is required for proliferative renewal of virus-specific memory CD8 T cells, *J. Exp. Med.* **195**, 1541–1548.

Bennett, S.R., Carbone, F.R., Karamalis, F., Flavell, R.A., Miller, J.F., Heath, W.R. (1998) Help for cytotoxic-T-cell responses is mediated by CD40 signalling, *Nature* **393**, 478–480.

Bird, J.J., Brown, D.R., Mullen, A.C., Moskowitz, N.H., Mahowald, M.A., Sider, J.R., Gajewski, T.F., Wang, C.R., Reiner, S.L. (1998) Helper T cell differentiation is controlled by the cell cycle, *Immunity* **9**, 229–237.

Bourgeois, C., Rocha, B., Tanchot, C. (2002) A role for CD40 expression on CD8+ T cells in the generation of CD8+ T cell memory, *Science* **297**, 2060–2063.

Bradley, L.M., Duncan, D.D., Tonkonogy, S., Swain, S.L. (1991) Characterization of antigen-specific CD4+ effector T cells in vivo: immunization results in a transient population of MEL-14-, CD45RB-helper cells that secretes interleukin 2 (IL-2), IL-3, IL-4, and interferon gamma, *J. Exp. Med.* **174**, 547–559.

Bui, J.D., Calbo, S., Hayden-Martinez, K., Kane, L.P., Gardner, P., Hedrick, S.M. (2000) A role for CaMKII in T cell memory, *Cell* **100**, 457–467.

Busch, D.H., Kerksiek, K.M., Pamer, E.G. (2000) Differing roles of inflammation and antigen in T cell proliferation and memory generation, *J. Immunol.* **164**, 4063–4070.

Busch, D.H., Pilip, I., Pamer, E.G. (1998) Evolution of a complex T cell receptor repertoire during primary and recall bacterial infection, *J. Exp. Med.* **188**, 61–70.

Busch, D.H., Pilip, I.M., Vijh, S., Pamer, E.G. (1998) Coordinate regulation of complex T cell populations responding to bacterial infection, *Immunity* **8**, 353–362.

Chase, M.W. (1945) The Cellular transfer of cutaneous hypersensitivity to tuberculin, *Proc. Soc. Exp. Biol. Med.* **59**, 134–135.

Cho, B.K., Wang, C., Sugawa, S., Eisen, H.N., Chen, J. (1999) Functional differences between memory and naive CD8 T cells, *Proc. Natl. Acad. Sci. U. S. A.* **96**, 2976–2981.

Crotty, S., Kersh, E.N., Cannons, J., Schwartzberg, P.L., Ahmed, R. (2003) SAP is required for generating long-term humoral immunity, *Nature* **421**, 282–287.

Garside, P., Ingulli, E., Merica, R.R., Johnson, J.G., Noelle, R.J., Jenkins, M.K. (1998) Visualization of specific B and T lymphocyte interactions in the lymph node, *Science* **281**, 96–99.

Gibson, T., Medawar, P.B. (1943) The fate of skin homografts in man, *J. Anat.* **77**, 299.

Goldrath, A.W., Bogatzki, L.Y., Bevan, M.J. (2000) Naive T cells transiently acquire a memory-like phenotype during homeostasis-driven proliferation, *J. Exp. Med.* **192**, 557–564.

Goldrath, A.W., Sivakumar, P.V., Glaccum, M., Kennedy, M.K., Bevan, M.J., Benoist, C., Mathis, D., Butz, E.A. (2002) Cytokine requirements for acute and basal homeostatic proliferation of naive and memory CD8+ T cells, *J. Exp. Med.* **195**, 1515–1522.

Gowans, J.L., Knight, E.J. (1964) The route of re-circulation of lymphocytes in the rat, *Proc. R. Soc. Lond. Ser. B. Biol. Sci.* **159**, 257–282.

Hamilton, S.E., Porter, B.B., Messingham, K.A., Badovinac, V.P., Harty, J.T. (2004) MHC class Ia-restricted memory T cells inhibit expansion of a nonprotective MHC class Ib (H2-M3)-restricted memory response, *Nat. Immunol.* **5**, 159–168.

Harrington, L.E., Galvan, M., Baum, L.G., Altman, J.D., Ahmed, R. (2000) Differentiating between memory and effector CD8 T cells by altered expression of cell surface O-glycans, *J. Exp. Med.* **191**, 1241–1246.

Hasegawa, K., Martin, F., Huang, G., Tumas, D., Diehl, L., Chan, A.C. (2004) PEST domain-enriched tyrosine phosphatase (PEP) regulation of effector/memory T cells, *Science* **303**, 685–689.

Hendriks, J., Gravestein, L.A., Tesselaar, K., van Lier, R.A., Schumacher, T.N., Borst, J. (2000) CD27 is required for generation and long-term maintenance of T cell immunity, *Nat. Immunol.* **1**, 433–440.

Homann, D., Teyton, L., Oldstone, M.B. (2001) Differential regulation of antiviral T-cell immunity results in stable CD8+ but declining CD4+ T-cell memory, *Nat. Med.* **7**, 913–919.

Hu, H., Huston, G., Duso, D., Lepak, N., Roman, E., Swain, S.L. (2001) CD4(+) T cell effectors can become memory cells with high efficiency and without further division, *Nat. Immunol.* **2**, 705–710.

Huster, K.M., Busch, V., Schiemann, M., Linkemann, K., Kerksiek, K.M., Wagner, H., Busch, D.H. (2004) Selective expression of IL-7 receptor on memory T cells identifies early CD40L-dependent generation of distinct CD8+ memory T cell subsets, *Proc. Natl. Acad. Sci. U. S. A.* **101**, 5610–5615.

Jacob, J., Baltimore, D. (1999) Modelling T-cell memory by genetic marking of memory T cells in vivo, *Nature* **399**, 593–597.

Janssen, E.M., Lemmens, E.E., Wolfe, T., Christen, U., Von Herrath, M.G., Schoenberger, S.P. (2003) CD4(+) T cells are required for secondary expansion and memory in CD8(+) T lymphocytes, *Nature* **421**, 852–856.

Jenner, E. (1798) *An Inquiry into the Causes and Effects of the Variolæ Vaccinæ, a Disease Discovered in Some of the Western Counties of England, Particularly Gloucestershire, and Known by the Name of the Cow Pox*, 1st edition, Low, Sampson, London, UK.

Kaech, S.M., Ahmed, R. (2001) Memory CD8+ T cell differentiation: initial antigen encounter triggers a developmental program in naive cells, *Nat. Immunol.* **2**, 415–422.

Kaech, S.M., Hemby, S., Kersh, E., Ahmed, R. (2002) Molecular and functional profiling of memory CD8 T cell differentiation, *Cell* **111**, 837–851.

Kaech, S.M., Tan, J.T., Wherry, E.J., Konieczny, B.T., Surh, C.D., Ahmed, R. (2003) Selective expression of the interleukin 7 receptor identifies effector CD8 T cells that give rise to long-lived memory cells, *Nat. Immunol.* **4**, 1191–1198.

Kassiotis, G., Garcia, S., Simpson, E., Stockinger, B. (2002) Impairment of immunological memory in the absence of MHC despite survival of memory T cells, *Nat. Immunol.* **3**, 244–250.

Kaufmann, S.H., Simon, M.M., Hahn, H. (1979) Specific Lyt 123 cells are involved in protection against Listeria monocytogenes and in delayed-type hypersensitivity to listerial antigens, *J. Exp. Med.* **150**, 1033–1038.

Kedl, R.M., Mescher, M.F. (1998) Qualitative differences between naive and memory T cells make a major contribution to the more rapid and efficient memory CD8+ T cell response, *J. Immunol.* **161**, 674–683.

Kerksiek, K.M., Busch, D.H., Pilip, I.M., Allen, S.E., Pamer, E.G. (1999) H2-M3-restricted T cells in bacterial infection: rapid primary but diminished memory responses, *J. Exp. Med.* **190**, 195–204.

Kieper, W.C., Tan, J.T., Bondi-Boyd, B., Gapin, L., Sprent, J., Ceredig, R., Surh, C.D. (2002) Overexpression of interleukin (IL)-7 leads to IL-15-independent generation of memory phenotype CD8+ T cells. *J. Exp. Med.*, **195**, 1533–1539.

Kitasato, S. (1977) *Collected Papers of Shibasaburo Kitasato*, Kitasato Institute, Tokyo, Japan.

Koch, R. (1878) *Untersuchungen über die Aetiologie der Wundinfectionskrankheiten*, F.C.W. Vogel, Leipzig, Germany.

Ku, C.C., Murakami, M., Sakamoto, A., Kappler, J., Marrack, P. (2000) Control of homeostasis of CD8+ memory T cells by opposing cytokines, *Science* **288**, 675–678.

Kursar, M., Bonhagen, K., Fensterle, J., Kohler, A., Hurwitz, R., Kamradt, T., Kaufmann, S.H., Mittrucker, H.W. (2002) Regulatory CD4+ CD25+ T cells restrict memory CD8+ T cell responses, *J. Exp. Med.* **196**, 1585–1592.

Landsteiner, K., Chase, M.W. (1942) Experiments on transfer of cutaneous sensitivity to simple compounds, *Proc. Soc. Exp. Biol. Med.* **49**, 688–689.

Lane, F.C., Unanue, E.R. (1972) Requirement of thymus (T) lymphocytes for resistance to listeriosis, *J. Exp. Med.* **135**, 1104–1112.

Lantz, O., Grandjean, I., Matzinger, P., Di Santo, J.P. (2000) Gamma chain required for naive CD4+ T cell survival but not for antigen proliferation, *Nat. Immunol.* **1**, 54–58.

Lauvau, G., Vijh, S., Kong, P., Horng, T., Kerksiek, K., Serbina, N., Tuma, R.A., Pamer, E.G. (2001) Priming of memory but not effector

CD8 T cells by a killed bacterial vaccine, *Science* **294**, 1735–1739.

Leishman, A.J., Naidenko, O.V., Attinger, A., Koning, F., Lena, C.J., Xiong, Y., Chang, H.C., Reinherz, E., Kronenberg, M., Cheroutre, H. (2001) T cell responses modulated through interaction between CD8alphaalpha and the nonclassical MHC class I molecule, TL, *Science* **294**, 1936–1939.

Mackaness, G.B., Blanden, R.V. (1967) Cellular immunity, *Prog. Allergy* **11**, 89–140.

Madakamutil, L.T., Christen, U., Lena, C.J., Wang-Zhu, Y., Attinger, A., Sundarrajan, M., Ellmeier, W., von Herrath, M.G., Jensen, P., Littman, D.R., Cheroutre, H. (2004) CD8alpha alpha-mediated survival and differentiation of CD8 memory T cell precursors, *Science* **304**, 590–593.

Marchesi, V.T., Gowans, J.L. (1964) The migration of lymphocytes through the endothelium of venules in the lymph nodes: an electron microscope study, *Proc. R. Soc. Lond. Ser. B. Biol. Sci.* **159**, 283–290.

Maruyama, M., Lam, K.P., Rajewsky, K. (2000) Memory B-cell persistence is independent of persisting immunizing antigen, *Nature* **407**, 636–642.

Masopust, D., Vezys, V., Marzo, A.L., Lefrancois, L. (2001) Preferential localization of effector memory cells in nonlymphoid tissue, *Science* **291**, 2413–2417.

Matsumoto, M., Mariathasan, S., Nahm, M.H., Baranyay, F., Peschon, J.J., Chaplin, D.D. (1996) Role of lymphotoxin and the type I TNF receptor in the formation of germinal centers, *Science* **271**, 1289–1291.

Medawar, P.B. (1944) The behaviour and fate of skin autografts and skin homografts in rabbits, *J. Anat.* **78**, 176.

Medawar, P.B. (1945) A second study of the behaviour and fate of skin homografts in rabbits, *J. Anat.* **79**, 157.

Mercado, R., Vijh, S., Allen, S.E., Kerksiek, K., Pilip, I.M., Pamer, E.G. (2000) Early programming of T cell populations responding to bacterial infection, *J. Immunol.* **165**, 6833–6839.

Merville, P., Dechanet, J., Desmouliere, A., Durand, I., de Bouteiller, O., Garrone, P., Banchereau, J., Liu, Y.J. (1996) Bcl-2+ tonsillar plasma cells are rescued from apoptosis by bone marrow fibroblasts, *J. Exp. Med.* **183**, 227–236.

Metchnikoff, E. (1905) *Immunity in Infective Diseases*, Cambridge University Press, Cambridge, UK.

Murali-Krishna, K., Lau, L.L., Sambhara, S., Lemonnier, F., Altman, J., Ahmed, R. (1999) Persistence of memory CD8 T cells in MHC class I-deficient mice, *Science* **286**, 1377–1381.

Murphy, J.B., Ellis, A.W.M. (1914) Experiments on the role of lymphoid tissue in the resistance to experimental tuberculosis in mice, *J. Exp. Med.* **20**, 397–403.

North, R.J. (1973) Importance of thymus-derived lymphocytes in cell-mediated immunity to infection, *Cell. Immunol.* **7**, 166–176.

Opferman, J.T., Ober, B.T., Ashton-Rickardt, P.G. (1999) Linear differentiation of cytotoxic effectors into memory T lymphocytes, *Science* **283**, 1745–1748.

Pasteur, L. (1880) *De l'attenuation du virus du cholera des poules, Vol. 91*, Comptes rendus de l'Academie des sciences, Paris, France.

Pihlgren, M., Dubois, P.M., Tomkowiak, M., Sjogren, T., Marvel, J. (1996) Resting memory CD8+ T cells are hyperreactive to antigenic challenge in vitro, *J. Exp. Med.* **184**, 2141–2151.

Reinhardt, R.L., Khoruts, A., Merica, R., Zell, T., Jenkins, M.K. (2001) Visualizing the generation of memory CD4 T cells in the whole body, *Nature* **410**, 101–105.

Ridge, J.P., Di Rosa, F., Matzinger, P. (1998) A conditioned dendritic cell can be a temporal bridge between a CD4+ T-helper and a T-killer cell, *Nature* **393**, 474–478.

Sallusto, F., Lenig, D., Forster, R., Lipp, M., Lanzavecchia, A. (1999) Two subsets of memory T lymphocytes with distinct homing potentials and effector functions, *Nature* **401**, 708–712.

Schoenberger, S.P., Toes, R.E., van der Voort, E.I., Offringa, R., Melief, C.J. (1998) T-cell help for cytotoxic T lymphocytes is mediated by CD40-CD40L interactions, *Nature* **393**, 480–483.

Seddon, B., Tomlinson, P., Zamoyska, R. (2003) Interleukin 7 and T cell receptor signals regulate homeostasis of CD4 memory cells, *Nat. Immunol.* **4**, 680–686.

Shedlock, D.J., Shen, H. (2003) Requirement for CD4 T cell help in generating functional CD8 T cell memory, *Science* **300**, 337–339.

Slifka, M.K., Antia, R., Whitmire, J.K., Ahmed, R. (1998) Humoral immunity due to long-lived plasma cells, *Immunity* **8**, 363–372.

Sourdive, D.J., Murali-Krishna, K., Altman, J.D., Zajac, A.J., Whitmire, J.K., Pannetier, C., Kourilsky, P., Evavold, B., Sette, A., Ahmed, R. (1998) Conserved T cell receptor repertoire in primary and memory CD8 T cell responses to an acute viral infection, *J. Exp. Med.* **188**, 71–82.

Sun, J.C., Bevan, M.J. (2003) Defective CD8 T cell memory following acute infection without CD4 T cell help, *Science* **300**, 339–342.

Sun, J.C., Williams, M.A., Bevan, M.J. (2004) CD4(+) T cells are required for the maintenance, not programming, of memory CD8(+) T cells after acute infection, *Nat. Immunol.* **5**, 927–933.

Suvas, S., Kumaraguru, U., Pack, C.D., Lee, S., Rouse, B.T. (2003) CD4+CD25+ T cells regulate virus-specific primary and memory CD8+ T cell responses, *J. Exp. Med.* **198**, 889–901.

Swain, S.L. (1994) Generation and in vivo persistence of polarized Th1 and Th2 memory cells, *Immunity* **1**, 543–552.

Swain, S.L., Hu, H., Huston, G. (1999) Class II-independent generation of CD4 memory T cells from effectors, *Science* **286**, 1381–1383.

Swanson, B.J., Murakami, M., Mitchell, T.C., Kappler, J., Marrack, P. (2002) RANTES production by memory phenotype-T cells is controlled by a posttranscriptional, TCR-dependent process, *Immunity* **17**, 605–615.

Sze, D.M., Toellner, K.M., Garcia de Vinuesa, C., Taylor, D.R., MacLennan, I.C. (2000) Intrinsic constraint on plasmablast growth and extrinsic limits of plasma cell survival. *J. Exp. Med.*, **192**, 813–821.

Tan, J.T., Ernst, B., Kieper, W.C., LeRoy, E., Sprent, J., Surh, C.D. (2002) Interleukin (IL)-15 and IL-7 jointly regulate homeostatic proliferation of memory phenotype-CD8+ cells but are not required for memory phenotype-CD4+ cells, *J. Exp. Med.* **195**, 1523–1532.

Teague, T.K., Hildeman, D., Kedl, R.M., Mitchell, T., Rees, W., Schaefer, B.C., Bender, J., Kappler, J., Marrack, P. (1999) Activation changes the spectrum but not the diversity of genes expressed by T cells, *Proc. Natl. Acad. Sci. U. S. A.* **96**, 12691–12696.

Torcia, M., Bracci-Laudiero, L., Lucibello, M., Nencioni, L., Labardi, D., Rubartelli, A., Cozzolino, F., Aloe, L., Garaci, E. (1996) Nerve growth factor is an autocrine survival factor for memory B lymphocytes, *Cell* **85**, 345–356.

Urdahl, K.B., Sun, J.C., Bevan, M.J. (2002) Positive selection of MHC class Ib-restricted CD8(+) T cells on hematopoietic cells, *Nat. Immunol.* **3**, 772–779.

Veiga-Fernandes, H., Rocha, B. (2004) High expression of active CDK6 in the cytoplasm of CD8 memory cells favors rapid division, *Nat. Immunol.* **5**, 31–37.

Veiga-Fernandes, H., Walter, U., Bourgeois, C., McLean, A., Rocha, B. (2000) Response of naive and memory CD8+ T cells to antigen stimulation in vivo, *Nat. Immunol.* **1**, 47–53.

Wherry, E.J., Teichgraber, V., Becker, T.C., Masopust, D., Kaech, S.M., Antia, R., Von Andrian, U.H., Ahmed, R. (2003) Lineage relationship and protective immunity of memory CD8 T cell subsets, *Nat. Immunol.* **4**, 225–234.

Wong, P., Pamer, E.G. (2001) Cutting edge: antigen-independent CD8 T cell proliferation, *J. Immunol.* **166**, 5864–5868.

Wong, P., Pamer, E.G. (2003) Feedback regulation of pathogen-specific T cell priming, *Immunity* **18**, 499–511.

Wong, P., Lara-Tejero, M., Ploss, A., Leiner, I., Pamer, E.G. (2004) Rapid development of T cell memory, *J. Immunol.* **172**, 7239–7245.

Yang, Y., Chang, J.F., Parnes, J.R., Fathman, C.G. (1998) T cell receptor (TCR) engagement leads to activation-induced splicing of tumor necrosis factor (TNF) nuclear pre-mRNA, *J. Exp. Med.* **188**, 247–254.

Zhang, X., Sun, S., Hwang, I., Tough, D.F., Sprent, J. (1998) Potent and selective stimulation of memory-phenotype CD8+ T cells in vivo by IL-15, *Immunity* **8**, 591–599.

Zimmermann, C., Prevost-Blondel, A., Blaser, C., Pircher, H. (1999) Kinetics of the response of naive and memory CD8 T cells to antigen: similarities and differences. *Eur. J. Immunol.*, **29**, 284–290.

# Part II
# Signaling in the Immune System

*Immunology. From Cell Biology to Disease.* Edited by Robert A. Meyers.

ISBN: 978-3-527-31770-7

# 6
# Molecular Mediators: Cytokines

*Jean-Marc Cavaillon*
*Institut Pasteur, Paris, France*

**1 Historical Record 140**

**2 A Universal Language of Cells 141**
2.1 Definitions 141
2.2 Families 143

**3 Receptors 144**
3.1 Families 144
3.2 Signaling 145
3.3 Soluble Receptors 146

**4 Functions 147**
4.1 Hematopoiesis 147
4.2 Immune Response 147
4.2.1 Innate Immunity 147
4.2.2 Adaptive Immunity 148
4.2.3 Th1 and Th2 Cytokines 149
4.3 Cell Survival and Cell Death 149
4.4 Link with the Central and Peripheral Nervous System 150
4.5 Reproduction 150
4.6 Inflammation 151

**5 Life without Cytokines (What Knockout Mice Tell Us) 153**

**6 Cytokine Synthesis 155**
6.1 Homeostasis 155
6.2 Induction 155
6.3 Parameters that Affect Production 156
6.4 Measurements 156

*Immunology. From Cell Biology to Disease.* Edited by Robert A. Meyers.

ISBN: 978-3-527-31770-7

**7 The Cytokine Network 157**

**8 Individual Heterogeneity 158**

**9 Cytokines and Infections 159**
9.1 Half Angel–Half Devil 159
9.2 The Strategies of Microbes against Cytokines 160

**10 Cytokines and Diseases 160**
10.1 Autoimmune Diseases 161
10.2 Allergy 161
10.3 Sepsis 161
10.4 Chronic Inflammatory Diseases 161
10.5 Cancer 162
10.6 Hepatitis C 162
10.7 Multiple Sclerosis 162
10.8 Transplantation 162
10.9 Neutropenia 163

**11 Conclusion 163**

**Bibliography 163**
Books and Reviews 163
Primary Literature 163

## Keywords

**Cytokines**
Constitutively released or produced upon cell activation, cytokines are soluble mediators – sometimes membrane bound – binding to receptors on target cells and inducing, modulating, or inhibiting cellular functions. Most cytokines are produced by immune cells and act on immune cells, but cytokines are produced by numerous cell types and act on a great variety of cells, accordingly displaying a large spectrum of activities. They are characterized by their redundancy, their synergistic action, and their regulatory loops. Those produced constitutively contribute to the tissue and systemic homeostasis. The cytokine are elements of a universal language used by cells to communicate.

**Chemokines**
A subfamily of cytokines with structural similarities and sharing the property to recruit cells within the extravascular space of the tissues (chemotaxis), either at homeostasis or during inflammatory processes.

**Interferons (IFN)**
A subfamily of cytokines sharing an antiviral bioactivity. Interferons are characterized by different subgroups (type I and II), depending on their origin. In addition to their common property, some (e.g. IFN$\gamma$) are involved at different stages of the onset and further development of the immune responses, or play a role during gestation in mammals (e.g. IFN$\tau$).

**Interleukin (IL-)**
A convenient name to classify cytokines. The numbers have been given as a reflection of their discovery through the years. This term does not define a family, since none are biochemically or biologically related. However, the most recent interleukins have been described after gene banks and are thus related to previous well-known interleukins. When the word was created (1979), there was only IL-1 and IL-2.

**Receptors**
The receptors of cytokines allow the cells to respond to the signal delivered by cytokines. They can involve 1, 2, or 3 different chains. Most chains are constituted with one extracellular domain, one transmembrane domain and one intracellular domain, except the chemokine receptors that are constituted by 7 transmembrane domains.

Cytokines are proteins or glycoproteins produced by cells and acting on other cells that display on their surface the specific cytokine receptor. Cytokines are words used by the cells to communicate. Alone or within a determined sequence, these mediators lead the responding cells to modify its function (e.g. secretion, proliferation, induction, inhibition, enhanced/reduced function, migration, apoptosis, etc.). Despite cytokines having been mainly discovered by immunologists as a product of cells of the leukocyte lineages, they are now recognized as elements of a universal language used by most cells of any other lineages. Accordingly, they are essential for many events through the life (reproductive tissue remodeling, embryogenesis, steady state and adaptive hematopoiesis, surveillance and maintenance of the tissue structure, functions by the immune system, cell survival, and cell death, etc). They also allow a dialogue with the central nervous system and some cytokines may modify different behaviors (fever, anorexia, sleep, etc). They usually act within their vicinity, but they can also act via an endocrine fashion. Their productions are tightly controlled within a complex network of positive and negative loops. They are a prerequisite to control infection by invasive microorganisms, but their exacerbated production can be deleterious at the local or systemic levels. Their half angel–half devil aspect has rendered them difficult to use with therapeutic purposes, although some recombinant cytokines or some cytokine-neutralizing strategies have been successfully used in different pathological conditions.

## 1 Historical Record

The history of cytokines can be divided in four main steps through the years. The first step corresponds to the identification of biological activities of factors present in cell culture supernatants or within the blood stream (late 40s–early 70s). During the second step (late 70s–early 80s), the purification and the biochemical characterization of these factors allowed to reach the stage of defined molecular entities. During the third step (mid-80s–mid-90s), the cloning of the coding and noncoding sequences, the production of recombinant cytokines, as well as the production of genetically manipulated mice extended the understanding of the complexity of the cytokine-mediated language. Since the late 90s (the fourth step), the discovery of new cytokines has been relying on the availability of different genome sequences and the different gene banks. In the later case, bioactivities of newly identified cytokines remained to be discovered. Probably, the first reported bioactivity was the induction of fever and the discovery of an endogenous pyrogen in the late 40s, early 50s. The purification of human endogenous pyrogen was published in 1977 by Charles Dinarello. This allowed the identification of endogenous pyrogen as interleukin-1 (IL-1). It definitively established that fever was due to endogenous factor(s) and not directly induced by microbial products. After the cloning of IL-1 in 1984, this was confirmed the following year, when injection of recombinant IL-1 was shown to induce fever in rabbits. Later, genes encoding different IL-1 homologues have been identified by means of their significant sequence similarities to IL-1, but further studies are required to know whether some members of the IL-1 family (IL-1F5–10) are pyrogenic. The other earlier-described bioactivity was reported in 1957 by Isaacs and Lindenmann. They showed that resistance to virus development could be conferred to a cell by another virus-infected cell. This bioactivity interfering with virus spreading was named "interferon." In the 60s, numerous bioactivities had been detected in supernatants of lymphocytes. Accordingly, identified as nonantibody mediators, D. Dumonde et al. proposed in 1969, the word "lymphokines." In the following years, the word "monokine" was employed to characterize other bioactivities found in monocyte/macrophage supernatants. In the 70s, the most famous lymphokine was the "macrophage migration inhibitory factor" (MIF). Owing its name to its *in vitro* property, MIF had been first identified in 1966 as a T-lymphocyte-derived product associated with delayed-type hypersensitivity. In the absence of cell lines producing high amounts of MIF, its biochemical characterization was delayed. The cloning of this molecule occurred in 1989, long after its discovery. In the meantime, MIF-like activity was reported in cultures of fibroblasts, and S. Cohen et al. found a MIF-like activity in the supernatants of virus-loaded cell lines. Then, it appeared evident that lymphokines were incorrectly named because many cell types other than lymphocytes could produce similar bioactivities. And, in 1974, S. Cohen coined the name "cytokine." Indeed, MIF is a good example of a cytokine produced by many cell types, including macrophages, dendritic leukocytes, endothelial cells, epithelial cells, and fibroblasts. Most fascinating was the rediscovery of MIF as a product of the pituitary gland, further reinforcing the concept of cytokine and illustrating that, sometimes, the border between cytokines and hormones may be difficult to draw.

Names of factors were traditionally reflecting their various biologic activities identified in different laboratories. This approach resulted in an imprecise and redundant nomenclature. In 1979, during the 2nd international Lymphokine Workshop in Ermatingen (Switzerland), it was decided that the name "interleukin" (IL-) be given to two different characterized factors. Under the new term IL-1 were gathered numerous names or acronyms defined according to one of the properties of this cytokine: osteoclast activating factor (OAF), lymphocyte activating factor (LAF), B-cell activating factor (BAF), epidermal thymocyte activating factor (ETAF), T-cell replacing factor III (TRFIII), leukocyte endogenous mediator, endogenous pyrogen, mitogenic protein, catabolin, hemopoietin-1, and so on. The name interleukin (between leukocytes) was obviously not the best choice since these factors were not restricted to leukocytes, but it had been mainly chosen by immunologists too self-absorbed, who were considering the immune system working independently of the other systems. When interviewed about it, Stanley Cohen specified: "It is a common misconception that the term interleukin was accepted for adoption by the nomenclature committee. When it was proposed, both Byron Waksman and I, as Co-chairs, felt it would turn out to be too restrictive (even then) and there was general agreement by the committee that it would be premature to refer to mediators as interleukins. This view was conveyed to the proposers. However, shortly thereafter, an editorial appeared in Journal of Immunology stating that the term 'interleukin' had been discussed and considered by the committee (true) and that henceforth the term would be used in scientific communications by the scientists writing the editorial. This was a true-true-unrelated kind of position to take, but it caught on in the general scientific community."

The last neologism, "chemokine," was created in 1992 to gather under one name a large family of small biochemically related cytokines, all possessing chemoattractant properties.

The availability of the first cloned molecules allowed major progresses, not only in fundamental research but also for clinical applications. While growth hormone was the first recombinant molecule used in humans in 1980, interferon-alpha was the first cytokine injected in humans in 1981, and IL-2 the second in 1985. Indeed, in 1984, 10 000 L of supernatants of activated Jurkat cell were required to prepare 30 mg of natural IL-2; the following year, 10 L of supernatants of recombinant *Escherichia coli* expressing the *IL-2* gene were sufficient to prepare 1 g of recombinant IL-2.

## 2 A Universal Language of Cells

### 2.1 Definitions

Cytokines have been defined by immunologists as cooperative factors produced by activated lymphocytes, dendritic leukocytes, macrophages that induce, modulate, favor, increase, inhibit cell proliferation, cell differentiation, or cell activities such as antibody production. The onset of the messages delivered by these soluble (or membrane-bound) molecules relies on the binding to their specific receptors displayed on the surface of the immune cells, a process leading to an intracellular signaling cascade. However, far from being limited to the leukocyte lineages, this is a universal language of cells of the whole body, independently of their lineages, their nature, and their localization. This is probably the main difference between cytokines and other soluble molecules such as neuromediators mainly produced by the brain and the neurons, or hormones mainly produced by the endocrine system (Table 1).

Like hormones, cytokines can act in an endocrine fashion, but their actions mainly occur within the cellular microenvironment in a paracrine fashion. However, since some cytokines can be expressed on the cell surface of the producer cell, cytokines can act in a juxtacrine fashion within a close contact between cells. Finally, cytokines can act as an autocrine factor for a cell that both produces a cytokine and displays its receptor (Fig. 1).

Cell interactions involve more than one cytokine. One can consider each cytokine as a word, and the full sentence that will finally emerge is the reflection of the delivery of different cytokine-mediated signals. Together, these signalings lead to orientate the activities of the target cell. Thus, the order of delivery of cytokines has major consequences on the nature of the message. Furthermore, the concentration of cytokines, their localization, the nature of the target cell, and the timing are important parameters that affect the exact nature of the cytokine-initiated process.

**Tab. 1** Comparison between hormones and cytokines.

| | *Sources* | *Targets* | *Activities* | *Action* |
|---|---|---|---|---|
| Hormones | Secreted by a unique cell lineage | Specificity rather limited to one single type of target cell | Single action | Endocrine |
| Cytokines | Produced by many cell types | Numerous target cells | Wide spectrum of activities (redundancy) | Paracrine<br>Juxtacrine<br>Autocrine<br>Endocrine |

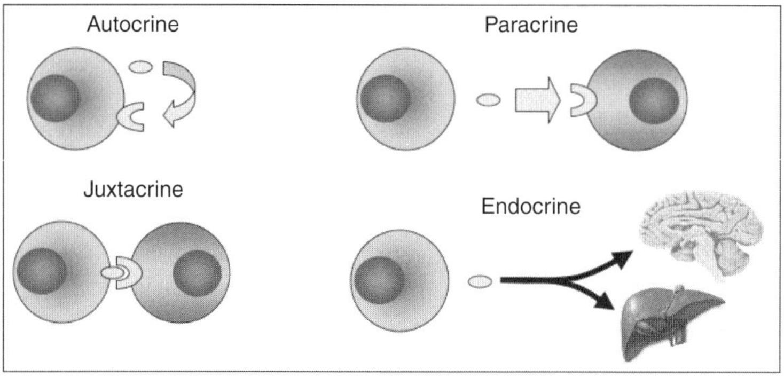


**Fig. 1** Mode of action of cytokines (see color plate p. xviii).

In most cases, constitutive production of cytokines is limited and most cytokines are produced upon activation of cells. However, highly sensitive readout assays have allowed to determine a constitutive production at homeostasis in numerous tissues and cells.

## 2.2 Families

The classification of cytokines can be achieved according to (1) a common bioactivity: antiviral activity (interferons), chemoattraction (chemokines), hematopoiesis (colony-stimulating factors, CSF); (2) a common biochemical structure, suggesting a common ancestral molecule (chemokines, tumor necrosis factor (TNF) family); or (3) the sharing of a common chain of a receptor (gp130 cytokine family) (Table 2). Different subfamilies of interferons are known. Twenty-six genes of IFN$\alpha$ have been identified, of which at least five are pseudo genes. Interferons-alpha are produced by leukocytes and interferon-beta is produced by fibroblasts. IFN$\gamma$ produced during immune response is a potent macrophage activator. IFN$\omega$ is induced by virus, IFN$\tau$ is produced by trophoblast, and IFN$\kappa$ is expressed by keratinocytes. Among chemokines, four different subfamilies can be identified on the basis of the highly conserved presence of the first two cysteine residues, which are either separated or are not separated by other amino acids: the CCL, CXCL, CX3CL, and the CL

**Tab. 2** Families of cytokines.

| | |
|---|---|
| Interferons | IFN$\alpha$, IFN$\beta$, IFN$\gamma$, IFN$\delta$, IFN$\kappa$, IFN$\tau$, IFN$\lambda$ |
| Interleukins[1] | [IL-1$\alpha$, IL-1$\beta$, IL-1Ra, IL-18, IL-1F5-10] [IL-2, IL-15, IL-21] |
| | [IL-10, IL-19, IL-20, IL-22, IL-26] [IL-12, IL-23, IL-27] |
| | [IL-17A-E, IL-25] [IL-28, IL-29] IL-3, IL-4, IL-5, IL-6, IL-7 |
| | IL-8 (CXCL8), IL-9, IL-11, IL-13, IL-14, IL-16 |
| Hematopoietic factors | M-CSF, G-CSF, GM-CSF, Stem cell factor |
| Chemokines | CCL1 ... CCL28 ; CXCL1 ... CXCL16 ; XCL1, XCL2 ; CX3CL1 |
| TNF superfamily | TNF, Lt$\alpha$, Lt$\beta$, NGF, CD27L, CD30L, CD40L, CD137L, APRIL, BAFF, FasL, GITRL, LIGHT, OPGL TRAIL, RANKL, TWEAK, VEGI. |
| Growth factors | TGF$\beta_{1,2,3}$ |
| Gp130 cytokine family | IL-6, IL-11, CNTF, LIF, Oncostatin-M, Cardiotrophin-1 |

[1] Biochemically related interleukins are framed.
APRIL: apoptosis-inducing ligand; BAFF: B cell-activating factor; CCL: chemokine CC ligand; CNTF: ciliary neurotrophic factor; CSF: colony-stimulating factor; CXCL: chemokine CXC ligand; GITRL: glucocorticoid-induced TNF receptor family related gene ligand; IL-1F: IL-1 family; LIF: leukemia inhibitory factor; LIGHT : lymphotoxins, inducible expression, competes with HSVglycoprotein D for HVEM, a receptor expressed on T-lymphocyte; Lt: lymphotoxin; OPGL: osteoprotegerin ligand; RANKL: receptor activator of NF-kappaB ligand; TGF$\beta$ : transforming growth factor-$\beta$; TRAIL: TNF-related apoptosis-inducing ligand; TWEAK: TNF-like weak inducer of apoptosis; VEGI: vascular endothelial cell growth inhibitor.

chemokines. The assigned numbers of interleukins only reflect the timing of their discovery. Thus, some interleukins are also hematopoietic factors (e.g. IL-3 previously named multi-CSF), chemokines (IL-8, also named CXCL8), or even interferon (IL-28 and IL-29 are identical to IFN$\lambda$). The first interleukins have been discovered as bioactive factors, and there are no similarities in their amino acid sequences. However, there are some homologies in tertiary and quaternary structures and similar helical structures are found. The more recently discovered interleukins have been identified in gene banks according to sequence homology to previously identified interleukins. The TNF ligand superfamily gathers 18 distinct members that include soluble molecules with biochemical homology as well as membrane-bound ones. There are few other cytokines that can be displayed at the plasma membrane of the producing cells: IL-1$\alpha$, IL-10, IL-15, and IFN$\gamma$. The TNF ligand superfamily also includes a growth factor (i.e. nerve growth factor). Usually, growth factors have a rather narrow spectrum of activity, are produced constitutively, are not acting on hematopoietic cells, and thus are distinct from cytokines. One exception is the transforming growth factor-$\beta$ family (TGF$\beta$), particularly because at least one member, namely TGF$\beta$1, displays anti-inflammatory properties.

Localization of genes on chromosomes is known, and often genes of a family are gathered on the same chromosome (e.g. IL-1$\alpha$, IL-1$\beta$, IL-1 receptor antagonist on human chromosome 2; most CXCL chemokines on human chromosome 4; most CCL chemokines on human chromosome 17). Most cytokines are glycosylated, their molecular weight ranging between 7 and 30 kDa. Most cytokines are monomers (e.g. IL-1, IL-2, IL-3, IL-4...), and few are homodimers (e.g. IL-8) or homotrimers (e.g. TNF). Some are heterodimers like IL-12, IL-23, and IL-27: IL-12 and IL-23 sharing the same p40 chain. Some are heterotrimers like lymphotoxin-$\beta$ (Lt$\beta$) that is the association of Lt$\alpha$ and Lt$\beta$ chains.

## 3
## Receptors

### 3.1
### Families

Most cytokine receptors had been characterized once their ligands had been identified, but there were few receptors, called *orphan receptors*, that were known before their ligands (e.g. c-kit, the receptor of stem cell factor; CD40, the receptor of CD40L; IL-1 receptor-related protein-1, part of the IL-18 receptor, etc). In most cases, receptors are the association of more than one transmembrane chain. Usually, one chain is essential for the binding ($\alpha$-chain), while the second chain is required to initiate the intracellular signaling ($\beta$- or $\gamma$-chain). Some chains are common for receptors of few cytokines; this is the case of the gp130 chain of the receptors of IL-6, IL-11, CNTF, LIF, Oncostatin-M, and Cardiotrophin-1, or the gamma chain of IL-2, IL-4, IL-7, IL-9, IL-15, and IL-21. Cytokine receptors are characterized by domains derived from ancestral common genes. In terms of evolution, it would be interesting to decipher which mutations in genes coding for cytokines or for cytokine receptors were first fixed. A common domain, called *hemopoietin receptor superfamily domain* characterized by the presence of four cysteine residues and a conserved sequence tryptophan-serine-x-tryptophan-serine, defines a family of numerous cytokine receptors (Fig. 2).

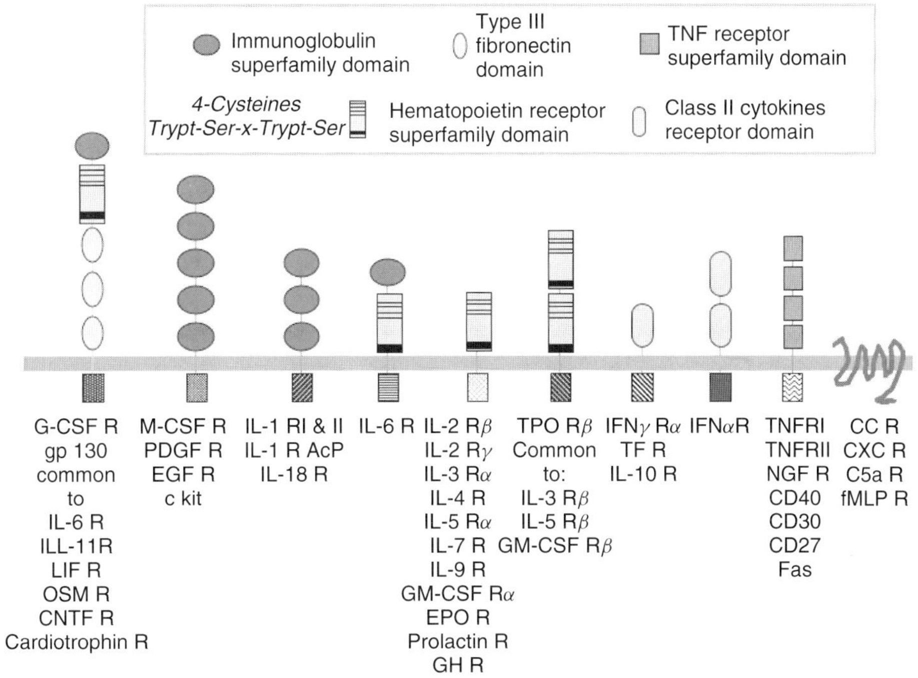


**Fig. 2** Families of cytokine receptors (see color plate p. xviii).

Most interestingly, this family of receptors also includes receptors for hormones (e.g. growth hormone, prolactin) or the hematopoietic factor (e.g. erythropoietin). This domain can be duplicated (the common $\beta$ chain of IL-3R, IL-5R, and GM-CSFR) or be associated with other domains like an immunoglobulin-like domain (IL-6R$\alpha$) or a fibronectin type III–like domain (gp130, G-CSFR). Another important family is the TNF receptor family. The activation by TNF is consecutive to the bridging of similar chains of the TNF receptor by the cytokine. Each cytokine receptor is highly specific for its ligand, except the TNFR p55 and p75 that bind both TNF and Lt$\alpha$, and the chemokine receptors that can bind various chemokines. Chemokine receptors, constituted by a 7-transmembrane chain, are similar to receptors for other chemoattractant molecules such as the anaphylatoxin C5a or the bacterial fMet-Leu-Phe peptide. Cytokine receptors are often internalized after ligation to their ligands, and their expression can be induced, down- or upregulated.

### 3.2 Signaling

Upon binding of the cytokine to the extracellular domain of receptors, the intracellular receptor domains become associated with a variety of signaling molecules. A cascade of phosphorylation of various adaptor proteins occurs following interaction of the intracellular domain of the receptors with cytoplasmic tyrosine kinases (e.g. Janus kinases, Jak). Phosphorylation of latent cytoplasmic transcriptional activators (e.g.

signal transducers and activators of transcription, STAT) allows their dimerization and their translocation into the nucleus where they bind to specific sequences present in promoters of certain genes. Other pathways involving the mitogen-activated protein kinase (MAPK) and other kinases lead to the activation of various transcription factors (e.g. c-fos/c-jun, NF-$\kappa$B, etc). Some signaling pathways are shared with other receptors. This is the case of the IL-1 and IL-18 receptors of which the intracellular domain is similar to the toll-like receptors (Toll–IL-1 Receptor, TIR domain). Thus, similar adaptors (e.g. MyD88, IRAK, etc) are involved after interaction of the respective receptors with their ligands, either members of the IL-1 family or pathogen-associated molecular patterns (PAMPs) such as endotoxin (lipopolysaccharide). Intracellular domain of some receptors (TNFR p55; Fas; TRAIL R) possesses a death domain that initiates a specific signaling cascade, leading to apoptosis.

## 3.3 Soluble Receptors

Receptors are integral part of the cytokine network (*cf* Sect. 7): not only can their membrane expression be modulated but they also exist as soluble molecules. Indeed, most cytokine receptors, except the chemokine receptors, can be shed from the cell surface after enzymatic cleavage following or not following a neosynthesis. As shown in Fig. 3, depending upon the nature of the receptor, its soluble form can behave as an inhibitor (e.g. soluble IL-1R, soluble TNFR) and can prevent the cytokine from reaching the membrane form. It can also be a carrier, protecting the cytokine within a complex, and increasing

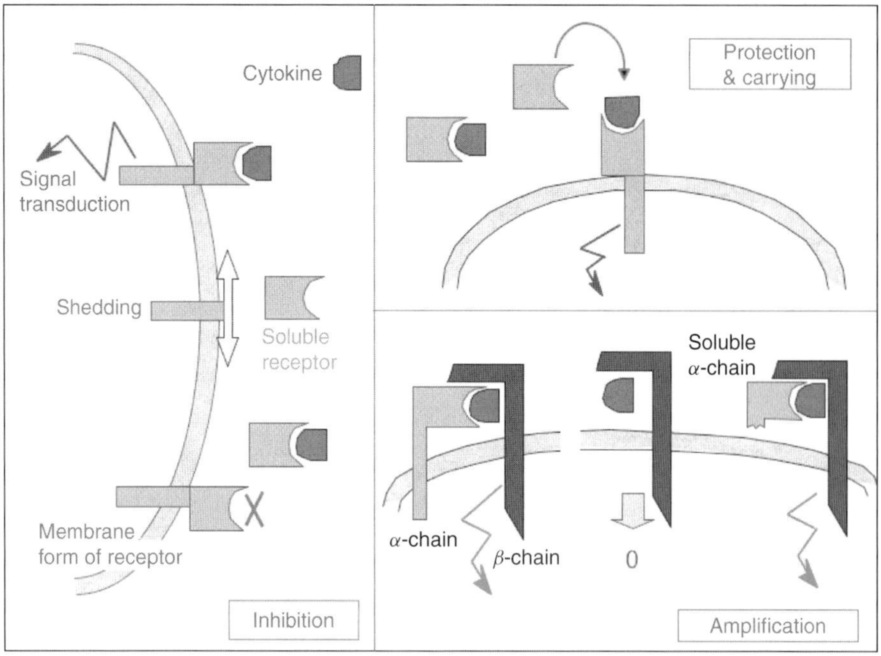


**Fig. 3** Different properties of soluble receptors (see color plate p. xix).

its half-life in the plasma or other biological fluids. Then, the dissociation of the cytokine-soluble receptor complex can allow the release of the cytokine close to a membrane receptor (e.g. TNFR). Lastly, in the case of a receptor constituted by one $\alpha$-chain that binds the cytokine and one $\beta$-chain that recognizes the bound cytokine, the soluble $\alpha$-chain receptor can enhance the responsiveness to the cytokine or can even allow the responsiveness of cells missing the $\alpha$-chain, but expressing the $\beta$-chain. One example of the latter case is provided by the soluble IL-6R that, together with IL-6, controls the switch of leukocyte recruitment during inflammation.

## 4 Functions

### 4.1 Hematopoiesis

Production of leukocytes within the bone marrow is a very active and efficient process ($4 \times 10^8$ white cells/leukocytes per hour in humans). Numerous cytokines, expressed at homeostasis in human bone marrow contribute to hematopoiesis, while others (like GM-CSF) are only active during infection to further increase the renewal of available leukocytes. Maturation, proliferation, and differentiation of bone marrow progenitor cells are under the control of different cytokines acting concomitantly. For example, IL-1 alone is unable to allow the differentiation of bone marrow precursors cells, but it acts in synergy with other hematopoietic cytokines. IL-3 and stem cell factor (also named *c-kit ligand*) are cytokines with a wide spectrum of activity, acting on pluripotent hematopoietic stem cells. GM-CSF favors the differentiation of myeloid progenitors cells, while M-CSF or G-CSF are specific of monocyte/macrophage and granulocyte lineages respectively. IL-6, IL-11, and mainly thrombopoietin favor thrombocytopoiesis, leading to megakaryocyte development and platelet production. Interleukin-5 has been identified as a major regulator of eosinophil development and function. Interleukin-7 is the main cytokine of lymphopoiesis required for the development of B, alpha-beta, and gamma-delta T-lymphocytes. Most of these hematopoietic cytokines also act as activators of mature cells and as antiapoptotic ligands.

### 4.2 Immune Response

#### 4.2.1 Innate Immunity

Innate immunity is characterized by a response localized within the site of infection (Fig. 4). Epithelial cells and leukocytes residing in the extravascular space of the tissue (mast cells, resident macrophages, and immature dendritic leukocytes), and endothelial cells of the microvessels are triggered by microorganisms and microbial-derived products and release inflammatory cytokines (e.g. IL-1, TNF, IL-12, IL-18, etc). These cytokines further activate the endothelial cells of the postcapillary venules, leading to the increased adherence of circulating leukocytes and to coagulation process. Other cytokines, such as IFN$\gamma$, are produced within an amplificatory loop. Local production of chemokines contributes to the recruitment of new leukocytes from the blood compartment to further prevent the infectious process. Cytokines contribute to enhance the antimicrobial activities of newly recruited neutrophils and monocytes/macrophages (e.g. production of free radicals, enhanced phagocytosis), and the

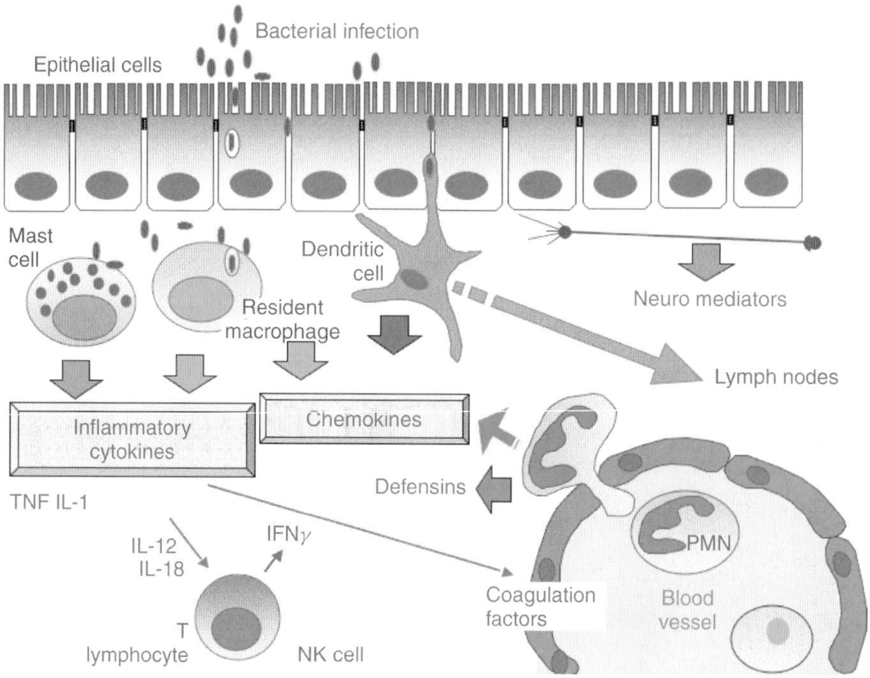


**Fig. 4** Cytokines and soluble mediators as coordinators of regulated processes taking place in a bacteria-loaded sites (see color plate p. xx).

antiviral activity of natural killer (NK) cells. Human leukocytes release defensins, a kind of natural antibiotics that contribute to limit bacterial dissemination. Neuromediators help control the inflammatory process. In addition to local responsiveness, a systemic response occurs, which includes fever, a reflection of the presence of pyrogenic cytokines within the plasma (e.g. IL-1, TNF, IL-6). Hyperthermia contributes to the anti-infectious process by enhancing some immune cell activities, and by reducing bacterial growth or viral replication. Finally, the production of hematopoietic cytokines (e.g. GM-CSF) further increases the number of leukocytes otherwise known to contribute to the clearance of the microorganisms and to the initiation of the healing process.

### 4.2.2 Adaptive Immunity

Adaptive immunity is characterized by processes relying on unique properties of the secondary lymphoid organs. Briefly, in secondary lymphoid organs such as the peripheral lymph nodes, professional antigen-presenting cells (dendritic leukocytes and mononuclear phagocytes), after processing of the native antigen, display on their membranes antigenic peptides and interact with T-lymphocytes, expressing a T-cell receptor able to specifically recognize these peptides. B-lymphocytes recognize native antigen. As a result of the cross talks between antigen-presenting cells, T- and B-lymphocytes, T-lymphocytes proliferate and differentiate as primed effector or memory T-cells, and B-lymphocytes proliferate and differentiate as antibody-secreting plasma cells. All these events are controlled by cytokines. For example, within the secondary lymphoid organs, IL-2 and IL-4 favor T-cell proliferation, IL-12 activates cytotoxic T-cells, IL-2 and IL-4 allow B-cell proliferation, IL-4, IL-5, IL-6, IL-13, and TGF$\beta$ influence plasma cell differentiation and favor different immunoglobulin classes synthesis. Within microorganisms-loaded tissues, the effector T-lymphocytes, once reactivated could release IFN$\gamma$ and GM-CSF that activate professional phagocytes conferring them microbicidal functions.

### 4.2.3 Th1 and Th2 Cytokines

In 1986, Tim Mosmann et al. described two types of murine helper CD4$^+$ T-cell (Th) clones defined according to the profile of released cytokines. Th1 cells produce IL-2, IFN$\gamma$, and Lt$\alpha$, and Th2 cells produce IL-4, IL-5, IL-6, IL-10, and IL-13. Both subpopulations produce IL-3 and GM-CSF. Their differentiation from a Th0 precursor occurs, depending upon the cytokine environment (Fig. 5). Th1 and Th2 cells control each other: IFN$\gamma$ derived from Th1 neutralizes Th2 cells, and IL-4 and IL-10 inhibit Th1 cells. Th1 cells are involved in cellular immunity and activate macrophages, Th2 cells are involved in humoral immunity and activate B-lymphocytes. Once stably differentiated as long-lived clones, some chemokine receptors allow to distinguish between these subpopulations: CXCR3 is mainly expressed on Th0 and Th1 cells, while CCR3 and CCR4 are expressed on Th2 cells. The Th1/Th2 dichotomy also exists in humans, as nicely illustrated by the early detection of Th2 cytokine mRNA expressed in response to an intradermal allergen challenge (immediate hypersensitivity) and the delayed detection of Th1 cytokine mRNA after tuberculin injection (delayed-type hypersensitivity). However, in most clinical settings, this dichotomy is not that obvious and mixed patterns are observed. More recently, other T-cell subsets with immunosuppressive properties have been characterized. Th3-type cells are a unique T-cell subset that primarily secretes TGF$\beta$, provides help for IgA switch, and has suppressive properties for Th1 and other immune cell. Type-1 regulatory T-cells (Tr1) inhibit Th1 and Th2-type immune responses through the secretion of IL-10. They are also able to secrete TGF$\beta$.

## 4.3 Cell Survival and Cell Death

Not only can cytokines be maturation and differentiation factors but they can

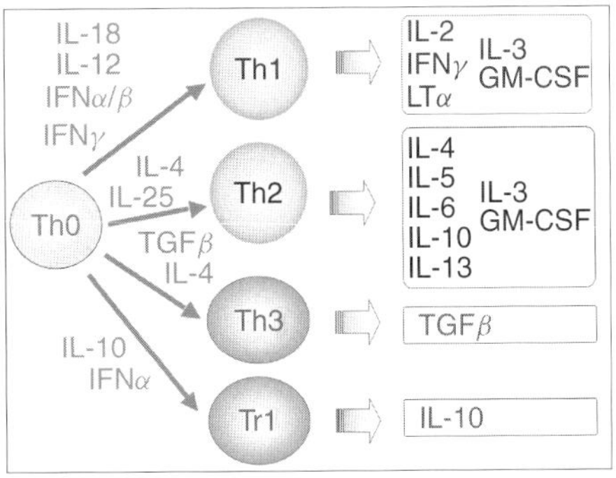


**Fig. 5** T-cell subpopulations and the nature of environmental cytokines required for their differentiation (see color plate p. xx).

also contribute to cell survival. For example, IL-2, IL-4, IL-6, IL-7, and IL-15 have all been shown to inhibit resting T-cell death *in vitro*. B-cell survival factors include IL-4 and B cell-activating factor (BAFF) of the TNF family. Plasma cell longevity depends on combination of IL-5, IL-6, stromal cell-derived factor-1$\alpha$ (SDF-1/CXCL12), and TNF$\alpha$. IL-15 can sustain NK cell survival in the absence of serum. Indeed, *in vitro* serum starvation often leads to apoptosis that can be prevented by cytokines like TGF$\beta$ (e.g. survival of epithelial cell). Indeed, these properties reflect an antiapoptotic mechanism initiated by cytokines like the suppression of caspase activity or the maintenance or induction of Bcl-2 protein expression. RANKL suppresses apoptosis of primary cultured endothelial cells; SDF-1/CXCL12 enhances the survival of myeloid progenitor cells *in vitro*; stem cell factor (SCF) regulates the survival of cellular lineages by suppressing apoptosis. In addition to SCF, most hematopoietic factors allow survival of hematopoietic progenitors. In addition to sustain cell survival, cytokines can allow cell growth. For example, this is the case of IL-2 for T-cells, or IL-8 and some other chemokines that favor angiogenesis. Cytokines can also be growth factors for tumor cells; this is, for example, the case of IL-6 for malignant myeloma plasma cells. In contrast, certain cytokines can induce apoptosis of a variety of tumor cells and normal cells. Indeed, TNF was first identified in 1984 as a cytokine with antitumor effects, and other members of the TNF superfamily can induce proliferation, survival, as well as cell death: specifically, TNF$\alpha$, FasL, TRAIL, and VEGI. The intracytoplasmic portion of their receptors contains a "death domain." Upon activation by their ligands, the death domain recruits intracytoplasmic adaptors expressing a death domain: TRAF (TNF receptor-associated factor) and FADD (Fas-associated death domain), which in turn recruit the proform of caspase-8. Autoactivation of caspase-8 leads to the subsequent activation of caspase-3, a proapoptotic enzyme.

## 4.4 Link with the Central and Peripheral Nervous System

A cross talk exists between immune leukocytes and the central nervous system. Proinflammatory cytokines, particularly IL-1 and TNF, induce fever, anorexia, and slow wave sleep. Part of their activities is conveyed by the vagal nerve. Fever involves local production of IL-6 and prostaglandins E2 (PGE2). Certain chemokines are also pyrogenic (e.g. IL-8, MIP-1$\alpha$, MIP-1$\beta$, RANTES), but do not induce PGE2. IL-1 and TNF also induce a neuroendocrine loop in response to their signal, hypothalamus produces a corticotropin-releasing factor (CRF), and CRF induces the release of adrenocorticotropin hormone (ACTH) by pituitary gland. ACTH induces the release of glucocorticoid by adrenals. Glucocorticoids inhibit the production of most cytokines and antagonize many of their activities. Vagal nerve releases acetylcholine that represses inflammatory cytokine production. In addition, other neural and neuronal mediators, adrenaline, vasoactive intestinal peptide (VIP), pituitary adenylate cyclase-activating polypeptide (PACAP) are inhibitors of IL-1 and TNF production, while noradrenaline and substance P are potentiators.

## 4.5 Reproduction

There is compelling evidence that cytokines are involved in spermatogenesis, ovogenesis, and gestation. Cytokines play an important regulatory role in the development and normal function of the testis. They are produced by Leydig cells, Sertoli cells, and germinal cells. Proinflammatory cytokines, including IL-1 and IL-6, have direct effects on spermatogenic cell differentiation and testicular steroidogenesis. Stem cell factor and leukemia inhibitory factor, cytokines normally involved in haematopoiesis, also play a role in spermatogenesis. Anti-inflammatory cytokines of the TGF$\beta$ family are involved in testicular development. TNF$\alpha$, a secretory product of round spermatids, increases endogenous androgen receptors expression in primary cultures of Sertoli cells. Given the requirement of testosterone for spermatogenesis and the importance of androgen receptors in mediating Sertoli cell responsiveness to testosterone, the stimulation of androgen receptors expression by TNF$\alpha$ may represent an important regulatory mechanism required to maintain efficient spermatogenesis. M-CSF is the principal growth factor regulating macrophage populations in the testis, male accessory glands, ovary, and uterus. Both male and female M-CSF deficient (op/op) mice have fertility defects. Males have low spermatozoid number and libido as a consequence of dramatically reduced circulating testosterone. Females have extended estrous cycles and poor ovulation rates.

Normal ovarian tissue is rich in cytokines. Cytokines and chemokines are important in the physiology of ovarian function and of ovulation. Actual rupture of the follicle during ovulation may be dependent on tissue remodeling that shares some features with an acute inflammatory process. TNF and IL-1 are implicated in ovarian follicular development and atresia, ovulation, steroidogenesis, and corpus luteum function (including formation, development, and regression). Chemokines such as MCP-1 are also involved in luteolysis.

A great number of cytokines are produced by endometrium and placenta. Cytokines released at the fetomaternal interface play an important role in regulating embryo survival, controlling not only the maternal immune system but also angiogenesis and vascular remodeling. For example, LIF plays a role in the embryo pre-, peri-, and postimplantation, as does IFN$\tau$ in bovine and ovine. IFN$\tau$ is produced constitutively by embryonic trophectoderm during the period immediately prior to implantation. It acts on uterine epithelium to suppress transcription of the genes for estrogen receptor and oxytocin receptor. It blocks development of the uterine luteolytic mechanism. TGF$\beta$ is involved in regulating placental development and functions. The abortive influences of TNF$\alpha$ and IFN$\gamma$ may terminate pregnancy during infection of uteroplacental unit. The slight dominance of proinflammatory cytokines in the fetal membranes and decidua suggest that inflammatory processes occur modestly with term labor, but much more robustly in preterm delivery, particularly in the presence of intrauterine infection.

## 4.6 Inflammation

"*Notae vero inflammationis sunt quatuor: rubor and tumor cum calore et dolore*" (Celsus, first century B.C.). Redness, swelling, heat, and pain are the four main parameters that characterize inflammation. They reflect the action of numerous inflammatory mediators and proinflammatory events that are orchestrated by inflammatory cytokines, mainly IL-1 and TNF, but also IL-12, IL-18, and IFN$\gamma$ (Fig. 6). These cytokines are produced following

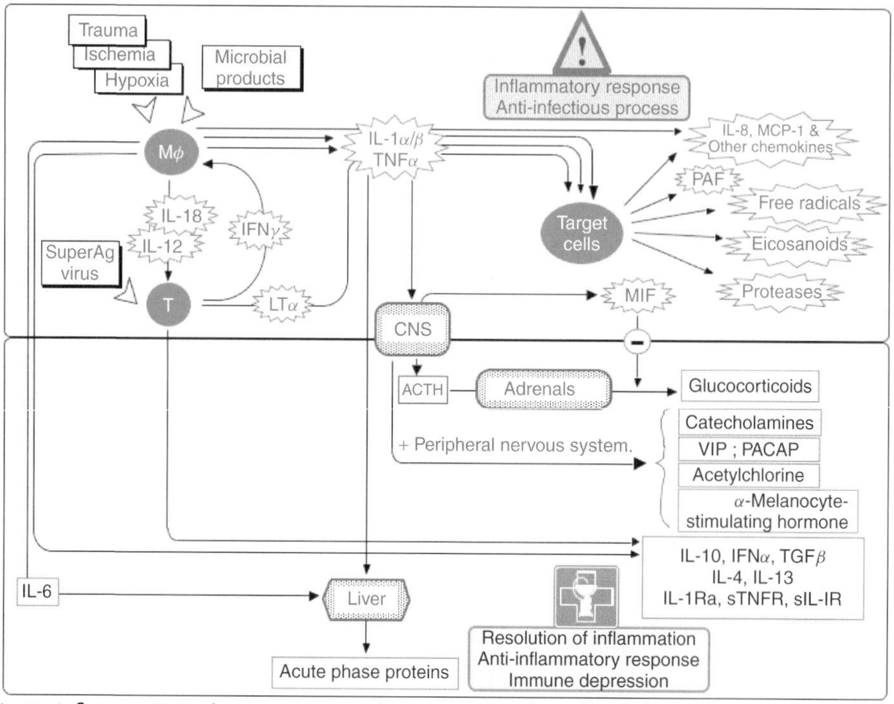


**Fig. 6** Inflammation is the consequence of a cascade of events initiated by IL-1 and/or TNF. The action of these cytokines on various target cells leads to the release of numerous inflammatory mediators. Anti-inflammatory cytokines, the activation of the neuroendocrine pathway and the effects of glucocorticoids as well as the enhanced production of acute-phase proteins control negatively the inflammatory process (ACTH: adrenocorticotropic hormone; CNS: central nervous system; PACAP: pituitary adenylate cyclase-activating polypeptide; PAF: platelet activating factor; VIP: vasoactive intestinal peptide) (see color plate p. xxi).

tissue or systemic steady state disruption like tissue loading by pathogens, ischemia, hypoxia, or trauma. Acting on target cells, they induce a cascade of mediators, including other cytokines and chemokines, proteases, lipid mediators (e.g. prostaglandins, platelet activating factor), free radicals (e.g. superoxide anion, nitric oxide) that contribute to the inflammatory process and tissue injury. Acting on endothelial cells, especially those of the postcapillary venules, they increase vascular permeability and plasma transudation, they increase circulating leukocyte adherence and margination, and they initiate the coagulation process by inducing tissue factor expression. Chemokines contribute to the recruitment of leukocytes that further maintain the inflammatory process. Some cytokines enhance the production of IL-1 and TNF (e.g. IFNγ, GM-CSF, IL-3), while others inhibit it (e.g. IL-4, IL-10, IL-13, TGFβ, IFNα). The latter contribute to attenuate the inflammatory process and to allow the tissue-repair process to occur. Other anti-inflammatory mediators include soluble IL-1 and TNF receptors, IL-1 receptor antagonist (IL-1Ra), some neuropeptides (e.g. adrenaline, acetylcholine), glucocorticoids induced after activation of the neuroendocrine loop by IL-1 and TNF, and the acute-phase proteins produced by the hepatocytes in response to IL-1, TNF, and particularly IL-6. Acute-phase proteins contribute to eliminate debris due to cell lysis, to favor phagocytosis, to neutralize toxic mediators, and to inhibit proteases. Recently, IL-1Ra was recognized as one of the acute-phase proteins. In vivo, in the endotoxin shock model, most of these molecules have been shown to be protective.

## 5
## Life without Cytokines (What Knockout Mice Tell Us)

Deletion of genes encoding for a cytokine or a cytokine receptor has led to the generation of many different knockout (KO) mice that further allowed to decipher the contribution of cytokines to different physiological events. In most cases, viable, fertile, and clinically healthy mice could be obtained. In few cases, the deletion of the genes led to death *in utero*, illustrating a major role of certain cytokines during embryogenesis. This was the case for SDF1 (CXCL12) and its receptor CXCR4, LIF, Cardiotrophin-1, and the gp130 chain of their receptors (see Table 3). Deletions of certain signaling molecules (e.g. STAT3) are also lethal during embryonic life. Other deletions reveal the role of certain cytokines for organogenesis during embryonic life. This is the case of lymphotoxin-$\alpha$, lymphotoxin-$\beta$, and Lt$\beta$R for the development of lymph nodes and Peyer's patches, and RANKL for the development of lymph nodes and mammary glands.

**Tab. 3** Cytokine or cytokine receptor KO mice reveal relative contribution of cytokines to life processes.

| *Biological events* | *Deleted gene* | *Consequences* |
|---|---|---|
| Embryogenesis | LIF | Absence of LIF in female prevents implantation of blastocytes |
| | Cardiotrophin-1 & gp130 receptor chain | Heart development blocked; death *in utero* on day 16 |
| | SDF-1 & CXCR4 | Defects in heart and brain development, intestinal vascularization, and B-cell hematopoiesis. Death *in utero* |
| Organogenesis | Lt$\alpha$, Lt$\beta$, Lt$\beta$R | Absence of lymph nodes and Peyer's patches |
| | TNF$\alpha$ | Altered spleen and lymph node organization |
| | RANKL | Absence of lymph nodes, mammary gland defects |
| | IL-8R | Splenomegaly, lymphoadenopathy, increased circulating neutrophils |
| Hematopoiesis | G-CSF | Chronic neutropenia |
| | Stem cell factor | Absence of mast cells |
| | M-CSF | Altered function of monocytes and osteoclast |
| | IL-7 | Reduced number of lymphocytes |

(*continued overleaf*)

**Tab. 3** (*Continued*)

| ***Biological events*** | ***Deleted gene*** | ***Consequences*** |
|---|---|---|
| Leukocyte differentiation | Ltα, Ltβ<br>IL-15<br>γ-chain | Absence of follicular dendritic cells<br>Absence of NK and NK-T cells<br>Impaired B, T and NK cell development |
| Osteoclastogenesis | RANK & RANKL | Defect in the differentiation of hematopoietic osteoclast progenitor |
| Pulmonary homeostasis | GM-CSF & β-chain of GM-CSF R | Accumulation of surfactant lipids and proteins in the alveolar space. Lymphoid hyperplasia. |
| Gut homeostasis | IL-10 | Intestinal mucosal hyperplasia, inflammatory cell infiltration, MHC Class II expression on colonic epithelium. |
| | IL-2<br>IL-15 | Colitis<br>Reduced intestinal epithelial B and T lymphocytes |
| Fever | IL-6 | No pyrogenic response |
| Acute-phase response | IL-6 | Profound alteration of acute-phase proteins production |
| Inflammation | TGFβ<br>IL-1-Ra | Systemic lethal inflammation (day 24)<br>Chronic arthropathy, spontaneous dermatosis, exacerbated delayed-type hypersensitivity |
| | IL-5 | Reduced eosinophilia |
| Antibody production | IL-4, IL-6, IL-13, CD40L | Altered antibody production |
| Immune function | IL-2, IL-4, IL-6, IL-12, IL-13, IL-15, IL-18, TNF, IFNγ, Stem cell factor, CSF, Chemokines, and their receptors. | Altered anti-infectious response |

Organization of hematopoietic compartments are under the control of numerous cytokines and absence of TNFα, or the γ-chain of the receptor of IL-3, -5, -7, -9, -15, and -21 have profound perturbations on tissue organization and leukocyte differentiation. This is also true for the deletion of chemokines that contribute to tissue colonization at homeostasis. While studies of KO mice often comforted the knowledge, there were some surprises. One of them was the discovery that the absence of the hematopoietic cytokine GM-CSF did not alter hematopoiesis (mice had normal numbers of peripheral leukocytes, bone marrow progenitors, and tissue hematopoietic populations), but led to a profound alteration of the lung status, demonstrating a role of GM-CSF in pulmonary homeostasis. Of major interest was the demonstration that the deletion of anti-inflammatory cytokines (IL-10, TGFβ, IL-1Ra) was deleterious at homeostasis in the absence of experimental exogenously delivered inflammatory signals, establishing their crucial role to prevent potential ongoing inflammatory processes at homeostasis. As expected, numerous deletions had major effects on the quality of the anti-infectious process. For example, in the absence of IL-6, mice produce lower levels of IgG antibodies against stomatitis vesicular virus, display a lower cytotoxic T-cell activity during vaccinia virus infection, leading to an increased number of virus in lungs, and failed to control *Listeria monocytogenes* load in the tissues they reached. In contrast, the same deficiencies may appear beneficial in certain experimental models of septic shock in response to endotoxin (e.g. TNF$\alpha$, TNFR p55, IL-1$\beta$ converting enzyme).

In humans, natural mutations occur that may affect cytokines, signaling molecules of receptors. In the latter case, the most famous mutation leads to the deletion of the $\gamma$-chain of the receptor of IL-3, -5, -7, -9, -15, and -21 and is associated with a severe combined immunodeficiency. This is for this clinical setting that occurred in 2000, the first successful gene therapy.

## 6 Cytokine Synthesis

### 6.1 Homeostasis

Owing to highly sensitive techniques (RT-PCR, *in situ* hybridization, ELISpot), it has been possible to show the presence of cytokine mRNA in various types of cells or to demonstrate the presence of cytokine-producing cells in the absence of activation. For example, IL-6 is spontaneously produced by 0.5% of bone marrow cells, 0.1% of spleen cells, and 0.01% of mesenteric lymph node cells. IL-6 is also produced in the absence of exogenous stimuli by enterocytes, eosinophils, neutrophils, epidermal cells, smooth muscle cells, bone marrow stromal fibroblasts, anterior pituitary cells, trophoblast cells, and so on. Furthermore, at homeostasis, certain cells contain cytokines. This is particularly the case for keratinocytes that contain large amounts of IL-1$\alpha$. Mast cells contain a large panel of preformed cytokines (e.g. IL-1, IL-4, IL-5, IL-6, IL-13, TNF, etc), and MIF is preformed in numerous leukocytes. Furthermore, the presence of cytokines in biological fluids has been reported in absence of any infection or inflammatory diseases: IL-1$\alpha$ and IL-8 are found in sweat; IL-1$\alpha$, IL-1$\beta$, IL-6, IL-8, GM-CSF, and TGF$\beta$ have been reported in tears; IL-2, IL-8, TNF$\alpha$, TGF$\beta$, and soluble TNFR are present in saliva; IL-2, IL-6, IL-8, IL-10, IL-12, TNF, and soluble IL-2R and IL-6R have been detected in seminal fluid; and a great number of different cytokines have been identified in human colostrum. In the later case, all of these cytokines probably act on the oropharyngeal and gut-associated lymphoid tissue of the newborn and favor the development and maturation of the immune system and may protect the newborn against invasive microorganisms delivered by the oral route. Finally, certain chemokines are present in tissues where they contribute to leukocyte recruitment at homeostasis (e.g. CCL17 & CCL25 in thymus, CCL21 & CXCL13 in lymph nodes, and Peyer's patches, CXCL12 in numerous tissues).

### 6.2 Induction

The production of cytokines by professional antigen-presenting leukocytes and activated T-cells occurs during the cellular cooperation required to initiate the adaptive immunity. This production is representative of the dialogue between immune leukocytes. In contrast, innate immunity is associated with the production of cytokines in response to exogenous activators, mainly the "pathogen associated molecular patterns" (PAMPs). Following their interaction with the "pattern recognition receptors" (PRR), various intracellular signaling cascades lead to the synthesis and the release of cytokines. The best-known PRRs are the toll-like receptors (TLR) that share with IL-1R and IL-18R a homologous Toll-IL-1R (TIR) domain. Ten different TLRs have been identified so far in humans that recognize bacterial, viral, parasitic, or fungal conserved structures expressed on the surface or within the microbes (e.g. DNA, single- or double-strand RNA). Endotoxin of Gram-negative bacteria (lipopolysaccharide, LPS) is among the most potent PAMPs and picograms of this molecule are sufficient to induce the production of a whole panel of cytokines by macrophages. LPS is trapped by the CD14 molecule on the cell surface and triggers the cell via a complex MD2/TLR4. In the case of Gram-positive bacteria, some of these bacteria release exotoxins, also known as superantigens that trigger the production of cytokines by both T-lymphocytes and macrophages.

## 6.3
## Parameters that Affect Production

In addition to the genetic polymorphism (see Sect. 8), there are numerous parameters that influence the level of the production of cytokines. Of course, *in vitro* experiments require careful sampling of the cells. The use of medium and agents free of endotoxin is very important. There are probably numerous wrong statements that suggest the capacity of a given molecule to induce the production of cytokines that are consecutive to their contamination by endotoxin, as it has been shown for heat shock proteins. *Ex vivo* studies of cytokine production by human cells is influenced by age, nutrition status, the use of drugs, alcohol, and smoking habit. Seasonal and circadian rhythms, physical exercise, altitude exposure, microgravity, social and psychological stress, physical stress, and gender affect cytokine production. In the later cases, neuromediators and sexual hormones modify cellular production of cytokines.

## 6.4
## Measurements

The measurement of cytokines in humans can be performed in natural biological fluids, in induced biological fluids (e.g. broncho-alveolar or peritoneal lavages), on tissue biopsies, and with blood leukocytes. The detection of cytokines in biological fluids may represent the "tip of the iceberg," and can be detected in the case of exacerbated production since, once produced, cytokines are efficiently trapped by their specific receptors on environmental cells. Cytokine mRNA analysis can be achieved by Northern blots, *in situ* hybridization or reverse transcriptase polymerase chain reaction (RT-PCR). Cytokine-producing cells can be monitored by ELISpot or by flow cytometry. Biological assays are not performed anymore and, nowadays, the measurement of cytokines is mainly achieved by enzyme-linked immunosorbent assays (ELISA). Cytokines constitute a tightly regulated network (see below). Thus, information concerning cytokines should be achieved through the analysis of simultaneous quantification of several cytokines. Indeed, it is the cytokine milieu that influences the cellular response rather than the action of a single cytokine. Techniques recently available such as microarrays combined to real-time PCR or multiplex immunological detection of cytokines might give precious information in a near future.

## 7 The Cytokine Network

As illustrated in Fig. 7, the interactions between cytokine-producing cells and target cells lead to define a cytokine network. Once produced in the context of a specific immune response or following microbial activation, cytokines act on target cells and induce synthesis of new cytokines. The induction by IL-1 of the release of IL-2 by lymphocytes, IL-6 by fibroblasts, G-CSF by macrophages, and IL-8 by endothelial cells are examples of cytokine cascades. Cytokines can induce their own synthesis, like IL-1, or further enhance the productions, leading to amplificatory loops, like IFN$\gamma$. Another key word to define this network is synergy. Owing to the great redundancy of cytokines, two cytokines can share similar activities. When acting together on the same target cell, the effect will be far greater than the only additive effect of each individual cytokine. For example, the production of complement factor C3 by endothelial cells in response to TNF and IL-1 is far higher than that induced by these cytokines alone. The synergy may be the consequence of the induction of cytokine receptors, allowing cells to respond to a second cytokine they would not do normally because they lack the specific receptor. The network also involves negative loops as a consequence of the action of cytokines, blocking the production or the effects of others. IL-10 is an example of a cytokine blocking the production and the action of other cytokines. However, the lineage of its target cell may lead to opposite observations. For example, IL-10 represses LPS-induced IL-8

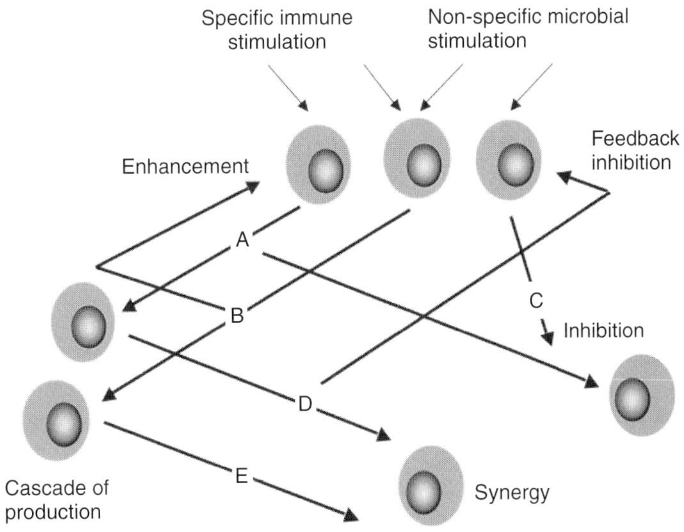


Fig. 7 The cytokine network (see Sect. 7). (See color plate p. xxii).

production by monocytes, but enhances that produced by endothelial cells.

## 8 Individual Heterogeneity

Among genetic predisposition to diseases, premature death due to uncontrolled invasion and growth by microorganisms has the higher relative risk linked to heritability. The individual heterogeneity reflects in part genetic polymorphisms of cytokines. Three levels can be dissociated (Fig. 8) as follows:

1. A genetic polymorphism exists for the receptors of PAMPs. These sensors are the first essential elements in initiating an intracellular signaling cascade that leads to cytokine production. Certain single nucleotide polymorphisms (SNP) or mutations can directly modify the intensity of the responsiveness and influence disease susceptibility, as recently shown by the existence of rare mutations of TLR4 among patients with meningococcal infection.
2. SNP or mutations can affect cytokine genes, particularly when expressed within gene promoters or other regulatory sequences. Accordingly, high, intermediate, or low producers are reported when assessing the levels of cytokines produced in response to an activator. Correlation between TNF genotypes and levels of released TNF in response to LPS has been shown.
3. Finally, depending upon the donors, target cells can react intensively, moderately, or weakly to a given amount of a cytokine. This has been nicely demonstrated with genetically distinct endothelial cells, which express various levels of adhesion molecules in response to similar amounts of IL-1 or TNF. In all these situations, the genetic polymorphism could also reflect SNP or mutations among the genes of the adaptors and signaling molecules involved in the signaling cascades.

As illustrated in Table 4, numerous cytokine gene polymorphisms have been associated with the occurrence or severity of diseases. The reported SNP or mutations can be associated with a protection against the disease or in contrast with a higher susceptibility. More recently, similar gene polymorphisms have been associated with the efficiency or the lack of efficiency of certain therapeutic approaches, particularly when the treatment targets cytokines.

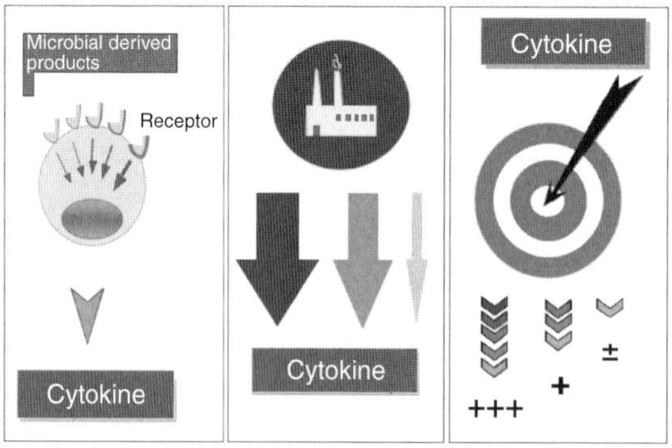


**Fig. 8** Individual heterogeneity for cytokine production and responsiveness (see color plate p. xxii).

**Tab. 4** Some examples of cytokine or cytokine receptor gene polymorphisms associated with diseases.

| *Disease* | *Gene polymorphism* |
|---|---|
| Acute renal failure mortality | TNF, IL-10 |
| Allergy | RANTES promoter; MIF promoter; IL-13 promoter IL-16 promoter; IL-1 gene complex; IL-18; IL-12B promoter |
| Asthma | IL4R $\alpha$-chain; IL-10; IL-15; IL-18; TGF$\beta$1 promoter; Eotaxin |
| Alzheimer | IL-1$\alpha$, TNFR2, IL-6 |
| Breast cancer | IL-6 |
| Chronic periodontitis | IL-10 promoter, IL-1B +3953 and TNF-A-308 allele 2 positive |
| Coronary disease | IL-1Ra |
| Crohn's disease | IL-10, MCP-1 (CCL2); IL-16 promoter; TNFR1 |
| HIV resistance | CCR5 deletion |
| Idiopathic pulmonary fibrosis | TNF-alpha (-308 A) allele |
| Infectious nephropathy | low CXCR1 expression |
| Infertility | functional mutations in the LIF gene |
| Lupus nephritis | MCP-1 (CCL2) promoter |
| Multiple sclerosis | microsatellite allele of TNF gene; IL-2 promoter |
| Parkinson's disease | promoter region of IL-8 |
| Rheumatoid arthritis | noncoding region of IFN$\gamma$ gene; TGF$\beta$1; MIF promoter; IL-1 gene cluster; IL-4 |
| Schizophrenia | IL-1 gene complex |
| Sepsis mortality | TNF |
| Systemic lupus erythematosus | IL-1$\alpha$; TNF$\alpha$ |
| Sudden infant death syndrome | IL-10 |
| Susceptibility to cerebral malaria | TNF |
| Transplantation | TNF, IL-10 |

## 9 Cytokines and Infections

### 9.1 Half Angel–Half Devil

As mentioned in Sect. 4.2.1, cytokines play a major role in innate immunity during the early processes that occur in sites of microorganisms' invasion. During viral infection, the antiviral properties of interferons have been established for more than half a century. After interaction of IFN$\alpha$ or IFN$\beta$ with its receptor, the activation of 2–5 A synthetase leads to the production of an endoribonuclease that degrades viral RNA, while activation of protein kinase P1 contributes to the capacity of the cell to inhibit viral protein synthesis. During bacterial infection, the beneficial effects of proinflammatory cytokines have been widely demonstrated. The injection of cytokines such as IL-1, IL-6, IL-12, IL-18, TNF, G-CSF, M-CSF, GM-CSF, and IFN$\gamma$ before the injection of bacteria have been shown to promote faster clearance of bacteria, and to enhance survival. The use of cytokine-neutralizing antibodies or antagonists molecules further demonstrated that endogenous cytokines produced in the course of infections play an essential role in preventing the microorganisms to establish themselves, or to grow, and their neutralization was highly deleterious. Finally, mice rendered deficient for the expression of a cytokine or a cytokine receptor display enhanced "susceptibility" to infection such as an enhanced microbial load in tissues, and reduced survival.

Despite their undoubtedly beneficial effects to clear microorganisms, some cytokines can also contribute to death in some clinical situations like the septic shock in response to endotoxin injection or after a lethal bacterial infection. Antibodies against TNF protect mice against lethal injection of LPS and baboons against lethal injection of *E. coli*. The ambivalent role of cytokines is illustrated in TNF receptor p55 knockout mice: these mice are more sensitive to *L. monocytogenes* infection than normal mice, but they are more resistant to toxic shock induced by endotoxin or Staphylococcal enterotoxin B. Furthermore, the synergy already described as a key word to define cytokine network is also valid for lethality and none lethal dose of TNF, injected with non-lethal dose of IL-1 or IFN$\gamma$ leads to lethality in mice.

### 9.2 The Strategies of Microbes against Cytokines

The fact that pathogenic microorganisms have developed numerous subterfuges to counteract the action of cytokines further illustrates that cytokines are crucial in the fight against the infectious process. Hijacking of genes for cytokine receptors has been perpetrated by virus to elaborated soluble molecules that neutralize cytokines of the host (e.g. IFN$\gamma$R and TNFR homologs of Myxoma virus, IL-1R homolog of Vaccinia virus). Another hijacking concerns the gene of IL-10, allowing viruses (Epstein–Barr virus, equine herpesvirus, parapoxvirus orf virus) to create an immunosuppressive environment. Another strategy has been developed by Vaccinia virus to prevent the maturation of biologically active IL-1 and IL-18. This virus produces an inhibitor (serpine) of caspase-1, the enzyme required for the cleavage of the proforms of these cytokines. Other mechanisms include the production of viral factors that downregulate cytokine mRNA expression (e.g. HTLV, adenovirus), the use of cytokines to further enhance the viral replication (e.g. HIV), the use of chemokine receptors to enter the cell (e.g. HIV), or the synthesis of viral chemokines that can behave as antagonists (e.g. Kaposi' sarcoma associated virus, human Herpesvirus 8, Steakth virus, etc.). Bacteria also display strategies to block cytokine production (e.g. *Yersinia enterocolitica*, enteropathogenic *E. coli*) and to use them as growth factors (*Staphylococcus aureus*, *Pseudomonas aeruginosa*, *Acinetobacter*).

## 10 Cytokines and Diseases

As the discovery of new cytokines were occurring, great hopes to cure diseases were mentioned, often followed by great disappointments. For example, while TNF was discovered for its antitumor effect, it was soon demonstrated that it was one of the most potent proinflammatory cytokines, obviously limiting its therapeutic use. One unique cytokine is rarely the unique key effector underlying the pathogenic process, and the complexity of the cytokine network has rendered difficult their use as therapeutic tools or as therapeutic target. In addition, the identification of an overexpressed cytokine at one peculiar stage of the disease process does not allow discriminating whether this cytokine is a marker or an actor. Nevertheless, there exist certain clinical situations in which the use of recombinant cytokines or the use of neutralizing antibodies has allowed significant therapeutic progresses. Few examples of diseases for which the involvement of cytokines is well established and clinical applications have been considered are summarized below.

## 10.1 Autoimmune Diseases

Among autoimmune diseases, systemic lupus erythematosus has been associated with an exacerbated production of IL-10. Considering the properties of IL-10, which activates B-lymphocytes and inhibits T-lymphocytes, its overproduction in lupus could explain the immune anomalies of this disease, which is characterized by anti-self antibodies production and by deficient T-responses. Thus, the use of IL-10 antagonists may be beneficial in the management of systemic lupus erythematosus, as suggested by a preliminary report. Human lupus patients also have elevated blood levels of BAFF that support differentiation of selected B-cells into mature long-lived B-cells and may be critical in generating deleterious autoimmune responses. Indeed, overexpression of BAFF in mice may lead to systemic lupus erythematosus–like disease. Thus, BAFF could be another potential cytokine to neutralize in systemic lupus erythematosus.

## 10.2 Allergy

Allergen specific IgE antibodies, mast cell degranulation releasing inflammatory mediators such as histamine, and influx of eosinophils in airway mucosa and airway lumen are a hallmark of atopy and allergic asthma. Thus, Th2 cytokines remain important candidates for a role in the pathogenesis of atopy and allergic asthma: IL-4 and IL-13 are required for IgE production, IL-9 is involved in mast cell growth, and IL-5 is required for eosinopoiesis. Humanized monoclonal antibodies (hMAbs) against IL-5, anti-IL-4, a recombinant soluble human IL-4 receptor, anti-IL-9, CCR3 antagonists (which block eosinophil chemotaxis), and CXCR2 antagonists (which block neutrophil and monocyte chemotaxis) have been developed as possible therapeutic interventions.

## 10.3 Sepsis

Animal models of sepsis have clearly shown that TNF, IL-1, or MIF are involved in sepsis-related lethality, and antibodies against these cytokines are highly protective. Unfortunately, anti-TNF antibodies and IL-1Ra have been used in numerous placebo double-blinded controlled assays in humans without any success. This failure may reflect that antibodies in animal models and in humans are not used with similar timings, and that mice, often used in experimental models, being $10^5$ less sensitive to Gram negative bacteria–derived endotoxin than humans, are not the most relevant laboratory animals to model the human settings!

## 10.4 Chronic Inflammatory Diseases

Fortunately for companies that had developed these anti-TNF antibodies, successful treatments have been achieved for Crohn's disease and rheumatoid arthritis. The significant improvement of patients allowed the marketing of these antibodies. However, in none of these studies, the successful improvement concerned all patients. For rheumatoid arthritis, the first successful treatment was reported in 1994. In responding patients, clinical improvements were associated with biological changes, reflecting the attenuation of the inflammatory process. There was no clinical response in 30% of the cases. While the effect of a single injection is long lasting (few months), it has to be repeated. As a consequence, the occurrence of microbial pathogenic processes (tuberculosis, *Pneumocystis*- triggered pneumonia, histoplasmosis, listeriosis, sepsis) has been deplored. Other approaches have been proposed to neutralize TNF with recombinant soluble TNF receptors. IL-1Ra has also been approved by the Food and Drug Administration (FDA) for treatment of rheumatoid arthritis. For this disease, the targeting of IL-17 has also been suggested to be of potential value. For Crohn's disease, the anti-TNF treatment was approved in 1998, and clinical response was still observed 12 weeks after the infusion. A remission could be obtained for half of the patients after 4 doses every 8 weeks.

## 10.5 Cancer

IL-2 has been extensively studied in cancer patients, either alone, or associated with chemotherapy, with activated NK cells (Lymphokine-activated killer cells, LAK), with tumor-infiltrating lymphocytes (TIL), with IFN$\alpha$, and, more recently, with tumor-pulsed dendritic cells. In some specific cases (melanoma, metastatic renal cancer), success remained modest. The major achievement has been obtained with IFN$\alpha$ in hairy cell leukemia and chronic myelogenous leukemia.

## 10.6 Hepatitis C

IFN$\alpha$ has been successful in a subset of hepatitis C virus–infected patients. More recently, once associated to ribavirin, a polymerized form of IFN$\alpha$, the half-life of which is prolonged, has been shown to lead to sustained virological response rates of $>50\%$ in chronic hepatitic C patients.

## 10.7 Multiple Sclerosis

In 1993, interferon-$\beta$ was approved in the United States for relapsing–remitting multiple sclerosis. The use IFN$\beta$ and glatiramer acetate for the treatment of multiple sclerosis has, to some extent, changed the course of the disease. The annual relapse rate of patients treated with these drugs is lower than that in placebo-treated patients, and more treated patients remain relapse-free as compared to untreated patients.

## 10.8 Transplantation

The alpha-chain of the IL-2R is a specific peptide against which monoclonal antibodies have been raised, with the aim of blunting the immune response by means of inhibiting proliferation and inducing apoptosis in primed lymphocytes. One of such antibodies has proved to be effective in reducing the episodes of acute rejection after kidney and pancreas transplantation. The use of this antibody was associated with a significant reduction in the incidence of any treated rejection episodes after kidney transplantation in the two major randomized European and US studies.

## 10.9 Neutropenia

Following chemotherapy for leukemia, neutropenia renders the patients more susceptible to develop infection. Treatment of patients with G-CSF and GM-CSF accelerates the recovery of normal neutrophil counts and significantly reduces the occurrence of documented infections.

## 11 Conclusion

Despite the huge number of molecules that belong to the cytokine family, the complexity of the cytokine network, the great number of parameters that may affect the biological properties of cytokines, tremendous amount of work has allowed to decipher the language of cells. Despite the difficulty, new promising use of cytokines or cytokine antagonists can be expected to master certain diseases.

## Bibliography

### Books and Reviews

Aggarwal, B.B. (1998) *Human Cytokines: Handbook for Basic and Clinical Research*, Blackwell, Oxford, UK.

Bona, C.A., Revillard, J-P. (2000) *Cytokines and Cytokine Receptors: Physiology and Pathological Disorders*, Harwood Academic Publishers, Amsterdam, Netherlands.

Cavaillon, J.M. (1994) Cytokines and macrophages, *Biomed. Pharmacother.* **48**, 445–453.

Cavaillon, J-M. (2001) Pro- versus anti-inflammatory cytokines: myth or reality, *Cell. Mol. Biol.* **47**, 695–702.

Fitzgerald, K.A. (2001) *Cytokine Facts Book*, Academic Press, San Diego, CA.

Giamila, F., Durum, S.K. (2003) *Cytokine Knockouts*, Contemporary immunology, 2nd edition, Humana Press, Totowa, NJ.

Kotb, M., Calandra, T. (2003) *Cytokines and Chemokines in Infectious Diseases Handbook*, Humana Press, Totowa, NJ.

Oppenheim, J.J., Feldmann, M., Durum, S.K. (2000) *Cytokine Reference: A Compendium of Cytokines and Other Mediators of Host Defense*, Academic Press, London, San Diego, CA.

Santamaria, P. (2003) *Cytokines and Chemokines in Autoimmune Disease*, Landes Bioscience/Eurekah.com, Georgetown, TX.

Thomson, A.W., Lotze, M.T. (2003) *The Cytokine Handbook*, 4th edition, Academic Press, Amsterdam, Netherlands.

### Primary Literature

Auron, P.E., Webb, A.C., Rosenwasser, L.J., Mucci, S.F., Rich, A., Wolff, S.M., Dinarello, C.A. (1984) Nucleotide sequence of human monocyte interleukin 1 precursor cDNA, *Proc. Natl. Acad. Sci. USA* **81**, 7907–7911.

Bartfeld, H., Atoynatan, T. (1969) Cytophilic nature of migration inhibitory factor associated with delayed hypersensitivity, *Proc. Soc. Exp. Biol. Med.* **130**, 497–501.

Bazan, J.F. (1990) Structural design and molecular evolution of a cytokine receptor superfamily, *Proc. Natl. Acad. Sci. USA.* **87**, 6934–6938.

Bender, J.R., Sadeghi, M.M., Watson, C., Pfau, S., Pardi, R. (1994) Heterogeneous activation thresholds to cytokines in genetically distinct endothelial cells: evidence for diverse transcriptional responses, *Proc. Natl. Acad. Sci. U S A.* **91**, 3994–3998.

Bernhagen, J., Calandra, T., Mitchell, R.A., Martin, S.B., Tracey, K.J., Voelter, W., Manogue, K.R., Cerami, A., Bvcala, R. (1993) MIF is a pituitary-derived cytokine that potentiates lethal endotoxaemia, *Nature* **365**, 756–759.

Bigazzi, P.E., Yoshida, T., Ward, P.A., Cohen, S. (1975) Production of lymphokine-like factors (cytokines) by simian virus 40-infected and simian virus 40-transformed cells, *Am. J. Pathol.* **80**, 69–78.

Burdin, N., Peronne, C., Banchereau, J., Rousset, F. (1993) Epstein-Barr virus transformation induces B lymphocytes to produce human interleukin 10, *J. Exp. Med.* **177**, 295–304.

Carswell, E.A., Old, L.J., Kassel, R.L., Green, S., Fiore, N., Williamson, B. (1975) An endotoxin-induced serum factor that causes necrosis of tumors, *Proc. Natl. Acad. Sci. USA* **72**, 3666–3670.

Cavazzana-Calvo, M., Hacein-Bey, S., de Saint Basile, G., Gross, F., Yvon, E., Nusbaum, P., Selz, F., Hue, C., Certain, S., Casanova, J.L., Bousso, P., Deist, F.L., Fischer, A. (2000) Gene therapy of human severe combined immunodeficiency (SCID)-X1 disease, *Science* **288**, 669–672.

Chai, Z., Gatti, S., Toniatti, C., Poli, V., Bartfai, T. (1996) Interleukin (IL)-6 gene expression in the central nervous system is necessary for fever response to lipopolysaccharide or IL-1 beta: a study on IL-6-deficient mice, *J. Exp. Med.* **183**, 311–316.

Coffman, R.L., Ohara, J., Bond, M.W., Carty, J., Zlotnik, A., Paul, W.E. (1986) B cell stimulatory factor-1 enhances the IgE response of lipopolysaccharide-activated B cells, *J. Immunol.* **136**, 4538–4541.

Cohen, S., Bigazzi, P.E., Yoshida, T. (1974) Similarities of T cell function in cell-mediated immunity and antibody production, *Cell. Immunol.* **12**, 150–159.

DeForge, L.E., Kenney, J.S., Jones, M.L., Warren, J.S., Remick, D.G. (1992) Biphasic production of IL-8 in lipopolysaccharide (LPS)-stimulated human whole blood. Separation of LPS- and cytokine-stimulated components using anti-tumor necrosis factor and anti-IL-1 antibodies, *J. Immunol.* **148**, 2133–2141.

de Waal Malefyt, R., Figdor, C.G., Huijbens, R., Mohan-Peterson, S., Bennett, B., Culpepper, J., Dang, W., Zurawski, G., de Vries, J.E. (1993) Effects of IL-13 on phenotype, cytokine production, and cytotoxic function of human monocytes. Comparison with IL-4 and modulation by IFN-gamma or IL-10, *J. Immunol.* **151**, 6370–6381.

Dinarello, C.A., Renfer, L., Wolff, S.M. (1977) Human leukocytic pyrogen: purification and development of a radioimmunoassay, *Proc. Natl. Acad. Sci. U S A.* **74**, 4624–4627.

Doherty, G.M., Lange, J.R., Langstein, H.N., Alexander, H.R., Buresh, C.M., Norton, J.A. (1992) Evidence for IFN-gamma as a mediator of the lethality of endotoxin and tumor necrosis factor-alpha, *J. Immunol.* **149**, 1666–1670.

Dranoff, G., Crawford, A.D., Sadelain, M., Ream, B., Rashid, A., Bronson, R.T., Dickersin, G.R., Bachurski, C.J., Mark, E.L., Whitsett, J.A. (1994) Involvement of granulocyte-macrophage colony-stimulating factor in pulmonary homeostasis, *Science* **264**, 713–716.

Dumonde, D.C., Wolstencroft, R.A., Panayi, G.S., Matthew, M., Morley, J., Howson, W.T. (1969) "Lymphokines": non-antibody mediators of cellular immunity generated by lymphocyte activation, *Nature* **224**, 38–44.

Echtenacher, B., Männel, D.N., Hultner, L. (1996) Critical protective role of mast cells in a model of acute septic peritonitis, *Nature* **381**, 75–77.

Elliott, M.J., Maini, R.N., Feldmann, M., Kalden, J.R., Antoni, C., Smolen, J.S., Leeb, B., Breedveld, F.C., MacFarlane, J.D., Bijl, H. (1994) Randomised double-blind comparison of chimeric monoclonal antibody to tumour necrosis factor alpha (cA2) versus placebo in rheumatoid arthritis, *Lancet* **344**, 1105–1110.

Fernandez-Botran, R. (1991) Soluble cytokine receptors: their role in immunoregulation, *FASEB J.* **5**, 2567–2574.

Frendeus, B., Godaly, G., Hang, L., Karpman, D., Lundstedt, A.C., Svanborg, C. (2000) Interleukin 8 receptor deficiency confers susceptibility to acute experimental pyelonephritis and may have a human counterpart, *J. Exp. Med.* **192**, 881–890.

Gabay, C., Smith, M.F., Eidlen, D., Arend, W.P. (1997) Interleukin 1 receptor antagonist (IL-1Ra) is an acute-phase protein, *J. Clin. Invest.* **99**, 2930–2940.

Gerard, C., Bruyns, C., Marchant, A., Abramowicz, D., Vandenabeele, P., Delvaux, A., Fiers, N., Goldman, M., Velu, T. (1993) Interleukin 10 reduces the release of tumor necrosis factor and prevents lethality in experimental endotoxemia, *J. Exp. Med.* **177**, 547–550.

Hapel, A.J., Lee, J.C., Farrar, W.L., Ihle, J.N. (1981) Establishment of continuous cultures of thy1.2+, Lyt1+, 2-T cells with purified interleukin 3, *Cell* **25**, 179–186.

Horai, R., Saijo, S., Tanioka, H., Nakae, S., Sudo, K., Okahara, A., Ikuse, T., Asano, M., Iwakura, Y. (2000) Development of chronic inflammatory arthropathy resembling rheumatoid arthritis in interleukin 1 receptor antagonist-deficient mice, *J. Exp. Med.* **191**, 313–320.

Howard, M., Farrar, J., Hilfiker, M., Johnson, B., Takatsu, K., Hamaoka, T., Paul, W.E. (1982) Identification of a T cell-derived B cell growth factor distinct from interleukin 2, *J. Exp. Med.* **155**, 914–923.

Hurst, S.M., Wilkinson, T.S., McLoughlin, R.M., Jones, S., Horiuchi, S., Yamamoto, N., Rose-John, S., Fuller, G.M., Topley, N., Jones, S.A.

(2001) IL-6 and its soluble receptor orchestrate a temporal switch in the pattern of leukocyte recruitment seen during acute inflammation, *Immunity* **14**, 705–714.

Isaacs, A., Lindenmann, J. (1957) Virus interference I. The interferon, *Proc. Roy. Soc. B.* **147**, 258–267.

Jacob, C.O., Fronek, Z., Lewis, G.D., Koo, M., Hansen, J.A., McDevitt, H.O. (1990) Heritable major histocompatibility complex class II-associated differences in production of tumor necrosis factor alpha: relevance to genetic predisposition to systemic lupus erythematosus, *Proc. Natl. Acad. Sci. USA.* **87**, 1233–1237.

Kopf, M., Baumann, H., Freer, G., Freudenberg, M., Lamers, M., Kishimoto, T., Zinkernagel, R., Bluethmann, H., Kohler, G. (1994) Impaired immune and acute-phase responses in interleukin-6-deficient mice, *Nature* **368**, 339–342.

Kuhn, R., Lohler, J., Rennick, D., Rajewsky, K., Muller, W. (1993) Interleukin-10-deficient mice develop chronic enterocolitis, *Cell* **75**, 263–274.

Llorente, L., Zou, W., Levy, Y., Richaud-Patin, Y., Wijdenes, J., Alcocer-Varela, J., Morel-Fourrier, B., Brouet, J.C., Alarcon-Segovia, D., Galanaud, P. (1995) Role of interleukin 10 in the B lymphocyte hyperactivity and autoantibody production of human systemic lupus erythematosus, *J. Exp. Med.* **181**, 839–844.

Moore, K.W., Vieira, P., Fiorentino, D.F., Trounstine, M.L., Khan, T.A., Mosmann, T.R. (1990) Homology of cytokine synthesis inhibitory factor (IL-10) to the Epstein-Barr virus gene BCRFI, *Science* **248**, 1230–1234.

Mosmann, T.R., Cherwinski, H., Bond, M.W., Giedlin, M.A., Coffman, R.L. (1986) Two types of murine helper T cell clone. I. Definition according to profiles of lymphokine activities and secreted proteins, *J. Immunol.* **136**, 2348–2357.

Nakano, Y., Onozuka, K., Terada, Y., Shinomiya, H., Nakano, M. (1990) Protective effect of recombinant tumor necrosis factor-alpha in murine salmonellosis, *J. Immunol.* **144**, 1935–1941.

Ohlsson, K., Bjork, P., Bergenfeldt, M., Hageman, R., Thompson, R.C. (1990) Interleukin-1 receptor antagonist reduces mortality from endotoxin shock, *Nature* **348**, 550–552.

Pfeffer, K., Matsuyama, T., Kundig, T.M., Wakeham, A., Kishihara, K., Shahinian, A., Wiegmann, K., Ohashi, P.S., Kronke, M., Mak, T.W. (1993) Mice deficient for the 55 kDa tumor necrosis factor receptor are resistant to endotoxic shock, yet succumb to L. monocytogenes infection, *Cell* **73**, 457–467.

Pociot, F., Briant, L., Jongeneel, C.V., Mölvig, J., Worsaae, H., Abbal, M., Thomsen, M., Nerup, J., Cambon-Thomsen, A. (1993) Association of tumor necrosis factor and class II major histocompatibility complex alleles with the secretion of TNF$\alpha$ and TNF$\beta$ by human mononuclear cells: a possible link to insulin-dependent diabetes mellitus, *Eur. J. Immunol.* **23**, 224–231.

Porat, R., Clark, B.D., Wolff, S.M., Dinarello, C.A. (1991) Enhancement of growth of virulent strains of Escherichia coli by interleukin-1, *Science* **254**, 430–432.

Rosenberg, S.A., Lotze, M.T., Muul, L.M., Leitman, S., Chang, A.E., Ettinghausen, S.E., Matory, Y.L., Skibber, J.M., Shiloni, E., Vetto, J.T. (1985) Observations on the systemic administration of autologous lymphokine-activated killer cells and recombinant interleukin-2 to patients with metastatic cancer, *N. Engl. J. Med.* **313**, 1485–1492.

Sallusto, F., Lenig, D., Mackay, C.R., Lanzavecchia, A. (1998) Flexible programs of chemokine receptor expression on human polarized T helper 1 and 2 lymphocytes, *J. Exp. Med.* **187**, 875–883.

Shalaby, M.R., Waage, A., Aarden, L., Espevik, T. (1989) Endotoxin, tumor necrosis factor-alpha and interleukin 1 induce interleukin 6 production in vivo, *Clin. Immunol. Immunopathol.* **53**, 488–498.

Shirai, A., Holmes, K., Klinman, D. (1993) Detection and quantitation of cells secreting IL-6 under physiologic conditions in BALB/c mice, *J. Immunol.* **150**, 793–799.

Shull, M.M., Ormsby, I., Kier, A.B., Pawlowski, S., Diebold, R.J., Yin, M., Allen, R., Sidman, C., Proetzel, G., Calvin, D. (1992) Targeted disruption of the mouse transforming growth factor-beta 1 gene results in multifocal inflammatory disease, *Nature* **359**, 693–699.

Smith, C.A., Davis, T., Wignall, J.M., Din, W.S., Farrah, T., Upton, C., McFadden, G., Goodwin, R.G. (1991) T2 open reading frame from the Shope fibroma virus encodes a soluble form of the TNF receptor, *Biochem. Biophys. Res. Commun.* **176**, 335–342.

Stuber, F., Petersen, M., Bokelmann, F., Schade, U. (1996) A genomic polymorphism within the tumor necrosis factor locus influences plasma tumor necrosis factor-alpha concentrations and outcome of patients with severe sepsis, *Crit. Care Med.* **24**, 381–384.

Supajatura, V., Ushio, H., Nakao, A., Akira, S., Okumura, K., Ra, C., Ogawa, H. (2002) Differential responses of mast cell Toll-like receptors 2 and 4 in allergy and innate immunity, *J. Clin. Invest.* **109**, 1351–1359.

Tracey, K.J., Fong, Y., Hesse, D.G., Manogue, K.R., Lee, A.T., Kuo, G.C., Lowry, S.F., Cerami, A. (1987) Anti-cachectin/TNF monoclonal antibodies prevent septic shock during lethal bacteraemia, *Nature* **330**, 662–664.

Tsicopoulos, A., Hamid, Q., Haczku, A., Jacobson, M.R., Durham, S.R., North, J., Barkans, J., Corrigan, C.J., Meng, O., Moqbel, R. (1994) Kinetics of cell infiltration and cytokine messenger RNA expression after intradermal challenge with allergen and tuberculin in the same atopic individuals, *J. Allergy Clin. Immunol.* **94**, 764–772.

Upton, C., Mossman, K., McFadden, G. (1992) Encoding of a homolog of the IFN-gamma receptor by myxoma virus, *Science* **258**, 1369–1372.

van Dullemen, H.M., van Deventer, S.J., Hommes, D.W., Bijl, H.A., Jansen, J., Tytgat, G.N., Woody, J. (1995) Treatment of Crohn's disease with anti-tumor necrosis factor chimeric monoclonal antibody (cA2), *Gastroenterology* **109**, 129–135.

Wang, H., Yu, M., Ochani, M., Amella, C.A., Tanovic, M., Susarla, S., Li, J.H., Wang, H., Yang, H., Ulloa, L., Al-Abed, Y., Czura, C.J., Tracey, K.J. (2003) Nicotinic acetylcholine receptor alpha7 subunit is an essential regulator of inflammation, *Nature* **421**, 384–388.

Yoshimura, T., Matsushima, K., Oppenheim, J.J., Leonard, E.J. (1987) Neutrophil chemotactic factor produced by lipopolysaccharide (LPS)-stimulated human blood mononuclear leukocytes: partial characterization and separation from interleukin 1 (IL 1), *J. Immunol.* **139**, 788–793.

# 7
# Interleukins

*Anthony Meager*
*Division of Immunobiology, National Institute for Biological Standards and Control, Blanche Lane, South Mimms, Potters Bar, Herts, UK*

**1 Background and Historical Perspective 170**

**2 Interleukin Proteins and Genes 173**
2.1 Interleukin Proteins 173
2.2 Interleukin Genes 175

**3 Interleukin Receptors 179**
3.1 Receptor Structure 179
3.2 Intracellular Signaling Pathways 183

**4 Biological Activities of Interleukins 185**
4.1 Interleukin-1 186
4.2 Interleukin-2 187
4.3 Interleukin-3 187
4.4 Interleukin-4 189
4.5 Interleukin-5 190
4.6 Interleukin-6 191
4.7 Interleukin-7 191
4.8 Interleukin-8 192
4.9 Interleukin-9 192
4.10 Interleukin-10 192
4.11 Interleukin-11 193
4.12 Interleukin-12 193
4.13 Interleukin-13 194
4.14 Interleukin-14 194
4.15 Interleukin-15 194
4.16 Interleukin-16 194
4.17 Interleukin-17 194

*Immunology. From Cell Biology to Disease.* Edited by Robert A. Meyers.

ISBN: 978-3-527-31770-7

4.18 Interleukin-18 195
4.19 Interleukin-19 195
4.20 Interleukin-20 195
4.21 Interleukin-21 195
4.22 Interleukin-22 195
4.23 Interleukin-23 196
4.24 Interleukin-24 196
4.25 Interleukin-25 196
4.26 Interleukin-26 196
4.27 Interleukin-27 196
4.28 Interleukin-28 196
4.29 Interleukin-29 197

**5 Interleukin Physiology 197**

**6 Pathophysiology and Disease Correlates 199**

**7 Clinical Uses of Interleukins 200**

**8 Concluding Remarks 202**

**Acknowledgments 202**

**Bibliography 202**
Books and Reviews 202


## Keywords

**Cytokine**
One of a class of inducible biologically active proteins that exercise specific, receptor-mediated effects in target cells or in the cytokine-producing cells themselves.

**Interleukin**
One of several different cytokines acting between leukocytes and other cell types, which has a variety of stimulatory activities that regulate immune, inflammatory, and hematopoietic responses.

**Hematopoietin**
One of a subgroup of interleukins having an $\alpha$-helical bundle structure.

**Hematopoietin Receptor**
One of a class of structurally related cell surface receptors for the hematopoietin subgroup of interleukins.

**Hematopoiesis**
The process of populating and replacing circulating erythrocytes and leukocytes from stem cells contained within the bone marrow.

**Macrophage**
Phagocytic leukocyte found in various tissues, which is important in nonspecific cellular immunity and antigen presentation, and which is a major producer of and responder to interleukins.

**T lymphocyte (T cell)**
A type of white blood cell capable of responding to foreign antigens and thus of mediating cellular immunity, which it does by secreting a variety of interleukins and other cytokines to activate leukocytes and other cells.

**Homolog**
A structurally related interleukin molecule whose activity is mediated by a receptor that is common to all homologs within an interleukin family.

**Paralog**
A structurally related interleukin molecule whose activity is mediated by a receptor distinct from those of other paralogs within an interleukin family.

The harmonious regulation of vital physiological processes, for example, replenishment of mature blood cells from bone marrow stem cells, termed *hematopoiesis,* and the activation of defense mechanisms against pathological microbes and injury has been shown to depend on the production and action of a variety of secreted biologically active proteins, collectively known as cytokines. Central among cytokines is a class of mediators, largely involved in the regulation of immune, inflammatory, and hematopoietic functions, designated interleukins primarily on the basis that interleukins are produced mainly by leukocytes and act locally on other leukocytes in surrounding tissues. Each interleukin, of which there are now 29 designated ones, exercises a spectrum of biological activities via specific cell surface receptors. Overlaps of biological activities have been found to be common among different interleukins and to result in many cases from the sharing of receptor components. On binding their cognate interleukins, receptors activate intracellular signaling pathways leading to the transcription of nuclear genes and expression of proteins necessary to commit the cell to a number of contingent events and responses according to the particular interleukin bound and cell type. Interleukins, however, probably rarely act alone and *in vivo* form complex interactive networks both among themselves and with other cytokines. Such complicated intercellular communications systems have made it difficult to precisely define the biological roles of interleukins in health and disease. In certain diseases, for example, cancer, some exogenously administered interleukins induce beneficial responses, but it has rarely been possible to dissociate their desirable pharmacological activities from their undesirable pharmacological activities, which often give rise to severe side effects. In fact, many disease symptoms, for example, fever, hypotensive shock, have been demonstrated to be strongly associated with the presence of endogenously produced interleukins, and this is leading to clinical evaluation of several interleukin antagonists in both acute and chronic diseases.

## 1 Background and Historical Perspective

From the beginning of the twentieth century, there has been a growing realization that many of the biological processes in multicellular organisms are regulated by extracellular factors. Together such factors constitute an elaborate, interactive communication system that governs the cellular and physiological responses at several levels. In particular, the immune and neuroendocrine systems of higher animals, such as mammals, are subject to regulation by secreted factors that can act both locally (paracrine action) and at a distance (endocrine action) on cells bearing cognate receptors. The word "cytokine" was coined in 1974 by Dr Cohen to describe any soluble factor produced by both lymphoid and nonlymphoid cells that exercise specific effects in its target cells. Since then, the definition of cytokine has become restricted to nonendocrine, biologically active, protein mediators involved in (1) cell proliferation and thus tissue development and repair and (2) cellular function, which is required for the maintenance of homeostatic and defense mechanisms. Cytokines may be perceived to be analogous to polypeptide hormones except, in contrast to hormones, cytokines mainly act locally rather than at a distance. However, in many respects the distinction between cytokines and hormones is blurred.

In today's classification, interferons, interleukins, colony stimulating factors, and polypeptide growth factors are considered to belong to the cytokine superfamily. The word "interleukin" was coined in 1979 at the 2nd International Lymphokine Workshop in Ermatingen, Switzerland (quite possibly interleukin was born from a corruption of Interlaken, another town in Switzerland!), to apply to soluble mediators produced by activated T lymphocytes that acted in a paracrine fashion on "responder" lymphocytes. However, the definition of interleukin was quickly widened to include soluble mediators produced by a wide variety of cell types that acted primarily on lymphocytes and other cell types within the immune system. Regrettably, prior designation of some of the earlier characterized cytokines, for example, interferon gamma (IFN$\gamma$), has effectively excluded them from the interleukin family to which they truly belong, and the inclusion/exclusion of newly discovered cytokines to the interleukin family has often been fairly arbitrary.

Historically, the biological activities that are now attributed to interleukins were first uncovered as early as the 1940s. Then, for example, a fever-inducing substance isolated from "neutrophils" was called *granulocyte pyrogen* (GP); it had similar properties to endogenous pyrogen (EP), a substance isolated from the blood of rabbits made febrile by the injection of bacteria. Since then and particularly in the late 1960s and early 1970s, many other biological activities have been defined, being originally ascribed to poorly characterized, soluble factors. Thus, lymphocyte activating factor (LAF) was the name given by immunologists to a macrophage-derived substance that enhanced mitogen-driven lymphocyte proliferation. A substance secreted by such mitogen (lectin)-driven lymphocytes was shown to specifically support the proliferation of T lymphocytes and was therefore named *T-cell growth factor* (*TCGF*). Subsequently, a number of other growth and differentiation factors acting on specific cell lineages were described.

By the late 1970s, it was beginning to be realized that some of these soluble factors, which had been separately designated, were in fact either similar or identical substances. For example, the production of EP and LAF could be shown to be stimulated by the same agents and their molecular weights (mw), isoelectric points (pI), and other properties were broadly similar. Their purification initially showed that they contained similar active, but heterogenous, proteins with pIs ranging between 5 and 8. Following the coining of "interleukin" in 1979, LAF was redesignated as interleukin-1. TCGF became interleukin-2. This naturally started the quest for "discovering" more interleukins and during the 1980s, with the major advance in rDNA technology, many "new" interleukins were cloned. Some of these were clearly identical to specific cell growth factors, which had been partially characterized earlier; others turned out to be novel biologically active proteins. The new and "old" names for interleukins are summarized in Table 1. New interleukins continue to be identified right up to the present day.

Once interleukins were cloned and made available in large quantities via production in recombinant bacteria, their biochemistry and biology could be and has been thoroughly investigated. A common characteristic that has emerged from such studies is that most, if not all, interleukins have more than one biological activity and there is frequently an overlap between the spectrum of activities of one particular interleukin and another interleukin or a member of the cytokine superfamily, such as an interferon or a colony stimulating factor. This indicates a significant redundancy in the biological functions of interleukins and of cytokines in general. Such redundancy suggests that interleukins trigger convergent signaling pathways in cells, and the more recent cloning of interleukin receptors has shed some light on why this should be so. For instance, it is now known that some interleukins share nonspecific receptor components besides those components that are necessary for specific interleukin binding. Furthermore, while interleukin receptors do not contain cytoplasmic kinase domains themselves, it appears increasingly likely that they are associated with a limited number of nonreceptor kinases, which mediate the phosphorylation of nuclear transcription factors. The latter are activated by this process to bind to interleukin-responsive elements of interleukin-inducible genes and activate their transcription. The subsequent expression of interleukin-inducible gene mRNAs and protein synthesis leads to cellular responses. If two (or more) interleukins trigger the induction of a common set of genes, then the observed cellular response, for example, mitogenesis, will probably be similar. However, cells have regulatory mechanisms that control the expression of interleukin receptors and thus their responsiveness to particular interleukins. Some interleukin receptors are restricted to specific cell types or lineages,

**Tab. 1** Interleukins and former nomenclature.

| ***Interleukin*** | ***Former names*** |
|---|---|
| Interleukin-1α (IL-1α) | Endogenous pyrogen (EP); lymphocyte activating factor (LAF); hemopoietin-1 (HP-1) |
| Interleukin-1ß (IL-1ß) | As for IL-1α; osteoclast-activating factor (OAF) |
| Interleukin-1 receptor antagonist (IL-1ra) | – |
| Interleukin-2 (IL-2) | T-cell growth factor-1 (TCGF-1); killer helper factor (KHF) |
| Interleukin-3 (IL-3) | Multiple colony stimulating factor (multi-CSF); hemopoietic cell growth factor (HCGF) |
| Interleukin-4 (IL-4) | B-cell stimulating factor-I (BSF-I); T-cell growth factor-II (TCGF-II); macrophage activation factor (MAF) |
| Interleukin-5 (IL-5) | T-cell replacing factor-I (TRF-I); eosinophil differentiation factor (EDF); IgA-enhancing factor |
| Interleukin-6 (IL-6) | B-cell stimulating factor-2 (BSF-2); hybridoma-plastocytoma growth factor (HPGF); hepatocyte stimulating factor (HSF) |
| Interleukin-7 (IL-7) | Lymphopoietin-1 (LP-1); thymocyte growth factor (THGF); pre-B-cell growth factor (PBGF) |
| Interleukin-8 (IL-8)[a] | Macrophage-derived neutrophil chemotactic factor (MDNCF); neutrophil activating factor-1 (NAF-1) |
| Interleukin-9 (IL-9) | T-cell growth factor-III (TCGF-III); p40; mast-cell enhancing activity |
| Interleukin-10 (IL-10) | Cytokine synthesis inhibitory factor (CSIF) |
| Interleukin-11 (IL-11) | Adipogenesis inhibitory factor (AGIF) |
| Interleukin-12 (IL-12) | Natural killer stimulatory factor (NKSF); cytotoxic lymphocyte maturation factor (CLMF) |
| Interleukin-13 (IL-13) | P-600 |
| Interleukin-14 (IL-14) | High molecular weight B-cell growth factor (HMW-BCGF) |
| Interleukin-15 (IL-15) | – |
| Interleukin-16 (IL-16) | Lymphocyte chemoattractant factor (LCF) |
| Interleukin-18 (IL-18) | Interferon gamma inducing factor |
| Interleukin-22 (IL-22) | IL-10-related T-cell-derived inducible factor (IL-TIF) |
| Interleukin-24 (IL-24) | Melanoma differentiation antigen-7 (MDA-7) |
| Interleukin-26 (IL-26) | AK155 |

[a] IL-8 is a member of the small cytokine or chemokine superfamily. There are two main subgroups distinguished on the basis of the positions of cysteine residues. IL-8 is a member of the CXC Ligand (CXCL) subgroup, which also includes platelet factor 4 (PF4), ß-thromboglobulin (ß-TG), gro-α, gro-ß, IP-10, and so on. The CC Ligand (CCL) subgroup includes monocyte chemoattractant protein-1 (MCP-1), RANTES, macrophage inflammatory proteins (MIP-1α MIP-1ß), I-309, and so on.

and their expression may also change according to the stage of cell development or differentiation.

Overall, the regulation of cellular responses by interleukins is complex, and probably even more so where they act as mixtures of interleukins or with other cytokines or other noncytokine mediators. The complex interaction of interleukins and other mediators, which is envisaged to occur *in vivo*, has often obscured the physiological roles of interleukins. However, do they have any major physiological roles or do they have one or more accessory or subordinate roles? The fact that most interleukins are inducibly, rather than constitutively, produced and mainly affect the working of the immune system, suggests they have been evolved primarily for triggering host defense mechanisms against infectious microorganisms. They have been clearly shown to regulate both cell-mediated and humoral immunity. As such, they could also be important in antitumor mechanisms and in chronic degenerative diseases.

The role(s) of many interleukins in preventing or combating infectious or invasive diseases has been supported by numerous studies carried out in experimental animal model systems. The antitumor effect of interleukin-2 (IL-2) has looked, for example, to be the most efficacious in causing tumor regression in allogenic and xenogenic tumors in mice. These studies have stimulated clinical interest in interleukins, and in recent years many clinical trials to evaluate interleukins as anticancer agents have been carried out. IL-2 has been used extensively, but as with IFN$\alpha$, its usefulness in treating cancer has appeared limited. It also causes severe side effects. Less is known generally about what clinical use other interleukins might offer; however, some such as IL-1 look to be too toxic. It is thus not at all certain that the therapeutic application of individual interleukins, which now number up to interleukin-29 (IL-29), will significantly affect the outcome or the management of clinical diseases.

## 2 Interleukin Proteins and Genes

### 2.1 Interleukin Proteins

There are now 29 distinct biologically active mediators that have been classified as interleukins (Table 1). However, the actual number of molecularly distinct proteins is larger than 29 because some interleukins are comprised of more than one molecular species, for example, IL-1, IL-8, IL-17. A comparable situation exists for the type I interferon (IFN) family, where for human IFN$\alpha$ in particular, there are 12 related molecular species known as subtypes. In the case of IL-1, initially three structurally related proteins, two of which are biologically active, that is IL-1$\alpha$ and IL-1ß, and the third, which is an inhibitor of IL-1$\alpha$/ß actions, known as IL-1 receptor antagonist (IL-1ra), were identified. The latter is the only known interleukin without agonistic activity and which behaves as a competitive inhibitor of an interleukin. More recently, seven more structural homologs of IL-1 have been described, including one that is more commonly known as IL-18. Other examples of structurally related interleukins that nevertheless are designated as differently numbered interleukins, for example, IL-10, IL-19, IL-20, IL-22, IL-24, and IL-26, (IL-28, IL-29), are now known. In one or two cases, the interleukin name has proved less appropriate, for example, IL-8, which is now clearly just one of a large family of chemokines.

The interleukins exhibit a wide variety of primary structures, sizes, and post-translational modifications. At the tertiary level, however, the interleukins generally fall into but a few categories. Excluding the IL-1 and IL-8 families, the majority of interleukins are $\alpha$-helical proteins, which although unrelated in amino acid sequences, fold up as 4-6-$\alpha$-helix bundles. This $\alpha$-helical bundle structure is common to IL-2, IL-3, IL-4, IL-5, IL-6, IL-7, IL-9, IL-10, IL-11, IL-12, IL-13, IL-15, and most of the newly described interleukins (Fig. 1). The structures of the

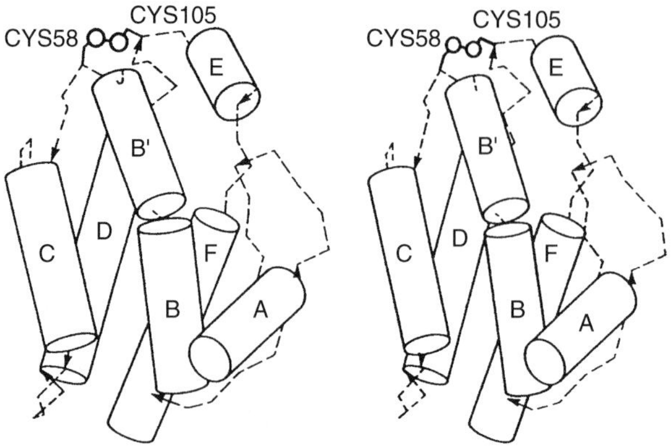


**Fig. 1** Schematic stereo drawing of IL-2; helices are represented as cylinders and are lettered sequentially from the N-terminus. (Reprinted with permission from Brandhuber et al. (1987) *Science* **238**, 1707. Copyright 1987 by American Association for the Advancement of Science.)

receptors for these interleukins also have common features (discussed in Sect. 3.1). These interleukins have sometimes been referred to as *hematopoietins* and their receptors as hematopoietin receptors (HR). Interestingly, the hematopoietin family extends to other noninterleukin mediators such as erythropoietin (EPO) and growth hormone (GH) as well as a number of cytokines, including granulocyte colony stimulating factor (G-CSF), granulocyte-macrophage colony stimulating factor (GM-CSF), leukemia inhibitory factor (LIF), oncostatin M (OSM), and ciliary neurotrophic factor (CNTF). In contrast, IL-1ß and other IL-1 family members are nonhelical proteins with about 60% of residues in ß-strands, in a compact configuration known as a ß-trefoil (Fig. 2). IL-1$\alpha$ and ß and IL-18 are also unlike the other interleukins in having long N-terminal leader sequences (prodomains) rather than the recognizable signal polypeptides of secreted interleukins. Their processing and the exit of mature IL-1$\alpha$, IL-1ß, or IL-18 from cells is therefore quite different from the other interleukins. A specific enzyme, a cysteine protease originally called IL-1ß converting enzyme but now designated caspase 1, which cleaves pro-IL-1ß to release the mature (active) IL-1ß protein, has been characterized. The IL-8 protein has a structure that incorporates both $\alpha$-helical and ß-strand domains. The three-dimensional structures of the latest additions to the interleukin family are not known in detail, but are predicted to be mainly $\alpha$-helical.

Another variable characteristic of interleukins is whether they are active in the monomeric or dimeric configuration. Most are active as monomers, but IL-5, IL-8, IL-10, (and IL-19, IL-20, IL-22, IL-24, IL-26) and IL-12 (and IL-23, IL-27) are dimers. Glycosylation is also a variable feature of interleukins; IL-1$\alpha$ and ß are not glycosylated, whereas most of the other interleukins have either O-linked and/or N-linked oligosaccharide side chains. The molecular characteristics of interleukins are summarized in Table 2.

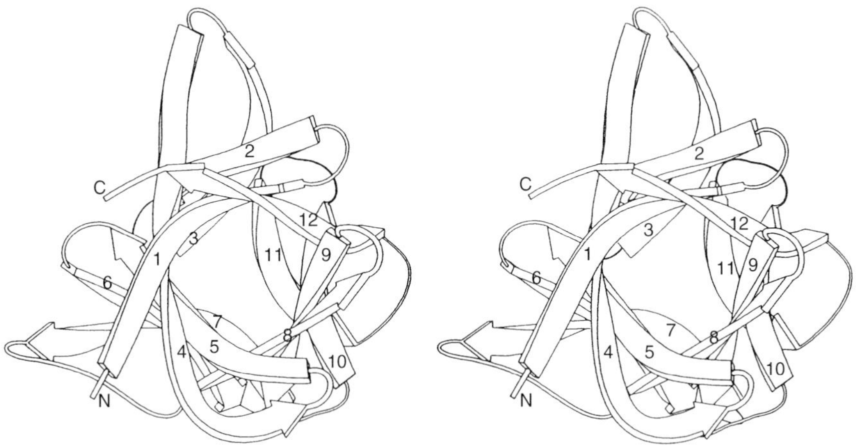


**Fig. 2** Stereo cartoon of IL-1ß. The twisted arrows represent ß-strands and they are numbered sequentially from the N-terminal. The view is down the axis of the barrel formed by six of the ß-strands. (Reprinted with permission from Priestle et al. (1988) *EMBO J*, **7**, 339. Copyright EMBO/IRL Press Ltd, UK.)

## 2.2 Interleukin Genes

The fact that interleukin-like molecules can be found in invertebrates indicates that interleukin genes probably arose very long ago (>500 million years). The precise ancestry of interleukin genes is not known, but it is clear that they have been conserved throughout evolution from an apparent ancient origin. The structures of interleukin genes are complex, being composed of several exons and introns (cf the intronless type I IFN genes). For example, the human IL-1ß gene, which is located on the long arm of chromosome 2, contains 7 exons and 6 introns. However, many of the other interleukin genes share a common 4-exon-3-intron structure, for example, IL-2, IL-4, IL-5, which is also found in the GM-CSF and IFN$\gamma$ genes. The gene for IL-6 has 5 exons and 4 introns, a feature it shares with the G-CSF gene. The interleukin genes are widely distributed among human chromosomes (Table 2), but there are several locations where interleukin families and their receptors are clustered, for example, the genes for IL-3, IL-4, IL-5, IL-9, IL-12p40, IL-13, GM-CSF, monocyte-colony stimulating factor (M-CSF) and its receptor are linked on the long arm (q) of chromosome 5. The equivalent site in the mouse genome is located on chromosome 11.

The expression of interleukin genes is dependent upon the activation of cellular transcription factors and the binding of these to the response elements located in the 5′ flanking regions of the coding DNA. The promoters and enhancers of interleukin gene transcription have been rather less studied than, for example, those of the IFNß gene. For instance, while IL-1$\alpha$, IL-1ß, and IL-1ra genes are activated by common inducers such as bacterial lipopolysaccharide (LPS) and phorbol esters, there are cellular mechanisms that give rise to differential expression of their respective mRNAs. These mechanisms are poorly understood, but are probably cell type specific. Inducers of IL-1 genes,

**Tab. 2** Molecular characteristics of human interleukins and their receptors.

| *Interleukin* | *Precursor, no. of AA* | *Mature protein, no. of AA.* | *Glycosylation* | *Disulfide bonds* | *Monomer (M) or dimer (D)* | *Chromosome assignment of gene and organization (exon/intron)* | *Receptors* |
|---|---|---|---|---|---|---|---|
| IL-1α | 271 | 159 | – | 0 | M | 2 (7/6) | 80 kDa type I receptor |
| IL-1ß | 269 | 153 | – | 1 | M | 2 (7/6) | 68 kDa type II receptor |
| IL-1ra | 177 | 152 | N-linked | | M | 2 | |
| IL-2 | 153 | 133 | O-linked | 1 | M | 4 (4/3) | 55 kDa α-chain<br>75 kDa ß-chain<br>65 kDa γ-chain |
| IL-3 | 152 | 133 | N-linked | 1 | M | 5 (5/4) | 80 kDa α-chain<br>120 kDa ß-chain |
| IL-4 | 154 | 129 | N-linked | 3 | M | 5 (4/3) | 130 kDa binding protein; other component? |
| IL-5 | 134 | 114 | N-linked | 1 | D | 5 (4/3) | 80 kDa α-chain<br>120 kDa ß-chain |
| IL-6 | 212 | 184 | N-linked | 2 | M | 7 (5/4) | 80 kDa α-chain<br>130 kDa ß-chain - (gp130) |
| IL-7 | 177 | 152 | N-linked | 3 | M | 8 | 68 kDa binding protein |
| IL-8 | 99 | 72 | – | 2 | D | 4 (c-x-c)<br>17 (c-c) (4/3) | 58 kDa binding proteins; two types at least |
| IL-9 | 144 | 126 | N-linked | 5 | M | 5 (5/4) | 64 kDa binding protein |

| | | | | | | | |
|---|---|---|---|---|---|---|---|
| IL-10 | 178 | 160 | Not glycosylated but contains one site of N-linked glycosylation | 2 (5-cysteines) | D | 1 (5/4) | 110 kDa binding protein |
| IL-11 | 199 | 178 | – | 0 (2-cysteines) | M | 19 | 150 kDa $\alpha$-chain<br>130 kDa ß-chain - (gp130) |
| IL-12 | | | | | | | Two similar chains |
| p35 | 253 | 197 | N-linked | 3 (7-cysteines) | D (Heterodimer) | 3<br>5 | related to gp130 named IL-12R$\beta$1 & IL-12R$\beta$2 |
| p40 | 326 | 306 | N-linked | 5 | | | |
| IL-13 | 132 | 112 | N-linked | 2 | M | 5 | Two IL-13R$\alpha$ chains (HR family) in combination with components shared with IL-4 |
| IL-14 | 498 | 483 | N-linked | ? | M | | |
| IL-15 | 162 | 114 | N-linked | 2 | M | 4 | Specific IL-15R$\alpha$ chain, with IL-2R$\beta$ & IL-2R$\gamma$ shared with IL-2 |

A growing number of additional interleukins have now been characterized; however, by and large, these belong to the previously characterized interleukin families. For instance, several IL-1 homologs have been identified, the genes of which are all on chromosome 2 close to IL-1$\alpha$, IL-1$\beta$, and IL-1ra genes; in contrast, the gene for IL-18, an IL-1 paralog, is located on chromosome 11. Genes for IL-10 paralogs IL-19, IL-20, and IL-24 are located at chromosome 1q32 along with the IL-10 gene, while those for IL-22 and IL-26 are located on chromosome 12q15, close to the IFN$\gamma$ gene. IL-10 and related paralogs are all V-shaped homodimers with topology resembling that of IFN$\gamma$. Receptors for IL-10 family members are constituted by specific pairs of class II cytokine receptor chains. IL-28 and IL-29 are related to both IL-10 and type I IFN families; however, their genes resemble the intron–exon structure of IL-10-type genes rather than the intronless type I IFN genes and are located on chromosome 19q13, away from either IL-10- and paralog genes, located as above, or type I IFN genes on chromosome 9. The receptor for IL-28/29 is a distinct class II cytokine receptor pair composed of IL-10R$\beta$ and CRF2-12 and different from type I IFN receptors. IL-21 is a 4-helix cytokine most closely to IL-15; IL-23 and IL-27 are heterodimers structurally related to IL-12, IL-23 sharing with IL-12 its p40 subunit and thus one of its receptors, IL-12R$\beta$1.

for example, LPS, phorbol myristate acetate (PMA), increase protein kinase C (PKC) activity, which in turn activates the nuclear transcription factors, NFκB and AP-1 (composed of the protooncogene products c-jun and c-fos). Binding motifs for NFκB and AP-1 are present in the promoter regions of IL-1 genes. It is probable that transcription of IL-1 genes is regulated by activated NFκB and AP-1, although presently it is not understood how their transcription is differentially regulated. Other factors such as intracellular levels of cyclic AMP (cAMP) could be important in this respect. Increased concentrations of cAMP have been found to enhance IL-1ß expression, both at the transcriptional and posttranscriptional level. Interestingly, it is clear that IL-1α and ß can themselves induce IL-1 gene expression, probably utilizing the same PKC signaling pathway and NFκB/AP-1 transcription factors as other IL-1 inducers. However, in macrophages, LPS is by far the strongest inducer of IL-1 gene expression.

In contrast, IL-2 gene expression is dependent on mitogenic or antigenic stimulation of T lymphocytes. Gene activation requires at least three transcription factors, AP-1, and two others designated NFAT-1 and NFIL-2A, which themselves are subject to regulatory mechanisms. The signaling pathway via PKC appears to be involved in IL-2 gene induction. IL-1, which also utilizes the PKC pathway and is known to enhance IL-2 synthesis, probably acts to augment/amplify the cooperation of nuclear transcription factors at the level of the IL-2 gene. The promoter region of the IL-6 gene contains the NFκB and AP-1 binding motifs, similar to those found in IL-1 genes, plus a cAMP-responsive element (cRE).

It is known that other interleukin genes are responsive to the same or similar inducers that lead to IL-1 gene activation, for example, IL-6, IL-8, while others are responsive to the same inducers that lead to IL-2 gene activation, for example, IL-3, IL-4, IL-5, IL-7, IL-9, IL-10, and IL-13. Synthesis of the latter interleukins is almost entirely restricted to T lymphocytes, but regulation at the transcriptional level can determine which interleukins are produced. Mature T-helper (h) $CD4^+$ lymphocytes, for example, can be separated into two or more functional subsets. The Th1 subset, which is responsible for initiating responses against intracellular pathogens and delayed-type hypersensitivity, produces IL-2 and IFNγ, whereas the Th2 subset, which preferentially induces antibody-mediated responses, produces IL-4, IL-5, IL-6, IL-10, and IL-13. Other interleukins, for example, IL-3, and cytokines, for example, GM-CSF, tumor necrosis factor alpha (TNFα) may be produced by either subset (Table 3). More immature Th lymphocytes, designated Th0, appear to be able to produce an unrestricted range of interleukins and cytokines. The mechanisms by which interleukin genes are switched on and off in Th lymphocytes are not fully understood, but are partly dependent on differentiation and maturation signals mediated by other interleukins and cytokines, for example, IL-12, IFNγ, and pathogenic triggers. For example, allergens and certain parasite antigens stimulate the development of Th2 lymphocytes.

The production of interleukins is also likely to be subject to posttranscriptional and posttranslational controls. As mentioned earlier, the processing of IL-1α and ß is complex, involving specific enzyme cleavage of IL-1 31 kDa precursors and transport of the mature, active 17.5 kDa IL-1α and ß proteins through the cytoplasm to cell membrane (neither IL-1 precursors

**Tab. 3** Interleukins and cytokines secreted by T-helper CD4+ cell subsets and cytotoxic CD8+ T lymphocytes.

| *Interleukin/cytokine* | *Th1* | *Th2* | *Th0* | *Cytotoxic T lymphocytes (CTL)* |
|---|---|---|---|---|
| Interferon-$\gamma$ (IFN$\gamma$) | ++ | − | ++ | ++ |
| IL-2 | ++ | − | ++ | ± |
| IL-3 | ++ | ++ | ++ | + |
| IL-4 | − | ++ | ++ | − |
| IL-5 | − | ++ | ++ | − |
| IL-6 | − | ++ | + | − |
| IL-9 | ? | ? | + | ? |
| IL-10 | − | ++ | ++ | ? |
| IL-13 | − | ++ | ? | ? |
| GM-CSF | ++ | + | + | ++ |
| Tumor necrosis factor-$\alpha$ (TNF$\alpha$) | ++ | + | ? | + |
| Tumor necrosis factor-ß (TNFß) | ++ | − | ? | + |

nor mature forms appear to enter the endoplasmic reticulum as would a "normal" secreted protein). Some IL-1$\alpha$ appears to remain "anchored" in the cell membrane or the cell associated for several hours before being released, whereas IL-1ß is quickly "secreted". The processing of IL-1$\alpha$ and ß is unusual and for all of the other interleukins, including IL-1ra, processing, posttranslational modification, for example, glycosylation, and secretion occurs through the regular route for a protein containing an N-terminal signal sequence, that is, via the endoplasmic reticulum and the Golgi apparatus.

The expression of all the interleukin genes is tightly regulated. Interleukin mRNAs are usually made for only short periods of time following induction. The interleukin genes then become hyporesponsive to further stimulation, probably due to increases in the level of repressor complexes. Synthesis of interleukins is also of a relatively short duration, but can depend on the stability of the mRNA. In addition, the level of synthesis will be affected (modulated) by the presence of other active regulatory agents [(for example, immunosuppressive hormones (prostaglandins, glucocorticoids) and cytokines (transforming growth factor beta TGF-$\beta$)].

The major cell sources and inducers of particular interleukins are detailed in Table 4.

## 3 Interleukin Receptors

### 3.1 Receptor Structure

As discussed in Sect. 2.1, many interleukins have similar $\alpha$-helical bundle structures and therefore it is not surprising that their cell surface receptors also share common structural features, particularly in their N-terminal extracellular binding domains. The receptors for IL-3, IL-4, IL-5, IL-6, IL-7, IL-9, IL-12, IL-13, IL-15, and IL-21 have been classified under the generic name of hematopoietin receptor (HR). Alternatively, they are known as type I cytokine receptors. Two

**Tab. 4** Major cell sources and inducers of interleukins.

| *Interleukin* | *Cell source* | *Inducers* |
|---|---|---|
| IL-1α/ß and IL-1ra | Monocytes/macrophages; endothelial cells; fibroblasts; most cell types | Bacterial lipopolysaccharide (LPS); phorbol esters; calcium ionophore; muramyldipeptide; IL-1α/ß; TNFα |
| IL-2 | T cells | Mitogens, e.g. plant lectins and bacterial enterotoxins; antigens |
| IL-3 | T cells; mast cells; thymic epithelial cells; keratinocytes | Mitogens and antigens (T cells) antibody cross-linking of FcR for IgE (mast cells) |
| IL-4 | T cells; mast cells; basophils; B cells; bone marrow stromal cells | Mitogens and antigens; antibody cross-linking of FcR for IgE |
| IL-5 | T cells; mast cells | Mitogens and antigens; antibody cross-linking of FcR for IgE |
| IL-6 | T cells; macrophages; fibroblasts, hepatocytes; endothelial cells | LPS; phorbol esters; IL-1α/ß; TNFα/ß; calcium ionophore |
| IL-7 | Bone marrow stromal cells; fetal liver cells | Cytokines; TGFß, PDGF and IL-1α/ß |
| IL-8 | Wide variety of cell types, including macrophages; T cells; endothelial cells | LPS; phorbol esters; IL-1α/ß; TNFα/ß; calcium ionophore |
| IL-9 | T cells | Mitogens and antigens |
| IL-10 | T cells; macrophages; B cells; keratinocytes | Mitogens and antigens; LPS; EBV (transformed B cells) |
| IL-11 | Stromal fibroblasts; trophoblasts | Cytokines; TGFß, PDGF and IL-1α/ß |
| IL-12 | Macrophages; dendritic cells | LPS; mitogens; phorbol esters |
| IL-13 | T cells | Mitogens and antigens |
| IL-14 | T cells | Mitogens and antigens |
| IL-15 | Macrophages;epithelial cells | Type I interferons |
| IL-16 | T cells; monocytes; eosinophils | ? |
| IL-17 | T cells | Mitogens and antigens |
| IL-18 | Macrophages | LPS |
| IL-19 | PBMC; T- and B-cell lines | LPS; GM-CSF |
| IL-20 | Keratinocytes; skin cells | ? |
| IL-21 | T cells | Mitogens and antigens |
| IL-22 | T cells | LPS; IL-9 |
| IL-23 | Macrophages; dendritic cells | LPS |
| IL-24 | Th2 cells; melanoma cell lines | IFNβ and phorbol esters |
| IL-25 | Th2 cells; mast cells | Mitogens; antigens; IgE |
| IL-26 | T cells | Herpesvirus transformation |
| IL-27 | Macrophages; dendritic cells | LPS |
| IL-28 | Dendritic cells | Virus infection; poly (I).(C) |
| IL-29 | Dendritic cells | Virus infection; poly (I).(C) |

components of the high-affinity IL-2 receptor also belong to the HR family as do the receptors for G-CSF, GM-CSF, EPO, GH, LIF, CNTF, and OSM. In addition, the two nonspecific receptor chains that are required for IL-3, IL-5, and GM-CSF receptor function and for IL-6, IL-11, LIF, CNTF, and OSM receptor function respectively, are HR family members.

The basic structural characteristics of HR are two extracellular domains of approximately 100 amino acids, the one closest to the cell membrane containing a WSXWS motif and the outer N-terminal domain containing a number of conserved cysteine residues (Fig. 3). As new members of the HR family were identified, it became clear that either (1) these two domains could be duplicated or (2) unrelated domains that were immunoglobulin-like or fibronectin-like could be added to extend the extracellular portion of the receptor. In the case of the GH receptor, which has the simplest two domain structure (Fig. 3), X-crystallography has recently shown that each of the 100 amino acid stretches consists of seven ß-strands that are folded to form a sandwich of two antiparallel sheets. However, only the N-terminal domain makes contact with the GH molecule. In fact, the GH molecule is able to bind to two GH receptors, thus effecting their dimerization, a process that is likely to be required for signal transduction via the C-terminal intracellular domains.

The G-CSF receptor probably functions in a similar manner. However, many of the interleukin receptors are more complex in that they are composed of more than one component (subunit). The IL-2 receptor, for example, has three subunits, the first (IL-2R$\alpha$) being a glycoprotein unrelated to the HR family and the second (IL-2Rß) and third (IL-2R$\gamma$) being HR family members. All three subunits can bind IL-2 with low affinity, the complexes of $\alpha$- and ß- and ß- and $\gamma$-subunits bind IL-2 with intermediate affinity, but only the complex of $\alpha$-, ß-, and $\gamma$-subunits can form a high-affinity receptor for IL-2, leading to signal transduction and internalization of the ligand–receptor complex (IL-2R$\gamma$ has been shown to be a subunit of IL-4, IL-7, IL-13, IL-15, and IL-21 receptors).

In the case of IL-3, IL-5, and GM-CSF, a different receptor system has evolved whereby ligand binding to a ligand-specific receptor subunit ($\alpha$-chain) induces the association of this complex with a nonspecific receptor subunit (ß-chain) that enables signal transduction to take place. Both $\alpha$- and ß- chains are members of the HR family (Fig. 3). The $\alpha$-chains have only small cytoplasmic domains, insufficient for signaling, and therefore the presence of the ß-chain, which does have a large cytoplasmic domain, is required for signal transduction. This receptor system explains why IL-3, IL-5, and GM-CSF share many biological activities since where cells express all three ligand-specific $\alpha$-chains, signaling can only take place by interaction with the single common, nonspecific, ß-chain yielding the same cellular response.

A similar receptor system exists for IL-6 and IL-11 and the cytokines LIF, CNTF, and OSM. Here the nonspecific ß-chain has been characterized as a 130-kDa transmembrane glycoprotein (gp130) that resembles the G-CSF receptor in overall structure (Fig. 3). Only in the presence of gp130, which itself cannot bind IL-6, can a high-affinity receptor complex be formed by interaction of the IL-6 binding $\alpha$-chain and signal transduction ensues. gp130 is also the ß-chain for IL-11, LIF, CNTF and OSM (gp130 can weakly bind OSM). The IL-6R$\alpha$ chain may be cleaved from the cell membrane to form a soluble receptor for IL-6 (sIL-6R$\alpha$), and this can also interact with IL-6 and gp130 to trigger signal transduction. Interestingly, IL-12 has two components, a p40 subunit that is homologous to sIL-6R$\alpha$ and a p35 subunit that resembles a typical helical cytokine. In effect, it is an interleukin that carries its own specific $\alpha$-chain around

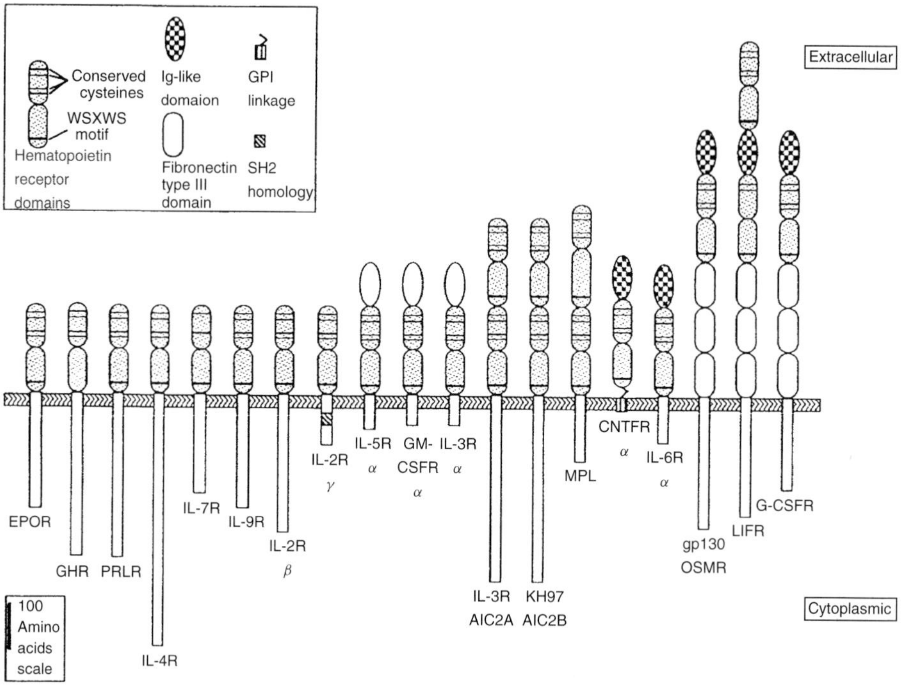


**Fig. 3** Schematic receptor structures of the hematopoietin superfamily. Modular elements are as shown in the box. (Reprinted with permission of Cosman (1993) *Cytokine* **5**, 95 Copyright Academic Press Ltd.)

with it. The high-affinity receptor for IL-12 is composed of two related ß-type receptor chains, IL-12Rß1 and IL-12Rß2, each independently having low affinity for IL-12. IL-12 p40 interacts primarily with IL-12Rß1, while p35 interacts with IL-12Rß2. IL-23, which contains the p40 subunit in common with IL-12, is able to bind only to IL-12Rß1, and probably a second, as-yet unidentified, receptor subunit is required to form the high-affinity IL-23 receptor. The receptors for IL-10 and related interleukins (IL-19, IL-20, IL-22, IL-24, and IL-26) also have two subunits, a long chain with a large intracellular domain and a smaller accessory chain, which have been classified on a structural basis as type II cytokine receptors. The latter, of which 12 members (cytokine receptor family 2, CRF2-1 to CRF2-12) are now characterized, include receptors for interferons.

Unlike the two or three component receptors of most interleukins, there are two distinct single-chain receptors for IL-1 molecules. The first of these to be characterized, now known as the type I IL-1 receptor (IL-1RI), has widespread tissue distribution and is found on T cells, fibroblasts, keratinocytes, endothelial cells, synovial lining cells, chondrocytes, and hepatocytes. The second receptor, the type II IL-1 receptor (IL-1RII), is found on B cells, neutrophils, and bone marrow cells. The two receptors are about 40% related in their N-terminal extracellular domains, contain three immunoglobulin (Ig) loop structures and are therefore included in the Ig superfamily rather than the hematopoietin superfamily. They have more recently been classified with Toll-like receptors, a family of 10 proteins involved in regulating innate immunity. The larger 80-kDa IL-1RI has an intracellular domain (217 amino acids) of sufficient size to effect signal transduction, whereas the smaller 68-kDa IL-1RII has only a short intracellular C-terminal tail and is probably not able to act as a signal transducer on its own. There is a large body of evidence demonstrating that IL-1RI is functional and is required to mediate IL-1 activities, but the biological significance of the "incomplete" IL-1RII is poorly understood. Recent evidence suggests IL-1RII acts as an inactive "decoy" receptor to bind excess IL-1ß. Expression of IL-1RI is inducible or highly regulated and this probably to some extent controls the responsiveness to IL-1$\alpha$ and ß, although it is known that low (5%) receptor occupancy is sufficient to trigger intracellular events. Both IL-1$\alpha$ and ß bind with similar affinity to IL-1RI and their binding is competitively inhibited by IL-1ra. Shortly after binding, the receptor-ligand complex is internalized and may be translocated to the nucleus.

The receptors for IL-8 and related molecules belong to a separate receptor family, that of the so-called G-protein-coupled receptor superfamily. These receptors contain seven membrane-spanning helices that couple to guanine nucleotide binding proteins (G-proteins). For IL-8, there are two distinct receptors, type I and type II (or type A and B), which are about 74% related in amino acids. It is probable that IL-8 and other members of the c-x-c subgroup of chemokines utilize these two receptors, whereas the c-c subgroup (includes MCP-1, RANTES, MIP-1$\alpha$/ß) have recently been shown to bind to another G protein–coupled receptor, which has approximately 33% homology to the IL-8 receptors.

## 3.2
## Intracellular Signaling Pathways

None of the interleukin receptors has an integral protein kinase. However, it is now known that they are associated through their intracellular domains, either directly or indirectly via adaptor proteins, with "nonreceptor" protein kinases to enable signal transmission to occur. For example, the intracellular domain of IL-1RI interacts with an adaptor molecule known as MyD88, which in turn couples to an "interleukin-1 receptor associated kinase" (IRAK). This serine kinase phoshorylates a signal transducer belonging to the tumor necrosis factor receptor associated factor (TRAF) family and this leads to the activation of nuclear transcription factors, such as AP-1 (jun/fos) and NF$\kappa$B; these in turn activate IL-1 responsive genes. It is known that IL-1 can potentially induce the expression of a whole catalog of genes ranging from those of other interleukins and cytokines, for example, IL-2 to IL-8, GM-CSF, TNF$\alpha$ and its own family genes, that is, IL-1$\alpha$/ß and IL-1ra to a wide variety of enzymes (for example, manganous superoxide dismutase, cyclooxygenase, tissue plasminogen activator, collagenase), oncogenes (*c-fos, c-jun, c-myc*), cell adhesion molecules (intercellular adhesion molecule-1 (ICAM-1), vascular cell adhesion molecule-1 (VCAM-1)) and the 25-kDa IL-2 receptor chain, IL-2R$\alpha$ (tac antigen). Many of these proteins are only specifically induced in certain cell types. However, mechanisms underlying such differential expression of IL-1 inducible genes remain not well understood.

It appears probable that individual interleukin receptors can be coupled to one or more intracellular signaling pathways, usually depending on the cell type. For example, IL-1α/ß through binding to IL-1R1 can evoke PKC activity in T cells and mouse NIH 3T3 fibroblasts, whereas in human foreskin fibroblasts, the cAMP-dependent protein kinase (PKA) is activated. In Th2 cells, both PKA and PKC are evoked by IL-1. It has been reported that IL-1 triggers the formation of several second messengers, including arachidonic acid metabolites via phospholipase $A_2$, prostaglandins via the cyclooxygenase pathway, and ceramide via sphingomyelin. Ceramide may be responsible for activating a serine kinase, distinct from PKC, which is involved in phosphorylating the epidermal growth factor (EGF) receptor and causing a lowering of the affinity of this receptor for EGF.

As mentioned above, one of the proteins induced by IL-1 is the IL-2Rα and this, when expressed at the cell surface in combination with the two other IL-2 receptor chains, IL-2Rß and IL-2Rγ, forms the high-affinity receptor for IL-2. In contrast to IL-1RI, high-affinity IL-2R appears not to connect to signaling pathways involving phosphatidyl inositol hydrolysis, $Ca^{2+}$ mobilization, PKC or PKA. However, IL-2 stimulation is known to result in tyrosine phosphorylation of several cytoplasmic proteins. It is now known that the intracellular domains of IL-2Rß and IL-2Rγ respectively, bind via specific tyrosine residues, the Janus tyrosine kinases JAK1 and JAK3. The JAKs constitute a family of receptor-activated kinases, which phosphorylate the members of a distinct class of transcription factors known as *signal transducers and activators of transcription* (*STATs*). Once activated, STATs translocate to the nucleus to activate transcription of IL-2 responsive genes. Increased expression of several oncogenes, including *c-myc, c-myb, c-jun, c-fos*, and src-related protein tyrosine kinase (PTK), which are probably involved in cell proliferation, are found in IL-2 activated cells. Notably, other interleukin and cytokine genes are also activated, but their expression depends on the phenotype of the lymphoid cell. For example, the Th-lymphocyte subsets Th1 and Th2 each express a defined spectrum of interleukins and cytokines (Table 3).

The high-affinity receptors for IL-4, IL-7, IL-9, IL-15, and IL-21 all share the IL-2Rγ chain, and therefore bind JAK3 and function in a similar way to the IL-2R. For example, high-affinity IL-15R also requires IL-2Rß and thus intracellular signaling pathways are likely to be identical to those triggered by IL-2. However, IL-15Rα is different from IL-2Rα and more widely expressed, suggesting a broader activity profile for IL-15. Most other high-affinity interleukin receptors are composed of two, rather than three chains – one α- and one ß-chain. In the case of IL-3 and IL-5 receptors, which share a common ß-chain, the latter is associated with JAK2, which activates STAT5. Similarly, for IL-6 and IL-11 receptors, the ß-chain associates primarily with JAK1, although other JAK family kinases might also be involved in signal transduction. In contrast, for IL-10 receptors, and those for the related interleukins IL-19, IL-20, IL-22, IL-24, IL-26, IL-28, and IL-29, both α- and ß-chains are associated with JAKs, in particular with TYK2 and JAK1 and 2; a similar association of JAK2 and TYK2 with the two receptor chains of IL-12R and IL-23R is also found. It is evident therefore that due to the sharing of JAKS and STATs, the intracellular pathways from the majority of interleukin receptors will have much in common, and that as a consequence activities can be shared among several interleukins. However, since interleukins are in the main locally produced intercellular mediators, such activities are dependent on the cells present in the vicinity of their release expressing appropriate receptors, having functional intracellular signaling pathways and having transcriptionally activatable "interleukin-responsive" genes.

For example, IL-6 is an inducer of acute-phase proteins in hepatocytes and this has proved to be a useful system for studying the regulation of transcription of acute-phase proteins. A nuclear transcription factor, NF-IL-6, has been identified in IL-6 stimulated hepatocytes that binds to the promoter regions of acute-phase protein genes. The NF-IL-6 transcription factor is highly homologous to a liver-specific transcription factor C/EBP, a member of the so-called basic leucine-zipper family. These leucine-zipper proteins are known to bind to DNA as dimers through a leucine-zipper structure that is required for dimerization and that an adjacent basic region makes direct contact with DNA. Serine phosphorylation of NF-IL-6 is required to activate it to bind to the IL-6-response elements in acute-phase protein genes. Most of these genes share a common consensus sequence (CTGGGAA(T)), which also appears to be important in IL-6-dependent acute-phase protein gene activation. The acute-phase proteins induced by IL-6 include C-reactive protein (CRP), serum amyloid A, haptoglobulin, $\alpha$1-antichymotrypsin, fibrinogen. These are largely restricted to hepatocytes, but as IL-6 is active in many different cell types, it is expected that IL-6 stimulation will lead to the enhanced transcription of a number of other genes, for example, *c-jun*, *c-fos*.

Members of IL-8 family bind to G-protein-coupled receptors. Binding of IL-8 to these receptors present on neutrophils triggers a large rise in intracellular $Ca^{2+}$ and activation of PKC, suggesting that these changes are involved in signal transduction leading subsequently to neutrophil activation and the chemoattractant response. G-proteins, or heterotrimeric GTP-binding regulatory proteins, have been clearly implicated in signal transduction; pertussis toxin (*Bordetella pertussis* islet-activating protein), which blocks the activation of certain G-proteins, inhibits the IL-8 stimulated chemoattractant response. The activated G-proteins have been found to activate phosphatidylinositol phospholipase C and $Ca^{2+}$ mobilization, subsequent PKC activation, secretion of granular enzymes, and activation of respiratory burst, and the generation of superoxide anion.

## 4 Biological Activities of Interleukins

There are now 29 recognized interleukins, three colony stimulating factors (G-CSF, GM-CSF, and M-CSF), and many more cytokines, for example, TNF$\alpha$, LIF, OSM, CNTF, transforming growth factor ß (TGF-$\beta$), stem cell or steel factor and interferons (IFN), to cite a noninclusive list, that can affect the proliferation, differentiation, and function of cells. With the number of these biologically active mediators now in the hundreds, the number of possible combinations is astronomic. *In vivo*, it is probable that interleukins and other cytokines form the basis of a complex interactive communication network with the overall cellular responses being dependent on integrated assimilation of multiple signals. *In vitro*, it has been rarely possible to mimic the *in vivo* situation and experimentally a highly reductionist approach using isolated molecules, isolated cells, and cell lines has formed the basis for studying the biological activities of individual interleukins. Such work has provided an important base, but is unsatisfactory in the long run as the true roles of interleukins cannot be accurately predicted. Only studies in whole animals and intact tissues will provide answers on the physiological importance of interleukins.

This section will first concisely review the biological activities of interleukins, starting with IL-1, that have been defined from *in vitro* experimentation. However, the vast amount of literature information precludes a totally comprehensive survey. Secondly, some examples of interleukin interactions will be briefly described and thirdly, the *in vivo* biological roles of interleukins, where information is available, will be outlined.

## 4.1 Interleukin-1

An unusually large number of biological activities have been attributed to IL-1. Broadly speaking, the activities of IL-1α and IL-1ß are qualitatively the same, and both IL-1α and ß are antagonized by IL-1ra. Currently, the activities of other members of the IL-1 superfamily, excluding IL-18, are not known. In general, IL-1α and ß behave as the primary activators of cells, readying them for secondary stimuli. As such, they have been included in a category of mediators called *competence factors*. These are considered to be elements of cell activation pathways, which commit cells to a series of stimulatable, contingent events. In this context, IL-1 has been shown to augment antigen activation of T lymphocytes and to potentiate the proliferation of hematopoietic progenitors. IL-1 has also been demonstrated to induce the synthesis of secondary acting mediators, the so-called *progression factors* that stimulate cells to undergo further proliferative or differentiating events. For example, IL-1 induces platelet-derived growth factor (PDGF), a potent mitogenic factor, in fibroblasts, GM-CSF in endothelial cells, IL-2 in T lymphocytes and IL-1, IL-6, IL-8, and TNFα in monocytes/macrophages. In addition, IL-1 can induce or regulate the expression of receptors and other cell surface molecules, for example, IL-2Rα in T lymphocytes, ICAM-1, and VCAM-1 in endothelial cells. IL-1, at least *in vitro*, can stimulate acute-phase protein synthesis in hepatocytes, although it is not as potent as IL-6 (NB possibly the effects of IL-1 on hepatocytes occur as the result of intermediate production of IL-6). At relatively high doses, IL-1 induces bone and cartilage resorption *in vitro* cell systems. In contrast, at low doses, IL-1 may promote osteoblast proliferation and transiently stimulate collagen synthesis. IL-1 induces prostaglandin ($PGE_2$) synthesis by synovial cells.

In hematopoiesis, that is, the generation of mature blood cells from bone marrow stem cells, IL-1 can act as activator of early progenitors, and in combination with other interleukins, for example, IL-3, IL-6, it probably stimulates proliferation and differentiation of the various cell lineages. It may synergize with IL-6 for IL-2 synthesis by activated T lymphocytes, with IL-4 for B-cell activation and Ig isotype regulation, and with IL-2 or IFN for augmenting natural killer (NK) cell activity.

*In vivo*, IL-1 is known to induce fever (IL-1 was originally called *endogenous pyrogen* (*EP*)), sleepiness, and anorexia. IL-1 also has been shown to affect the neuroendocrine system, principally by increasing pituitary adrenocorticotropic hormone (ACTH) and endorphin levels as well as glucocorticoids. Administration of IL-1 causes neutrophilia, probably due to IL-1-mediated induction of GM-CSF from endothelial cells. The latter are activated and sticky for lymphocytes and monocytes due to increased ICAM-1 expression. There are effects on liver function leading to increased amino acid turnover and hyperlipidemia. By itself, but particularly in combination with TNFα and/or bacterial LPS, high levels of IL-1 can lead to profound hypotension, myocardial suppression, shock, and death.

The biological activities of IL-1 and other interleukins are summarized in Table 5.

## 4.2 Interleukin-2

In contrast to IL-1, IL-2 is much more limited in its biological activities, due largely to the restricted expression of high-affinity IL-2 receptors to a relatively few cell types. Mitogen- or antigen-activated mature T lymphocytes express high-affinity IL-2R and subsequent interaction with IL-2 leads to clonal proliferation. Many T-cell lines and clones, specific for particular antigens, remain wholly or partially dependent on the presence of exogenous IL-2 for proliferation. In addition, antigen-independent murine cytotoxic T-lymphocyte line (CTLL) requires IL-2 for continuous growth. NK cells, contained within the large granular lymphocyte (LGL) population, express IL-2Rß and not only proliferate in response to IL-2 but also exhibit enhanced cytolytic activity. In the presence of high IL-2 concentrations, the so-called lymphokine-activated killer (LAK) cells emerge from resting populations of lymphoid cells. LAK cells, which are cytolytic for some tumor cells, can be induced *in vitro* and *in vivo*. IL-2 can also act on activated B lymphocytes to stimulate both their proliferation and the induction of Ig synthesis.

Besides its growth-promoting activity, IL-2 stimulates T cells to secrete a range of other interleukins and cytokines (Table 3). While the principal role of these interleukins and cytokines is to act as "helper factors" for the growth and differentiation of leucocytes other than T lymphocytes, they could also be involved in the differentiation of the latter. Thus, IL-2 may indirectly regulate T-lymphocyte differentiation, for example, in the maturation of CTL or LAK to express cytolytic activity.

*In vivo*, IL-2 induces lymphoid hyperplasia, for example, increases in mature T cells, neutrophilia, and eosinophilia; the latter are probably indirectly caused by IL-2 induced interleukin/cytokine synthesis by activated T cells. LAK cells appear rapidly following the infusion of IL-2. However, high doses of IL-2 induce undesirable side effects such as hypotension, oliguria, fluid retention, progressive dyspnoea (difficulty in breathing), atrial arrhythmias, thrombocytopenia, and vascular leak syndrome.

## 4.3 Interleukin-3

IL-3 has been called the pan-specific interleukin because of its highly pleiotropic activities. Like IL-2, IL-3 acts mainly as a specific growth factor, principally affecting the proliferation and differentiation of erythroid and myeloid lineages. It has been shown to be particularly effective in stimulating the proliferation of early erythroid/myeloid progenitors, for example, in the formation of granulocyte-erythroid-macrophage-megakaryocyte colony-forming units

**Tab. 5** The biological activities of interleukins.

| *Interleukin* | *Principal activities* |
|---|---|
| IL-1α/ß | – Activates mature T cells to produce IL-2 and express IL-2R.<br>– Costimulant of proliferation and activation of B cells and hematopoietic progenitor cells.<br>– Induces proinflammatory cytokines, e.g. TNFα, IL-6, and other mediators depending upon cell type.<br>– Increases expression of endothelial cell adhesion molecules.<br>– Induces neutrophil accumulation *in vivo*.<br>– Induces fever and hypotension. |
| IL-1ra | – Inhibits the activities of IL-1α/ß by competing for IL-1R. |
| IL-2 | – Stimulates T-cell proliferation and differentiation.<br>– Activates NK and LAK cells.<br>– Promotes proliferation of and Ig secretion in activated B cells.<br>– Causes vascular leak syndrome *in vivo* at high doses. |
| IL-3 | – Acts in combination with lineage-restricted cytokines to stimulate proliferation and differentiation of hematopoietic progenitors of macrophages, neutrophils, basophils, eosinophils, mast cells, megakaryocytes, and erythrocytes.<br>– Supports the proliferation of early multipotential bone marrow stem cells. |
| IL-4 | – B-cell growth and differentiating factor, induces IgE synthesis.<br>– Generates Th2 cell subset from naive Th0 cell population.<br>– Antagonist of IFNγ. |
| IL-5 | – Stimulates generation of eosinophils from hematopoietic precursors.<br>– Possible involvement in B-cell differentiation and Ig class switching. |
| IL-6 | – Growth factor for many transformed and tumor cells.<br>– Costimulant of proliferation in thymocytes and IL-3-dependent hematopoietic progenitors, and of IL-2 production in mature T cells.<br>– Terminal B-cell differentiation factor.<br>– Hepatocyte stimulating factor; inducer of acute-phase protein synthesis. |
| IL-7 | – Supports the proliferation of immature B- and T cells.<br>– Possible involvement in proliferation and activation of mature T cells. |
| IL-8 | – Chemoattractant for neutrophils and T cells.<br>– Stimulates neutrophil activation and degranulation. |
| IL-9 | – Augments mast-cell proliferation response to IL-2.<br>– Sustains antigen-independent growth of certain Th cells. |
| IL-10 | – Inhibitor of cytokine/interleukin synthesis by Th1 cells.<br>– Suppressant of macrophage functions, including down regulation of inflammatory cytokine production.<br>– Enhances B-cell proliferation and Ig synthesis.<br>– Growth factor for B-cell tumors. |
| IL-11 | – Synergistic factor for IL-3-dependent proliferation of bone marrow progenitors (similar to IL-6 in this respect), particularly megakaryocyte colonies.<br>– Can act on hepatocytes to induce acute-phase protein synthesis. |
| IL-12 | – Stimulates the differentiation of naive Th0 cells into the Th1 cell subset and thereby is an initiator of cell-mediated immunity.<br>– Induces IFNγ production with IL-18 as costimulant.<br>– Stimulates the proliferation and activity of NK cells and mature T cells. |
| IL-13 | – Induces B-cell proliferation and differentiation (similar to IL-4).<br>– Inhibits inflammatory cytokine synthesis by monocytes/macrophages. |

Tab. 5 (continued)

| Interleukin | Principal activities |
|---|---|
| IL-14 | – Inducer of proliferation of activated B cells (but not resting B cells).<br>– Inhibits Ig secretion by mitogen-stimulated B cells. |
| IL-15 | – Inducer of proliferation of T cells and T-cell lines.<br>– Activates NK and LAK cells. |
| IL-16 | – Lymphocyte chemoattractant factor; stimulates migratory response in $CD4^+$ lymphocytes, monocytes and eosinophils. |
| IL-17 | – Induces IL-6 and IL-8 production in fibroblasts; neutrophil recruitment.<br>– Enhances cell surface antigen and adhesion molecule expression. |
| IL-18 | – Induces T-cell IFN$\gamma$ production in combination with IL-12.<br>– Enhances T-cell cytotoxicity and inflammatory responses. |
| IL-19 | – Similar to IL-10, but poorly defined as yet. |
| IL-20 | – Involved in skin differentiation and keratin expression. |
| IL-21 | – Regulation of NK cell activation and differentiation. |
| IL-22 | – Activates synthesis of acute-phase proteins in liver cells. |
| IL-23 | – Similar to IL-12, but poorly defined as yet. |
| IL-24 | – Induction of apoptosis and inhibition of proliferation in tumor cells. |
| IL-25 | – Induces expression of Th2 type immunosuppressive interleukins. |
| IL-26 | – Not known. |
| IL-27 | – Similar to IL-12, but poorly defined as yet. |
| IL-28/29 | – Induce antiviral activity. |

(GEMM-CFU) and in the stimulation of erythroid burst-forming units (BFU-E) and megakaryocyte progenitors. There is some evidence that IL-3 also acts on very early multipotential progenitor cells. IL-3 may be considered to act as a progression factor following cell activation with competence factors, such as IL-1. IL-3 may also stimulate the proliferation/differentiation of more mature cells of the myeloid lineage, for example, by inducing macrophage precursors to express cytokine receptors, such as M-CSF-R. There is experimental data showing that IL-3 has a growth-supporting activity for murine mast cells, but it is not clear that this is so for human mast cells.

*In vivo*, release of IL-3 by injected WEHI-3B tumor cells results in the stimulation of all of the various types of hematopoietic cells predicted from the *in vitro* investigations, that is, increases in numbers of myeloid cells, erythroid cells, mast cells, and megakaryocytes. Administration of IL-3 to mice, primates, and humans yields broadly similar effects; the progenitors of mast cells, neutrophils, and macrophages are increased.

## 4.4 Interleukin-4

IL-4 was originally discovered by its action on B lymphocytes and was the first B-cell stimulating factor (BSF-1, Table 1) to be characterized. B lymphocytes, which express surface Ig, were found only to proliferate in response to anti-Ig if IL-4 (BSF-1) was present. It would appear that IL-4 is a B-lymphocyte competence factor, perhaps the counterpart of IL-1 for T lymphocytes. As such, IL-4 is involved in B-lymphocyte activation, rather than proliferation, resulting in, for example, the increased expression of cell surface molecules, such as major histocompatibility complex (MHC) class II antigens, low-affinity IgE receptor (CD23) and surface IgM. IL-4 also regulates which Ig isotype is synthesized and secreted by mature B cells; it appears to be directly responsible for switching on IgG1 and IgE synthesis *in vitro*. This action is opposed by IFN$\gamma$, which preferentially induces the synthesis of IgG2a.

Additionally, IL-4 can act on other cell lineages. In particular, it drives the differentiation of naive T lymphocytes into the Th2-lymphocyte subset, possibly by blocking the synthesis of its own natural antagonist, IFN$\gamma$. In some antigen-independent T-cell lines, IL-4 can act as a proliferation signal, similar to IL-2. In contrast, IL-4 has been reported to antagonize the stimulatory actions of IL-2, for example, in LAK cells. IL-4 has also been shown to affect hematopoiesis, especially the development of mast cells and eosinophils from progenitor cells. It has a variety of actions in monocytes/macrophages, for example, upregulation of low-affinity IgE receptor (CD23), enhancement of MHC class II antigens, inhibition of spontaneous and induced IL-1, IL-6, IL-8, and TNF$\alpha$ synthesis, but stimulation of G-CSF and M-CSF, and in granulocytes, fibroblasts, epithelial, and endothelial cells. IL-4 has many activities in common with IL-13 with which it shares receptor components (see Sect. 4.13).

*In vivo*, IL-4 release results in the generation of large numbers of eosinophils (through induction of IL-5 production), mast cells, and macrophages. In transgenic mice, expressing IL-4 in the thymus and T cells, thymic hypoplasia, eosinophilia, and eye inflammation (possibly due to activated macrophages) have been found.

## 4.5 Interleukin-5

It is known that the control of eosinophil numbers is T cell dependent. In steady state hematopoiesis, relatively few eosinophils are found in the circulation. In the presence of IL-3 and GM-CSF, only small numbers of eosinophil colonies develop from GEMM-CFU progenitor cells. However, infection by parasites, such as helminths, dramatically increases eosinophil numbers and the factor responsible for eosinophil proliferation and differentiation has been identified as T-cell-derived IL-5. The main action of IL-5 is to stimulate the proliferation and maturation of eosinophil precursors (Eo-CFU); IL-5 appears to be lineage specific and does not induce the proliferation of other myeloid lineage precursors. This could be linked to the restricted IL-5R$\alpha$ chain expression to eosinophils. IL-5 activity in eosinophils is probably enhanced by the actions of other interleukins, such as IL-3 and IL-4.

In addition to its growth-promoting and differentiating effects in immature eosinophils, IL-5 also acts on mature eosinophils, for example, increasing phagocytosis, killing of antibody-coated tumor cells, stimulation of superoxide ($O_2^-$) production. IL-5 has been demonstrated to induce the differentiation of murine- but not human-activated B lymphocytes. This activity is only apparent if the B lymphocytes have been activated by a priming stimulant such as LPS, anti-immunoglobulin, or specific antigen. IL-5 alone preferentially enhances IgA production, while specific combinations of IL-5, IL-2, and IL-4 appear to regulate the amount of IgG1 isotype synthesis.

*In vivo*, IL-5 causes eosinophilia, an expected result from its *in vitro* activities. In transgenic mice that constitutively express IL-5 in their T cells, massive eosinophilia with eosinophil infiltration of tissues was observed, although in all other respects such mice remained normal and healthy.

### 4.6 Interleukin-6

The biological activities that are associated with IL-6 are very wide ranging – it has been shown to be very pleiotropic, indicating that IL-6R are expressed on many cell types. Originally, it was described as a B-cell stimulating factor (BSF-2, Table 1) and it is clear that IL-6 is a growth and differentiation factor for a variety of B-lymphoid cells. It is a growth factor for many murine antibody-secreting hybridomas, for some human myelomas, and for low-density cultures of human B cells transformed with Epstein–Barr virus (EBV). IL-6 is also a late-acting differentiating factor in mature B lymphocytes; their terminal differentiation into Ig-secreting plasma cells appears to be upregulated by IL-6. For T cells, IL-6 probably acts as a costimulant (secondary signal) for proliferation of thymocytes and IL-2 production by mature T cells. In hematopoiesis, IL-6 acts as a costimulant for IL-3-dependent proliferation of multipotential cells and induces maturation of megakaryocytes leading to increased platelet number. It may also stimulate the proliferation of myeloid leukemic blast cells and keratinocytes.

A major activity of IL-6 is that of a "hepatocyte stimulating factor", principally as an inducer of acute-phase proteins. IL-6 induces a variety of acute-phase proteins, including fibrinogen, $\alpha$-1-antichymotrypsin, $\alpha$1 acid glycoprotein and haptoglobulin, from the hepatoma cell line, Hep2G, and primary hepatocytes (which also produce C-reactive protein (CRP), serum amyloid A and $\alpha$-1-antitrypsin)). *In vivo*, levels of IL-6 are positively correlated with levels of CRP. Serum levels of IL-6 also correlate well with fever. In transgenic mice that constitutively express IL-6 in their B cells, IgG1 plasmacytosis is developed with the infiltration of plasma cells into lung, spleen, and kidney. In contrast, infection of bone marrow cells with an IL-6 expressing recombinant retrovirus resulted in a fatal myeloproliferative disease with massive neutrophil infiltration of lungs, liver, and lymph nodes, but no plasmacytosis.

### 4.7 Interleukin-7

IL-7, which is produced by bone marrow stromal cells, appears to be principally involved in stimulating the growth and differentiation of cells of the lymphoid lineages. In particular, it has been shown to induce the proliferation of precursors of B lymphocytes, the so-called pro-B-cells, *in vitro*. In addition, IL-7 can stimulate the proliferation of thymocytes; it has been demonstrated to act synergistically with IL-1 in this respect. Probably, IL-7 occupies a pivotal role in T-cell development, but it also may be involved in differentiation of mature T lymphocytes. Here IL-7 may act rather like IL-6 as a costimulant of IL-2 production by T lymphocytes. Possibly IL-7 can also substitute IL-2 as a mitogen in the generation of CTL and LAK cells.

*In vivo*, it is likely that IL-7 activities mirror those found *in vitro*. For instance, in transgenic mice expressing IL-7 in lymphoid organs, increased numbers of both B and T cells were observed. In some transgenic mice, the continuous secretion of IL-7 has led to lymphoproliferative disorders, including B and T lymphomas.

### 4.8
### Interleukin-8

IL-8 is the name given to just one of a number of related chemokines that are induced by inflammatory stimuli, for example, LPS, IL-1, and that act as chemoattractants for particular cell types, neutrophils, in the case of IL-8. Thus, release of IL-8 from infiltrating or tissue macrophages specifically recruits neutrophils to sites of injury or infection. IL-8 also acts on neutrophils to induce respiratory burst responses and degranulation, resulting in the superoxide ion ($O_2^-$) production and the release of lysosomal enzymes. In addition, IL-8 has been demonstrated to be a chemotactic and activating agent for about 10% of T cells and some basophils *in vitro*. However, it is not chemotactic for blood monocytes, the role for this being assigned to another member of the IL-8 superfamily, namely, the monocyte chemoattractant protein-1 (MCP-1). (NB IL-8 is a member of the c-x-c intercrine $\alpha$-subgroup, while MCP-1 is a member of the c-c intercrine ß-subgroup).

IL-8 has been recovered from inflammatory sites such as psoriatic lesions, the synovial fluid of rheumatoid arthritis patients, and the bronchial lavage fluid taken from patients with acute respiratory distress syndrome. Injection of IL-8 leads to local neutrophil infiltration in both rabbits and humans. If repeatedly injected, IL-8 appears to cause neutrophil accumulations in joints and lungs, and may be associated with damage to cartilage and pulmonary inflammation.

### 4.9
### Interleukin-9

IL-9 is produced by Th1 cells and has been shown to sustain the antigen-independent growth of certain Th-cell clones and lines *in vitro*. Subsequently, mouse IL-9 has also been shown to enhance the growth of bone marrow–derived mast-cell lines in response to IL-3, and that of fetal thymocytes in response to IL-2. In addition, human and mouse IL-9 act to enhance erythroid burst-forming (BFU-E) activity and human IL-9 acts to stimulate proliferation of the human megakaryoblastic leukemic cell line, MO7E. Human IL-9 can synergize with stem cell factor (SCF) and/or EPO in the stimulation of growth of this megakaryoblastic cell line and megakaryocytic progenitors.

An IL-9-dependent cell line that was transfected with the IL-9 gene to give deregulated IL-9 expression became highly tumorigenic in mice. This suggests that uncontrolled expression of IL-9 can support T-cell proliferation *in vivo*, and may be a transforming event involved in the development of certain T-cell lymphomas.

### 4.10
### Interleukin-10

IL-10 is produced by activated Th2 lymphocytes, macrophages, and B lymphocytes and was originally characterized as a cytokine synthesis inhibitory factor (CSIF) because it inhibited IFN-$\gamma$ synthesis in T cells. It also appears that IL-10 actively suppresses the development of the Th1 subset of Th lymphocytes. This probably occurs as a result of the immunosuppressive activity of IL-10 on macrophages leading to the inhibition of IL-12 production. IL-12, produced by activated macrophages and dendritic cells, stimulates Th1 lymphocyte proliferation. IL-10, in possible synergy with IL-4 and transforming growth factor ß, also blocks the production of IL-1 and TNF$\alpha$ by macrophages and the development of macrophage cytotoxicity and thus can be considered as an anti-inflammatory agent. In contrast, IL-10 enhances B-cell proliferation and Ig synthesis.

Th1 cells are generally host-protective for several infectious diseases. Therefore, the suppression of Th1 cell activities will favor the growth of the infecting agent. In this context, it is interesting that EBV and several other large DNA viruses encode (viral) IL-10 molecules with limited homology to IL-10, but which act through IL-10R and thus have the same activities as IL-10. These viral IL-10 homologs appear to induce immunosuppression, thus providing an immune-evasion mechanism for such viruses. In contrast, the paralogous IL-10-related proteins, IL-19, -20, -22, -24 -26, -28, and -29 with similar limited homology to IL-10, act through distinct heterodimeric receptors composed of receptor chains belonging to the type II cytokine receptor family, which also includes receptors for interferons.

## 4.11 **Interleukin-11**

IL-11, produced by bone marrow–derived stromal cells, appears to be intimately involved in hematopoiesis and, in particular, the generation of megakaryocytes. *In vitro* studies suggest that IL-11 is capable of directly supporting the proliferation of committed murine myeloid progenitors and, like IL-6 and G-CSF, acts synergistically with IL-3 to shorten the Go phase of the cell cycle in early progenitors. Although IL-11 has no inherent megakaryocytic colony stimulating activity, it can synergize with IL-3 in stimulating human and murine megakaryocyte colony formation.

In many respects, the activities of IL-11 are very similar to those of IL-6. For example, like IL-6, IL-11 has been found to promote an increase in the number of Ig-secreting B cells. It has also been shown to stimulate the proliferation of an IL-6-dependent murine plasmacytoma cell line, and to act as an autocrine growth factor for human megakaryoblastic cell lines. Lastly, IL-11 appears to stimulate the synthesis of hepatic acute-phase proteins, but is less effective than IL-6.

## 4.12 **Interleukin-12**

The actions of IL-12 are primarily on T cells and NK cells. It appears to be a necessary factor in the generation of Th1 lymphocytes from naive or uncommitted T cells. In addition, there are results that indicate that IL-12 stimulates IFN$\gamma$ production by T cells and NK cells by cooperation with either IL-1, or IL-18, or TNF$\alpha$. The presence of IFN$\gamma$ may also favor the development of the Th1 subset by inhibiting the production of IL-10 by macrophages. IL-12 appears to be an important factor for the differentiation and maturation of dendritic cells.

The emerging experimental data for IL-12 therefore strongly suggest that it plays an important part in the initiation of immune responses by providing a link between natural resistance mediated by phagocytic cells, for example, macrophages, and NK cells and adaptive immunity mediated by Th cells, CTL, and B cells. In support of this proposed role, recombinant IL-12 has recently been shown to cure mice infected with the parasite, Leishmania major. A neutralizing antibody against IFN$\gamma$ abrogated the curative effect of IL-12 indicating that IL-12 was acting by stimulating T-cell and/or NK-cell IFN$\gamma$-production *in vivo*, and the development of a protective Th1-cell immune response. Conversely, generation of Th2 cells and associated immune responses was inhibited.

4.13
## Interleukin-13

IL-13 bears approximately 30% homology with IL-4 and appears to act very similarly to IL-4. In particular, IL-13, like IL-4, induces IgE and IgG4 synthesis by human B cells. In addition, IL-13 inhibits inflammatory cytokine production by macrophages and monocytes, another activity characteristic of IL-4. Recent studies indicate that IL-13 and IL-4 share receptor components, that is, the IL-4R$\alpha$ and IL-13R$\alpha$1 chains, which together form high-affinity receptors for either IL-4 or -13, thus offering an explanation of their similar activities. Emerging experimental evidence suggests a more complex picture with receptors for IL-13 being constituted from among four components, including IL-13R$\alpha$1, IL-13R$\alpha$2, IL-4R$\alpha$, and IL-2R$\gamma$.

4.14
## Interleukin-14

IL-14 is a high molecular weight B-cell growth factor. However, it appears to act as a costimulant, that is, it induces the proliferation of activated B cells, but not resting B cells. It has been speculated that human IL-14 fulfills a similar role to murine IL-5 for B cells. IL-14 has also been demonstrated to inhibit Ig synthesis by mitogen-activated B cells.

4.15
## Interleukin-15

IL-15 is a 14 to 15-kDa glycoprotein with shared bioactivities, but no sequence homology, with IL-2. Its biological activities are remarkably like those of IL-2 and include the induction of proliferation of the established T-cell line CTLL.2 and mitogen-stimulated peripheral blood mononuclear cells (PBMC) as well as the generation of cytolytic NK and LAK cells. The overlap of IL-15 activities with those of IL-2 is probably largely accounted for by their sharing both the IL-2Rß and IL-2R$\gamma$ chains as components of their respective high-affinity receptors.

4.16
## Interleukin-16

IL-16 (or lymphocyte chemoattractant factor) is an atypical interleukin, possibly produced as a degradation product of a larger protein of 42 kDa present in lymphocytes, which probably utilizes CD4 receptors. Nevertheless, IL-16 has been shown to stimulate a migratory response in CD4$^+$ lymphocytes, monocytes, and neutrophils, upregulate IL-2 receptors and class II MHC antigens in resting T lymphocytes and exhibit some anti-HIV activity.

4.17
## Interleukin-17

IL-17 is the prototypic member of a family of structurally related IL-17 proteins (6 human members A–E) produced by T lymphocytes. IL-17 appears to act like IL-1 by inducing IL-6, IL-8, and G-CSF synthesis and enhancing expression of the cellular adhesion molecule ICAM-1 in human fibroblasts. Thus, IL-17 is representative of those cytokines involved in initiating inflammatory responses and innate host defense mechanisms, for example, the pulmonary recruitment of neutrophils and microbial host defense in response to the lung pathogen *Klebsiella pneumoniae*. It may also promote the development of hematopoietic precursors.

### 4.18 Interleukin-18

IL-18, formerly known as *interferon gamma inducing factor* (*IGIF*) and now known to be structurally related to IL-1, appears to be mainly involved in the augmentation of Th1 cellular immune responses to invading pathogens. Its capacity to induce IFN$\gamma$ is reliant on the costimulation of T lymphocytes by IL-12, whose synthesis is induced by bacterial LPS in monocytes/macrophages. It has also been shown to enhance Fas ligand-mediated cytotoxicity of cloned murine Th1 cells, but not Th0 or Th2 cells. Its activities are mediated by IL-1 receptor-related protein (IL-1Rrp), an Ig superfamily member related to IL-1RI, and can be inhibited by a naturally occurring, specific IL-18 binding protein.

### 4.19 Interleukin-19

IL-19, one of several IL-10 paralogs, is produced mainly by LPS-stimulated monocytes and macrophages and in common with IL-10 is expected to exert immunoregulatory functions that influence the activities of many of the cell types in the immune system. It appears probable that IL-19 will, like IL-10, act as a feedback inhibitor of proinflammatory cytokine, for example, IL-1, TNF$\alpha$ synthesis and thus initiate immunosuppressive effects.

### 4.20 Interleukin-20

IL-20, one of several IL-10 paralogs, is in contrast to IL-10 and IL-19 produced mainly by skin cells, and is thus expected to target distinctive activities on typical skin cells such as keratinocytes, which express its specific heterodimeric receptor, IL-20R1/IL-20R2. Overexpression of IL-20 has been shown to result in severe skin abnormalities, including hyperkeratosis, hyperproliferation, and aberrant epidermal differentiation.

### 4.21 Interleukin-21

IL-21, which is most closely related to IL-15, is one of the several structurally related cytokines, including IL-2, which is secreted by activated CD4$^+$ T cells as part of the normal response to foreign antigens. Its main role appears to be as a differentiation factor for NK cells. It has been shown to promote IFN$\gamma$ synthesis, which acts to enhance macrophage killing of microbial pathogens, and to prevent apoptosis of precursor NK cells, allowing them to become terminally differentiated and fully activated.

### 4.22 Interleukin-22

IL-22, one of several IL-10 paralogs, formerly known as IL-10-related T-cell-derived inducible factor (IL-TIF) is primarily produced by activated CD4$^+$ T cells, but unlike immunoregulatory IL-10, its target cell types include mesangial, neuronal, and liver cells (hepatocyctes). Its action on hepatocytes (and hepatoma cell lines) appears similar to that of IL-6 in that it results in the upregulation of a number of acute-phase proteins. It should therefore probably be included with other proinflammatory cytokines as a prime mediator of the acute-phase response.

4.23
**Interleukin-23**

IL-23, produced mainly by macrophages and dendritic cells, is a novel heterodimeric cytokine constituted by the p40 subunit of IL-12 and a p19 subunit that is most closely related to the p35 subunit of IL-12. It is evident that IL-23, besides sharing one structural subunit with IL-12, also shares IL-12R$\beta$1 of the high-affinity IL-12R as part of IL-23R. However, a second specific IL-23R component is indicated for the recognition of p19 suggesting IL-23 has distinct cell targets from IL-12. For instance, although IL-23 in common with IL-12 stimulates IFN$\gamma$ production, IL-23 is a strong inducer of memory T-cell proliferation, while IL-12 is not. In addition, IL-23, in contrast to IL-1, for example, appears to act more broadly as an end-stage effector cytokine via direct actions on macrophages.

4.24
**Interleukin-24**

IL-24, one of several IL-10 paralogs, formerly known as melanoma differentiation-associated antigen 7 (MDA7), was originally found to be expressed in differentiated melanoma cells, but is now also known to be expressed by a variety of cell types, including T lymphoblasts. IL-24 appears to act through IL-20R and thus is expected to have similar cell targets to IL-20. It has been demonstrated to have antiproliferative activity against cultured tumor cell lines, including those derived from melanomas.

4.25
**Interleukin-25**

IL-25, also known as SF20, is a novel cytokine belonging to the IL-17 family, which is mainly produced by Th2 cells. It has been shown to induce expression of the immunosuppressive interleukins, IL-4, IL-5 and IL-13, required for Th2-like responses and may thus be an important mediator of allergic disease.

4.26
**Interleukin-26**

IL-26, one of several IL-10 paralogs, also previously known as AK155, is expressed at high levels in activated T cells, monocytes, and various T-derived leukemic cell lines. However, little is yet known about its cell targets and biological activities.

4.27
**Interleukin-27**

IL-27 is a newly identified "IL-12-like" heterodimeric cytokine, also similar to IL-23. It is constituted by Epstein–Barr virus-induced gene 3 (EBI3) protein, a protein that is widely expressed and related to IL-12 p40 subunit and a further homolog of IL-12 p35 known as p28. IL-27 is the product of activated antigen-presenting cells (macrophages and dendritic cells) and stimulates the rapid clonal expansion of naïve but not memory $CD4^{+}$ T cells. It also strongly synergizes with IL-12 to trigger IFN$\gamma$ production by naïve $CD4^{+}$ T cells.

4.28
**Interleukin-28**

IL-28 consists of two near identical proteins, IL-28A and B, which are distantly related to both the IL-10 and IFN$\alpha$ families, which are induced by viral infection. They interact with a heterodimeric class II cytokine receptor that consists of IL-10R$\beta$ chain and an orphan class II receptor chain designated IL-28R$\alpha$. IL-28 induces interferon-stimulated response elements (ISRE), but not via IFN receptors, and mediates moderate antiviral activity in response to virus infection. This has led to IL-28A and B being also named as IFNs, IFN-lambda2, and -lambda3, respectively. In common with IFNs, IL-28 upregulates several known IFN-responsive genes, including MxA, 2-5A synthetase, and class I MHC antigen.

### 4.29 Interleukin-29

IL-29 shows 81% sequence homology to IL-28 and acts in a very similar way. It also has been shown to induce IFN-responsive genes and exert antiviral activity, and thus is alternatively designated IFN-lambda1.

## 5 Interleukin Physiology

Interleukins are mainly, but not exclusively, inducible mediators of nonspecific host resistance and adaptive immunity, that is, they are stimulators of host-protective mechanisms against infectious and invasive diseases. The biological effects of most interleukins are readily observable *in vivo*. However, the administration of high doses of a single interleukin by abnormal routes may also create imbalances that perturb homeostatic mechanisms resulting in aberrant biological and physiological responses. Therefore, studies in which deficiencies of interleukin production or action can be investigated can provide perhaps a better guide to the physiological role(s) of individual interleukins. Such studies can either be done by blocking interleukin action, for example, with anti-interleukin antibodies, antireceptor antibodies, or soluble receptors, or by examining the effects of genetic deficiencies in interleukin protein or receptor. The former are probably the less satisfactory of the two approaches since interleukin production needs to be stimulated before an effect of an inhibitor of interleukin action can be demonstrated. Generally speaking, it is not possible to induce the production of a single interleukin; several interleukins and cytokines are normally produced in response to a stimulant. In addition, most antibodies used will be foreign (xenogenic) to the responding animal, and this may result in problems of an immunological nature.

There have been very few natural genetic deficiencies in interleukin genes and interleukin-receptor genes identified so far. However, it is now possible to artificially create knockout mutations in mice by using the strategy of gene targeting in embryonic stem (ES) cells. Increasing numbers of different knockout mice are being created, and it is perhaps surprising that some of them apparently develop normally, are born alive, and can grow and survive for weeks, if not months. Possibly, compensatory molecules or mechanisms are induced in these knockout mice. In the IL-2 knockout or null mouse, for example, development is normal with no effects being observed on the early development of thymocytes and inactivation of T lymphocytes within the thymus. However, a lack of response of mature T cells to polyclonal activators together with a lack of T cell–mediated "help", that is, interleukin production, for B-cell growth and differentiation were manifest in these mice. There were also excessively high levels of IgG1 and IgE found. About half of these mice became severely immunocompromised and died between four and nine weeks from birth. The remainder developed an inflammatory bowel disease, probably resulting from an abnormal immune response to a normal antigenic stimulus. In man, a child with a defect in IL-2 mRNA production (but not due to a defect in the IL-2 gene) was found to be more susceptible to infections. In addition, a few patients suffering from a particular form of severe combined immunodeficiency disease (SCID), which has been linked to the X chromosome (thus XSCID), are characterized by a lack of expression of IL-2R$\gamma$. XSCID patients are therefore unable to form high-affinity IL-2 (or IL-4, -7, -9, -15, -21) receptors and their immunocytes cannot respond to IL-2 or the other interleukins that require IL-2R$\gamma$. In total, these studies provide strong evidence for IL-2 playing a key role in the regulation of immune and inflammatory responses. In contrast, either IL-15 or IL-15R$\alpha$ knockout mice are lymphopenic with selective loss particularly of NK and $CD8^+$ T-cell subsets, rendering these mice susceptible to vaccinia virus infection. Thus, this phenotype is distinctive from that of IL-2 knockouts and indicates a role different to IL-2 for IL-15, despite the latter sharing two of the high-affinity IL-2R components (see Sect. 4.15).

Other knockout mice that are interesting in this respect are those of the IL-4 and IL-10 null phenotypes. In IL-4 knockout mice, it has been found that they generally are unable to mount a Th2 cell type of immune response. For instance, IgE was undetectable in these mice. In addition, when these mice were parasitized with the nematode *Nippostrongylus brasiliensis*, only a weak eosinophilia was observed. Since eosinophil proliferation is dependent on IL-5 (see Sect. 4.5), this finding indicates that IL-4 is required for inducing Th2 cell interleukin production, that is, IL-4, IL-5, IL-10, and so on. In the case of IL-10 knockout mice, it might be expected that these animals are also less able to mount a Th2 cell immune response, since loss of IL-10, which actively suppresses macrophage "help" (IL-12) for Th1 cell generation, would relieve the inhibition on Th1 cell development. However, the main observable effect in IL-10 knockout mice is chronic bowel inflammation where mice are not kept under specific pathogen-free conditions. This finding indicates that IL-10 acts as a regulator of immune responses stimulated by enteric antigens in the intestine. In the absence of IL-10, such immune responses are uncontrolled leading to bowel inflammation, probably due to a loss of IL-10-mediated suppression of macrophage cytotoxicity and release of inflammatory cytokines, for example, IL-1, TNF$\alpha$. In support of this hypothesis, it has been shown that normal mice become more sensitive to LPS-induced endotoxic shock by treatment with anti-IL-10 antibodies, while lethal endotoxemia and elevated serum TNF$\alpha$ levels were suppressed by injected IL-10.

Another interesting cohort of interleukin knockouts that is currently emerging is that with knockouts of IL-12 or IL-23. Knockouts of IL-12 p40 (which lack both IL-12 and IL-23) are more immunocompromised than knockouts of IL-12 p35 (which lack only IL-12). However, knockouts of IL-23 p19 (which lack only IL-23) manifest a phenotype that can be distinguished from the p35 knockout. The p19 null mice can still generate Th1 cells and IFN$\gamma$, but are susceptible to experimental autoimmune encephalitis (EAE), whereas p35 null mice cannot generate Th1 cells, but are highly susceptible to EAE. Such studies indicate the importance of the interleukin recognition components of their cognate receptors in determining their functional roles.

An alternative method for determining *in vivo* roles of interleukins involves neutralizing their biological activity with specific antibodies, soluble receptors, and other antagonists, that is, essentially preventing the interaction of the interleukin with its cognate receptors, also generates interesting results. To date, however, the abrogation of interleukin activities has mostly been carried out under conditions in which pathological conditions have been induced, for example, microbial challenge, tissue injury, graft-versus-host disease, or autoimmune disease. From these studies, for example, for IL-1, it has been shown that neutralizing antibodies suppress cell-mediated immunity and increase susceptibility to pathogens such as *Listeria monocytogenes*. In this case, the IL-1ra and soluble IL-1R can also produce similar effects. Such work indicates a protective role for IL-1. Interestingly, supportive evidence for this has been revealed by a series of recent findings that members of certain virus families, particularly the poxvirus family, encode soluble homologs of interleukins or their receptors, suggesting that some viruses have evolved means of countering host-defense mechanisms that would otherwise inhibit their replication. For example, vaccinia and cowpox viruses encode a soluble form of IL-1RII that can compete for IL-1 binding to cell surface receptors and thus reduce IL-1 actions. Besides this, the IL-1RII homolog, cowpox virus has also been found to encode for an inhibitor, a serpin, of the IL-1ß converting enzyme, and this too would be expected to inhibit an IL-1ß-driven immune response from being mounted. The example of the Epstein–Barr virus encoded homolog of IL-10 has been previously discussed under Sect. 4.10.

## 6 Pathophysiology and Disease Correlates

There are a large number of examples of interleukins having pathophysiological roles or disease correlates, regrettably far too many to be covered here, which may help understand the complex and versatile nature of interleukins in health and disease. Starting with IL-1, it is clear that when produced at low levels it has a protective role, but when produced at high levels, which spill into the circulation, it produces a predominance of proinflammatory effects that are correlated with disease states such as hypotensive shock and sepsis. High levels of IL-1 can induce bone and cartilage resorption and degradation and IL-1ß has been found in the synovial fluid and serum of rheumatoid arthritis (RA) patients. Other inflammatory cytokines such as TNF$\alpha$ can also be found in RA synovial fluids. Serum or plasma IL-6 levels can be greatly increased by bacterial infections and result in acute-phase responses. IL-8 levels are raised in psoriasis scales. IL-10 levels can be raised in malignant B-cell lymphomas in AIDS patients.

Studies in transgenic mice where expression of interleukin genes is under the control of constitutive promoters have also contributed evidence of pathophysiological effects. For example, IL-2 expression in many body organs of transgenic mice leads to baldness and interstitial pneumonia due to an inflammatory infiltrate. In contrast, IL-2 expression in the pancreas leads to a lethal pancreatitis, but not diabetes. Constitutive expression of IL-5, IL-6, and IL-7 can result in the expected hyperplasias, that is, eosinophilia (IL-5), plasmacytosis (IL-6), lymphocytosis (IL-7). However, the site and control of IL-6 expression, for example, is important in determining the outcome. When IL-6 is expressed under the control of the human Ig heavy chain promoter in B cells, the result is IgG1 plasmacytosis and infiltration of plasma cells into lung, spleen, and kidney. If, however, IL-6 is expressed under the control of the human keratin (K14) promoter in skin and tongue, there is growth retardation, poor hair growth, and epidermal scaliness (tail, paws), but no changes typical of psoriasis. By contrast, if bone marrow cells are infected with an IL-6-expressing recombinant retrovirus, a fatal myeloproliferative disease with massive neutrophil infiltration of lungs, liver, and lymph nodes develops within four weeks of engraftment, but no plasmacytosis occurs. As a further example of the diverse outcomes of constitutive interleukin expression, it has been found that IL-4 expression in the thymus and T cells results in thymic hypoplasia, inflammatory ocular lesions (blepharitis), and raised IgE and IgG1 levels, whereas IL-4 expression in B cells results in raised IgE and IgG1, but no thymic hypoplasia or blepharitis. Such differences indicate an importance of where or in which cells an interleukin is expressed, together with the quantitative and temporal elements, on the outcome.

## 7 Clinical Uses of Interleukins

The biological activities of interleukins have suggested that some may be of clinical use as therapeutic agents in human diseases as shown in Table 6. For example, the stimulating activity of IL-2 on the cytotoxic function of NK and LAK cells against tumor cells *in vitro* suggested that IL-2 had potential as an anticancer agent. Thus, once adequate amounts of recombinant IL-2 (rIL-2) became available, clinical trials to evaluate rIL-2 in different human malignancies were carried out. It was quickly established that high-dose IL-2 therapy caused severe side effects, for example, hypotension, oliguria, fluid retention, breathing difficulties, heart problems, with the dose-limiting side effect being the so-called *vascular leak syndrome*. In the latter, fluid extravasation and subsequent edema (swelling) takes place in the pleural and peritoneal cavities. The underlying cause appears to be the adherence of IL-2-activated lymphocytes to vascular endothelial sites, that is, the linings of blood vessels, resulting in holes in the endothelial cell layer being produced through which fluids leak. Despite these severe side effects, some tumor responses were found in a limited number of malignancies, including renal cell carcinoma, melanoma, and non-Hodgkins lymphoma. However, overall response rates with IL-2 as a single agent have been disappointingly low. New strategies involving combinations of IL-2 with other cytokines (e.g. IFN$\alpha$, TNF$\alpha$), antitumor monoclonal antibodies, cytotoxic drugs, or LAK cells removed by leukopheresis, activated by IL-2 and grown *ex vivo* before reinfusing back into the patient, have or are being tested in order to improve efficacy. The most "successful" approach has been that of adoptive cellular therapy where activated LAK cells are combined with high doses of IL-2. However, this is a highly aggressive antitumor therapy, which has serious complications for patients, with major life-threatening side effects. Despite initial tumor regressions following this therapy in "hard-to-treat" cancers such as colorectal cancer and malignant melanoma, in most instances remissions were not durable. A more sophisticated approach in which tumor-infiltrating lymphocytes (TIL) are obtained, IL-2

**Tab. 6** Disease correlates and possible scope for therapeutic/clinical intervention using interleukin inhibitors.

| *Interleukin* | *Disease correlates* | *Possibilities for therapeutic/ clinical intervention* |
|---|---|---|
| IL-1 | – Gram-negative septacemia, | IL-1ra, soluble IL-1R (other anti-endotoxic shock, hypotension inflammatory agents, e.g. anti-TNF$\alpha$) |
| | – Rheumatoid arthritis | As above |
| | – Multiple sclerosis | ,, |
| | – Kawasaki syndrome (inflammation of veins) | ,, |
| | – Myeloid leukemia | ,, |
| | – Insulin-dependent diabetes mellitus (IDDM) | ,, |
| IL-2 | – Lymphoid leukemia | Soluble IL-2R (also anti-TNF$\alpha$ antibody) |
| | – Systemic lupus erythematosus (SLE) | ,, |
| | – Graft-versus-host disease | ,, |
| IL-3 | – Cerebral malaria | Anti-IL-3, anti-GM-CSF, anti-IFN$\gamma$ |
| IL-4 | – Allergy/asthma | Anti-IL-4, soluble IL-4R |
| | – IDDM | ,, |
| | – Blepharitis (inflammatory eye lesion) | ,, |
| IL-5 | – Allergy/asthma | Anti-IL-5 |
| | – Eosinophilia | ,, |
| IL-6 | – Gram-negative septicemia | IL-1ra, soluble IL-1R, anti-TNF$\alpha$ |
| | – Multiple myeloma | Anti-IL-6, soluble IL-6R |
| IL-7 | – Lymphoid leukemia | Anti-IL-7, soluble IL2R |
| IL-8 | – Psoriasis | Signal transduction inhibitors |
| | – Erythroderma | ,, |
| IL-10 | – Burkitt's lymphoma | Anti-IL-10 |
| | – Malignant B-cell lymphomas in AIDS patients | ,, |
| IL-11 | – Megakaryocytic leukemia | Anti-IL-11 |
| IL-12 | – Multiple sclerosis | Anti-IL-12 |
| | – Autoimmune diseases | ,, |
| IL-20 | – Psoriasis | Anti-IL-20 |

activated and expanded *ex vivo*, and then reinfused back into the patient with the expectation that the TIL would home back into tumors has shown some promise, but again is complicated by side effects and less-than-durable responses. Novel gene therapy strategies involving the transduction of patients' lymphocytes or fibroblasts with the IL-2 gene leading to constitutive IL-2 production are now being tested.

So far, other interleukins despite their proven immunoregulatory activities *in vitro*, have not yet found much favor among clinicians for the treatment of human diseases. IL-1, for example, has been shown to enhance cellular proliferation and immune responses, but given in high-dose schedules it is probably too profoundly toxic for safe use. Similarly, IL-4 can act as a stimulator of lymphocytes, but its augmentation of IgE secretion and thus the potential for inducing allergic responses may disqualify it from being used clinically.

IL-12 has been proposed as an antitumor agent, but has proved to be quite toxic when administered systemically to cancer patients.

In patients with chronic disease, the presence of certain interleukins and their potential involvement in inflammatory and degradative processes suggests that countering their activities could produce clinical benefits. For example, IL-1 is frequently found in the synovial fluids of RA patients and is quite possibly associated with inflammatory symptoms. Thus, intervention using anti-IL-1 antibodies, IL-1ra, or soluble IL-1R is suggested as a therapeutic approach for RA. For autoimmune diseases such as type 1 diabetes and multiple sclerosis, a similar strategy employing IL-1 antagonists is also suggested. These antagonists could additionally be potentially useful in the treatment of acute illness, for example, bacterial sepsis and shock, as well as perhaps in other conditions such as asthma and acute respiratory distress syndrome (ARDS). So far, clinical phase II trials of IL-1ra in sepsis and RA patients have led to encouraging results and, although IL-1ra was administered in high doses, there were few, if any, side effects.

In several types of hematological malignancies, tumor cells secrete interleukins or other cytokines that may act as autocrine growth factors. For example, IL-2 and IL-7 may be produced by lymphoid leukemia cells, IL-6 by multiple myeloma cells and IL-10 by B-lymphoma (e.g. Burkitt's lymphoma) cells. Such interleukin production therefore suggests a case for investigating whether antagonists of these interleukins would inhibit tumor cell proliferation. There is as yet little clinical work done in the cancer area with interleukin antagonists, and it will be for future clinical trials to determine if they will be successful.

## 8
## Concluding Remarks

Interleukin biology is highly complex due to biochemical redundancy, pleiotropic activities, and the numerous interactions among interleukins themselves and with other biological effector molecules. While some biological activities of individual interleukins shown *in vitro* can be demonstrated *in vivo*, there remain questions about types of expressing and responding cells, sites, levels and duration of expression, developmental stages, and genetic background. Advances in molecular genetics, for example, knockout mice, have provided means for identifying physiological roles for some interleukins, but in most instances it is still not certain that such roles can be compensated for by other interleukins or cytokines, or how such roles are integrated within the host intercellular communication network as a whole. The present fragmentary evidence for defined physiological roles for the majority of interleukins remains an obstacle for translating the activities of these powerful biological molecules into clinically useful treatments of human diseases.

## Acknowledgments

I am indebted to Miss Deborah Kirk for the excellent typing of this chapter.

## Bibliography

### Books and Reviews

Baxter, A., Ross, R. (Eds.) (1991) *Cytokine Interactions and Their Control*, John Wiley & Sons, Chichester, UK.

Callard, R.E., Gearing, A.J.H. (Eds.) (2001) *The Cytokine Factsbook*, 2nd edition, Academic Press, London, UK.

Cosman, D. (1993) The hematopoietin receptor superfamily, *Cytokine* **5**, 95.

Fickenscher, H., et al. (2002) The interleukin-10 family of cytokines, *Trends Immunol.* **23**, 89.

Horst Ibelgaufts Cytokines Online Pathfinder Encyclopaedia. A COPE free servive on the World Wide Web. (2003).

Ibelgaufts, H. (1995) *Dictionary of Cytokines*, VCH, Weinheim (Germany) and New York (USA).

Meager, A. (1998) *The Molecular Biology of Cytokines*, John Wiley & Sons, Chichester, UK.

Mire-Sluis, A., Thorpe, R. (Eds.) (1998) *Cytokines*, Academic Press, San Diego (USA) and London (UK).

Stahl, N., Yancopoulos, G.D. (1993) The alphas, betas, and kinases of cytokine receptor complexes, *Cell* **74**, 587.

Thomson, A.W. (Ed.) (1992) *The Molecular Biology of Immunosuppression*, John Wiley & Sons, Chichester, UK.

Thompson, A.W. (Ed.) (1998) *The Cytokine Handbook*, 3rd edition, Academic Press, London, San Diego.

# 8
# Viral Inhibitors and Immune Response Mediators: The Interferons

*Anthony Meager*
*Division of Immunology and Endocrinology, National Institute for Biological Standards and Control, South Mimms, Herts, UK*

**1 Historical Perspective 208**

**2 Interferon Genes and Proteins 209**
2.1 Identification and Nomenclature 209
2.2 Interferon Genes 210
2.3 Interferon Proteins 211
2.3.1 IFN-$\alpha$ 211
2.3.2 IFN-ß 211
2.3.3 IFN-$\kappa$ 214
2.3.4 IFN-$\omega$ 214
2.3.5 IFN-$\tau$ 214
2.3.6 IFN-$\gamma$ (type II IFN) 214
2.3.7 IFN-$\lambda$ (type III IFNs) 215

**3 Regulation of IFN Synthesis 215**
3.1 Type I IFNs 215
3.1.1 IFN-$\alpha$ 215
3.1.2 IFN-ß 217
3.1.3 IFN-$\kappa$ 218
3.1.4 IFN-$\omega$ 218
3.1.5 IFN-$\tau$ 218
3.2 Type II IFN 218
3.2.1 IFN-$\gamma$ 218
3.3 Type III IFNs 219
3.3.1 IFN-$\lambda$s 219

**4 Regulation of Cellular Gene Expression by IFNs 219**
4.1 IFN Receptors 219

*Immunology. From Cell Biology to Disease.* Edited by Robert A. Meyers.

ISBN: 978-3-527-31770-7

4.1.1 Molecular Characteristics 219
4.1.2 Receptor Signaling 220
4.2 IFN-response Gene Sequences 222

**5 Proteins Induced by IFNs 222**

**6 Biological Activities Associated with IFNs 223**
6.1 Antiviral Activity 223
6.1.1 Molecular Mechanisms 223
6.1.2 Defense Mechanisms of Viruses 225
6.2 Antiproliferative Activity 226
6.3 Antimicrobial Activity 226
6.4 Immunoregulatory Activity 227

**7 Potential Physiological and Pathophysiological Roles of IFNs 228**
7.1 Do IFNs have Physiological Roles? 228
7.2 Ways of Studying IFN "Physiology" 229
7.3 IFNs and Pathophysiological Phenomena 230
7.4 IFNs and Pathogenesis 230

**8 Clinical Uses of IFNs 232**

**9 Final Remarks 236**

**Bibliography 236**
Books and Reviews 236
Primary Literature 237

## Keywords

**Cytokine**
One of a class of inducible biologically active proteins that exercise specific, receptor-mediated effects in target cells or in the cytokine-producing cells themselves.

**Homology**
Degree of structural similarity.

**Interferon**
One of a group of cytokines with antiviral activity, also displaying antimicrobial, antiproliferative, and immunomodulatory activities.

**Interferon Receptors**
Transmembrane glycoproteins specifically binding IFN and capable of signal transduction across the cell surface membrane.

**Interferon/Type I**
Acid-stable species of interferon, including, IFN-$\alpha$, -ß, -$\omega$, -$\kappa$, and -$\tau$.

**Interferon/Type II**
Acid-labile species of interferon, IFN-$\gamma$.

**Interferon/Type III**
A small family of interferon-like molecules known as IFN-$\lambda$s. Alternatively designated interleukin-28A and B and interleukin-29.

**Subtype**
One of two or more molecularly and antigenically related IFNs.

**Tumor Cell**
Cell that is capable of producing a tumor and that is unresponsive to mediators of normal cell growth control.

The interferons (IFNs), a category of biologically active proteins or cytokines with common antiviral activity, are induced by a variety of pathological stimuli, including viruses, bacteria, protozoa, and foreign antigens, as well as by endogenous cellular interactions. IFNs are secreted locally from induced cells to stimulate, via specific cell surface receptors, host defense mechanisms in the surrounding tissues. Such mechanisms are manifested as a wide spectrum of antiviral, antimicrobial, antitumor, and immunomodulatory actions, and have led to the belief that IFNs could have therapeutic potential in the treatment of many infections and invasive diseases.

Clinical investigations, begun in earnest in the 1980s with the abundant availability of "recombinant" IFNs, have demonstrated that IFNs, used as single therapeutic agents, induce beneficial responses in a select number of virus-mediated and malignant diseases, but not in the major human cancers, such as breast, lung, and colon. IFNs also induce significant, undesirable, but reversible, side effects. Nevertheless, IFNs continue to be widely used, for example, in the cancer area, and increasingly so for disease indications such as chronic hepatitis, caused by hepatitis-B and -C viruses, and multiple sclerosis, a chronic degenerative disease of unknown etiology.

## 1 Historical Perspective

The phenomenon of "interference" in virus infections was studied from the end of the 1940s. In 1957, Isaacs and Lindemann discovered a biological substance produced in virus-infected cells that in turn produced an antiviral response in cell cultures infected by the same virus. The word *interferon* (IFN) was coined for this substance. Initially, there were difficulties in characterizing IFN because under laboratory conditions it was made in only minute quantities. Gradually however, with the introduction of large-scale production facilities and the development of efficient purification procedures, the nature of IFN began to emerge. First, it was established that IFN was protein in nature. Second, it was discovered that there were different molecular species (types) of IFN. Third, it became evident that IFNs were associated with a growing number of biological activities beside their antiviral activity. One of these activities, the antitumor action, was the spur to increased efforts during the 1970s to develop IFN as an anticancer agent. Progress was made particularly in the area of purification and in the sequencing of the N-terminal regions of pure IFN proteins. However, the development of rDNA technology in the late 1970s was to revolutionize both the production of IFNs and their characterization. This technology was also to reveal that one type of human IFN, designated IFN-$\alpha$, was composed of many different, but structurally related, IFN proteins, now known as subtypes. Using nucleotide-sequencing techniques, the amino acid sequences of IFN molecules could be deduced once complementary DNA (cDNA) copies of individual IFN messenger RNA (mRNA) were cloned. Genes for individual IFN proteins were also isolated, characterized, and their chromosomal location established. By comparing the IFN genes of other animal species to those of humans, evolutionary relationships and the ancestry of IFN genes were deduced.

The early 1980s were times of intensive research investigations into all aspects of IFN biochemistry and biology and the period that saw the instigation of many clinical trials to evaluate IFN as a potential therapeutic agent in cancer patients. Reports of initial successes in the clinic were followed by more sober analyses of IFN's worth as an anticancer agent. IFN clearly had potential, but tumor responses were seldom large enough to cure cancer patients. In addition, IFN produced a number of undesirable side effects in patients. The cloning in the mid-1980s of the second major type of IFN, now designated IFN-$\gamma$, led to a further expansion of biochemical and biological investigations and more clinical trials.

The period from around 1985 to date has witnessed many new discoveries about IFNs and their biological actions and featured the cloning of their cell surface receptors, responsible for triggering cellular responses following IFN binding. There has been a rationalization of the clinical results leading to the now-accepted view that IFN administrated as a single anticancer agent is beneficial in only a handful of malignancies, for example, hairy cell leukemia (which is extremely rare), renal cell carcinoma, AIDS-related Kaposi's sarcoma and chronic myelogenous leukemia (CML). Nevertheless, further attempts to exploit IFN's potential have and are being made by combining it with other anticancer drugs. IFN-$\alpha$ is now having some success as an antiviral agent in the treatment of hepatitis-B and -C virus infections, while IFN-$\beta$ has emerged as a therapeutic agent for multiple sclerosis.

## 2
## Interferon Genes and Proteins

### 2.1
### Identification and Nomenclature

Originally two types of human IFN were defined on the basis of the capacity of the biological activity of IFN preparation to withstand acidification to pH 2. These were designated type I (acid-stable) IFN and type II (acid-labile) IFN. On the basis of the cell source, type I IFN was subdivided into two subclasses: *leukocyte IFN* and *fibroblast IFN*. These were subsequently shown to be antigenically distinct. Type II IFN was often referred to as *immune IFN* to denote its production by stimulated T lymphocytes.

In 1978, an international IFN nomenclature committee recommended that *leukocyte IFN* should become IFN-$\alpha$, *fibroblast IFN*, IFN-ß, and *immune IFN*, IFN-$\gamma$. These designations were widely accepted, but nowadays with increased knowledge of IFN genes, protein structures, and receptors, IFN-$\alpha$ and IFN-$\beta$ are often classified as type I IFNs and IFN-$\gamma$ as type II IFN. However, complications arose when it was discovered that IFN-$\alpha$ was heterogeneous and contained many different, but structurally and antigenically related species, now commonly referred to as subtypes. Some groups labeled these subtypes $\alpha$A, $\alpha$B, $\alpha$C, $\alpha$D, and so on, while others adopted the numerical system of $\alpha_1$, $\alpha_2$, $\alpha_4$, and so on. The latter nomenclature now prevails. However, there is no general correspondence between letters and numbers, for example, $\alpha A = \alpha 2$; $\alpha B = \alpha 8$; $\alpha D = \alpha 1$. When an IFN preparation is a mixture of IFN-$\alpha$ subtypes, these are often designated IFN-$\alpha n$. The discovery in the mid-1980s of additional IFN-$\alpha$-like proteins antigenically distinct from the main family of IFN-$\alpha$ subtypes then created another problem. At first, the new IFN-$\alpha$-like proteins were classified as IFN-$\alpha$ subclass II, with members of the original IFN-$\alpha$ family being referred to as IFN$\alpha$ subclass 1. However, a radical departure from the sequence of letters in the Greek alphabet resolved this complex nomenclature. IFN-$\alpha$ subclass II has become IFN omega ($\omega$)! Recently, another human type I IFN distantly related to either IFN-$\alpha$ or IFN-$\omega$ and expressed only in epidermal keratinocytes has been characterized; this is designated *keratinocyte-derived IFN*, or IFN-$\kappa$. The most recent addition to the human type I IFN family is IFN-$\varepsilon$, which appears distantly related to IFN-$\beta$ but whose function remains uncertain.

In mice, similar IFN-$\alpha$ subtypes, IFN-$\beta$, and IFN-$\varepsilon$ are present, but IFN-$\kappa$ and IFN-$\omega$ are absent. However, a type I IFN-like molecule called limitin, which is 30% related to human IFN-$\omega$, has recently been identified in mice. In other animal species (e.g. bovines), besides multiple IFN-$\alpha$ subtypes similar to those in humans, there are frequently several different IFN-ß and/or IFN-$\omega$ subtypes, although besides mice IFN-$\omega$ is also absent in canines. Further, other classes of IFN-$\alpha$-like proteins have been identified in certain animal species, but not in humans; the *trophoblast* ($\tau$) and *delta* ($\delta$) IFNs are produced by the trophectoderm and are involved in the maternal recognition of pregnancy in ruminants (e.g. bovines) and pigs, respectively.

All members of the type I ($\alpha$, $\beta$, $\delta$, $\varepsilon$, $\kappa$, $\tau$, and $\omega$) and II ($\gamma$) IFNs induce antiviral activity. A further small family of human cytokines with limited homology to type I

IFNs have recently been characterized as also having antiviral activity. They have been designated IFN-λ1 (alternatively, interleukin-29 (IL-29)), IFN-λ2 (IL-28A), and IFN-λ3 (IL-28B) and collectively as type III IFNs. Their counterparts probably exist in other animal species, but IFN-λ1 (IL-29) is absent in mice.

The nomenclature system for designating IFNs according to their animal species of origin is reasonably straightforward. The first two letters of the animal name are most often used: human, Hu; murine, Mu; bovine, Bo, and so on. However, Mo is sometimes used for mouse IFN and Ra for rat IFN to distinguish between these murine species. In other cases, no abbreviations are used, as in, for example, chick IFN and rabbit IFN.

## 2.2 Interferon Genes

The number of IFNs varies among different animal species, but those of the IFN-α family are usually the most numerous. In humans, there are 13 nonallelic IFN-α genes (Note that these genes are designated as *IFNA1, IFNA2,* and so on, so that they correspond exactly to the IFN-α proteins they encode), plus four nonallelic *IFNA-like* pseudogenes (i.e. genes that cannot give rise to functional IFN-α molecules). In addition, allelic variants of some *IFNA* subtype genes (e.g. for *IFNA2*) have been found. There is one HuIFN-ß gene (designated *IFNB1*), one IFN-ε gene (designated *IFNE1*) one HuIFN-κ gene (designated *IFNK*), one HuIFN-ω gene (designated *IFNW1*) and several *IFNW*-like pseudogenes. All of these genes and pseudogenes are located on the short arm of human chromosome 9 where they are arranged in tandem. The equivalent site for mouse *IFNA* genes is in mouse chromosome 4. Each of the genes specifying IFN-α, -ß, -ε and -ω proteins has no introns, but the *IFNK* gene contains one intron in the 3′ noncoding region. Lack of introns is an unusual feature among mammalian cytokine genes, in which there are usually three or more introns (IFNs are considered to belong to the cytokine superfamily of multifunctional, intercellular biological mediators). The lack of introns suggests a very ancient origin of the common ancestral gene. It has been predicted that the primordial type I IFN gene arose some 500 million years ago. A split about 400 million years ago appears to have given rise to the first *IFNA* and *IFNB* genes. Since then, the *IFNA* gene has evolved and duplicated many times to give the multiple *IFNA* genes found in present day mammals. Around 100 million years ago, an *IFNA* gene apparently diverged enough to give rise to the *IFNW* gene family, which is found in most mammals. In ruminants, the *IFNW* gene family appears to have split around 60 to 80 million years ago to give the *IFNT* gene family, which is now only found in the Ruminantia suborder Artiodactyls (sheep, cattle). The relationship of the pig IFN-δ gene and the mouse limitin gene to the *IFNA, W, T* gene family remains uncertain. However, *IFNE1* and *IFNK* genes are more likely related to the *IFNB* gene than to the *IFNA, W, T* genes.

In contrast, throughout evolution there appears to have been no duplication of the IFN-γ gene (designated *IFNG*). Only a single *IFNG* gene has been found in all animal species tested. This gene has four exons and three introns, a common organizational feature of many cytokine genes, and is located on chromosome 12 in humans and chromosome 10 in the mouse. The *IFNG* gene also shares other structural features in common with cytokine genes such as an inducible transcriptional enhancer in the 5′ flanking region found, for example, in the interleukin-2 gene.

The genes for IFN-λ1-3 (type III IFNs) are located on human chromosome 19 and contain introns. These genes appear to have emerged about the same time as type I IFN genes and are widespread in mammals, although the mouse lacks a functional IFN-λ1 gene.

All the active IFN genes have regulatory regions in their 5′ noncoding promoter sequences, which control transcription. In all coding regions, the part of the gene that encodes the mature IFN protein is preceded by a nucleotide sequence specifying an N-terminal signal peptide.

## 2.3 Interferon Proteins

### 2.3.1 IFN-α

All IFN-α's are secreted proteins and, as mentioned earlier, are transcribed from mRNA as precursor proteins, pre-IFN-α, containing N-terminal signal polypeptides of about 23 amino acids. The signal polypeptide is cleaved off before the mature IFN-α molecule is secreted from the cell. The mature form of most HuIFN-α subtypes contains 166 amino acids (except HuIFN-$\alpha_2$, 165 amino acids: (Fig. 1). IFN-α subtypes from the mouse are of a similar size with 166 or 167 amino acids (except mouse IFN-$\alpha_4$, 162 amino acids). The calculated molecular weight of HuIFN-α subtypes is approximately 18.5 kDa, although apparent molecular weights in sodium dodecyl sulfate (SDS)-polyacrylamide gels vary between 17 and 26 kDa. The amino acid sequences of HuIFN-α subtypes are highly related, with complete identity at 85 of the 166 (amino acid) positions. Many of the positions where amino acids differ from subtype to subtype are conservative substitutions. Interestingly, HuIFN-α subtypes contain four cysteine residues whose positions (1, 29, 99, and 139) are highly conserved (Fig. 1). These four cysteines form two disulphide bridges (1–99, 29–139), which induce folding of the IFN-α molecule and whose integrity is essential for biological activity. IFN-α subtypes are predicted to have a high proportion (ca 60%) of α-helical regions and are folded to form globular proteins. The three-dimensional structure of HuIFN-$\alpha_2$ revealed by X-ray crystallographic analysis comprises 5α-helices (A–E) connected by loops in the "classic" up-up-down-down configuration. This α-helical "bundle" structure IFN-αs share with many other cytokines, particularly those of the hematopoietin family. Only one of the HuIFN-α subtype sequences contain sites for N-linked glycosylation; however, limited O-linked glycosylation may occur in several. Natural HuIFN-$\alpha_2$, for example, appears to contain some O-linked oligosaccharides (N.B. Recombinant IFN-α subtypes produced by *Escherichia coli* are nonglycosylated since bacteria lack the biosynthetic machinery to add sugar residues to polypeptides).

Although the mouse IFN-α subtypes, of which 10 have been cloned and expressed, are only approximately 40% related in amino acid sequence to their HuIFN-α subtype counterparts, they are likely to be structurally similar. However, in contrast to HuIFN-α subtypes, most mouse IFN-α subtypes contain one N-linked glycosylation site and are probably glycoproteins when produced from mouse somatic cells.

### 2.3.2 IFN-ß

Pre-HuIFN-ß contains 187 amino acids, of which 21 comprise the N-terminal

| | |
|---|---|
| α1 | $^{-23}$M A S P F A L L M V L V V L S C K S S C S L G$^{1}$C D L P |
| α2 | M A L T F A L L V A L L V L S C K S S C S V G$^{1}$C D L P |
| αII(w) | M V L L L P L L V A L P L C H C G P C G S L S$^{1}$C D L P |
| β | $^{-21}$M T N K C L L Q I A L L L C F S T T A L S$^{1}$M S Y N L L |
| γ | $^{-23}$M K Y T S Y I L A F Q L C I V L - G S L G C Y C Q$^{1}$D P |
| α1 | E T H S L D N R R T L M L L A Q M S R I S P S S C L M D R |
| α2 | Q T H S L G S R R T L M L L A Q M R K I S L F S C L K D R |
| αII(w) | Q N H G L L S R N T L V L L H Q M R R I S P F L C L K D R |
| β | G F L Q R S S N F Q C Q K L L W Q L N G R L E Y C L K D R |
| γ | Y V K E A E N L K K Y F N A G H S D V A D N G T L F L G I |
| α1 | H D F G F P Q E E F D G N Q F Q K A P A I S V L H E L |
| α2 | H D F G F P Q E E F - G N Q F Q K A E T I P V L H E M |
| αII(w) | R D F R F P Q E M V K G S Q L Q K A H V M S V L H E M |
| β | M N F D I P E E I K Q L Q Q F Q K E D A A L T I Y E M |
| γ | L K N W K E E S D R K I M Q S Q I V S F Y F K L - - - |
| α1 | I Q Q I F N L F T T K D S S A A W D E D L L D K F C T E L |
| α2 | I Q Q I F N L F S T K D S S A A W D E T L L D K F Y T E L |
| αII(w) | L Q Q I F S L F H T E R S S A A W N M T L L D Q L H T E L |
| β | L Q N I F A I F R Q D S S S T G W N E T I V E N L L A N V |
| γ | F K N F K D D Q S I Q K S V E T I K E D M N V K F F N S N |
| α1 | Y Q Q L N D L E A C V M Q E E R V G E T P L M N A D S I |
| α2 | Y Q Q L N D L E A C V I Q G V G V T E T P L M K E D S I |
| αII(w) | H Q Q L Q H L E T C L L Q V V G E G E S A G A I S S P A |
| β | Y H Q I N H L K T V L E E K L E K E D F T R G K L M S S |
| γ | K K K R D D F E K L T N Y S V T D L N V Q R K A I H E L |
| α1 | L A V K K Y F R R I T L Y L T E K K Y S P C A W E V V R A |
| α2 | L A V R K Y F Q R I T L Y L K E K K Y S P C A W E V V R A |
| αII(w) | L T L R R Y F Q G I A V Y L K E K K Y S D C A W E V V R M |
| β | L H L K R Y Y G R I L H Y L K A K E Y S H C A W T I V R V |
| γ | I Q V M A E L - - - - - - - - - - - - S P A A K T G K R K |
| α1 | E I M R S L S L S T N L Q E R L R R K E$^{166}$ |
| α2 | E I M R S F S L S T N L Q E S L R S K E$^{165}$ |
| αII(w) | E I M K S L F L S T N M Q E R L R S K D A D L G S S$^{172}$ |
| β | E I L R N F Y F I N R L T G Y L R N$^{166}$ |
| γ | R - - - - - - - S Q M L - F R G R R A S Q$^{143}$ |

**Fig. 1** Amino acid sequences of human IFNs, $\alpha_1$, $\alpha_2$, $\alpha$II ($\omega$), ß and $\gamma$. Sequences are aligned to show maximum structural relatedness. Sites of N-linked glycosylation (Asn = N) are underlined.

signal polypeptide. The signal polypeptide is cleaved on secretion to yield the mature HuIFN-ß protein of 166 amino acids, the same size as most HuIFN-$\alpha$ subtypes (Fig. 1). However, HuIFN-ß shows only approximately 30% amino acid sequence relatedness with HuIFN-α and is antigenically distinct from it. The mature HuIFN-ß molecule lacks the N-terminal cysteine residue at position 1 found in HuIFN-α subtypes, but contains three other cysteines at position 17, 31, and 141. The cysteines at position 31 and 141 are equivalent to those found at positions 29 and 139 in HuIFN-α and are joined to form a single disulphide bond. Replacement of the cysteine residue at position 17 by serine does not result in any loss of biological activity, whereas serine substitution of $Cys^{141}$ does. Unlike most HuIFN-α subtypes, HuIFN-ß's amino acid sequence contains one site, asparagine at position 80, for N-linked glycosylation and it is known that HuIFN-ß produced by human fibroblasts is a glycoprotein of approximately 23 kDa. Like HuIFN-α subtypes, HuIFN-ß is predicted to have a high α-helical content, and folds up as an α-helical bundle of 5 domains.

Mouse IFN-ß contains 161 amino acids and shows some 60% relatedness to HuIFN-ß: in contrast, however, mouse IFN-ß contains only one cysteine residue and thus intramolecular disulphide bonds are impossible. There are three N-glycosylation sites present in mouse IFN-ß, and the rather high apparent molecular weight of 35 kDa probably reflects extensive glycosylation of this molecule. The three-dimensional crystal structure of recombinant mouse IFN-ß has been solved, revealing that the structure consists of five α-helices folded into a compact α-helical bundle (Fig. 2). There are many structural similarities between mouse IFN-β and HuIFN-$\alpha_2$. Subsequently, the three-dimensional structure of HuIFN-β has also been shown to be similar, but dimers may form through a zinc-binding site present at the interface between two protomers. On the basis of comparative sequence analyses, it is predicted that in all mammalian IFN-α, IFN-ß, and IFN-ω proteins, the five α-helical domains are conserved. Pig IFN-δ, mouse limitin, IFN-ε, and ruminant IFN-τ are

**Fig. 2** Schematic drawing of the "side" view of a recombinant MuIFN-ß molecule. Helices A-E and an intervening loop AB are labeled N- and C-termini are marked. The three shaded areas along loop AB, helix D, and loop DE appear to comprise a spatially continuous binding site for the IFN receptor(s). (Reprinted with permission of Senda, et al. (1992) *EMBO J.* **11**, 3193; Copyright EMBO/IRL Press Ltd., UK).

also predicted to have the same basic structure.

#### 2.3.3 IFN-$\kappa$

HuIFN-$\kappa$, made principally by keratinocytes, is a recently identified novel member of the type I IFN family displaying around 30% homology to HuIFN-$\alpha$, HuIFN-$\beta$, or HuIFN-$\omega$. In distinction to the latter, HuIFN-$\kappa$ contains an additional cysteine at its C-terminus following the four conserved cysteines, and an insertion of 12 amino acids between the C and D helices. Its full amino acid sequence is therefore predicted to comprise 180 amino acids. However, sequence alignment with other type I IFNs predicts that HuIFN-$\kappa$ has an overall structure typical of this family. Molecular modeling of HuIFN-$\kappa$ has indicated a structure consisting of five $\alpha$-helices as in other type I IFNs.

#### 2.3.4 IFN-$\omega$

The structure of IFN-$\omega$ is similar to that of IFN-$\alpha$ subtypes, but there have been sufficient amino acid substitutions to make it only about 60% related and thus antigenically distinct from them (HuIFN$\omega$ is only 30% related to HuIFN-ß). Like pre-IFN-$\alpha$, pre-IFN-$\omega$ has a signal sequence of 23 amino acids. However, the mature form of HuIFN-$\omega$ contains 172 amino acids, six more at the C-terminal end than HuIFN-$\alpha$ subtypes (Fig. 1). The four cysteine residues occur in precisely the same positions, 1, 29, 99, and 139, as they do in HuIFN-$\alpha$ subtypes. With regard to glycosylation, HuIFN-$\omega$ resembles HuIFN-ß more than HuIFN-$\alpha$ in having a single N-glycosylation site at Asn78.

IFN-$\omega$ is not found in mice or dogs, but three functional IFN-$\omega$ subtypes have been identified in sheep, cattle, and horses.

#### 2.3.5 IFN-$\tau$

Over a decade ago, it was discovered that ovine trophoblastic protein-1 (OTP-1), a hormonelike molecule having antiluteolytic activity produced by the preimplantation ruminant conceptus during the second and third weeks of pregnancy, has antiviral activity. This led to the finding that OTP-1 was itself structurally related to type 1 IFN, in particular to IFN-$\omega$. As a consequence, OTP-1 and related proteins have been designated by one group of researchers as trophoblast IFNs or IFN-$\tau$. Preovine IFN-$\tau$ contains 195 amino acids; a signal sequence of 23 amino acids followed by the mature polypeptide of 172 amino acids. IFN-$\tau$ molecules all contain the precisely conserved cysteines at positions 1, 29, 99, and 139 found both in IFN-$\alpha$ and IFN-$\omega$ molecules. Like IFN-$\omega$, some IFN-$\tau$ subtypes possess a potential site for N-glycosylation on Asn78. It is predicted that IFN-$\tau$ will have a similar three-dimensional structure to IFN-$\alpha$ and IFN-ß (i.e. a bundle configuration of five $\alpha$-helices).

IFN-$\tau$ subtypes are only found in the Ruminantia suborder Artiodactyla (e.g. sheep, goats, cattle), where they appear to have evolved to perform a special antiluteolytic function in early pregnancy. In man and other mammals, the fertilized egg, the conceptus, immediately attaches to the uterine wall and therefore there is no special need to maintain the corpus luteum as in ruminants, where the conceptus undergoes extensive preimplantation development.

#### 2.3.6 IFN-$\gamma$ (type II IFN)

The nucleotide sequence of HuIFN-$\gamma$ mRNA contains an open-reading frame of 166 codons. At first, it was believed that the mature HuIFN-$\gamma$ protein contained 146 amino acids, the 20 N-terminal amino acids of pre-HuIFN-$\gamma$ forming the signal peptide. However, this proved incorrect, inasmuch as cleavage of the signal peptide and the mature HuIFN-$\gamma$ occurred between residues 23 and 24 of the pre-HuIFN-$\gamma$ (Fig. 1). It was also shown that the N-terminal amino acid of HuIFN-$\gamma$ was pyroglutamic acid and that several of the C-terminal amino acids of the natural 143 amino acid polypeptide could be cleaved off without loss of biological activity. The predicted molecular weight of the unmodified polypeptide is approximately 17 kDa, but there are two N-glycosylation sites (Asn25 and Asn97), both of which are used to yield fully glycosylated HuIFN-$\gamma$ molecules of 25 kDa (Note that HuIFN-$\gamma$ preparations derived from T lymphocytes often contain minor amounts of a 20-kDa molecule representing glycosylation at one site only.). In its native state, HuIFN-$\gamma$ is not only heavily glycosylated but is also a homodimer of two 25-kDa molecules (subunits). The role of the oligosaccharide side chains is not known. Since unglycosylated HuIFN-$\gamma$ produced from *E. coli* containing IFN-$\gamma$ cDNA is biologically active, they appear to be dispensable.

The amino acid sequence of HuIFN-$\gamma$ shows very little homology with those of type I IFNs ($\alpha$, ß, $\kappa$, $\omega$, or $\tau$). Despite this, rather interestingly the three-dimensional structure of HuIFN-$\gamma$ (illustrated in Fig. 3), which was reported in 1991, contains some features in common with IFN-ß. Like the type I IFNs, IFN-$\gamma$ molecules are essentially $\alpha$-helical in content, with no ß-sheet. IFN-$\gamma$ subunits have six $\alpha$-helical regions as opposed to the five present in the IFN-ß structure. On dimerization of IFN-$\gamma$ subunits, five of the 12 helices are folded to form a structural domain that resembles the folded arrangement of the five helices in IFN-ß. Other similarities in folding topology between IFN-$\gamma$ and IFN-ß are apparent, suggesting that despite their present low degree of homology, an ancient precursor to both types might have existed.

Mouse IFN-$\gamma$ containing 133 amino acids is smaller than HuIFN-$\gamma$ and is only 40% related in sequence. It also is heavily glycosylated and the molecular weight of the natural dimeric molecule is around 38 kDa. Its structure, and those of IFN-$\gamma$ from other species, are expected to be similar to that worked out for the three-dimensional structure of HuIFN-$\gamma$.

#### 2.3.7 IFN-λ (type III IFNs)

The IFN-λs contain 200 amino acids: in sequence IFN-λ1 is 81% related to IFN-λ2; IFN-λ2 is 96% related to IFN-λ3. IFN-λ1 contains an N-glycosylation site that is not present in IFN-λ2 or IFN-λ3. The IFN-λs are distantly related to type I IFNs (15–19% homology) and to the interleukin-10 (IL-10; 11–13% homology) family of cytokines. The 3-D structures of IFN-λs are not yet known, but since cysteine patterns are conserved on alignment of amino acid sequences with those of IFN-$\alpha$ and IL-10, it is predicted that IFN-λs are helical cytokine family members.

## 3 Regulation of IFN Synthesis

### 3.1 Type I IFNs

#### 3.1.1 IFN-$\alpha$

IFN-$\alpha$ can be induced by a whole variety of stimuli, including viruses, bacteria, xenogeneic or allogeneic tumor cells, virus-infected cells, B-cell mitogens, and so on. IFN-$\alpha$ subtypes are produced by stimulated or activated null lymphocytes,

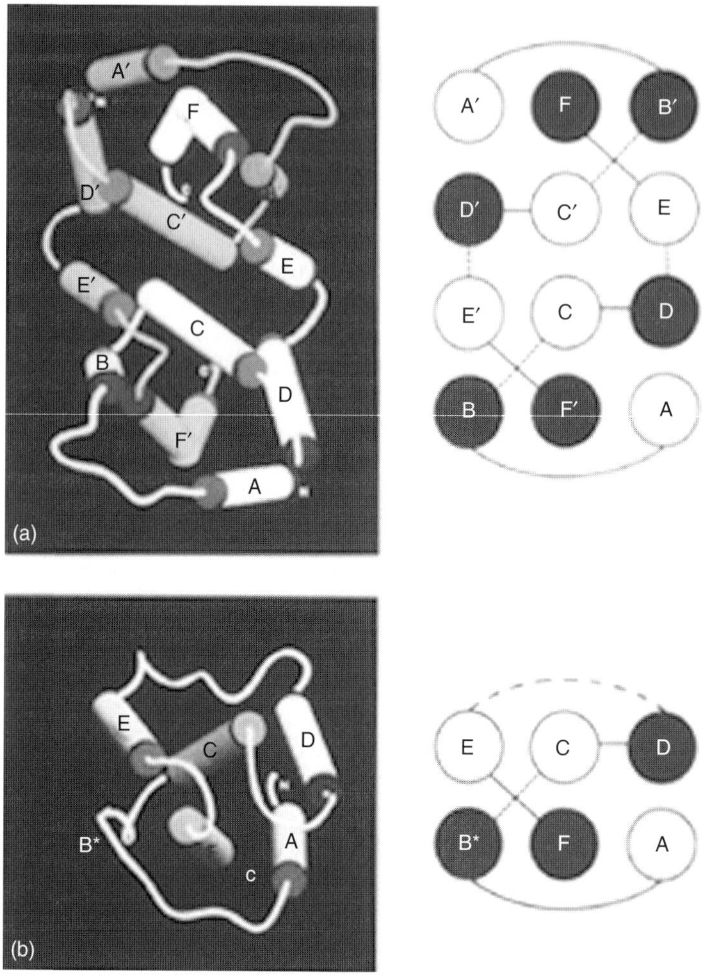


**Fig. 3** Schematic drawings of (a) the recombinant human IFN-$\gamma$ D′ dimer and (b) IFN-ß. On the left-hand side, the $\alpha$ helices are represented as cylinders, and the nonhelical regions are shown as tubes. The N- and C-terminal ends are marked. On the right-hand side, the $\alpha$-helices are shown as circles, and the nonhelical regions are shown as solid or dashed lines, depending on whether they are above or below the plane of the figure. The regions labeled B* represent a kinked region of IFNß that is not helical, but shows some helical features. (Reprinted with permission of Ealick, et al. (1991) *Science*, **252**, 698. Copyright 1991 by American Association for the Advancement of Science).

monocytes, and lymphoblastoid cell lines of B-cell origin. More recently, plasmacytoid dendritic cells, a subset found mainly in lymph nodes and thymus, have been characterized as IFN producer cells (IPC) that produce large amounts of IFN-$\alpha$ following virus stimulation. The proportion of each IFN-$\alpha$ subtype in the mixtures of subtypes produced as a consequence of stimulation/activation varies according to the nature of the stimulus and the producer cell type. It is still unclear why so many IFN-α subtypes are required and how their proportions are regulated at the level of their genes (but see below).

Normally, *IFNA* genes are tightly repressed, but when cells are infected by a virus, for example, "virus-responsive elements" in the *IFNA* genes are activated leading to derepression. *IFNA* genes contain several IFN gene regulatory elements (IREs), termed positive-response domains (PRDs) and negative-response domains (NRDs), in their 5′ noncoding promoter region whose sequences appear to govern inducibility by various inducers. Such IREs are acted upon by host cell interferon regulatory factors (IRF), a family of 9 known structurally related proteins, which act as transcriptional enhancers (e.g. IRF-1, −3, −5, −7) or repressors (e.g. IRF-2). It is now believed that IRF-3, which is constitutively present and phosphorylated by a serine kinase early after virus infection, plays a key role in IFN-α induction. Phosphorylated IRF-3 dimerizes, translocates to the nucleus, and activates the transcription of an "early" *IFNA* (most likely *IFNA4*) gene together with *IFNB*. The newly made "early" IFNs act in an autocrine/paracrine manner to activate type I IFN signaling pathways (see under Sect. 4.1.2) leading to transcription of IFN responsive genes, including IRF-7, an inducible transcription factor, which, following phosphorylation in virus-infected cells, activates several other *IFNA* genes plus many host defense genes (e.g. cytokines, chemokines). Thus, IFN-α production appears to be amplified in a controlled manner subject to the virus-dependent, sequential activation of specific IRFs. However, *IFNA* gene transcription leading to synthesis of IFN-α mRNAs continues only for a short interval before the genes are shut down again, even if the presence of the inducer is maintained, though IFN-α mRNA expression can be extended by certain drugs (e.g. cycloheximide, which blocks protein synthesis and hence the formation of labile repressors such as IRF-2).

### 3.1.2 IFN-ß

The regulation of transcription of the *IFNB* gene has been more closely studied than that of *IFNA* genes. Here, it is known that within the 120 bp 5′ to the transcription initiation site, there are at least six IRE, four PRD (I–IV), and two NRD (I, II). PRDI and PRDII are involved in both positive control and in postinduction repression. Several protein factors have been identified, which bind to PRDI including IRF-1 and IRF-3, transcriptional enhancers, and IRF-2, a repressor. PRDII resembles an NFκB transcription factor binding site and agents that activate NFκB (e.g. viruses, double-stranded RNA, phorbol esters), therefore promote IFN-ß induction. In contrast to *IFNA* gene regulation, active NFκB is probably a requirement for efficient *IFNB* gene transcription.

Although the major regulatory elements within the *IFNB* gene promoter have been identified, the mechanisms by which the gene is differentially induced and shut off are still not completely understood. However, research into the signaling pathways involved in following virus, double-stranded RNA (dsRNA), or cytokine stimulation has increased our knowledge of potential mechanisms for induction. For example, it is apparent that dsRNA, often an intermediate in viral genome-directed nucleic acid synthesis, can act as a "proximal" inducer (e.g. via Toll-like receptor-3) of IRF-3 and NF$\kappa$B activation, which are then able to activate transcription of the *IFNB* gene. The production of IFN-$\beta$ appears to be the first stage in the cooperative activation of other IRFs (e.g. IRF-5, IRF-7), which appear to be crucial for transcriptional induction of *IFNA* and *IFNW* genes. In the absence of IRF-2, viral induction of IFN-ß is increased.

#### 3.1.3 IFN-$\kappa$

Unlike other type I IFN genes, the *IFNK* gene appears to be constitutively active to some extent, particularly in IFN-$\kappa$-producer cells such as keratinocytes. However, IFN-$\kappa$ mRNA expression is reported to be upregulated by viral infection, by other type I IFNs, and characteristically, by IFN-$\gamma$ (type II IFN) in monocytes and dendritic cells. Currently, there is no information about which transcription factors are involved in *IFNK* gene activation.

#### 3.1.4 IFN-$\omega$

The *IFNW* gene is virus-inducible, like *IFNA* and *IFNB* genes, and, since there is much sequence identity with many of the *IFNA* subtype genes, it is assumed that its regulation will also be similar to that applying to *IFNA* genes. As a proportion of the antiviral activity present in the IFN mixture secreted by virally induced leukocytes, IFN-$\omega$ constitutes approximately 15% of the total.

#### 3.1.5 IFN-$\tau$

The *IFNT* genes are not virus-inducible and are therefore regulated in a different manner to the other type I IFN genes. IFN-$\tau$ is produced by trophoblast cells and the temporal pattern of expression depends on the time frame of maternal recognition of pregnancy, which varies from one to three weeks in ruminants. For instance, ovine (Ov) IFN-$\tau$ is only produced in large amounts between the 13th and 21st days of pregnancy, although OvIFN-$\tau$ mRNA can be detected as early as 8 to 10 days. It is believed that a factor(s) in the luteal phase environment of the uterus is required to trigger and maintain expression of IFN-$\tau$ genes. The promoter regions of *IFNT* genes appear to be under different regulatory control than other type I IFN genes, despite a high degree of superficial similarity in nucleotide sequences across all members of this gene family.

### 3.2 Type II IFN

#### 3.2.1 IFN-$\gamma$

IFN-$\gamma$ is produced mainly by antigen- or mitogen-stimulated mature, T lymphocytes and natural killer (NK) cells. Some T-cell lymphoblastoid lines, especially those chronically infected with HLTV-I, also can be stimulated to produce IFN$\gamma$. Like IFN-$\tau$, IFN-$\gamma$ does not appear to be virus-inducible. The gene structure of *IFNG* is quite apart from the intron-less type I IFN genes and resembles more the four exon–three intron arrangement found in cytokine genes (e.g. IL-2, IL-4, and GM-CSF). There are also similarities in the 5′ flanking region, upstream from the initiation site, of the *IFNG* gene and the 5′ flanking regions of such cytokine genes. *IFNG* gene transcription is probably controlled by binding of nuclear transcription factors to regulatory sequences in the 5′ flanking region. Since phorbol esters can act as (co)-stimulators of IFN-$\gamma$ induction, it is probable that activation of NF$\kappa$B is involved in gene derepression.

## 3.3 Type III IFNs

### 3.3.1 IFN-λs

Transcription of IFN-λ genes is stimulated by the same inducers of IFN-$\alpha$, -$\beta$, and -$\omega$ genes, that is, several viruses, dsRNA (e.g. poly I.C.), in a variety of cell lines and dendritic cells. Virus-infected cells produce IFN-λ activity. The regulation of mRNA expression of IFN-λ genes is likely to involve a positive feedback mechanism, similar to that operating for IFN-$\alpha/\beta$ genes, but details are not yet available.

# 4 Regulation of Cellular Gene Expression by IFNs

## 4.1 IFN Receptors

### 4.1.1 Molecular Characteristics

IFNs exercise their actions in cells via IFN-specific cell surface receptors. These receptors bind IFNs with high affinity and transduce the signal occasioned by ligand (IFN)-binding across the cell membrane into the cytoplasm. All type I IFNs appear to share a common receptor, whereas the sole type II IFN, IFN-$\gamma$, has a distinct receptor. This division of receptors has been corroborated by the cloning of both type I and type II IFN cell surface binding proteins, which revealed little apparent relatedness of their respective amino acid sequences. However, these IFN binding proteins, two for type I IFN and one for type II IFN, share structural features that are frequently found in cytokine receptors; each has a large N-terminal extracellular domain, which is glycosylated, and which forms the binding site for IFN, a comparatively small transmembrane region and an intracellular domain of variable size (Fig. 4). The genes for the type I IFN binding proteins are located on human chromosome 21 and that for the type II IFN binding protein on human chromosome 6. It seems probable that the type I IFN binding proteins are intimately associated and together form high-affinity receptors. The type II IFN binding protein also appears to require a second component or accessory factor (AF), coded for by a gene located on chromosome 21, to form a functional receptor. One group has deduced that possibly the two types of receptor are evolutionarily connected and in their extracellular domains resemble the structures of cytokine receptors of the hematopoietin family (Fig. 4). It is now widely accepted that IFN receptors together with the individual receptors for IL-10 and its paralogs constitute a distinct subclass of hematopoietin receptors, currently referred to as *type II cytokine receptor family* (CRF2).

Comparison of protein sequences of the IFN receptors of different species has indicated that these have evolved with their cognate IFNs in a concerted fashion, and this has resulted in some degree of species specificity for most IFNs. For example, most HuIFN-$\alpha$ subtypes, HuIFN-ß and HuIFN-$\gamma$ are poorly bound by mouse IFN receptors and vice versa. However, in some instances some IFNs are able to act across species; for example, HuIFN-$\alpha$ subtypes bind as well or better to bovine type I IFN receptors and can trigger antiviral activity in bovine cells.

It has recently been shown that the receptors for IFN-λs are contained within the CRF2 family. All IFN-λs bind to a unique receptor chain known as IL-28R$\alpha$ (CRF2-12), which forms a functional heterodimeric receptor by combining with the accessory receptor chain of IL-10,

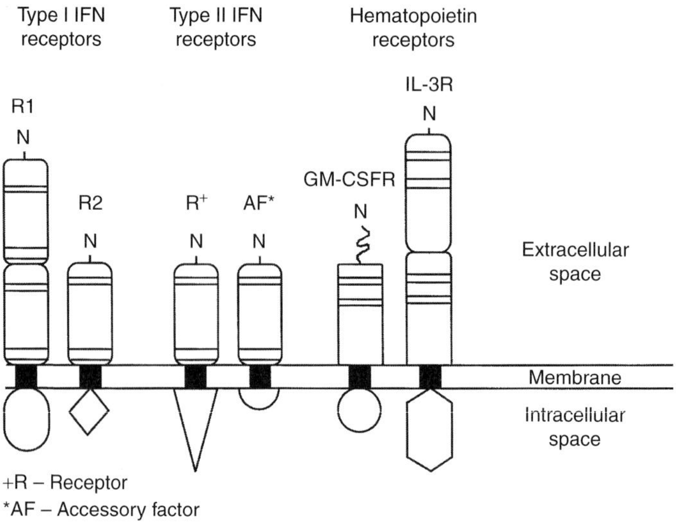


**Fig. 4** Schematic drawing of the domain organization of the human type I IFN and type II IFN$\gamma$ receptors and representative, analogous hematopoietin receptors. The extracellular segments of the granulocyte-macrophage colony-stimulating factor receptor (GM-CSFR) and interleukin-3 receptor (IL-3R) are drawn to linear scale, and conserved cysteines are marked by black bands. Adapted from Bazan, Shared architecture of hormone binding domains in type I and type II interferon receptors, *Cell* (1990) **61**, 753. Copyright Cell Press.

designated IL-10R2. It is not unusual for different cytokine ligands to share receptor components; besides IL-10 and IFN-$\lambda$s, IL-10R2 is also an accessory receptor chain of the IL-22 and IL-26 receptors.

### 4.1.2 Receptor Signaling

The intracellular domains of the type I and type II IFN receptors are not predicted to have kinase activity of any sort: many growth factor receptors (e.g. epidermal growth factor receptor) contain integral tyrosine kinase, which is essential for phosphorylation of downstream, intracellular mediators of the signaling pathway. Nevertheless, IFN receptors do associate with intracellular, nonreceptor kinases. The type I IFN-receptor chains are linked with nonreceptor tyrosine kinases, designated TYK2 (R1) and JAK1 (R2), whereas the type II IFN-receptor chains are associated with JAK1 (R1) and JAK2 (R2 or AF). In type I IFN signaling, TYK2 and JAK1 tyrosine kinases further link to what is known as interferon-stimulated gene factor-3 (ISGF-3), a complex of three/four distinct proteins that contains the primary transcriptional activator for IFN-induced genes. ISGF-3 is normally present in an inactive form being composed of inactive ISGF-3$\alpha$ (three related proteins: p84/91, two alternatively spliced transcripts of the same gene, and p113) and a constitutive and IFN-inducible functional component ISGF-3$\gamma$ (now synonymous with IRF-9) that is associated with p113. The ISGF-3$\alpha$ components have been redesignated as "Signal Transducers and Activators of Transcription" (STATs); p84/91 and p113 are thus STAT1 and STAT2, respectively. STAT1 contains a DNA-binding domain and a nuclear location signal (NLS), which is absent in STAT2; IRF-9 does have an NLS and thus by binding to STAT2 confers ability to be translocated to the nucleus. Following phosphorylation by TYK2 and JAK1, STAT1 and STAT2 dimerize and the ISGF-3 complex (STAT1/STAT2/IRF-9) translocates to the nucleus, where it acts as a transcription factor for the activation of type I IFN-inducible genes (Fig. 5). In contrast, type II IFN (IFN-$\gamma$) stimulation leads to phosphorylation of only p91 STAT1: homodimers of STAT1 act as transcription factors for the activation of IFN-$\gamma$ inducible genes.

For type III IFNs, the IFN-$\lambda$s, it appears that the same signal transduction pathway operating for type I IFNs is activated, despite the distinct receptors for IFN-$\lambda$s. This similarity in signaling is probably due to JAK1 and TYK2 association with IL-28R$\alpha$ and IL-10R2, respectively, the two receptor chains of the IFN-$\lambda$ (IL-28/-29) receptor complex, and the contingent activation of ISGF3 (see above).

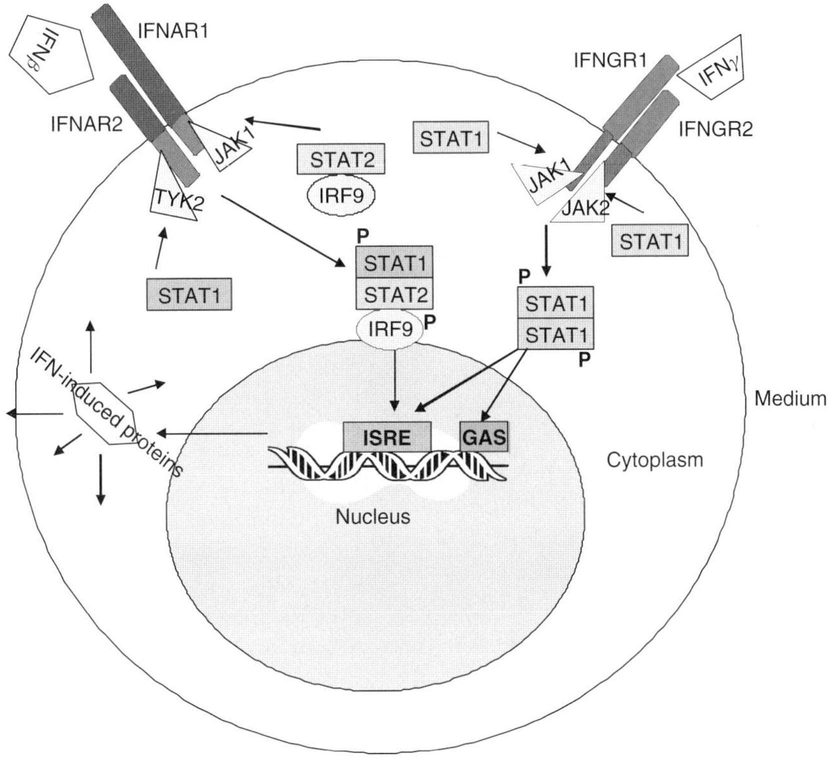


**Fig. 5** Schematic drawing depicting signal transduction pathways from type I and type II IFN receptors at the cell membrane to the cell nucleus. ISGF3, interferon-stimulated gene factor-3. The ISGF3$\alpha$ complex contains three structurally related proteins (p84), p91[STAT1] and p113 [STAT2], and the ISGF3$\gamma$ subunit is composed of a single protein, p48/IRF-9. ISRE, interferon-stimulated response element, present in IFN-inducible genes. GAS, $\gamma$-activated sequences, present in IFN-$\gamma$-inducible genes. JAK1, JAK2 and TYK2, nonreceptor protein tyrosine kinases involved in the phosphorylation of ISGF3$\alpha$ proteins, p91 [STAT1] and p113 [STAT2] (see color plate p. xxiii).

While the "JAK-STAT pathway" is the major intracellular signaling mechanism activated in IFN-treated cells, it may not be the only one. There has been some, albeit somewhat controversial, evidence to implicate protein kinase C (PKC) pathways as well. In particular, the type II IFN receptor appears to trigger PKC translocation from membrane to cytoplasm upon IFN-$\gamma$ binding. However, possibly other non-PKC serine/threonine kinases are activated and are responsible for further signal amplification and activation of transcription factors. IFN-$\gamma$ is a known inducer of IRF-9, which is necessary for the translocation of type I IFN-activatable ISGF-3 from cytoplasm to nucleus.

### 4.2
### IFN-response Gene Sequences

Soon after the binding of type I IFNs to their cognate receptors, a number of nuclear genes are activated and transcripts (mRNA) of a variety of proteins are synthesized. The type I IFN-inducible genes have in common a regulatory nucleotide sequence in their 5′ flanking promoter region, and this type of sequence, which resembles the IRE present in IFN genes, has been designated IFN-stimulated response element (ISRE). The resemblance between ISRE and IRE could account for the fact that many IFN-inducible genes are transcriptionally activated by virus infection or dsRNA, which also activate the transcription of IFN genes (but not *IFNT* genes). The ISRE for type II IFN (IFN-$\gamma$), are similar, but generally known as $\gamma$-activation sequences (GAS). In contrast to the almost consensus ISRE sequences in type I IFN-inducible genes, it appears that GAS are likely to be heterogeneous. It is evident that signaling pathways of both type I and type II IFN systems share common signal transduction components (e.g. STAT1) leading to "convergent" transcriptional activation of a common set of IFN-inducible genes (Fig. 5). However, IFN-$\gamma$ also induces a unique set of protein genes not induced by type I IFNs suggesting the involvement of other transcription factors. For example, IRF-1 (formerly known as ISGF-2) forms complexes with ISRE in both type I and type II IFN-stimulated cells, but is most particularly involved in transcription of type II IFN-inducible genes, for example, class II major histocompatibility (MHC) antigen genes. Current experimental evidence strongly indicates that type III IFNs, the IFN-$\lambda$s, induce the same genes as type I IFNs, although probably more weakly.

## 5
## Proteins Induced by IFNs

It is well established that the biological activity of IFNs depends on selective protein synthesis. Thus, it is clear that at least some of the set of IFN-inducible proteins must mediate the biological activity of IFN. IFN-inducible proteins, whose number probably runs into the 100s, include both those proteins induced early after IFN stimulation and those proteins that are produced at later times, often in response to the actions of early IFN-inducible proteins. Many IFN-inducible proteins are induced by both type I IFN and type II IFN, although some may be preferentially induced by either type of IFN. Yet other proteins are exclusively induced by either type I or type II IFN. In some cases, IFN-inducible proteins may be completely absent from the cell before IFN stimulation, but in other cases the effect of IFN causes an overall increase in production of proteins that are already being expressed. Table 1 shows an incomplete list of IFN-inducible proteins together with their likely functions and their relative inducibility by type I and type II IFNs. Note that since IFN-inducible proteins tend also to be cell type specific, not all of the proteins listed will be expressed in all cell types.

## 6 Biological Activities Associated with IFNs

### 6.1 Antiviral Activity

#### 6.1.1 Molecular Mechanisms

As mentioned earlier, viruses often induce IFN responsive genes, including type I IFNs, *IFNA, IFNB,* and *IFNW* and type

**Tab. 1** IFN-inducible proteins.

| ***Protein*** | ***Function*** | ***Induction by*** | |
|---|---|---|---|
| | | ***Type I IFN*** | ***Type II IFN*** |
| 2,5 A-synthetase | dsRNA-dependent synthesis of ppp-(A2p)n-A (2-5A); activator of RNase L | + | + |
| Protein (p68) kinase (PKR), dsRNA-activatable protein kinase | Phosphorylation of peptide initiation factor eIF-2$\alpha$ | + | + |
| MHC class I (HLA-A, B, C) | Antigen presentation to CTL | + | + |
| MHC class II (HLA-DR) | Antigen presentation to Th-lymphocytes | ± | + |
| Indoleamine 2, 3-dioxygenase | Tryptophan catabolism | − | + |
| Guanylate-binding proteins (GBP; $\gamma$67) | GTP, GDP binding | ± | + |
| Mx | Specific inhibition of influenza virus | + | − |
| IRF1/ISGF2 | Transcription factor | ± | + |
| IRF2 | Transcription factor | + | − |
| IP-10 | Related to chemotactic IL-8-like cytokines | ± | + |
| Metallothionein | Metal detoxification | + | + |
| (TNF) receptors | Mediate TNF action | ± | + |
| IL-2 receptors | Mediate IL-2 action | − | + |
| Intercellular adherence molecule-1 (ICAM-1) | Endothelial cell adhesion protein | − | + |
| Immunoglobulin Fc-receptor (FcR) | Ig binding by macrophages/neutrophils | ± | + |
| Thymosin ß4 | Induction of terminal transferase in lymphocytes | + | ? |
| ß2-microglobulin | Immune function | + | + |
| Nitric oxide synthetase | Production of nitric oxide from arginine. Increased microbicidal activity in macrophages | − | + |
| 200 family | From cluster of 6 genes; p204 is located in nucleolus | + | − |
| 6–16 | Unknown; extracellular | + | ± |
| 1–8/9–27 | Cell surface proteins | + | + |

III IFNs. The corresponding IFNs probably act in a paracrine manner on adjacent cells to induce antiviral mechanisms. The vast number of viruses with different replication strategies notwithstanding, it appears that these agents can be countered by relatively few IFN-inducible "antiviral" proteins. One of the best characterized of these proteins is an enzyme known as 2-5A synthetase which, in the presence of dsRNA (often an intermediate of viral RNA synthesis), catalyzes the formation of an unusual oligonucleotide, ppp (A2′p)nA (2-5A), which in turn activates a latent endonuclease, RNase L. This endonuclease degrades viral (and cellular) mRNA and therefore inhibits viral protein synthesis. Small RNA viruses (*Picornaviridae*), such as Mengo virus or murine encephalomyocarditis virus (EMCV), whose replication is cytoplasmic are most affected by the induction of 2-5A synthetase. A second important antiviral protein is a dsRNA-activatable protein kinase, now designated PKR, which in the active form phosphorylates the peptide initiation factor, eIF2, involved in polyribosomal translation of mRNA. Phosphorylated eIF2 is inactive and thus protein synthesis is inhibited. This inhibition has been associated with the loss of replicating capacity of rhabdoviruses such as vesicular stomatitis virus (VSV). Both type I and type II IFNs have the potential to induce 2-5A synthetase and PKR, but induction occurs more slowly in type II IFN-treated cells. A third "antiviral" protein, Mx, is however primarily induced by type I IFNs; type III IFNs probably weakly induce Mx too. The Mx protein inhibits the replication of influenza viruses in rodent cells, specifically by inhibiting the nuclear stage of influenza viral nucleic acid synthesis.

Probably there are other antiviral proteins and mechanisms by which virus replication is inhibited. For instance, some viruses (e.g. herpesvirus and certain retroviruses) appear to be inhibited at the relatively late stage of virus particle maturation and budding. Table 2 summarizes the antiviral mechanisms applying to different virus families.

**Tab. 2** Antiviral mechanisms for different virus groups.

| *Virus family* | *Principal antiviral mechanism* |
|---|---|
| Small RNA viruses (picornaviruses), including rhinoviruses, poliovirus, encephalomyocarditis virus, Coxsackie viruses | 2-5A synthetase system, inhibition of viral polypeptide synthesis by RNase L degradation of viral mRNA. |
| Rhabdoviruses, e.g. vesicular stomatitis virus (VSV) | dsRNA-activatable protein kinase, phosphorylation of initiation factor eIF-2 to inhibit viral polypeptide synthesis |
| Influenza viruses (orthomyxoviruses) | Mx protein, inhibition of early nuclear phase transcription of viral mRNAs (type I IFN only) |
| Retroviruses, e.g. Rous sarcoma virus, human immunodeficiency viruses (HIV-1, 2) | Acute infections: early inhibition of proviral DNA synthesis or integration. Chronic infections: late inhibition of virion (virus particle) assembly or maturation |
| Herpesviruses, e.g. herpes simplex viruses 1 and 2, varicella virus | Possibly early inhibition of viral protein synthesis by 2-5A synthetase system. Late stage inhibition of viral glycoprotein assembly |

### 6.1.2 Defense Mechanisms of Viruses

In the course of evolution, many viruses have developed countermechanisms by which they can disrupt the antiviral mechanisms induced by IFNs. The IFN-inducible, dsRNA-dependent protein kinase PKR is one of the main targets, and different viruses overcome this by various means. For example, adenoviruses and Epstein–Barr virus (a member of the herpesvirus family) produce small RNA molecules, VAI-, and EBER-RNAs, respectively, which bind to PKR and block its activation by dsRNA. Reoviruses and vaccinia virus (a member of the poxvirus family) produce proteins, sigma 3 and SKI respectively, that bind to dsRNA and thus impair the activation of PKR. Interestingly, if IFN-treated VSV-infected cells are coinfected by vaccinia virus, VSV-replication is rescued, presumably by the inhibitory effect of SKI on PKR. Vaccinia also produces a nonfunctional protein analog of eIF2, which competes with eIF2 for phosphorylation by PKR and thus dilutes out the antiviral effect of activated PKR. Other viruses (Table 3) may "defeat" the kinase by yet other means.

The 2-5A synthetase-RNase L system can also be attacked. For example, encephalomyocarditis virus (EMCV), a picornavirus, can inactivate RNase L in several cell lines, but this inactivation is usually blocked by IFN treatment. Herpesviruses in contrast appear to impair RNase L activation by producing competing analogs of 2-5A.

Some viruses even have the capability to block the transcription of (normally) IFN-inducible genes. The E1A protein of adenoviruses, for example, which is produced early in viral replication, prevents activation of ISGF3 by IFN$\alpha$ (type I IFN). Hepatitis C virus uses a viral protease to block IRF-3 activation and IFN synthesis, while in hepatitis-B virus-infected cells, the so-called virus-specified "terminal protein" inhibits both type I and type II IFN-inducible gene expression.

**Tab. 3** Viral defense mechanisms against IFN antiviral effects.

| *Virus family* | *Defense mechanism* |
|---|---|
| Small RNA viruses, e.g. encephalomyocarditis virus | Inactivation of RNase L of 2-5A synthetase system |
| Herpesvirus | Viral analogs of 2-5A block 2-5A synthetase |
| Epstein–Barr virus (a herpesvirus) | Production of small RNA molecules, EBER-DNA, which block activity of dsRNA-dependent protein kinase (PKR) |
| Adenoviruses | Production of small RNA molecules, VAI-RNA, which block activity of dsRNA-dependent protein kinase (PKR). Early adenoviral E1a protein blocks activation of ISGF-3 by type I IFN |
| Reoviruses and poxviruses, in particular vaccinia virus | Production of viral proteins that bind dsRNA and therefore inhibit activation of PKR. Vaccinia also produces a nonfunctional analog of the initiation factor eIF-2, which competes for PKR-mediated phosphorylation with the cellular PKR |
| Influenza viruses | Activate preexisting inhibitor of PKR |
| Human immunodeficiency virus-1 (HIV-1) | HIV-1 TAT gene mediated inhibition of PKR synthesis. HIV-1 TAR RNA may also inhibit PKR activity |

Finally, members of the poxvirus family, including vaccinia, encode IFN binding proteins, which are soluble and thus able to act as decoy receptors for either type I or type II IFNs. These viral decoy IFN receptors competitively bind IFNs and therefore reduce their antiviral activity.

### 6.2 Antiproliferative Activity

The antiviral actions wrought by IFNs are mediated by enzymes whose mechanisms have broad implications for cell growth and proliferation. Virus replication may be regarded as a form of pathological growth of a foreign, cell-like, entity at the cell's expense. In the presence of IFN, enzymes that generally curtail protein synthesis are activated, but because viral protein synthesis is normally rapid, the inhibitory effect on virus replication appears more dramatic than on the slower and more complex cellular growth. Possibly, the type I IFN system was evolved as part of a complex, interactive network of intercellular mediators that regulate cell growth and proliferation.

The dsRNA-dependent protein kinase, PKR, is constitutively produced in low amounts in a variety of cell types and is made in larger amounts following IFN induction. If a mutant form of this enzyme, which is unable to phosphorylate eIF2, is introduced into cells they undergo neoplastic transformation. This result suggests that normal PKR acts as a tumor-suppressor gene product whose activity is abrogated in the presence of dominant-negative (catalytically inactive) PKR mutants. The 2-5A synthetase-RNase L system could also have tumor-suppressor activity. For example, the levels of these two enzymes are high in growth-arrested cells. Introduction of 2-5A-like oligoadenylates into proliferating cells causes growth impairment, probably through activation of RNase L.

The foregoing observations indicate that enzymes such as PKR, 2-5A synthetase, and RNase L are involved in negative growth control mechanisms. That these enzymes are increased and activated in IFN-treated cells correlates well with IFN's antiproliferative activity. The increase in synthesis of these enzymes is however dependent on IFN-inducible transcription factors. Two of these transcription factors, IRF-1 and IRF-2, behave antagonistically; IRF-1 is a transcriptional activator while IRF-2 is a transcriptional inhibitor of the same set of IFN-inducible genes. IRF-1 has a short half-life and thus transcription of IFN-inducible genes is normally inhibited by the longer lasting IRF-2. After IFN induction, IRF-1 is transiently increased and leads to derepression of IFN-inducible genes. If IRF-2 is overexpressed, cells become transformed, probably because even low-level constitutive production of IFN-inducible gene products, such as PKR and 2-5A synthetase, is blocked. This transformation can be reversed by overexpressing IRF-1.

### 6.3 Antimicrobial Activity

A variety of microorganisms, including bacteria, mycoplasma, rickettsiae, chlamydia, protozoa, and fungi, have been cited as IFN inducers. Since, however, they lack IFN receptors, they are not directly susceptible to IFN actions. Nevertheless, IFNs tend to inhibit the growth of many of these microorganisms, particularly those that are intracellular parasites. It is unlikely, though, that the actions of the enzymes (e.g. 2-5A synthetase, PKR) that inhibit viral replication will affect the replication of intracellular parasites, since their biosynthetic machinery is protected by their cell membrane. Other indirect mechanisms are therefore implicated. For example, type II IFN induces the enzyme indoleamine 2,3-dioxygenase (IDO), which catalyzes the conversion of tryptophan to N-formyl-kynurenine, resulting in tryptophan starvation, an activity that could also contribute to IFN-mediated antiproliferative effect. The growth of certain intracellular parasites (e.g. the protozoan *Toxoplasma gondii*) is inhibited by type II IFN, and tryptophan deprivation appears to be one of the inhibitory mechanisms involved. In phagocytic cells such as macrophages, which can be invaded by protozoans (e.g. *Leishmania*), IFNs, particularly type II IFN, inhibit parasite growth, probably by the induction of synthesis of reactive oxygen intermediates (ROI) and reactive nitrogen intermediates (RNI, e.g. $NO^{\bullet}$), which are toxic. The production of ROI following type II IFN induction may also contribute to the killing of intraerythrocytic malarial parasites.

There is rather less evidence to support a protective role of IFNs in bacterial and mycobacterial infections. Few bacteria appear to be growth inhibited by IFNs, but one exception is *Legionella pneumophilia*, whose replication inside macrophages is inhibited by type II IFN. Again, the production of ROI appears to be involved. It is noteworthy that patients with chronic granulomatous disease (CGD), who are inherently deficient in the respiratory burst oxidase system required for ROI generation, suffer from repeated severe bacterial infections of the deep tissues. At least some CGD patients benefit from type II IFN therapy.

## 6.4
## Immunoregulatory Activity

Emerging evidence strongly suggests the involvement of IFNs not only in innate defense mechanisms but also in shaping adaptive immune responses to microbial attack. For example, type I IFN production by PDC following viral stimulation probably mediates T and NK cell maturation. However, the most crucial action is the increased expression of MHC antigens in leukocytes and other cell types triggered by IFNs. (The MHC antigens are cell surface glycoproteins essential for recognition of foreign, nonself, antigens). Both type I and type II IFNs induce expression of class I MHC antigens (i.e. HLA-A, B, C), but only type II IFN can induce *de novo* synthesis of class II MHC antigens (i.e. HLA-DR, DP, DQ). Class II MHC antigen expression is critically required for presentation of foreign antigen by macrophages and dendritic cells to T-helper ($T_h$) ($CD4^+$) lymphocytes leading to their activation. $T_h$-lymphocytes then secrete a range of cytokines that regulate B-lymphocyte proliferation and immunoglobulin (antibody) synthesis. Class I MHC antigen expression occurs more widely and is required for recognition of foreign antigen by cytotoxic T lymphocytes (CTL, $CD8^+$). For example, recognition of virus-infected cells by CTL depends on class I MHC antigen presentation of viral antigens at the cell membrane. Another category of leukocytes composed of large granular lymphocytes and known as NK cells are also activated by IFN, particularly type I IFN. It is not presently known how type I IFN augments the cytotoxic function of NK cells whose recognition of virus-infected cells or tumor cells appears to be independent of MHC antigen expression.

In the presence of antigen-specific antibodies, macrophages can effect cell-mediated cytotoxicity. Such antibody-dependent cell-mediated cytotoxicity (ADCC) can be enhanced by IFN through augmentation of immunoglobulin G-Fc receptor (IgG-FcR) expression.

Other cell surface proteins whose expression is increased by IFNs include cell adhesion molecules. Intercellular adhesion molecule 1 (ICAM-1) in particular is highly induced by type II IFN, but not by type I IFN, in vascular endothelial cells. The increased expression of ICAM-1 and other cell adhesion molecules leads to lymphocytes and monocytes bearing counterreceptors, for example, those expressing the integrin LFA-1, sticking to the endothelial cell lining of blood vessels from where they can migrate into injured or infected tissues.

Apart from the upregulation of cell surface molecules, IFNs also appear to be involved in the proliferation and functions of B lymphocytes, particularly immunoglobulin (antibody) synthesis. For example, type I IFNs inhibit antibody production when present at high concentrations but may stimulate antibody production at low concentrations. In comparison, type II IFN may act as a costimulant for B-cell proliferation as well as an immunoglobulin isotype switch in antibody-producing cells. Type II IFN (IFN-$\gamma$) enhances $IgG_{2a}$ synthesis and inhibits $IgG_1$ and IgE production, an action that is opposed to that of interleukin-4 (IL-4). Type II IFN may also act by an unknown mechanism as a maturation factor for certain classes of cytotoxic T lymphocytes (CTL). IFNs, in general, are inhibitory for the proliferation of hematopoietic progenitor cells and therefore can depress hematopoiesis and the production of leukocytes.

## 7 Potential Physiological and Pathophysiological Roles of IFNs

### 7.1 Do IFNs have Physiological Roles?

All of the biological activities ascribed to IFNs were discovered through *in vitro* experimentation. Here it is possible to pick and choose conditions that favor particular outcomes (e.g. the antiviral response) by adjusting doses of IFN, times of incubation, levels of virus challenge, and so on. Experiments are frequently performed in cell systems where IFNs are the only exogenous, biologically active reagents present. The whole range of IFN-inducible activities is not demonstrable in a single-cell type; thus *in vitro* experimentation, often involving multiple cell "targets," including tumor cell lines, tends to illustrate, and perhaps exaggerate, the potential of IFNs as multifunctional regulators rather than define their "true" physiological roles. *In vivo*, IFNs together with other cytokines are often coinduced; therefore, it can be difficult to deduce whether IFN, or rather a particular type of IFN, has a major physiological role or one or more subordinate or accessory roles. Furthermore, although IFNs can be shown to have actions that regulate cell growth and differentiation *in vitro*, it is much harder to show unequivocally that IFNs are involved in physiological processes such as embryogenesis, organ and tissue development, and homeostatic mechanisms. That IFNs have the potential for inducing antiviral, antimicrobial, and antitumor activity suggests that their true physiological worth is as regulators of host defense mechanisms, to prevent pathophysiological events.

## 7.2
## Ways of Studying IFN "Physiology"

Probably the best approach to determine how IFNs work *in vivo* is to investigate animals in which a deficiency of the IFN system exists or can be induced. For the IFN system, however, natural genetic deficiencies, either in the IFN genes or in the IFN-receptor genes, are probably very rare. Nevertheless, it is now possible to knockout the function of particular genes by specific site-directed disruption of the gene, for example, in mice. This approach involves targeting a nonfunctional gene by means of a vector into the mouse embryonic stem (ES) cell genome resulting in disruption of gene expression. Subsequently, ES cell lines are selected for the disrupted gene and injected into blastocysts. A program of breeding of the resulting chimeric animals and selection of offspring carrying the disrupted gene in the germline eventually leads, as long as this is not prenatally lethal, to the generation of viable null or knockout mice, which are homozygous for the disrupted gene.

To date, knockout mice with disrupted IFN, IFN-receptor, IRF, JAK, and other IFN-stimulated genes have been generated. For example, knockout mice with either a disrupted type II IFN gene or a disrupted type II IFN-receptor gene have been produced. Both kinds of mice developed normally and were healthy, but they exhibited a defect in natural resistance when challenged with pathogens (e.g. vaccinia virus or *Listeria monocytogenes* or mycobacteria). Macrophages in these mice showed impaired ROI and nitric oxide production and expression of class II MHC antigen expression; NK cell activity, but not CTL activity, was reduced while splenocytes proliferated in an uncontrolled manner in response to mitogen or alloantigen. Knockout mice deficient in the IFN receptor lacked an antigen-specific $IgG_{2a}$ response. It thus has been clearly demonstrated that type II IFN (IFN-$\gamma$) is essential for immune responses to pathogens by several cell types.

Production of macrophage nitric oxide is crucially dependent on type II IFN-inducible nitric oxide synthetase (iNOS), whose gene transcription is regulated by IRF-1, a transcriptional enhancer. Thus, mice deficient in IRF-1 are also deficient in iNOS and susceptible to mycobacterial (e.g. *M. tuberculosis*) infections, which they are unable to clear. IRF-1 knockout mice also share other phenotypic similarities, such as reduced NK activity, with IFN-$\gamma$R knockout mice.

Although IFN-$\gamma$R knockout mice were susceptible to vaccinia virus, they were no more susceptible to other viruses (e.g. Semliki Forest virus (SFV), VSV) than their normal littermates. In contrast, knockout mice defective in type I IFN-receptor R1 gene, while displaying no overt phenotypic abnormalities, were unable to counter infection with several different viruses, including SFV and VSV. These findings indicate that type I and type II IFNs exert their antiviral effects through different, partially nonredundant pathways.

Certain mouse strains are more susceptible to infection by viruses or microorganisms, and in some cases this can be attributed to low type I IFN production. In addition, injection of mice with antitype I IFN antibody has been shown to increase their susceptibility to a range of virus infections. Children who suffer repeated rhinovirus (common cold) infections also have been shown to secrete lower than average levels of type I IFNs.

There are many molecularly different species of type I IFN (e.g. IFN-$\alpha_1$,

$\alpha_2$, ...$\alpha_n$, IFN-ß, IFN-$\kappa$, IFN-$\omega$, IFN-$\tau$). Except for IFN-$\tau$, which has a special role as an antiluteolytic agent in ruminant pregnancy, there is no good evidence why such a large diversity of type I IFN molecules is necessary. Qualitatively, there are no differences in the *in vitro* biological activities of IFN-$\alpha$, IFN-ß, and IFN-$\omega$. Quantitative differences among these IFNs can be demonstrated, but these probably reflect their differing affinities for the type I IFN receptor. Possibly, differences in their hydrophobicity causes individual type I IFNs to locate more efficiently to particular tissues or anatomical sites but, as yet, there is little to support this hypothesis.

## 7.3 IFNs and Pathophysiological Phenomena

IFN genes are normally switched off and only expressed transiently during intervals of acute illness or injury. IFNs are thus induced to contribute to the host's specific and nonspecific defense mechanisms to combat infectious diseases and resolve tissue injuries. They are part of a network of mediators, including many other cytokines, responsible for regulatory immune responses and inflammatory reactions. However, if immunity and inflammation are excessive, pathophysiological phenomena often arise. IFNs are usually secreted locally, but can, if overexpressed, find their way into the circulation and produce symptoms such as fever, malaise, and headache, often associated with such acute viral infections as influenza. Such symptoms can be reproduced in patients receiving therapeutic doses of exogenous IFN-$\alpha$ or IFN-ß for malignancies and so it is clear that it is the IFN molecule itself that has the capacity to induce undesirable "side effects."

IFNs, and particularly type II IFN (IFN-$\gamma$), are proinflammatory. By increasing recognition of infected cells through augmentation of MHC antigen expression and enhancing cell adhesion molecules, leukocytes (neutrophils, lymphocytes, and macrophages) are recruited to sites of infected or injured tissues, are activated by antigen and cytokines to become cytotoxic or phagocytic, and synthesize a range of inflammatory and vasoactive mediators. In damaged skin, for instance, this results in local reddening and swelling before healing and resolution of the damaged area occurs. Infection or injury occurring a second time at the same site sometimes leads to a more severe inflammatory reaction as leukocytes are "primed" to infiltrate the site. This phenomenon is known as delayed hypersensitivity (DH). The severe complications (e.g. hemorrhagic shock) observed in children suffering from a second dengue virus infection (normally of a different serotype to the one that caused the primary infection) probably result from DH reactions. In a murine model, DH has been shown to be inhibited by neutralizing antibody against IFN-$\gamma$, thus strongly implicating IFN-$\gamma$ as a proximal mediator of DH.

## 7.4 IFNs and Pathogenesis

On some occasions, the presence of IFNs correlates well with disease progression. Again, in a murine model, *Plasmodium berghei* induces cerebral malaria, often a severe complication of human infection with *Plasmodium falciparum*. Neutralizing antibodies to IFN-$\gamma$ injected prior to infection with *P. berghei*, however, prevented the development of cerebral malaria. IFN-$\gamma$ and other cytokines (e.g. TNF-$\alpha$) are also partly responsible for the formation of granulomas, nodular-like tissue masses characteristic of infections caused by some microorganisms, particularly mycobacteria (e.g. *Mycobacterium leprae*, the microbe that causes leprosy). Granulomas prevent spread of the mycobacteria, and since they contain large numbers of macrophages and neutrophils, mycobacteria can be killed by IFN-$\gamma$ induction of ROI and nitric oxide synthesis. If, for example, ROI production fails to clear the mycobacteria and persists over an extended period, however, this cell-mediated response may lead to chronic inflammation and tissue damage. In such cases, chronic disease might result not from excessive IFN-$\gamma$ levels, but from continuous, low-level production, which is insufficient to induce resolution of the disease. Other types of acute illnesses (e.g. viral hepatitis) may also become chronic due to a failure of the IFN system to clear the virus or microorganism. In both hepatitis-B and -C virus infections, viral mechanisms actively inhibit type I IFN production and action, and probably contribute to the establishment of chronic virus infection.

IFNs have also been associated with autoimmune diseases, but while levels of certain IFN types may be raised in such diseases, it is difficult to distinguish whether the IFNs are contributing to pathogenesis or whether they are by-products of underlying pathological processes. Autoimmune diseases usually result from a breakdown in immunological tolerance to self-antigens, but the causative agents are mostly unknown. It is, however, widely believed that viruses, bacteria, or mycoplasma, often infecting individuals years prior to the onset of disease, are responsible for lymphocyte sensitization to self-antigens. There are some genetic factors also (e.g. MHC antigen haplotypes) that increase susceptibility to particular autoimmune diseases. The inbred New Zealand Black (NZB) mice, for example, "spontaneously" develop autoantibodies and glomerulonephritis; this autoimmune disease progresses more rapidly if the mice are injected with type I or type II murine IFN and less rapidly if the mice are given neutralizing antitype II murine IFN antibody. Type I IFN-receptor deficiency in homozygous IFN-$\alpha/\beta$R gene deleted NZB mice also resulted in reduced lupus-like disease. These findings suggest that IFNs have a pathogenic role in this autoimmune disease of NZB mice.

A small number of people suffer from systemic lupus erythematosus (SLE), an autoimmune disease characterized by high titers of autoantibodies of a range of specificities, including nuclear components, DNA, and nucleosomes leading to immune complex induced vasculitis and kidney failure, a feature in common with the NZB mice's autoimmune disease. Circulating IFN-$\alpha$ can be found in some SLE patients, with levels correlating to disease activity. Its role remains enigmatic, but there is increasing evidence of IFN-$\alpha$ being involved on both the initiation and maintenance of the central autoimmune processes of SLE. It is hypothesized that a specific class of dendritic cells is stimulated by viruses to secrete IFN-$\alpha$ in the first phase and by immune complexes in the second phase, thus creating a vicious cycle mechanism driving autoimmune processes. The unabated IFN-$\alpha$ production might contribute to autoimmune disease in a number of ways (e.g. by stimulating B-cell production of autoantibodies, by increasing MHC antigen expression and thus stimulating autoreactive T and B cells. In support of this hypothesis, an IFN-$\alpha$ "signature" of many IFN-induced genes has been demonstrated in white blood cells from patients with active SLE.

**Tab. 4** IFNs and disease correlations.

| *Disease* | *IFN type* | *Putative pathogenic mechanism(s)* |
|---|---|---|
| Systemic lupus erythematosus (SLE) | IFN-$\alpha$ | Stimulation of autoantibodies, autoreactive T and B cells |
| Insulin-dependent diabetes mellitus (IDDM) | IFN-$\alpha$/IFN-$\gamma$ | Upregulation of MHC class I and class II antigens in ß-islet cells, generation of autoreactive cytotoxic T lymphocytes |
| Behçets disease (eye lesions) | "Acid labile" – IFN-$\alpha$/IFN-$\gamma$ | Increased MHC class I and class II antigen expression. Increased autoantibody production |
| Grave's ophthalmopathy (immunological cytotoxicity against eye muscle cells) | IFN-$\gamma$ | Increased MHC class II antigen expression. Increased ADCC of eye muscle cells |
| Multiple sclerosis | IFN-$\gamma$ | Generation of autoreactive CTL in central nervous system |
| Sjögren's syndrome (chronic inflammatory disease of exocrine glands) | IFN-$\gamma$ | Downregulation of type I IFN receptors in NK cells leading to loss of NK cell activity |
| Aplastic anemia (primary disorder of hematopoietic system) | IFN-$\alpha$/IFN-$\gamma$ | Inhibition of hematopoietic precursor proliferation |

In tissue-restricted autoimmune diseases, such as insulin-dependent diabetes mellitus (IDDM), localized IFN secretion in a particular tissue or tissues may also be involved. For example, the ß-islet cells in the pancreas of IDDM patients hyperexpress class I and class II MHC antigens suggesting the presence of types I and II IFN, respectively. This has been confirmed by immunohistochemical staining of sections of pancreases. There are several other autoimmune or autoimmune-like diseases for which IFNs have been associated and these are summarized in Table 4.

## 8
## Clinical Uses of IFNs

The potent antiviral activity of IFNs together with their potential antitumor actions provided the spur to the large-scale manufacture of IFNs for the purpose of clinical evaluation in a variety of viral and malignant diseases. In the early 1970s, large-scale production was limited by the quantities of human buffy coats (leukocytes) which could be used to make type I IFN. Later in that decade, human lymphoblastoid cells (e.g. Namalwa cell line), which could be grown to large culture volumes, became available for type I IFN production. By the 1980s, following the cloning of IFN-$\alpha$ subtype genes, some IFN species were mass-produced in *E. coli* or, in some instances, in mammalian cells, leading to abundant availability of certain subtypes (e.g. IFN-$\alpha_2$). Production of IFN-ß, IFN-$\omega$ and IFN-$\gamma$ by this means followed. Synthesis or construction of consensus sequence IFNA genes and various hybrid IFNA genes (e.g. A1/A2, A1/A8) has led to the production of nonnatural IFN-$\alpha$ molecules with enhanced activities or stability.

As a result of the anecdotal or often uncontrolled clinical studies of the 1970s, there was great enthusiasm in the 1980s, both from manufacturers of IFNs and clinicians, to test the therapeutic potential of interferons. It was only after many controlled or randomized studies had been conducted that it became obvious that IFN administered as a single agent was not beneficial for the treatment of the majority of malignant diseases, including the major cancers, lung, breast, colon. The initial optimism all but vanished and was replaced in the mid-1980s by a more sober and realistic appreciation of the potential therapeutic value of IFNs.

One major problem remains, that is: there is little real understanding of how IFNs were having their antitumor effects. Although IFNs had been shown to have antiproliferative activity in many tumor cell lines *in vitro*, it was also known that other tumor cell lines that were resistant to IFNs *in vitro* were nevertheless destroyed by IFN treatment when grown as transplantable tumors in mice. This indicated that IFNs probably acted on tumors *in vivo* by indirect mechanisms (e.g. augmentation of NK cell activity or enhancement of MHC antigen expression and antitumor immunity). In other words, their antitumor effects had more to do with modifying biological responses than direct antiproliferative actions. IFNs and cytokines in general are sometimes referred to as biological response modifiers (BRMs). However, it is still not clear what biological and pharmacological responses are necessary for IFN-induced tumor regression to occur.

In patients IFNs are administered in high doses, often close to the maximum tolerated dose (MTD): IFN, a "biological drug" is thus being used as if it were a chemical drug, a practice that sometimes leads to severe side effects (Table 5) and toxicity. Therefore, it is difficult to establish whether IFN is acting on tumors in the manner of a chemotherapeutic agent (i.e. direct action) or as a BRM (i.e. indirect action) or as both. It has become apparent nonetheless that IFNs administered at low doses are generally less effective in causing tumor regression. In addition, high-dose IFN therapy over a long period causes leukopenia (i.e. reduced numbers of leukocytes that respond to IFN and mediate indirect IFN antitumor effects), relative immunosuppression, and autoimmune phenomena. It is therefore difficult, against a background of generalized toxic effects, to "harness" whatever antitumor activity IFNs may have for particular tumor types.

So far IFNs have only been partially successful in the treatment of cancers, and only for a very limited number of types of malignancy. Clinical trials have shown that the efficacy of individual IFN-$\alpha$ subtypes, in particular IFN-$\alpha_2$, is more or

**Tab. 5** Side effects of type I IFN therapy.

| |
|---|
| *Observed clinical responses* (frequent) |
| Fever |
| Chills |
| Malaise/Fatigue |
| Myalgia |
| Headache |
| Anorexia/weight loss |
| *Pharmacological toxicity* (infrequent) |
| Liver and kidney toxicity |
| Inhibition of hematopoiesis, leukopenia |
| Central nervous system toxicity, e.g. confusion, altered mental states |
| Heart disturbances, e.g. arrhythmia, ischemia |
| *Immunological responses* |
| Depressed natural killer cell activity and ADCC on prolonged therapy schedules |
| Induction of autoantibodies |

less the same as heterogeneous, multicomponent, type I IFN preparations such as lymphoblastoid and leukocyte IFN, both of which contain mixtures of IFN-$\alpha$ subtypes. IFN-ß is probably as efficacious, but it has not been used as widely as IFN-$\alpha$ preparations in antitumor therapies. IFN-$\gamma$, despite its apparent greater immunoregulatory activity, has been found to be rather less potent than IFN-$\alpha$. All IFNs in high doses cause significant, but readily reversible, side effects (Table 5). The most responsive cancer to IFN-$\alpha$ therapy is a very rare form of leukemia known as hairy cell leukemia (HCL), where up to an 80% response rate has been reported. In HCL patients, the "hairy cells" invade the spleen and bone marrow. It has been convincingly shown that IFN-$\alpha$ therapy continued over several months leads to clearance of "hairy cells" and in some patients, a long-term remission is achieved. Some patients develop neutralizing antibodies against IFN, particularly IFN-$\alpha_2$, which may cause resistance to further treatment. In these cases, beneficial responses maybe rescued by switching to a multicomponent IFN-$\alpha$ preparation. On the whole, IFN-$\alpha$ therapy of HCL appears at least as effective and durable as chemotherapy with the drug pentostatin (2-deoxycoformycin).

IFN-$\alpha$ therapy has also been shown to slow down the progression of CML. In this cancer, leukemic cells grow slowly in the initial chronic phase and persist for 2 to 4 years, but there follows a dramatic "blast crisis," producing rapidly proliferating myeloid leukemia cells and a fatal outcome. CML patients can be treated with IFN-$\alpha$ in the chronic phase, and some achieve complete hematological remission (i.e. disappearance of leukemic cells, and often increased survival). Recently, IFN-$\alpha$ therapy of CML has been replaced by the administration of tyrosine kinase inhibitors such as Gleevec®, which block c-abl activity in leukemic cells, and thus their proliferation.

Other malignancies in which IFN-$\alpha$ therapy seems to work, although with a lower percentage of patients responding than in HCL and CML, include low-grade non-Hodgkin's lymphoma, cutaneous T-cell lymphoma, AIDS-related Kaposi's sarcoma, carcinoid tumors, renal cell carcinoma, squamous epithelial tumors of the head and neck, multiple myeloma, and malignant melanoma (Table 6). In most of these cancers, complete responses are low compared to partial responses, but IFN-$\alpha$ may help with maintenance therapy of disease in some cases. In preclinical systems, the combination of IFN therapy and conventional chemotherapy has appeared to offer greater chances of producing effective treatment of many cancers, but in clinical trials this strategy has produced mostly disappointing results, probably because optimal combinations of IFNs with cytotoxic drugs cannot be deduced owing to a lack of understanding of the physiological interaction of these agents.

As mentioned earlier, IFN therapy induces leukopenia and generalized immunosuppression. This effect may be useful for the treatment of diseases in which there is uncontrolled leukocytosis (e.g. thrombocytosis: markedly elevated platelet numbers) associated with various myeloproliferative diseases. Parenteral (subcutaneous or intramuscular) injection of IFN-ß appears to benefit multiple sclerosis (MS) patients in the early stages of this disease by leading to a reduced rate of exacerbations. This, however, could be a result of the relative immunosuppression caused by IFN-ß (e.g. by inhibiting IFN-$\gamma$ actions and thereby

Tab. 6 Responses seen in human malignancies following IFN$\alpha$ therapy.

| | | ***Approximate response rate[a] [%]*** |
|---|---|---|
| *Hematological malignancies:* | Hairy cell leukemia | 80 |
| | Chronic myelogenous leukemia (preblast crisis) | 79 |
| | Cutaneous T-cell lymphoma | 52 |
| | Non-Hodgkin's lymphoma | 46 |
| | Multiple myeloma | 20 |
| *Solid tumors:* | Squamous tumors of head and neck | 91% |
| | Carcinoid tumors | 55 |
| | Bladder cancer | 65 |
| | Cervical cancer | 43 |
| | Kaposi's sarcoma (AIDS-related) | 30 |
| | Ovarian cancer | 19 |
| | Renal cell carcinoma | 17 |

Notes: Intermediate and high-grade lymphomas, Hodgkin's disease, and chronic lymphocytic leukemia, all less than 20%. Malignant melanoma, breast cancer, colorectal carcinoma, lung cancer, osteogenic sarcoma, all around 10% or less.
[a]Response rate includes complete, partial, and minor responses.

suppressing the growth and activity of autoreactive T lymphocytes in the central nervous system.

Type II IFN, IFN-$\gamma$, has been rather less successful in the treatment of cancers, but uniquely it has been found to benefit some chronic granulomatous disease (CGD) patients, probably by boosting and thus correcting the impaired ROI producing capacity of macrophages and neutrophils. IFN-$\gamma$ may also be useful in the adjunctive treatment of leishmaniasis and mycobacterial diseases.

In the treatment of viral diseases, IFNs, despite having proven antiviral activity *in vitro,* have not proved to be the panacea for most common viral infections of man. IFNs prevent the replication of rhinoviruses in cell cultures when administered to volunteers in the form of a nasal spray, but IFNs cannot "cure" colds once established. IFNs are only partially effective in preventing influenza infection. Topical applications of IFN in the form of creams or ointments to herpesvirus lesions (e.g. in herpes zoster) and genital warts caused by papilloma viruses have been tried but have given limited beneficial effects. It has been found that parenteral administration (i.e. intramuscular or intravenous injection) has greater beneficial effects of IFN-$\alpha$ on virus-caused lesions and warts, although not to the extent that IFN therapy has become the treatment of choice. Another wartlike disease, juvenile laryngeal papilloma (JLP), which can severely obstruct the airways of young children, also has been found to respond beneficially to IFN-$\alpha$ therapy. Nevertheless, it appears neither curative nor of substantial value as an adjunctive agent in the long-term management of JLP.

Probably, the most successful application of IFN-$\alpha$ therapy in virus-mediated diseases is in the treatment of chronic active hepatitis, caused by either hepatitis-B or -C viruses. A certain proportion of chronic active hepatitis-B patients respond positively to IFN-$\alpha$ treatment. For instance, viral infectivity markers disappear and seroconversion and cure follows. It is interesting that in the case of hepatitis-B virus, viral activity is probably responsible for inhibiting the endogenous IFN system, and therefore the administration of exogenous IFN-$\alpha$ constitutes a replacement therapy. In hepatitis C virus infection, the most common cause of transfusion-linked hepatitis, IFN-$\alpha$ is the treatment of choice. Recent improvements in combination therapies using "pegylated" IFN-$\alpha_2$ (the pegylation markedly increases the half-life of IFN in the circulation) and the antiviral drug ribovarin have resulted in cure rates approaching 70%. However, some strains of the virus, in particular genotype 1, which is the most prevalent worldwide, are strongly resistant to IFN therapy. This resistance has been recently demonstrated to be associated with the capacity of the virus to block IRF-3 activation, which is essential for antiviral activity.

Type I IFN, both IFN-$\alpha$ and IFN-ß, has been shown to inhibit HIV-1 replication *in vitro*. However, *in vivo* there is little evidence showing IFN-$\alpha$ has any long-term beneficial effect in HIV-1 positive individuals. IFN-$\alpha$ can cause tumor regression in some AIDS patients with Kaposi's sarcoma, but this possibly has more to do with IFNs antitumor or antiangiogenic activity than with its antiviral one. Combination therapies for HIV-1 infections involving IFN-$\alpha$, antiviral drugs such as 3′-azido-2′,3′-dideoxythymidine (AZT) and cytokines, such as granulocyte-macrophage colony-stimulating factor (GM-CSF), appeared promising, but have not been followed up.

## 9 Final Remarks

Since their discovery over four decades ago, IFNs have ceased to be poorly characterized antiviral factors; both at the molecular and cell biological levels; they are now among the best defined and better understood biologically active intercellular mediators. However, despite the large body of information concerning IFNs and their actions, there are still many issues to be resolved. Why, for example, is the type I IFN family so heterogeneous? There is a need to find more ways to translate the potent antiviral, antimicrobial, and antitumor activities of IFNs into the means for combating human disease. Future progress will depend on further understanding not only of IFN-mediated actions at the molecular and cellular level but also of the numerous ways IFNs can interact with other biologically active mediators, including cytokines, hormones, and chemical drugs.

## Bibliography

### Books and Reviews

Barnes, B., Lubyova, B., Pitha, P.M. (2002) On the role of IRF in host defense, *J. Interferon Cytokine Res.* **22**, 59–71.

Baron, S., Coppenhaven, D.H., Dianzani, F., et al. (Eds.) (1992) *Interferon: Principles and Medical Applications.* International Society for Interferon Research, University of Texas Medical Branch, Galveston, TX.

Billiau, A. (1996) Interferon-$\gamma$: biology and role in pathogenesis, *Adv. Immunol.* **62**, 61–90.

Brierley, M.M., Fish, E.N. (2002) IFN-$\alpha/\beta$ receptor interactions to biologic outcomes:

understanding the circuitry, *J. Interferon Cytokine Res.* **22**, 835–845.
Byrne, G.I., Turco, J. (Eds.) (1988) *Interferon and Non-viral Pathogens*, Dekker, New York.
Corssmit, E.P.M., De Metz, J., Sauerwein, H.P., Romijn, J.A. (2000) Biologic responses to IFN-$\alpha$ administration in humans, *J. Interferon Cytokine Res.* **20**, 1039–1047.
De Maeyer, E., De Maeyer-Guignard, J. (1988) *Interferons and other Regulatory Cytokines*, Wiley, New York.
De Maeyer, E., DeMaeyer-Guignard, J. (1998) Interferon gamma, in: Mire-Sluis, A., Thorpe, R. (Eds.) *Cytokines*, Academic Press, San Diego, CA, p. 391.
Gutterman, J.U. (1994) Cytokine therapeutics: lessons from interferon $\alpha$, *Proc. Natl. Acad. Sci. U.S.A.* **91**, 1198.
Ihle, J., Kerr, I. (1995) JAKs and STATs in signaling by the cytokine receptor superfamily, *Trends Genet.* **11**, 69–74.
Langer, J.A., Cutrone, E.C., Kotenko, S. (2004) The class II cytokine receptor (CRF2) family: overview and patterns of receptor-ligand interactions, *Cytokine Growth Factor Rev.* **15**, 33–48.
Meager, A. (1995) Interferons in clinical use: an appraisal, *J. Biotechnol. Health Care* **2**, 281–317.
Meager, A. (1998) *The Molecular Biology of Cytokines*, John Wiley, Chichester, UK.
Meager, A. (2002) Biological assays for interferons, *J. Immunol. Methods* **261**, 21–36.
Mogensen, K.E., Lewernz, M., Reboul, J., et al. (1999) The type I interferon receptor: structure, function, and evaluation of a family business, *J. Interferon Cytokine Res.* **19**, 1069–1098.
Pestka, S., Krause, D.K., Walter, M.R. (2004) Interferons, interferon-like cytokines, and their receptors, *Immunol. Rev.* **202**, 8–32.
Roberts, R.M., Cross, J.C., Leaman, D.W. (1992) Interferons as hormones of pregnancy, *Endocr. Rev.* **13**, 432–452.
Ronnblom, L., Alm, G.V. (2002) The natural interferon-$\alpha$ producing cells in systemic lupus erythematosus, *Hum. Immunol.* **63**, 1181–1193.
Sen, G.C., Lengyel, P. (1992) The interferon system: a bird's eye view of its biochemistry, *J. Biol. Chem.* **267**, 5017–5020.
Stark, G.R., Kerr, I.M., Williams, B.R., et al. (1998) How cells respond to interferons, *Annu. Rev. Biochem.* **67**, 227.
Stewart, W.E. II, (1979) *The Interferon System*, Springer-Verlag, Vienna.
Wadler, S., Schwartz, E.L. (1990) Antineoplastic activity of the combination of interferon and cytotoxic agents against experimental and human malignancies: a review, *Cancer Res.* **50**, 3473–3486.

## Primary Literature

Aboud, M., Hassan, Y. (1983) Accumulation and breakdown of RNA-deficient intracellular virus particles in interferon-treated NIH 3T3 cells chronically producing Moloney murine leukaemia virus, *J. Virol.* **45**, 489–495.
Adolf, G.R. (1990) Monoclonal antibodies and enzyme immunoassays specific for human interferon (IFN) $\omega$1: evidence that IFN-$\omega$1 is a component of human leukocyte IFN, *Virology* **175**, 410–417.
Aguet, M., Dembic, Z., Merlin, G. (1988) Molecular cloning and expression of the human interferon-$\gamma$ receptor, *Cell* **55**, 273–280.
Allen, G., Diaz, M. (1996) Nomenclature of the human interferon proteins, *J. Interferon Cytokine Res.* **16**, 181–185.
Barber, G.N., Jagus, R., Meurs, E.F., et al. (1995) Molecular mechanisms responsible for malignant transformation by regulatory and catalytic domain variants of the interferon-induced enzyme RNA-dependent protein kinase, *J. Biol. Chem.* **270**, 17423–17428.
Borden, E.C., Hogan, T.F., Voelkel, J.G. (1982) Comparative antiproliferative activity in vitro of natural interferons $\alpha$ and $\beta$ for diploid and transformed human cells, *Cancer Res.* **42**, 4948–4953.
Bouillon, M., Audette, M. (1993) Transduction of retinoic acid and $\gamma$-interferon signal for intercellular adhesion molecule-1 expression on human tumour cell lines: evidence for the late-acting involvement of protein kinase C inactivation, *Cancer Res.* **53**, 826–832.
Capon, D.J., Shepard, H.M., Goeddel, D.V. (1985) Two distinct families of human and bovine interferon-$\alpha$ genes are co-ordinately expressed and encode functional polypeptides, *Mol. Cell. Biol.* **5**, 768–779.
Coccia, E.M., Severa, M., Giacomini, D., et al. (2004) Viral infection and Toll-like receptor agonists induce a differential expression of type I and $\lambda$ interferons in human

plasmacytoid and moncyte-derived dendritic cells, *Eur. J. Immunol.* **34**, 796–805.

Da Silva, A.J., Brickelmaier, M., Majeau, G.R., et al. (2002) Comparison of the gene expression patterns induced by treatment of human umbilical vein endothelial cells with IFN-$\alpha$2b vs. IFN-$\beta$1a: understanding the functional relationship between distinct type I interferons that act through a common receptor, *J. Interferon Cytokine Res.* **22**, 173–188.

Deblandre, G.A., Marinx, O.P., Evans, S.S., et al. (1995) Expression cloning of an interferon-inducible 17-kDa membrane protein implicated in the control of cell growth, *J. Biol. Chem.* **270**, 23860–23866.

Diaz, M.O., Ziemin, S., Le Beau, M.M., et al. (1988) Homozygous deletion of the $\alpha$- and $\beta$-genes in human leukaemia and derived cell lines, *Proc. Natl. Acad. Sci. U.S.A.* **85**, 5259–5263.

Dumoutier, L., Tounsi, A., Michaels, T., et al. (2004) Role of the interleukin-28 receptor tyrosine residues for the antiproliferative activity of IL-29/IFN-$\lambda$1: similarities with type I interferon signaling, *J. Biol. Chem.* **279**, 32269–32274.

Ealick, S.E., Cook, W.J., Vijay-Kumar, S., et al. (1991) Three-dimensional structure of recombinant interferon-$\gamma$, *Science* **252**, 698–702.

Fellous, M., Kamoun, M., Gresser, I., Bono, R. (1979) Enhanced expression of HLA antigens and $\beta$2-microglobulin on interferon-treated human lymphoid cells, *Eur. J. Immunol.* **9**, 446–449.

Gray, P.W., Leung, D.W., Pennica, D., et al. (1982) Expression of human immune interferon cDNA in *E. coli* and monkey cells, *Nature (Lond.)* **295**, 503–508.

Gupta, S.L., Holmes, S.L., Mehra, L.L. (1982) Interferon action against reovirus: activation of interferon-induced protein kinase in mouse L929 cells upon reovirus infection, *Virology* **120**, 495–499.

Hauptmann, R., Swetly, P. (1985) A novel class of human type I interferons, *Nucleic Acids Res.* **13**, 4739–6219.

Heron, I., Hokland, M., Berg, K. (1978) Enhanced expression of $\beta$2-microglobulin and HLA antigens on lymphoid cells by type I interferons, *Proc. Natl. Acad. Sci. U.S.A.* **75**, 6215–6219.

Hokland, M., Berg, K. (1981) Interferon enhances the antibody-dependent cellular cytotoxicity (ADCC) of human polymorphonuclear leukocytes, *J. Immunol.* **127**, 1585–1588.

Horisberger, M.A. (1992) Interferon-induced human protein MxA is a GTPase which binds transiently to cellular proteins, *J. Virol.* **66**, 4705–4709.

Hovanessian, A.G., Wood, J.N. (1980) Anticellular and antiviral effects of ppp(2′5′A)$_n$, *Virology* **101**, 81–90.

Karpusus, M., Nolte, M., Benton, C.B., et al. (1997) The crystal structure of human interferon $\beta$ at 2.2-A resolution, *Proc. Natl. Acad. Sci. U.S.A.* **94**, 11813–11818.

Kotenko, S.V., Gallagher, G., Baurin, V.V., et al. (2003) IFN-$\lambda$s mediate antiviral protection through a distinct class II cytokine receptor complex, *Nat. Immunol.* **4**, 69–77.

Kroger, A., Koster, M., Schroeder, K., et al. (2002) Activities of IRF-1, *J. Interferon Cytokine Res.* **22**, 5–14.

Kumar, R., Choubey, D., Lengyel, P., Sen, G.C. (1988) Studies on the role of the 2′-5′-oligoadenylate synthetase-RNase L pathway in $\beta$-interferon-mediated inhibition of encephalomyocarditis virus replication, *J. Virol.* **62**, 3175–3181.

LaFleur, D.W., Nardelli, B., Tsareva, T., et al. (2001) Interferon-kappa, a novel type I interferon expressed in human keratinocytes, *J. Biol. Chem.* **276**, 39765.

McNurlan, M., Clemens, M. (1986) Inhibition of cell proliferation by interferons: relative contributions of changes in protein synthesis and breakdown to growth control of human lymphoblastoid cells, *Biochem. J.* **237**, 871–876.

Meurs, E.F., Galabru, J., Barber, G.N., et al. (1993) Tumor suppressor function of the interferon-induced double-stranded RNA-activated protein kinase, *Proc. Natl. Acad. Sci. U.S.A.* **90**, 232–236.

Müller, U., Steinhoff, U., Reis, L., et al. (1994) Functional role of type I and type II interferons in antiviral defense, *Science* **264**, 1918–1921.

Nagata, S., Mantei, N., Weissmann, C. (1980) The structure of one of the eight or more distinct chromosomal genes for human interferon-$\alpha$, *Nature (Lond.)* **287**, 401–408.

Pellegrini, S., John, J., Shearer, M., et al. (1989) Use of a selectable marker regulated by alpha interferon to obtain mutations in the signalling pathway, *Mol. Cell. Biol.* **9**, 4605–4612.

Pestka, S., Meager, A. (1995) Interferon standardization and designations, *J. Interferon Cytokine Res.* **17**(Suppl. 1), S9.

Qin, X.-Q., Runkel, L., Deck, C., et al. (1997) Interferon-$\beta$ induces S phase accumulation selectively in human transformed cells, *J. Interferon Cytokine Res.* **17**, 355–367.

Radhakrishnan, R., Walter, L.J., Hruza, A., et al. (1996) Zinc mediated dimer of human interferon-$\alpha_{2b}$ revealed by X-ray crystallography, *Structure* **4**, 1453–1463.

Resnitzky, D., Tiefenbrun, N., Berissi, H., Kimchi, A. (1992) Interferons and interleukin-6 suppress phosphorylation of the retinoblastoma protein in growth-sensitive hematopoietic cells, *Proc. Natl. Acad. Sci. U.S.A.* **89**, 402–406.

Rice, A.P., Duncan, R., Hershey, J.W.B., Kerr, I.M. (1985) Double-stranded RNA-dependent protein kinase and 2-5A system are both activated in interferon-treated, encephalomyocarditis virus-infected HeLa cells, *J. Virol.* **54**, 894–898.

Rigby, W.F., Ball, E.D., Guyre, P.M., Fanger, M.W. (1985) The effect of recombinant-DNA-derived interferons on the growth of myeloid progenitor cells, *Blood* **65**, 858–861.

Ronni, T., Melén, K., Malygin, A., Julkunen, J. (1993) Control of IFN-inducible MxA gene expression in human cells, *J. Immunol.* **141**, 1715–1726.

Sheppard, P., Kindsvogel, W., Xu, W., et al. (2003) IL-28, IL-29 and their class II cytokine receptor IL-28R, *Nat. Immunol.* **4**, 63–68.

Stojdl, D.F., Lichty, B., Knowles, S., et al. (2000) Exploiting tumor-specific defects in the interferon pathway with a previously known oncolytic virus, *Nat. Med.* **6**, 821–825.

Tan, S.-L., Katze, M.G. (1999) The emerging role of the interferon-induced PKR protein kinase as an apoptotic effector: a new face of death? *J. Interferon Cytokine Res.* **19**, 543–554.

Taniguchi, T., Mantei, N., Schwarzstein, M., et al. (1980) Human leukocyte and fibroblast interferons are structurally related, *Nature (Lond.)* **285**, 547–549.

Wang, E., Pfeffer, L.M., Tamm, I. (1981) Interferon $\alpha$ induces a protein kinase C-$\varepsilon$ (PKC-$\varepsilon$) gene expression and a 4.7 kb PKC-$\varepsilon$ related transcript, *Proc. Natl. Acad. Sci. U.S.A.* **78**, 6281–6285.

Wong, L.H., Kramer, K.G., Hatzinisiriou, I., et al. (1997) Interferon-resistant human melanoma cells are deficient in ISGF3 components, STAT1, STAT2, and p48-ISGF3 gamma, *J. Biol. Chem.* **272**, 28779–28784.

Zhang, K., Kumar, R. (1994) Interferon-$\alpha$ inhibits cyclin E- and cyclin D1-dependent CDK-2 kinase activity associated with RB and E2F in Daudi cells, *Biochem. Biophys. Res. Comm.* **200**, 522–528.

Zhou, A., Hassel, B.A., Silverman, R.H. (1993) Expression cloning of 2-5A-dependent RNAase: a uniquely regulated mediator of interferon action, *Cell* **72**, 753–765.

# 9
# Signaling Through JAKs and STATs: Interferons Lead the Way

*Christian Schindler and Jessica Melillo*
*Columbia University, New York, USA*

**1 Introduction 244**

**2 Discovery of the JAK-STAT Signaling Paradigm 246**

**3 The Janus Kinases (JAK) Family 249**
3.1 JAK Structure 249
3.2 JAK Family Members 250
3.2.1 Jak1 250
3.2.2 Jak2 250
3.2.3 Jak3 250
3.2.4 Tyk2 251

**4 The STAT Family of Transcription Factors 251**
4.1 STAT Evolution 251
4.2 STAT Structure 252
4.2.1 $NH_2$ Domain 252
4.2.2 Coiled-coil Domain 253
4.2.3 DNA Binding Domain 253
4.2.4 Linker Domain 253
4.2.5 SH2 and Tyrosine Activation Domain 253
4.2.6 Transcription Activation Domain 254
4.3 STAT Family Members 254
4.3.1 Stat1 254
4.3.2 Stat2 255
4.3.3 Stat3 255
4.3.4 Stat4 255
4.3.5 Stat5a and Stat5b 255
4.3.6 Stat6 256

*Immunology. From Cell Biology to Disease.* Edited by Robert A. Meyers.

ISBN: 978-3-527-31770-7

**5 Hematopoietin Family of Cytokine Receptors 256**
5.1 The Cytokine Type II or IFN Receptor Family 256
5.1.1 Type I IFN Receptor 256
5.1.2 Type II IFN Receptor 258
5.1.3 IFN-$\lambda$ Receptor 258
5.1.4 IL-10 Receptor Family 258
5.2 Extended gp130 Receptor Family 259
5.2.1 IL6-gp130 Receptor Subfamily 259
5.2.2 gp130 Receptor-like Subfamily 260
5.2.3 IL-12 Family of gp130-Related Receptors 260
5.3 $\gamma$C Receptor Family 261
5.3.1 IL-2 Receptor Family 261
5.3.2 IL-4 Receptor Family 261
5.4 IL-3 Receptor Family 262
5.5 Single-chain Receptor Family 262
5.6 Noncytokine Receptors 262

**6 Regulation of JAK-STAT Signaling 262**
6.1 Phosphatases 263
6.2 Nuclear Export 263
6.3 The SOCS Family 263

**7 Concluding Comments 264**

**Bibliography 264**
Books and Reviews 264
Primary Literature 265

## Keywords

### Cytokines
A large family of secreted peptide ligands that bind to specific receptors on target cells to mediate potent biological responses. This includes immunomodulatory peptides from a number of important families including the TGF-$\beta$, TNF, and hematopoietin families.

### GAS (Gamma Activation Site)
The enhancer element recognized by most activated STAT homodimers. It was first identified as an IFN-$\gamma$ response element.

**Hematopoietins**
These four-helix-bundle cytokines mediate potent biological responses on target cells (see Table 1). They are predominantly involved in regulating inflammation and immunity.

**IFN (Interferon)**
A family of cytokines initially identified for their antiviral activity. This family includes the type I, type II, and λ IFNs.

**IRFs (Interferon Response Factors)**
A family of transcription factors that is important in immune response. They bind to the family of IFN response elements (IREs).

**ISGF-3**
An IFN-$\alpha$-activated transcription factor consisting of Stat1, Stat2, and IRF-9.

**ISRE (IFN-stimulated Response Element)**
A specific enhancer that is recognized by ISGF-3. They constitute a subset of the IRE enhancer family.

**JAK (Janus kinase)**
Tyrosine kinase that associates with cytokine receptors and provides catalytic activity, which is essential to initiate signal transduction. Members include Jak1, Jak2, Jak3, and Tyk2.

**Ligand**
The specific peptide that binds to a receptor.

**NK Cells (Natural Killer Cells)**
A family of lymphocytes that are important in innate immunity. They are potently activated by IFN-I and secrete IFN-$\gamma$.

**Protein Tyrosine Kinases**
Enzymes that catalyze phosphorylation of protein tyrosine residues, which serve an important regulatory role in higher eukaryotes.

**Receptors**
Transmembrane-spanning proteins that specifically bind extracellular ligands, like hematopoietins, to mediate a biological response.

**SH2 (Src Homology 2)**
The domain that mediates protein–protein interactions through its capacity to recognize and bind to phosphotyrosines in the context of a specific sequence of amino acids, referred to as a *tyrosine motif* (see final entry). SH2-phosphotyrosine interactions play an important role in signal transduction.

**SOCS (Suppressors of Cytokine Signaling)**
A family of rapidly induced proteins that feed back inhibit JAK-STAT signaling.

**STAT (Signal Transducers and Activators of Transcription)**
A family of latent cytoplasmic transcription factors that transduces signals for cytokines. Members include Stat1, Stat2, Stat3, Stat4, Stat5a, Stat5b, and Stat6.

**T cells**
Thymic derived lymphocyte family whose members are activated by specific antigen-MHC complex. They constitute an important limb of adaptive immunity. They can be divided into several subsets including Th1 and Th2 cells.

**Tyrosine Motif**
A phosphotyrosine and the four-carboxy-terminal flanking amino acids that constitute an SH2 domain recognition/binding site.

Characterization of the "viral interfering activity" secreted by infected tissues led to the identification of interferons, the founding members of the cytokine family. Subsequent studies, directed at understanding how IFNs mediated their biological response then led to identification of the JAK-STAT signaling cascade. Today, the important role this pathway plays in transducing signals for the over 50 members of the hematopoietin family of cytokines is widely appreciated. After providing a brief historical perspective, this review will provide an overview of the JAK-STAT signaling cascade, with emphasis on some of the more recent developments in this field.

## 1 Introduction

In 1957, Isaacs and Lindenmann first described the potent antiviral properties of interferons (IFNs). It was not until 20 years later, when the IFNs had been purified to homogeneity, that the size and diversity of this family was recognized. They were segregated into two distinct families, now referred to as type I (i.e. IFN-$\alpha/\beta$) and type II (i.e. IFN-$\gamma$) IFNs. The early generation IFN cDNA clones, circa 1980, served to propel them as one of the first products of a nascent biotechnology industry. Concomitantly, other secreted peptides, also with potent biological activity, were identified. These were initially referred to as *monokines* or *lymphokines*, reflecting their cellular source (i.e. monocytes and lymphocytes respectively). This confusing terminology was later replaced by more generic terms, interleukins (i.e. between leukocytes) and cytokines. The cytokine family of biological response modifiers has subsequently grown to number in the hundreds, including members of the four-helix bundle, TNF, TGF-$\beta$, growth factor, and chemokines families of ligands.

The early availability of relatively large quantities of purified recombinant IFNs provided an opportunity to determine how these cytokines mediate their potent biological responses. Initial studies determined that the rapid induction of genes were an important feature of this response. Subsequent characterization of the ability of IFN-$\alpha$ to drive the expression of new genes led to the identification of Stat1, Stat2, and the JAK-STAT signaling cascade (STAT refers to signal transducers and activators of transcription and JAK refers to Janus kinase). Further studies determined that IFN-$\gamma$ also signals through an analogous, but distinct JAK-STAT pathway.

Within three years, five additional members of the STAT family had been cloned and found to transduce signals for other members of the cytokine family of ligands. Subsequent gene targeting studies have confirmed the pivotal role the STATs play in transducing the biological response for each four-helix-bundle cytokines (see Table 1). More recent efforts have focused on understanding how the transient nature of STAT signaling is regulated. These studies have revealed an elaborate system of controls, including tight regulation

**Tab. 1** Hematopoietin dependent JAK-STAT signaling. Hematopoietins transduce their signals through specific sets of JAKs and STATs as indicated. The assignments for which there is the most confidence (i.e. based on knockout and biochemical studies) are shown in bold. Those with less confidence are shown in plain lettering, and those with the least confidence are shown in brackets.

| *Ligands* | *JAKs* | *STATs* |
|---|---|---|
| **IFN family** | | |
| **IFN-I** (Type I)[a] | **Jak1**, Tyk2 | **Stat1**, **Stat2**, Stats3, Stat4 (Stats5-6) |
| **IFN-$\gamma$** (Type II) | **Jak1**, **Jak2** | **Stat1** |
| **IFN-$\lambda$** (IL-28a,b,-29) | Jak1, Tyk2 | Stat1, Stat2, Stat3 |
| **IL-10** | **Jak1**, Tyk2 | **Stat3**, Stat1 |
| **IL-19** | ? | ? |
| **IL-20** | ? | Stat3 |
| **IL-22** | Jak1, Tyk2 | Stat3, (Stat5) |
| **IL-24** | ? | Stat3 |
| **IL-26** (AK155) | ? | Stat1/Stat3 |
| **Extended gp130 family** | | |
| **IL-6** | **Jak1**, (Jak2) | **Stat3**, Stat1 |
| **IL-11** | Jak1 | **Stat3**, Stat1 |
| **LIF** | **Jak1**, (Jak2) | **Stat3**, Stat1 |
| **CNTF** | Jak1, (Jak2) | **Stat3**, Stat1 |
| **NNT-1/BSF-3** | Jak1, (Jak2) | **Stat3**, Stat1 |
| **NP** | Jak1, (Jak2) | Stat3 |
| **CT-1** | Jak1, (Jak2) | **Stat3** |
| **OSM** | Jak1, (Jak2) | **Stat3**, Stat1 |
| **IL-31** | Jak1, (Jak2) | Stat3, Stat5 |
| **G-CSF** | Jak1, (Jak2) | **Stat3** |
| **Leptin** | Jak2 | **Stat3** |
| **IL-12** | **Tyk2**, Jak2 | **Stat4** |
| **IL-23** | ? | **Stat4** |
| **IL-27** | ? | (Stat1), Stat3, Stat4, (Stat5) |
| **IL-30** | ? | (Stat1), Stat3, Stat4, (Stat5) |

*(continued overleaf)*

**Tab. 1** (*Continued*)

| *Ligands* | *JAKs* | *STATs* |
|---|---|---|
| $\gamma$C family | | |
| **IL-2** | **Jak1**, **Jak3** | **Stat5**, (Stat3) |
| **IL-7** | **Jak1**, **Jak3** | **Stat5**, (Stat3) |
| TSLP[b] | ? | Stat5 |
| **IL-9** | Jak1, Jak3 | **Stat5**, Stat3 |
| **IL-15** | **Jak1**, Jak3 | **Stat5**, (Stat3) |
| **IL-21** | Jak3, (Jak1), | **Stat3**, Stat5, (Stat1) |
| **IL-4** | **Jak1**, **Jak3** | **Stat6** |
| IL-13[b] | Jak1, Jak2 | **Stat6**, (Stat3) |
| IL-3 family | | |
| **IL-3** | **Jak2** | **Stat5** |
| **IL-5** | **Jak2** | **Stat5** |
| **GM-CSF** | **Jak2** | **Stat5** |
| Single-chain family | | |
| **Epo** | **Jak2** | **Stat5** |
| **GH** | Jak2 | **Stat5**, (Stat3) |
| **Prl** | Jak2 | **Stat5** |
| **Tpo** | **Jak2** | **Stat5** |

[a] In humans this family consists of 12 IFN-$\alpha$s, IFN-$\beta$.$\omega$ and Limitin.
[b] Bind to a related, but $\gamma$c independent receptor.

of nuclear import–export, phosphorylation, as well as novel regulatory proteins, most notably the SOCS (suppressors of cytokine signaling) family. As there are many excellent reviews on cytokines (i.e. hematopoietins) and JAK-STAT signaling, this review will focus on the more recent developments and the important role IFN studies have played in elucidating hematopoietin biology.

## 2 Discovery of the JAK-STAT Signaling Paradigm

Studies with early preparations of recombinant IFN-Is led to the identification of a set of rapidly induced genes. Classic promoter analysis of these IFN-stimulated genes (ISGs), identified a single conserved IFN-stimulated response element (ISRE), with a consensus of AGTTTN$_3$TTTCC. Three unique ISRE binding complexes were identified in nuclear extracts prepared from IFN-$\alpha$–stimulated cells. They were named *IFN-stimulated gene factor 1* (ISGF-1; aka IRF-2), ISGF-2 (aka IRF-1), and ISGF-3. Kinetic and biochemical studies suggested that ISGF-3 was the critical factor directing the rapid expression of IFN-$\alpha$ stimulated genes. Purification of ISGF-3 led to the identification of four component proteins, with molecular weights of 113 kDa, 91 kDa, 84 kDa, and 48 kDa. The 48-kDa protein was recognized as the ISRE binding component of this complex. Analysis of the three larger proteins determined that they were members of a novel family of proteins, for which the term STAT was coined. The 84 and 91 kDa proteins turned out to be alternative mRNA splice products of a single gene and were renamed Stat1$\alpha$ and Stat1$\beta$. The 113-kDa product became known as Stat2 and the 48-kDa protein, with its homology to interferon response factors (IRFs) was named IRF-9.

With the development of antibodies against Stat1 and Stat2, it was possible to demonstrate that both proteins resided in the cytoplasm of resting cells. Upon IFN-$\alpha$ stimulation, Stat1 and Stat2 (both isoforms) were rapidly phosphorylated on a single conserved tyrosine, leading to the formation of stable Stat1–Stat2 heterodimer. Subsequent biochemical and structural studies determined that STAT dimerization is driven by the stable interaction between the phosphotyrosine residue of one STAT with the SH2 (Src homology 2) domain of the corresponding STAT. In contrast to the inactive STATs, Stat1–Stat2 dimers are competent for translocation into the nucleus, where they associate with IRF-9 to form a stable ISRE DNA binding complex, and direct the expression of target genes (see Fig. 1). Numerous ISRE-driven genes have been characterized, including PKR (the dsRNA dependent protein kinase), OAS (2′-5′ oligoadenylate synthetase), Mx, PML, the

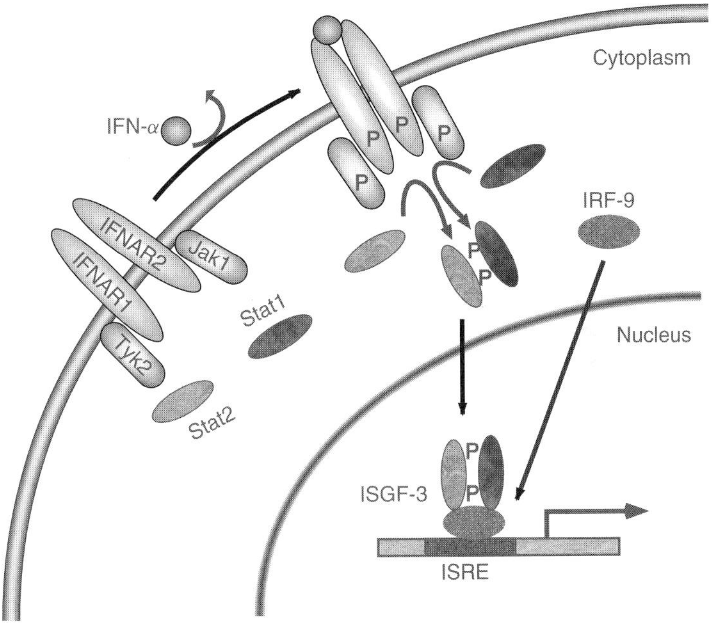


**Fig. 1** The IFN-I signaling paradigm. Upon binding to its dimeric receptor, type I IFN promotes the apposition of two receptor associated JAKs, which then activate each other by transphosphorylation. The activated JAKs in turn phosphorylate receptor tyrosine(s), directing the SH2 domain dependent recruitment of Stat1 and Stat2. At the receptor, Stat1 and Stat2 are activated by phosphorylation, whereupon they heterodimerize, translocate into the nucleus, and associate with IRF-9 to bind to the ISRE enhancers. This DNA binding complex, referred to as ISGF-3, directs the expression of ISRE target genes. IFN-I also promotes the formation of Stat1 homodimers, which directs the expression of GAS-driven genes as outlined in Fig. 2 (see color plate p. xxiv).

Ifi 200 cluster, GBP, iNOS (inducible nitric oxide synthetase, aka NOS2), IP-10, IRF-7, 9–27, and ISG-15.

Contemporaneous with the discovery of STATs, was the recognition that members of the JAK family of tyrosine kinases were important in IFN-I signaling. This provided the missing link as to how STATs were activated and the term JAK-STAT Pathway was coined. Parallel efforts to understand how IFN-$\gamma$ mediated potent but distinct biological responses led to the discovery of an analogous yet different JAK-STAT signaling pathway (see Fig. 2). Notable similarities with the IFN-$\alpha$ signaling cascade included requirement of a unique receptor, JAKs and Stat1. As was the case with IFN-$\alpha$, elucidation of this pathway began with the identification of an IFN-$\gamma$ responsive element (i.e. the gamma activation site; GAS) in the promoters of GBP and Fc$\gamma$RI. Subsequently, a unique IFN-$\gamma$-induced GAS binding activity was recognized and determined to consist of Stat1. Moreover, IFN-$\gamma$ was found to stimulate the phosphorylation of the same conserved tyrosine residue on Stat1 as IFN-$\alpha$. However, in the absence of Stat2 activation, a Stat1 homodimer was formed that translocated into the nucleus and directly bound DNA, culminating in the expression of GAS-driven genes. Subsequent studies determined that all STATs, with the exception of Stat2, form homodimers upon activation and then directly bind members of the GAS family of enhancers. This family of conserved enhancers features a palindrome, where the

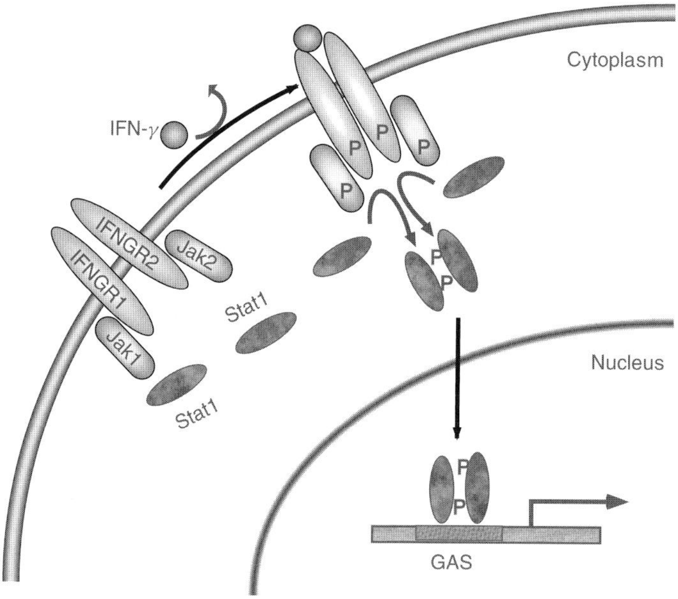


**Fig. 2** The IFN-II signaling paradigm. Upon binding to its dimeric receptor, IFN-$\gamma$ promotes the activation of two receptor associated JAKs. The activated JAKs in turn activate the receptor. Stat1 is recruited to the receptor, whereupon it is activated and forms homodimers. Stat1 homodimers translocate to the nucleus and bind to the GAS family of enhancers, culminating in the expression of a distinct set of genes (see color plate p. xxv).

optimal sequence is $TTCN_{2-4}GAA$ (Stat1 homodimers bind to an element where $N = 3$; Stat3 homodimers favor $N = 2$; and Stat6 homodimers favor $N = 4$). IFN-$\gamma$ stimulated GAS-driven genes include a number of important immunoregulatory transcription factors (e.g. IRF-1, ICSBP and CIITA), chemokines (e.g. Mig), and cell surface molecules (e.g. ICAM-1, Ly6E and Fc$\gamma$RI). CIITA is critical for adaptive immunity, driving the expression of major histocompatibility complex (MHC) class II and the Invariant chain, whereas IRF-1 has more broad effects, directing the expression of immune response genes like *iNOS, MHC I, LMP2, TAP2,* and *IL-12.* The ability of IRF-1 also to drive ISRE gene expression is likely to account for some of the overlap in the biological responses to IFN-$\alpha$ and IFN-$\gamma$.

The rapid identification of additional members of the STAT family provided an opportunity to determine that the other four-helix-bundle cytokines (i.e. hematopoietins) also transduce signals through this pathway. As the number of four-helix-bundle cytokines has grown, so has the recognition that they all transduce pivotal signals through JAK-STAT signaling cascade. The critical role hematopoietins play in directing immune response has directed further attention to this important pathway.

## 3
## The Janus Kinases (JAK) Family

Studies directed at understanding how IFNs mediated biological response directly implicated three known members of the JAK family, Jak1, Jak2, and Tyk2 in signal transduction. Subsequent homology screens identified a fourth member, Jak3, whose expression was restricted to hematopoietic tissues. However, each of these JAKs shared important structural and functional features that enable them to serve as an essential kinase for cytokine receptors. This is also the case for a JAK homolog, Hop, which has been identified in *Drosophila melanogaster* (see Sect. 4.1).

### 3.1
### JAK Structure

JAKs range in size from 120 to 130 kDa. Comparison of these ~1000 residue kinases reveals seven high homology regions, JH1–JH7 (see Fig. 3a). JH1 encodes the tyrosine kinase, whose activity is regulated by an activation loop. Thus, in resting cells when the activation loop sits in the catalytic pocket, activity is limited to a low or basal state. However, cytokine dependent ligation brings the two receptor associated JAKs into close apposition, enabling them to activate each other through a transphosphorylation event. This entails exploiting basal activity to phosphorylate the opposing kinases activation loop. The loops move out of the catalytic site, rendering these kinases fully active, whereupon they sequentially phosphorylate critical receptor and STAT tyrosine residues (see Figs. 1 and 2). The second JAK homology domain, JH2, represents a pseudokinase domain (i.e. catalytically inactive) that is both a signature feature of this kinase family and appears to be required for full catalytic activity. The five amino-terminal JAK homology domains, JH3–JH7, constitute a FERM (four point one, ezrin, radixin, moesin) domain, which is responsible for association with hematopoietin family of cytokine receptors. Specifically, JAKs associate with proline rich, membrane proximal box1/box2 domain found in these receptors. As outlined in Sect. 5,

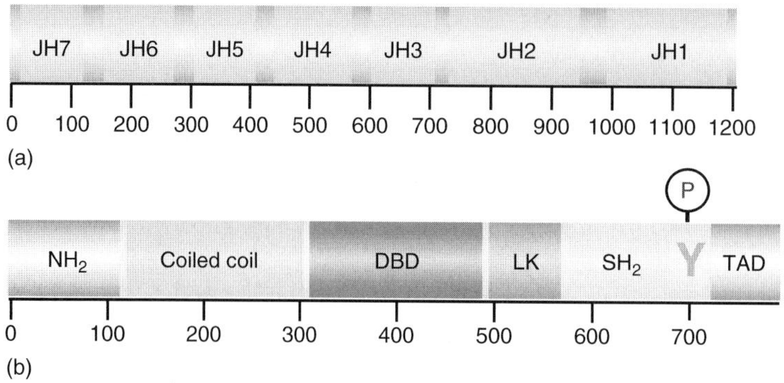


**Fig. 3** Structural models of the JAK and STAT family of signaling proteins. (a) The four members of the JAK family share seven homology domains regions, JH1–JH7. JH1 serves as the catalytic domain, whereas JH2 represents a pseudokinase domain, which is a characteristic feature of this family of tyrosine kinases. JH3–JH7 comprise a FERM domain that is responsible for association with cytokine receptors. (b) The seven members of the STAT family of transcription factors share seven functionally conserved domains. This includes the amino-terminal domain ($NH_2$), the coiled-coiled domain (coiled coil), the DNA binding domain (DBD), the Linker domain (LK), the SH2 domain, the tyrosine activation domain, and the transcriptional activation domain (TAD), which is conserved in function but not in sequence. Domain colors correspond with those in Stat1 Crystal structure in Fig. 4 (see color plate p. xxv).

the cytokine receptors depend on JAKs to transduce signals that can be functionally divided into several groups, each exhibiting a unique pattern of JAK association and STAT activation (see Table 1).

## 3.2 JAK Family Members

### 3.2.1 Jak1

This kinase was initially recognized for its association with the second chain of the IFN-$\alpha$ receptor (IFNAR2) and then the first chain of the IFN-$\gamma$ receptor (IFNGR1). Subsequent biochemical studies demonstrated association with receptors for the gp130 and the common $\gamma$ ($\gamma$C) families. Jak1 knockout mice die perinatally, secondary to neurological defects attributed to a loss in gp130 signaling. Jak1$^{-/-}$ mice also exhibit reduced lymphocyte counts due to a lack of T cells and NK (natural killer) cells. Biochemical studies on Jak1$^{-/-}$ cells have underscored a failure to respond to cytokines from the gp130, $\gamma$C, and IL-10 families.

### 3.2.2 Jak2

This kinase was initially determined to associate with IFNGR2 and the gp130 component of the IL-6 receptor. Targeted deletion of Jak2 yielded embryonic lethal phenotype, at day 12.5, due to failure in definitive erythropoiesis. Tissues collected from Jak2 knockout mice fail to respond to IFN-$\gamma$ and most of the cytokines from the IL-3 and single-chain cytokine receptor families (see Table 1).

### 3.2.3 Jak3

The expression in this JAK is limited to hematopoietic cells, where it exclusively associates with the $\gamma$C chain. Hence, the Jak3 knockout mice exhibit a severe combined immunodeficiency disease (SCID) like syndrome, analogous to what has been reported for $\gamma$C deficient mice.

#### 3.2.4 Tyk2

Although the first member of the family to be associated with cytokine signaling, it was the last to be knocked. Earlier biochemical studies implicated Tyk2 in IFN-I, IL-10, IL-12, and IL-13 signaling. Yet, Tyk2 knockout mice exhibit surprisingly modest defects in their response to IFN-I and IL-10. They do, however, exhibit significant defects in their response to IL-12, and unexpectedly profound defects in their responses to LPS. While the mechanism behind this latter defect is not fully elucidated, it underscores the intimate relationship between toll-like receptor (TLR) dependent signaling (LPS signals through TLR-4), IFN-I production, and the IFN-I autocrine loop (see also Fig. 5).

## 4 The STAT Family of Transcription Factors

The rapid identification of seven members of the STAT family facilitated a functional classification of the growing family of cytokine receptors (see the following and Table 1). It also provided an opportunity to explore STAT function. The seven mammalian STATs (Stats 1–6 with two Stat5 genes), range in size from 750 to 900 amino acids. They consist of six distinct domains (Fig. 3b), and except for Stat2, they all transduce signals through the paradigm illustrated in Fig. 2. STATs have also been identified in numerous more primitive model eukaryotes, including *D. melanogaster*, *Caenorhabditis elegans*, and *Dictyostelium*. Growth in the number of STATs during evolution that appears in these lower eukaryotes reflect an increasing need for cell-to-cell communication as eukaryotes have become more complex.

### 4.1 STAT Evolution

STATs have been remarkably well conserved during eukaryotic evolution and may encode the primordial SH2 domain (see Sect. 4.2.5). Homologs have been identified in *Dictyostelium*, *C. elegans*, *Drosophila*, *Anopheles*, *Xenopus*, and zebrafish. The JAK-STAT pathway has been genetically characterized in *Drosophila*, where it plays important roles in developmental segmentation, larval hematopoiesis, sexual identity, and the development of eye polarity. The single *Drosophila* JAK exhibits 27% identity to Jak2 and the single STAT exhibits 37% identity to Stat5. Also present in *Drosophila* are a distant homolog of a cytokine receptor (Upd) and ligand (Dom). The three STAT homologs in *Dictyostelium* are distantly related to mammalian Stat5 and Stat3, but the signaling pathways they are involved in are unrelated to the one found in *Drosophila* and vertebrates. STAT homologs in *C. elegans* are also more distantly related. Zebrafish encode two STATs, homologous to Stat1 and Stat3. The Stat3 homolog is expressed in early development and is syntenic with both murine and human Stat3. The zebrafish Stat1 homolog is more divergent and expressed later in life.

The seven murine and human STATs are segregated into three syntenic genetic clusters, each consisting of two tandem STATs and suggesting evolution from a single tandem cluster (see the following). Specifically, in mice, Stat1 and Stat4 map to chromosome 2; Stat2 and Stat6 map to chromosome 12; and Stat3 and Stat5 map to chromosome 17. The Stat5 gene, however, has undergone an additional evolutionarily more recent duplication, yielding tandem Stat5a and Stat5b genes. Analysis of these clusters, the recent duplication of Stat5, and the sequence conservation of Stat3 and Stat5 amongst lower eukaryotes, all serve to suggest that Stat3–Stat5 gene cluster represents the primordial tandem array mammalian STAT genes. This may also explain the relative functional pleiotropy of these two STATs (see the following).

## 4.2 STAT Structure

Biochemical, genetic, and structural studies have provided clear evidence that each of the seven mammalian STATs consist of six domains, as well as a tyrosine near residue 700 that serves as the activation site (see Figs. 3b and 4). These functionally conserved domains include: the amino-terminal domain ($NH_2$), the coiled-coiled domain (CCD), the DNA binding domain (DBD), the linker domain, the SH2 domain and the carboxy-terminal transcriptional activation domain (TAD; conserved in function but not sequence).

### 4.2.1 $NH_2$ Domain

The amino-terminal domain, about 130 residues in length, forms a unique dimeric structure when crystallized as an independent domain. Likewise, biochemical and genetic studies of full-length inactive Stat1, Stat3, Stat4, and Stat5 indicate that the amino-terminal domain drives homotypic

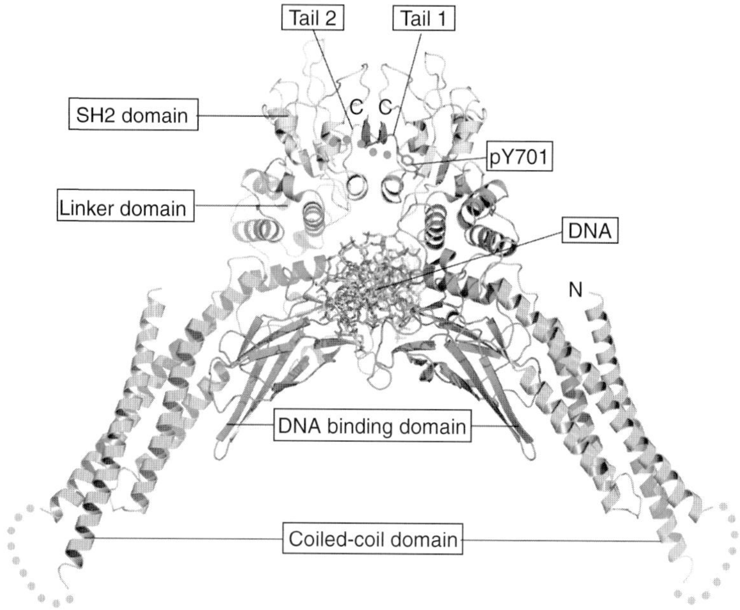


**Fig. 4** Stat1 crystal structure. The structure represents two activated Stat1 fragments (amino acids 135–710) that are dimerized and bound to a GAS palindrome. Each domain is shown in a distinct color and labeled (see Fig. 3). Note that the CCD forms a four-$\alpha$-helix bundle and the DBD forms a $\beta$-barrel with an immunoglobulin fold (see color plate p. xxvi).

(i.e. to self) dimerization. This may assure delivery of "packets" of STAT dimers to the receptor in a conformation that renders the tyrosine, near residue 700, exposed. Upon activation, STATs undergo a conformation change, in which this phosphotyrosine becomes buried in an activated structure. Earlier studies also implicated the ability of the amino termini to dimerize, thereby promoting cooperativity of DNA binding to tandem GAS elements. Finally, the amino terminus may also serve to regulate STAT nuclear import and export.

#### 4.2.2 Coiled-coil Domain

This amino proximal domain, ~150 residues in length (amino acids 135 to 315 for Stat1), folds up into a four-$\alpha$-helix bundle (see Fig. 4). This structural feature is likely to afford the ability to dynamically regulate interactions with regulatory proteins (e.g. karyophilins) and a large surface area for protein–protein interactions. Consistent with this, the coiled-coil domain protrudes 80 Å laterally from the core structure of both Stat1 and Stat3, providing a hydrophilic surface that is known to associate with other proteins including IRF9, c-Jun, N-myc interacting protein (Nmi), and StIP.

#### 4.2.3 DNA Binding Domain

This highly conserved ~200 amino acid domain (residues 320–480 in Stat1) forms a $\beta$-barrel with an immunoglobulin fold, reminiscent of the NF-$\kappa$B and p53 DBDs. This affords each STAT component in the dimer the ability to recognize bases in the proximal half of the GAS palindrome. The number of direct contact sites between STAT residues and DNA are modest, yielding a dissociation constant in the nanomolar range, and suggesting that cooperative DNA binding activity may be an important aspect for STAT function. Recall, that Stat2 does not appear to form active homodimers or bind DNA directly.

#### 4.2.4 Linker Domain

As the name suggests, this short domain (amino acids 488–576 for Stat1), links the DBD with the SH2 domain. Detailed analysis of the linker structure suggests that this domain serves to appropriately "translate" the SH2 dependent dimerization to the DBD.

#### 4.2.5 SH2 and Tyrosine Activation Domain

The STAT SH2 domain is the most highly conserved and arguably the most important domain (residues 580–680 in Stat1). It plays critical roles in signal transduction through its capacity to recognize specific receptor phosphotyrosine motifs (i.e. affecting specific receptor recruitment). After activation it specifically recognizes the phosphotyrosine motif on the partner STAT (i.e. directing active STAT dimerization). Moreover, the SH2 domain in *Dictyostelium* STAT appears to represent the "earliest" identified SH2 domain, raising the possibility that it is the primordial SH2 domain. Although the STAT SH2 domain is well conserved evolutionarily, it is surprisingly divergent from other SH2 domains in sequence, but not structure. The structure consists of an antiparallel $\beta$-sheet, flanked by two $\alpha$-helices, forming a pocket. An absolutely conserved arginine (Arg-602 for Stat1) resides at the bottom of this pocket and mediates the interaction with phosphotyrosine. As is the case with all SH2 domains, flanking residues recognize the other amino acids that are characteristic of a tyrosine motif (i.e. phosphotyrosine + four carboxy-terminal amino acids). Of note, the residues most important in defining the specificity of binding to the STAT tyrosine motif (i.e. the tyrosine activation domain) during dimerization are located at positions +1 and +3 in this motif. In Stat1 residues, +5, +6, and +7 also contribute to the specificity of dimerization by associating with Ala-641 and Val-642 of the SH2 domain. The proximity of the SH2 to the corresponding STAT tyrosine activation domain precludes an intramolecular association between these domains. Thus, the STAT SH2 domain is only able to bind to tyrosine motifs on other proteins.

#### 4.2.6 Transcription Activation Domain

The critical TAD resides at the STAT carboxy terminus. Consistent with the ability of this domain to regulate unique transcriptional responses for each STAT, it is the most poorly conserved STAT domain. The function of this domain can be regulated by either deletion or serine phosphorylation. The former feature was recognized with the initial isolation of Stat1$\beta$ (i.e. the 84 kDa ISGF-3 component), which is an alternate splice isoform. The *Stat3* gene gives rise to a similar carboxy terminally deleted Stat3 isoform, Stat3$\beta$. Intriguingly, recent studies suggest that truncated isoforms are not necessarily transcriptionally inert, but rather can also direct the expression of a unique gene program through the association with distinct sets of transcriptional regulators. Regulation of STAT transcriptional activity by serine kinases has also received considerable attention. These studies have identified numerous candidate serine kinases, potentially reflecting the ability of multiple "stress" pathways to modulate STAT activity. Elegant studies, in which the critical serine residue in Stat1 and Stat3 have been mutated by a knockin approach, have demonstrated that serine phosphorylation is important for the expression of a subset of STAT target genes.

## 4.3 STAT Family Members

As indicated above, seven STAT genes have been identified in both mouse and man. They include *Stats1–6* and two Stat5 genes, *Stat5a* and *Stat5b*.

### 4.3.1 Stat1

Both biochemical and gene targeting studies have determined that Stat1 is critical for both type I and type II IFN signaling (see Figs. 1 and 2). Stat1 forms transcriptionally active homodimers in response to stimulation with both classes of IFNs (studies with λ-IFNs are under way; see Sect. 5.1.3). IFN-Is and IFN-λs also direct the formation of active Stat1–Stat2 heterodimers. Intriguingly, the relative role that Stat1 homodimers versus Stat1–Stat2 heterodimers play in the biological response to IFN-Is is tissue specific, with Stat1–Stat2 heterodimers playing a considerably more prominent role in fibroblasts and epithelial cells. Finally, gene targeting studies have also underscored a modest role for Stat1 in the biological response to IL-6 (see Table 1). Specifically, Stat1 appears to alter the nature of the response to Stat3 (the major gp130 activated STAT; see Sects. 4.3.3 and 5.2.2), highlighting the seemingly antagonistic roles Stat1 and Stat3 play in cancer and immune response. That is to say, numerous studies have implicated Stat3 as an oncogene, whereas Stat1 appears to antagonize cell proliferation. Likewise, Stat1 drives the expression of proinflammatory genes (aka "danger signals"), whereas Stat3 activation can be associated with the suppression of inflammation.

#### 4.3.2 Stat2

Stat2 is a most unique member of the STAT family by virtue of its large size, inability to homodimerize, and inability to directly bind DNA. Sequence comparison of the murine and human homologs also reveals an uncharacteristic level of carboxy-terminal sequence divergence. Both biochemical and gene targeting studies have determined that Stat2 is critical for the biological response to IFN-Is. Upon activation, Stat2 heterodimerizes with its exclusive partner Stat1. This heterodimer is competent for nuclear translocation, but must associate with IRF-9 to form a stable complex on ISRE elements, a signaling paradigm that is unique to Stat2 (see Fig. 1). Biochemical studies have suggested a similar role in the response to λ-IFNs. Although gene targeting studies have highlighted an essential role for Stat2 in IFN-I signaling, there is growing evidence for a regulatory role in some IFN-γ responses.

#### 4.3.3 Stat3

Biochemical and gene targeting studies have demonstrated that Stat3 plays a critical role in mediating the biological response to members of the IL-6/gp130 and IL-10 families of cytokines. Underscoring an important developmental role for gp130 dependent signals, deletion of the *Stat3* gene yields an early embryonic lethal phenotype (i.e. day 6.5–7.5). However, tissue-specific deletions of Stat3 are largely associated with abnormal inflammation and/or decreased susceptibility to cancer. As suggested by the latter phenotype, there is a burgeoning interest in understanding the role Stat3 plays in tumorigenesis. This contrasts the biological activities associated with Stat1, pointing to the seemingly antagonistic relationship between Stat1 and Stat3 with respect to the control of growth and inflammation alluded to in Sect. 4.3.1.

#### 4.3.4 Stat4

Stat4 is predominately involved in signaling for the IL-12 family (see Table 1 and Sect. 5.2.3), which plays a critical role in activating the effector functions of both Th1 and NK cells (e.g. enhanced IFN-γ production). Studies on Stat4 and Stat6 (see Sect. 4.3.6) knockout mice have underscored the important role these two STATs play in maintaining a healthy balance between Th1 and Th2 dependent immune responses. More recent studies have also implicated Stat4 in the ability of IFN-Is to synergize with IL-18 in the activation of Th1 cells. For these responses, IFN-I has been shown to drive Stat4 tyrosine phosphorylation and IL-18 to stimulate Stat4 serine phosphorylation, both of which are required for full activity. Although Stat4 is also important in the biological response to IL-23, a potential role in the biological response to the two newest members of the family, IL-27 and IL-30, remains more controversial.

#### 4.3.5 Stat5a and Stat5b

Consistent with their location in the primordial STAT locus, Stat5a/Stat5b, like Stat3, are relatively pleiotropic in their effects. They transduce important signals for all members of the IL-2 family (all γC receptors except for IL-4 and IL-13), the IL-3 family, and the single-chain receptor family (see Table 1). Underscoring their homology, Stat5a and Stat5b play overlapping roles in mediating the responses to IL-2 and IL-3 ligand families.

Surprisingly however, the responses to members of the single-chain family are far more specific. Stat5a knockout mice exhibit a profound defect in the response to prolactin (PRL), and Stat5b knockout mice exhibit a profound defect in the response to growth hormone (GH). Reproductive defects and an autoimmune phenotype, attributed to a deficient IL-2 response, have made Stat5a/5b double knockout mice difficult to propagate and study.

#### 4.3.6 Stat6

Biochemical and gene targeting studies have determined that Stat6 transduces signals for two closely related and distinct members of the $\gamma$C family of receptors, the IL-4 and IL-13 receptors. Both receptors, and their corresponding ligands, play an essential role in the development of an effective Th2 dependent response. Thus, as outlined above, Stat4 and Stat6 serve as protagonists in maintaining a critical balance between the Th1 and Th2 subsets of T-helper cells.

## 5 Hematopoietin Family of Cytokine Receptors

To date, over 50 members of the four-helix-bundle family of cytokines, also referred to as *hematopoietins*, have been identified. They each bind to a conserved family of receptors, referred to as *hematopoietin receptors*, or more commonly known as *type I* and *type II cytokine receptors*, based on the presence or absence, respectively, of a conserved WSXWS amino acid sequence motif in their extracellular domain. Members of the cytokine receptor family type I can be further subdivided into four functionally related groups that transduce signals through analogous JAK-STAT signaling pathways (see Sects. 5.2–5.5; see also Table 1). Members of the cytokine receptor family type II (CRF2) include receptors for the many members of the IFN and IL-10 ligand families. Consistent with this, they also signal through a characteristic, albeit more pleiotropic set of JAK-STAT signaling pathways (see Sect. 5.1; see also Table 1).

### 5.1 The Cytokine Type II or IFN Receptor Family

IFNs represent the largest single family of cytokines that bind to and signal through receptors composed of CRF2 chains. Until recently, IFNs were divided into two families, type I and type II, on the basis of distinct receptors they bind. However, recently a third family of "λ-IFNs" has been identified, which signal through a unique receptor, consisting of an IL-10 receptor chain and an "orphan" CRF2 chain. The related IL-10 family of ligands also binds to a partially overlapping set of receptors consisting of CRF2 chains.

#### 5.1.1 Type I IFN Receptor

Ligands binding to the type I IFN receptor represent the largest single family of four-helix-bundle cytokines, numbering close to 20 in man and mouse. Referred to as type I IFNs (IFN-Is), they are encoded in a large locus of single exon genes. IFN-Is include numerous IFN-$\alpha$s, IFN-$\beta$, IFN-$\delta$, IFN-$\varepsilon$, IFN-$\kappa$, IFN-$\tau$, IFN-$\omega$, and Limitin. Their receptor, which is usually referred to as the IFN-$\alpha$ receptor (IFNAR), is expressed in all cells. It consists of two CRF2 chains, IFNAR1 and IFNAR2. The cytoplasmic domain of IFNAR1 is physically associated with Tyk2, whereas the cytoplasmic domain of IFNAR2 binds Jak1. As outlined in Fig. 1, upon binding IFN-I, these receptor associated tyrosine kinases become activated by transphosphorylation, whereupon they phosphorylated receptor tyrosine(s) to affect STAT recruitment (see Sect. 3.1). At the receptor, Stat1 and Stat2 become phosphorylated (i.e. activated) and form both Stat1–Stat2 heterodimers and Stat1 homodimers, which in turn direct the expression of ISRE- and GAS-driven genes. The relative balance of these two pathways differs between tissues, with Stat1–Stat2 dimers playing a more important role in fibroblasts and macrophages relying more heavily on Stat1 homodimers. Although IFN-Is were initially recognized for their important role in the innate immune response to viral infection, they are now also recognized for their capacity to regulate the interface between innate and adaptive immunity. Consistent with their important role in immune response, the regulation of IFN-I expression is quite complex (see Fig. 5). During infection with either a viral or bacterial pathogen, pattern recognition receptors, often from the TLR family, become activated. These receptors trigger activation of IKK-$\varepsilon$ (aka IKK-i) and TBK-1, culminating in the activation of IRF-3. Activated IRF-3 drives the expression of IFN-$\beta$, the prototypical immediate early IFN-I. IFN-$\beta$ is secreted, and binds to the IFN-$\alpha$ receptor directing the activation of ISGF-3 and expression of ISRE-driven genes. One such target gene, *IRF-7*, either independently or along with *IRF-3* (and potentially other IRFs) directs the robust secretions of the many delayed IFN-Is (e.g. IFN-$\alpha$s, IFN-$\delta$, IFN-$\omega$, etc.). Intriguingly, the IFN-I autocrine/paracrine loop appears to be active at low levels in resting cells (i.e. basal activity), potentially underscoring its role as a sentry for infection. More recently however, studies have implicated this autocrine loop in the regulation of noninfectious cellular stresses, including cancer and autoimmune disease. Finally, while it is clear that both Stat1 and Stat2 play a critical role in mediating IFN-I response, the role other STATs play in this response has remained more controversial. Recent studies suggest that Stat3, which is activated by IFN-Is in most tissues, may serve to antagonize Stat1 activity, a theme that was outlined in the preceding sections. (Sects.4.3.1 and 4.3.3). In contrast, the IFN-I dependent activation

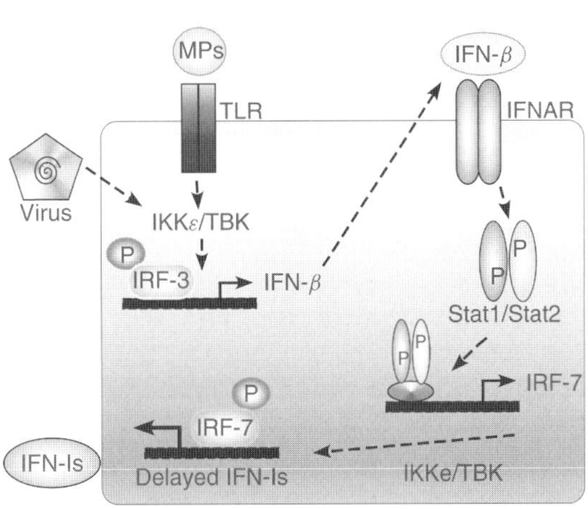


**Fig. 5** IFN-I autocrine/paracrine loop. Viral and bacterial pathogens include molecular patterns (MPs) that are recognized by toll-like receptors (TLRs). Once bound to these receptors, they promote the activation of IKK$\varepsilon$/TBK1 kinases, which in turn phosphorylate IRF-3. Activated IRF-3 then drives the expression and secretion of IFN-$\beta$, an immediate early IFN-I. IFN-$\beta$ binds the IFN-$\alpha$ receptor (IFNAR) initiating an important autocrine loop. This leads to the expression of IRF-7, an important ISGF-3 target gene, which is responsible for driving the expression/secretion of the delayed IFN-Is (see color plate p. xxvi).

of Stat4 is limited to leukocytes, where it appears to play an important role in promoting the expression of IFN-$\gamma$.

#### 5.1.2 Type II IFN Receptor

There is only one ligand that binds to the type II IFN receptor, IFN-$\gamma$. In contrast to the type I IFNs, the gene encoding IFN-$\gamma$ consists of three exons and resides in a distinct locus. The type II IFN receptor, which is usually referred to as the IFN-$\gamma$ *receptor*, consists of two CRF2 chains, IFNGR1 and IFNGR2. They associate with Jak1 and Jak2 respectively. In contrast to type I IFNs, IFN-$\gamma$ secretion is largely restricted to T cells and NK cells, underscoring the need to carefully regulate the activity of this potent proinflammatory cytokine. In T cells, IFN-$\gamma$ is the signature cytokine of effector Th1 cells, which play a critical role in adaptive immunity. However, NK cell dependent IFN-$\gamma$ production assures an important role in innate immunity as well. As outlined in Fig. 2, upon binding, ligand two receptor associated JAKs are activated by transphosphorylation, whereupon IFNGR1 becomes phosphorylated on a single tyrosine residue that specifically serves to recruit Stat1 to the receptor complex. Stat1 is activated at the receptor, dimerizes, translocates to the nucleus, and directly binds GAS elements to drive the expression of target genes. Notable GAS-driven target genes include chemokines and transcription regulators. One such transcription regulator is IRF-1, which then directs the expression of ISRE-driven genes, accounting for some of the functional overlap between type I and type II IFNs. Another important transcriptional regulator is CIITA, which coordinates the expression of MHC class II, a signature response associated with IFN-$\gamma$ stimulation, and is essential for adaptive immunity.

#### 5.1.3 IFN-λ Receptor

The ligands that bind to the IFN-$\lambda$ receptor are either referred to as $\lambda$ IFNs or IL-28a, IL28b, and IL-29. Both receptor and ligands were recently identified through genome mining. Although the IFN-$\lambda$s are functionally quite similar to IFN-Is, they reside in a distinct genomic cluster and are encoded for by multiexon genes, more reminiscent of the gene for IFN-$\gamma$. In contrast, the genes encoding IFN-$\lambda$ receptor chains reside in IFN-I/IL-10 locus (aka CRF2 locus), consisting of CRF2–12 (provisionally named IFN-$\lambda$R1 or IL-28R$\alpha$) and CRF2–4 (a component of the IL-10 receptor; see the following). Although many functional and mechanistic details of this family have not yet been fully elucidated, $\lambda$-IFNs do appear to be induced in response to viral infection and predominately signal (at least in fibroblasts) through Stat1–Stat2 heterodimers. The relative role of this unique family during viral or bacterial infections is an area of current investigation.

#### 5.1.4 IL-10 Receptor Family

This family of receptors and their corresponding ligands has grown considerably in size over the last few years. There are five distinct receptor chains that alternatively pair to form at least 5 unique receptors (not including the IFN-$\lambda$ receptor). They bind and transduce signals for six ligands, including IL-10, IL-19, IL-20, IL-22 (IL-TIF), IL-24 (MDA-7; FISP), and IL-26 (AK-155). IL-10, the founding member of this family, is widely expressed and recognized for its potent and clinically important anti-inflammatory activity. The first evidence that IL-10 might be related to the IFNs came when the sequence of the first IL-10 receptor chain (IL-10R1) was found to be homologous to the IFN-$\gamma$ receptor. Subsequently, several orphan *CRF2* genes were found to serve as receptors for IL-10 family members. First was CRF-2–4, identified as the second chain of the IL-10 receptor (IL-10R2). More recently, three additional receptor chains, IL-20R1, IL-20R2, and IL-22R, have been identified. IL-19, IL-20, and IL-24 have been shown to signal through a receptor consisting of IL-20R1 and IL-20R2. However, both IL-20 and IL-24 can also signal through a second receptor consisting of IL-22R and IL-20R2. In contrast, IL-22 appears to signal exclusively through a receptor consisting of IL-22R and IL-10R2, whereas IL-26 appears to signal through a receptor consisting of IL-20R1 and IL-10R2. Of these ligands and receptors, IL-10 and its receptor remain the best characterized. The IL-10 receptor is expressed in most tissues. Its cytoplasmic domains associate with Jak1 and Tyk2 respectively, and serve to recruit and activate Stat3, the same STAT that transduces signals for the gp130 family (Sect. 5.2). Recent studies have provided important insight as to how IL-6 and IL-10 can both transduce signals through Stat3, and yet achieve very different biological responses. It has been determined that because the IL-10 receptor fails to recruit SOCS-3, a potent Stat3 antagonist (see Sect. 6.3), IL-10 directs a significantly prolonged pattern of Stat3 activation, culminating in a distinct pattern of gene expression. In addition, in contrast to gp130 responses (Sects. 4.3.3 and 5.2.1), Stat1 does not appear to play a significant role in directing the biological response to IL-10. With the recent growth in IL-10 like ligands and receptors, much remains to be learned about their biological responses. Initial studies have suggested that IL-19 and IL-20 are preferentially expressed in monocytes, and that IL-20 may play an important role in dermal inflammation. IL-22 has been reported to induce the expression of acute phase response genes in the liver, and IL-24 appears to be a Th2 secreted effector cytokine. Intriguingly, IL-24 appears to selectively antagonize the growth of some tumors. Likewise, IL-26 may also function as a T-cell effector cytokine.

## 5.2 Extended gp130 Receptor Family

The extended family of gp130 receptors includes a large group of receptors that either employs gp130, or a gp130 homolog as one of their receptor chains. The first, and therefore prototypical member of this family is the receptor for IL-6, consisting of a unique IL-6 ligand-binding chain (IL-6R$\alpha$) and gp130. The extended gp130 receptor family can be subdivided into those receptors that are functionally distinct from the IL-6 receptor (i.e. IL-12 family of gp130-related receptors; Sect. 5.2.3) and those that are functionally similar. The latter group can also be further subdivided into two groups, those receptors in which gp130 participates (i.e. IL6-gp130 receptor subfamily; Sect. 5.2.1), and those in which the receptor solely employs gp130 homologs (i.e. gp130-like receptors; Sect. 5.2.2). The discussion of this extended family will start with the prototypical IL6-gp130 receptor subfamily.

### 5.2.1 IL6-gp130 Receptor Subfamily

gp130 was first identified as the signaling component of the IL-6 receptor. It associates predominantly with Jak2 and encodes four STAT recruitment motifs (referred to as the YXXQ amino acid motifs), which exhibit a differential affinity for Stat1 and Stat3. gp130 also encodes fifth and distinct tyrosine motif (Y759 in mice) that exhibits a unique affinity for SHP2 and SOCS-3. Gene targeting studies creating gp130 chains that either only express a single YXXQ motif or Y759, have elegantly served to elucidate their roles in mediating the biological response to IL-6. Specifically, the Y759 motif has been shown to direct the activation of an SHP2-MAP kinase signaling cascade that serves to promote cytoprotective (i.e. anti-inflammatory) responses, as well as antagonizing Stat3-dependent signals through the recruitment of SOCS-3. In contrast, YXXQ motifs direct recruitment/activation of Stat3 (and Stat1) to drive the expression of a number of proinflammatory genes, as well as SOCS-3, a Stat3-specific antagonist (Sect. 6.3). These genes are associated with both inflammation and cancer progression. Other IL-6 related ligands that transduce signals through gp130 include, IL-11, CNTF, LIF, OSM, CT-1, NP, and NNT-1/BSF-3. In each case, the corresponding receptors employ a unique chain that either only serves as a ligand-binding chain (like IL-6R$\alpha$), or both as a ligand-binding and signal transducing chain. The latter group includes specific receptor chains for OSM (i.e. OSM-R$\alpha$) and LIF (i.e. LIF-R$\beta$), which both exhibit significant homology to gp130. Of note, OSM appears to straddle this and the following group (Sect. 5.2.2), because it binds to two receptors, one that consists of gp130 and OSM-R$\alpha$, and the other consisting of OSM-R$\alpha$ and OSM-R$\beta$ (also a gp130 homolog).

### 5.2.2 gp130 Receptor-like Subfamily

Although most members of the extended gp130 receptor family include a gp130 chain, several appear to function independent of gp130. As alluded to above, this includes one of the OSM receptors, as well as receptors for IL-31, G-CSF, and Leptin. Curiously, the IL-31 receptor consists of both a unique chain (IL-31R$\alpha$) and OSM-R$\beta$. The receptors for both G-CSF and Leptin function as homodimers (see also section on single-chain receptors). The G-CSF receptor is important for neutrophil maturation, whereas the Leptin receptor is important in weight homeostasis. Importantly however, like in the IL6-gp130 subfamily, each of these ligand–receptor pairs transduces pivotal signals through Stat3.

### 5.2.3 IL-12 Family of gp130-Related Receptors

This group includes three receptors that are distantly related to, but functionally distinct from the IL-6-gp130 family. They bind to and transduce signals for a unique family of related ligands, which include IL-12, IL-23, IL-27. IL-12, the founding and most fully characterized member of this family is composed of two disulfide-linked peptides of 35 kDa (i.e. p35) and 40 kDa (p40). p35 has homology to IL-6, whereas p40 has homology to the IL-6R$\alpha$ chain (also IL-11R$\alpha$ and CNTFR$\alpha$). p40 can also associate with a distinct 19-kDa subunit (i.e. p19) to form a novel cytokine, IL-23. Another recently identified and more poorly characterized member of this family is IL-27, which consists of EBI3, an IL-12 p40 homolog (recently renamed IL-30), and p28, an IL-12 p35 homolog. Not unexpectedly, this family of related ligands binds to and transduces signals through a family of related receptors. IL-12 binds to a receptor consisting of IL-12R$\beta$1 and IL-12R$\beta$2. This receptor is known to direct the sequential activation of Tyk2 and Stat4, both of which have been shown to be important in the biological response to IL-12. The IL-23 receptor consists of the p40 binding IL-12$\beta$1 chain and a unique p19 binding chain, now known as IL-23R. Although, one chain of the IL-27 receptor has recently been identified as the orphan WSX-1 receptor chain, the nature of the second chain and the STAT-dependent signal downstream from this receptor remains controversial. However, the important role these ligands play in regulating the maturation of naïve T cells indicates that their signaling pathways will be similar.

## 5.3 γC Receptor Family

This family of receptors exploits the common hematopoietic receptor gamma chain (i.e. γC) and plays an important role in directing development and activity of lymphocytes. These receptors can be divided into two functionally distinct subgroups, the IL-2 family and the IL-4 family.

### 5.3.1 IL-2 Receptor Family

The receptor for IL-2 is a founding member of the cytokine receptor family. The role this receptor and its ligand play in lymphocyte development has been studied extensively. Additional family members that also play an important role in lymphocyte regulation, and their corresponding ligands have been identified over the ensuing years. They include the receptors for IL-7, TSLP, IL-9, IL-15, and IL-21. Each receptor consists of a ligand specific chain (the α-chain) and the common gamma chain (γC). In most cases, α-chains are associated with Jak1 and serve to recruit and mediate the activation of Stat5. The γC, whose expression is limited to hematopoietic tissues, serves to bring Jak3 to the receptor complex. The critical role these latter two components play in lymphocyte function is underscored by the severe combined immunodeficiency (SCID) phenotype that develops when γC or Jak3 are defective. The IL-2 family exhibits a number of additional noteworthy features. First, two closely related members of this receptor family (i.e. the IL-2 and IL-15 receptors) employ a third receptor chain, the β-chain, which serves as the signal transducing chain. In the absence of the α-chain, this β-chain directs the activity of a low affinity IL-2 receptor. Also, the receptor for TSLP does not use γC, rather, it consists of the IL-7R α-chain and a unique TSLP receptor chain. Finally, as mentioned to earlier, each of these receptors, except perhaps the IL-21 receptor, rely on Stat5 to transduce important ligand specific signals.

### 5.3.2 IL-4 Receptor Family

The receptors for IL-4 and IL-13 consist of an overlapping set of chains and are functionally distinct in the IL-2 family. These receptors and their corresponding ligands have received considerable attention for the role they play in regulating humeral immunity and allergic response. IL-4 binds two distinct receptors. One is composed of the IL-4 receptor α-chain (IL-4Rα), which associates with Jak1, and γC chain, which is associated with Jak3. The second receptor is a heterodimer between IL-4Rα (associated with Jak1) and the IL-13 receptor α-chain (IL-13Rα), which associates with Tyk2. IL-13 binds and signals through the second of these two receptors (i.e. IL-4Rα and IL-13Rα). For both receptors, the IL-4Rα chain serves as the signaling chain, directing the recruitment and activation of Stat6. This, along with the ability to signal through IRS adaptor proteins, functionally distinguishes this family from the IL-2 family. While this small family has overlapping function, gene targeting studies have identified important differences. IL-4 plays a more important role in lymphocyte maturation, whereas IL-13 is more important in directing lymphocyte effector functions.

### 5.4
### IL-3 Receptor Family

The IL-3 family includes receptors for IL-3, IL-5, and GM-CSF (granulocyte macrophage-colony-stimulating factor), all of which play important roles in the development of myeloid lineages. The IL-3 and GM-CSF receptors, and their corresponding ligands, play a more important general role, directing development of granulocytes, macrophages, and dendritic cells (DCs). IL-5 and its receptor appear more specialized, promoting the development of mast cells and eosinophils. Each receptor in this family is composed of two chains: a ligand specific $\alpha$-chain (i.e. IL-3R$\alpha$, IL-5R$\alpha$, and GM-CSFR$\alpha$); and a shared signal transducing $\beta$-chain (i.e. $\beta_{\text{common}}$). The $\alpha$-chains associate with Jak1 and the $\beta$-chain associates with Jak2. The $\beta$-chain also serves to recruit and activate Stat5. In mice, however, the receptor for IL-3 is a bit more complex. There are two functionally homologous IL-3 receptor $\beta$-chains, $\beta_{\text{IL3}}$ and $\beta_{\text{common}}$. Analogous to human system, $\beta_{\text{common}}$ serves as the $\beta$-chain for GM-CSF and IL-5 receptors, as well as some IL-3 receptors. But the remaining IL-3 receptors employ $\beta_{\text{IL3}}$, providing redundancy in the ability to respond to IL-3.

### 5.5
### Single-chain Receptor Family

The single-chain receptors are the final family of structurally and functionally related hematopoietin receptors. As their name suggests, these receptors consist of a single chain that both binds ligand and transduces signals. Members of this family include receptors for traditional hematopoietins, like Epo (erythropoietin) and Tpo (thrombopoietin), as well as the receptors for GH and PRL (prolactin). In each case, the ligand binds to and activates a receptor homodimer, culminating in the sequential activation of receptor associated Jak2 and then Stat5. It is worth noting that two receptors from the gp130-related family, the G-CSF and Leptin receptors also function as homodimers.

### 5.6
### Noncytokine Receptors

A number of receptors that are not from the cytokine family also appear to direct a subset of signals through STAT activation. Well-known examples include several receptor tyrosine kinases (RTks), like the receptors for EGF (epidermal growth factor), PDGF (platelet-derived growth factor), FGF (fibroblast growth factor) and potentially Flt3. There are also numerous reports implicating STAT signaling in the response to G proteins and G-protein-coupled receptors. The best-characterized example is the receptor for Angiotensin II (i.e. $AT_1$). This should not be surprising, since some of the receptors implicated in STAT activation in *Dictyostelium* are G-protein-coupled receptors.

## 6
## Regulation of JAK-STAT Signaling

Two characteristic features of the JAK-STAT signals are their rapid onset and transient nature. With respect to rapid onset, STATs are briskly activated at the receptor, whereupon they quickly translocate to the nucleus and bind DNA. Like for many other nuclear proteins, the process of nuclear import is dependent on a well-characterized set of proteins. Within a period of hours, however, the cytokine stimulated signals decay and the STATs are reexported back to the cytoplasm, resetting the cell for its next round of stimulation. This signal decay entails both downregulation of the upstream signaling components (i.e. the receptors and JAKs), as well as terminating STAT transcription activity. This section will quickly review three mechanisms that are known to play an important role in STAT signal decay: dephosphorylation, active nuclear export, and inhibition by members of the SOCS family of regulators.

### 6.1 Phosphatases

It is not surprising that several phosphatases, which regulate all kinase-based signals, have been implicated in the control of STAT signal transduction. Genetic and biochemical approaches have been exploited to identify several cytoplasmic phosphatases that promote the decay of activated cytokine receptors and potentially JAKs. They include SHP-1, SHP-2 and potentially CD45. In contrast, studies directed at characterizing the decay of activated STATs have suggested that STATs are dephosphorylated in the nucleus prior to nuclear export. Although TC-PTP has been implicated as a STAT nuclear phosphatase, it is likely that other STAT phosphatases remain to be identified.

### 6.2 Nuclear Export

Within hours of having accumulated in the nucleus, STATs are dephosphorylated and reexported back to the cytoplasm, resetting the cell for the next round of stimulation. The process of nuclear export, just like that of nuclear import, is dependent on Ran-GTPases, but the sensitivity of export to leptomycin B has greatly facilitated the analysis of the export process. As expected, leptomycin B has been shown to effectively block poststimulation nuclear export. Unexpectedly however, treatment of resting cells with leptomycin B also promotes STAT nuclear accumulation. This observation highlights an intriguing aspect of STAT biology, not only are STATs rapidly translocated into and out of the nucleus in response to activation, but they also shuttle into and out of the nucleus in resting cells. Moreover, this process is independent of tyrosine phosphorylation! Thus, the predominately cytoplasmic localization of STATs in resting cells represents a steady state balance between nuclear import and export. While the reason for this continuous flux has not been elucidated, distinct nuclear export sequence (NES) elements, which direct either poststimulation or basal nuclear export, have been identified. The distribution of NES elements suggests that their activity (i.e. access to nuclear export machinery) is regulated by conformational changes STATs undergo as they become activated, bind DNA, and then become dephosphorylated.

### 6.3 The SOCS Family

The SOCS proteins were identified as a family of STAT target genes that directly antagonize STAT activation, establishing a classic "feedback loop." CIS-1 (cytokine-inducible SH2 containing protein), the founding member, was identified as an Epo induced, Stat5 dependent target gene that blocked subsequent Stat5 recruitment to the Epo receptor. SOCS-1 was identified next through several approaches. Comparison of the CIS-1 and SOCS-1 sequences identified a conserved SH2 domain and SOCS box, now recognized as characteristic features of this family. Subsequent gene targeting studies determined that SOCS-1 knockout mice die perinatally from a profound IFN-$\gamma$ – Stat1 dependent inflammatory process, underscoring a critical role for SOCS-1 regulating these potent cytokine. In contrast, SOCS-2 has been shown to antagonize the ability of GH to activate Stat5. Consistent with this, the SOCS-2 knockout mice develop gigantism. The final SOCS protein found to regulate cytokine activity is SOCS-3. (SOCS4-7 appear to regulate responses to different classes of stimuli.) Careful studies have determined that SOCS-3 is a Stat3 antagonist. Furthermore, its specificity for IL-6 dependent Stat3 activation appears to be the major determinant in distinguishing between the differing Stat3 dependent responses to IL-6 and IL-10. This level of specificity is achieved through a single SOCS-3 recruitment motif found in gp130 receptor chain (i.e. Y759; see Sect. 5.2.1), but not either of the IL-10 receptor chains. An analogous SOCS-2 recruitment motif has been identified in the GH receptor. Finally, the mechanism by which the SOCS proteins regulate STAT signaling appears to entail more than competition for receptor recruitment. A number of studies have determined that the conserved SOCS box promotes association with the elongin B/C complex, which in turn binds an E3-like ubiquitin ligase, cullin-2, to promote proteosome dependent degradation. Thus, SOCS proteins appear to direct the degradation of at least some of the components in the JAK-STAT signaling cascade. The existence of SOCS proteins underscores the need to counterregulate STAT based signals.

## 7 Concluding Comments

Studies over the last decade have provided an important insight into how members of the JAK and STAT family direct signal transduction for all four-helix-bundle cytokines (i.e. hematopoietins). While much remains to be learned, significant advances in the future are likely to entail the identification of additional molecules that serve to fine tune this signaling cascade, as well as provide an insight into how these signals integrate with other important signaling pathways. Consistent with this prediction, a number of regulatory molecules, whose role in STAT signaling are not yet fully elucidated, have been identified, including, members of the PIAS (protein inhibitors of activated STATs) family, StIP, Nmi, several phosphatases, and several serine kinases. Recent studies, which have highlighted the role covalent modifications play in the regulation of transcription factors, beget the question of whether STAT signaling will also be found to be regulation by acetylation, methylation, SUMOylation, or nitrosylation.

## Bibliography

### Books and Reviews

Akira, S., Takeda, K. (2004) Toll-like receptor signalling, *Nat. Rev. Immunol.* **4**, 499–511.

Barton, G.M., Medzhitov, R. (2003) Linking Toll-like receptors to IFN-alpha/beta expression, *Nat. Immunol.* **4**, 432–433.

Dearolf, C.R. (1999) JAKs and STATs in invertebrate model organisms, *Cell. Mol. Life Sci.* **55**, 1578–1584.

Decker, T., Kovarik, P. (2000) Serine phosphorylation of STATs, *Oncogene* **19**, 2628–2637.

Decker, T., Kovarik, P., Meinke, A. (1997) GAS elements: a few nucleotides with a major impact on cytokine-induced gene expression, *J. Interferon Cytokine Res.* **17**, 121–134.

Ehret, G.B., Reichenbach, P., Schindler, U., Horvath, C.M., Fritz, S., Nabholz, M., Bucher, P. (2001) DNA binding specificity of different STAT proteins. Comparison of in vitro specificity with natural target sites, *J. Biol. Chem.* **276**, 6675–6688.

Fickenscher, H., Hor, S., Kupers, H., Knappe, A., Wittmann, S., Sticht, H. (2002) The interleukin-10 family of cytokines, *Trends Immunol.* **23**, 89–96.

Heinrich, P.C., Behrmann, I., Haan, S., Hermanns, H.M., Muller-Newen, G., Schaper, F. (2003) Principles of interleukin (IL)-6-type cytokine signalling and its regulation, *Biochem. J.* **374**, 1–20.

Horvath, C.M. (2000) STAT proteins and transcriptional responses to extracellular signals, *Trends Biochem. Sci.* **25**, 496–502.

Ihle, J.N. (2001) The Stat family in cytokine signaling, *Curr. Opin. Cell Biol.* **13**, 211–217.

Kile, B.T., Nicola, N.A., Alexander, W.S. (2001) Negative regulators of cytokine signaling, *Int. J. Hematol.* **73**, 292–298.

Kisseleva, T., Bhattacharya, S., Braunstein, J., Schindler, C.W. (2002) Signaling through the JAK/STAT pathway, recent advances and future challenges, *Gene* **285**, 1–24.

Kovanen, P.E., Leonard, W.J. (2004) Cytokines and immunodeficiency diseases: critical roles of the gamma(c)-dependent cytokines interleukins 2, 4, 7, 9, 15, and 21, and their signaling pathways, *Immunol. Rev.* **202**, 67–83.

Levy, D.E., Darnell, J.E. Jr. (2002) Stats: transcriptional control and biological impact, *Nat. Rev. Mol. Cell Biol.* **3**, 651–662.

Miyoshi, K., Cui, Y., Riedlinger, G., Lehoczky, J., Zon, L., Oka, T., Dewar, K., Hennighausen, L. (2001) Structure of the mouse stat 3/5 locus: evolution from drosophila to zebrafish to mouse, *Genomics* **71**, 150–155.

O'Shea, J.J., Gadina, M., Schreiber, R.D. (2002). Cytokine signaling in 2002: new surprises in the Jak/Stat pathway, *Cell* **109**, S123–S131.

Pestka, S., Krause, C.D., Sarkar, D., Walter, M.R., Shi, Y., Fisher, P.B. (2004a) Interleukin-10 and related cytokines and receptors, *Annu. Rev. Immunol.* **22**, 929–979.

Pestka, S., Krause, C.D., Walter, M.R. (2004b) Interferons, interferon-like cytokines, and their receptors, *Immunol. Rev.* **202**, 8–32.

Shuai, K. (2000) Modulation of STAT signaling by STAT-interacting proteins, *Oncogene* **19**, 2638–2644.

Zeidler, M.P., Bach, E.A., Perrimon, N. (2000) The roles of the Drosophila JAK/STAT pathway, *Oncogene* **19**, 2598–2606.

## Primary Literature

Akira, S., Nishio, Y., Inoue, M., Wang, X.-J., Wei, S., Matsusaka, T., Yoshida, K., Sudo, T., Naruto, M., Kishimoto, T. (1994) Molecular cloning of APRF, a novel IFN-stimulated gene factor 3 p91-related transcription factor involved in the gp130-mediated signaling pathway, *Cell* **77**, 63–71.

Alexander, W.S., Starr, R., Fenner, J.E., Scott, C.L., Handman, E., Sprigg, N.S., Corbin, J.E., Cornish, A.L., Darwiche, R., Owczarek, C.M. et al. (1999) SOCS1 is a critical inhibitor of interferon $\gamma$ signaling and prevents the potentially fatal neonatal actions of this cytokine, *Cell* **98**, 597–608.

Argetsinger, L.S., Carter-Su, C. (1996) Mechanism of signaling by growth hormone receptor, *Physiol. Rev.* **76**, 1089–1107.

Azam, M., Erdjument-Bromage, H., Kreider, B.L., Xia, M., Quelle, F., Basu, R., Saris, C., Tempst, P., Ihle, J.N., Schindler, C. (1995) Interleukin-3 signals through multiple isoforms of Stat5, *EMBO J.* **14**, 1402–1411.

Becker, S., Groner, B., Müller, C.W. (1998) Three-dimensional structure of the Stat3$\beta$ homodimer bound to DNA, *Nature* **394**, 145–151.

Begitt, A., Meyer, T., van Rossum, M., Vinkemeier, U. (2000) Nucleocytoplasmic translocation of Stat1 is regulated by a leucine-rich export signal in the coiled-coil domain, *Proc. Natl. Acad. Sci. U.S.A.* **97**, 10418–10423.

Bhattacharya, S., Schindler, C. (2003) Regulation of Stat3 nuclear export, *J. Clin. Invest.* **111**, 553–559.

Braunstein, J., Brutsaert, S., Olson, R., Schindler, C. (2003) STATs dimerize in the absence of phosphorylation, *J. Biol. Chem.* **278**, 34133–34140.

Bromberg, J.F., Wrzeszczynska, M.H., Devgan, G., Zhao, Y., Pestell, R.G., Albanese, C., Darnell, J.E. (1999) Stat3 as an Oncogene, *Cell* **98**, 295–303.

Chen, X., Vinkemeier, U., Zhao, Y., Jeruzalmi, D., Darnell, J.E. Jr., Kuriyan, J. (1998) Crystal structure of a tyrosine phosphorylated STAT-1 dimer bound to DNA, *Cell* **93**, 827–839.

Chen, X., Bhandari, R., Vinkemeier, U., Van Den Akker, F., Darnell, J.E. Jr., Kuriyan, J. (2003) A reinterpretation of the dimerization interface of the N-terminal domains of STATs, *Protein Sci.* **12**, 361–365.

Copeland, N.G., Gilbert, D.J., Schindler, C., Zhong, Z., Wen, Z., Darnell, J.E. Jr., Mui, A.L., Miyajima, A., Quelle, F.W., Ihle, J.N., et al. (1995) Distribution of the mammalian Stat gene family in mouse chromosomes, *Genomics* **29**, 225–228.

Croker, B.A., Krebs, D.L., Zhang, J.G., Wormald, S., Willson, T.A., Stanley, E.G., Robb, L., Greenhalgh, C.J., Forster, I., Clausen, B.E., et al. (2003) SOCS3 negatively regulates IL-6 signaling in vivo, *Nat. Immunol.* **4**, 540–545.

Durbin, J.E., Hackenmiller, R., Simon, M.C., Levy, D.E. (1996) Targeted disruption of the mouse Stat1 gene results in compromised innate immunity to viral disease, *Cell* **84**, 443–450.

Gupta, S., Yan, H., Wong, L.H., Ralph, S., Krolewski, J., Schindler, C. (1996) The SH2 domains of Stat1 and Stat2 mediate multiple interactions in the transduction of IFN-alpha signals, *EMBO J.* **15**, 1075–1084.

Heim, M.H., Kerr, I.M., Stark, G.R., Darnell, J.E. (1995) Contribution of STAT SH2 groups to specific interferon signaling by the JAK-STAT pathway, *Science* **267**, 1347–1349.

Hou, J., Schindler, U., Henzel, W.J., Ho, T.C., Brasseur, M., McKnight, S.L. (1994) An Interleukin-4 induced transcription factor: IL-4 Stat, *Science* **265**, 1701–1705.

Kaplan, M.H., Schindler, U., Smiley, S.T., Grusby, M.J. (1996a) Stat6 is required for mediating responses to IL-4 and for the development of Th2 cells, *Immunity* **4**, 313–319.

Kaplan, M.H., Sun, Y.-L., Hoey, T., Grusby, M.J. (1996b) Impaired IL-12 responses and enhanced development of Th2 cells in Stat4-deficient mice, *Nature* **382**, 174–177.

Karaghiosoff, M., Neubauer, H., Lassnig, C., Kovarik, P., Schindler, H., Pircher, H., McCoy, B., Bogdan, C., Decker, T., Brem, G., et al. (2000) Partial impairment of cytokine responses in Tyk2-deficient mice, *Immunity* **13**, 549–560.

Kawata, T., Shevchenko, A., Fukuzawa, M., Jermyn, K.A., Totty, N.F., Zhukovskaya, N.V., Sterling, A.E., Mann, M., Williams, J.G. (1997) SH2 signaling in a lower eukaryote: a STAT protein that regulates stalk cell differentiation in dictyostelium, *Cell* **89**, 909–916.

Kieslinger, M., Woldman, I., Moriggl, R., Hofmann, J., Marine, J.-C., Ihle, J.N., Beug, H., Decker, T. (2000) Anti-apoptotic activity of Stat5 required during terminal stages of myeloid differentiation, *Genes Dev.* **14**, 232–244.

Klingmuller, U., Lorenz, U., Cantley, L.C., Neel, B.G., Lodish, H.F. (1995) Specific recruitment of SH-PTP1 to the erythropoietin receptor causes inactivation of JAK2 and termination of proliferative signals, *Cell* **80**, 729–738.

Marie, I., Durbin, J.E., Levy, D.E. (1998) Differential viral induction of distinct alpha-interferon genes by positive feedback through interferon regulatory factor-7, *EMBO J.* **17**, 6660–6669.

Marine, J.-C., Topham, D.J., McKay, C., Wang, D., Parganas, E., Stravopodis, D., Yoshimura, A., Ihle, J.N. (1999a) SOCS1 deficiency causes a lymphocyte-dependent perinatal lethality, *Cell* **98**, 609–616.

Marine, J.C., McKay, C., Wang, D., Topham, D.J., Parganas, E., Nakajima, H., Pendeville, H., Yasukawa, H., Sasaki, A., Yoshimura, A., Ihle, J.N. (1999b) SOCS3 is essential in the regulation of fetal liver erythropoiesis, *Cell* **98**, 617–627.

Maritano, D., Sugrue, M.L., Tininini, S., Dewilde, S., Strobl, B., Fu, X., Murray-Tait, V., Chiarle, R., Poli, V. (2004) The STAT3 isoforms alpha and beta have unique and specific functions, *Nat. Immunol.* **5**, 401–409.

McBride, K.M., Banninger, G., McDonald, C., Reich, N.C. (2002) Regulated nuclear import of the STAT1 transcription factor by direct binding of importin-alpha, *EMBO J.* **21**, 1754–1763.

Meraz, M.A., White, J.M., Sheehan, K.C., Bach, E.A., Rodig, S.J., Dighe, A.S., Kaplan, D.H., Riley, J.K., Greenlund, A.C., Campbell, D., et al. (1996) Targeted disruption of the Stat1 gene in mice reveals unexpected physiological specificity in the JAK-STAT pathway, *Cell* **84**, 431–442.

Metcalf, D., Greenhalgh, C.J., Viney, E., Willson, T.A., Starr, R., Nicola, N.A., Hilton, D.J., Alexander, W.S. (2000) Gigantism in mice lacking suppressor of cytokine signalling-2, *Nature* **405**, 1069–1073.

Moriggl, R., Sexl, V., Piekorz, R., Topham, D., Ihle, J.N. (1999) Stat5 activation is uniquely associated with cytokine signaling in peripheral T cells, *Immunity* **11**, 225–230.

Neubauer, H., Cumano, A., Mueller, M., Wu, H., Huffstadt, U., Pfeffer, K. (1998) Jak2 deficiency defines an essential developmental checkpoint in definitive hematopoiesis, *Cell* **93**, 397–409.

Nosaka, T., van Deursen, J.M., Tripp, R.A., Thierfelder, W.E., Witthuhn, B.A., McMickle, A.P., Doherty, P.C., Grosveld, G.C., Ihle, J.N. (1995) Defective lymphoid development in mice lacking Jak3, *Science* **270**, 800–802.

Ohtani, T., Ishihara, K., Atsumi, T., Nishida, K., Kaneko, Y., Miyata, T., Itoh, S., Narimatsu, M., Maeda, H., Fukada, T., et al. (2000) Dissection of signaling cascades through gp130 in vivo: reciprocal roles for STAT3- and SHP2-mediated signals in immune responses, *Immunity* **12**, 95–105.

Parganas, E., Wang, D., Stravopodis, D., Topham, D., Marine, J.-C., Teglund, S., Vanin, E.F., Bodner, S., Colamonici, O.R., van Deursen, J.M., et al. (1998) Jak2 is essential for signaling through a variety of cytokine receptors, *Cell* **93**, 385–395.

Park, C., Li, S., Cha, E., Schindler, C. (2000) Immune response in Stat2 knockout mice, *Immunity* **13**, 795–804.

Park, S.Y., Saijo, K., Takahashi, T., Osawa, M., Arase, H., Hirayama, N., Miyake, K., Nakauchi, H., Shirasawa, T., Saito, T. (1995) Developmental defects of lymphoid cells in Jak3 kinase-deficient mice, *Immunity* **3**, 771–782.

Remy, I., Wilson, I.A., Michnick, S.W. (1999) Erythropoietin receptor activation by a ligand-induced conformation change, *Science* **283**, 990–993.

Rodig, S.J., Meraz, M.A., White, J.M., Lampe, P.A., Riley, J.K., Arthur, C.D., King, K.L., Sheehan, K.C., Yin, L., Pennica, D., et al. (1998) Disruption of the Jak1 gene demonstrates obligatory and nonredundant roles of the Jaks in cytokine-induced biologic responses, *Cell* **93**, 373–383.

Schindler, C., Fu, X.Y., Improta, T., Aebersold, R., Darnell, J.E. Jr. (1992a) Proteins of transcription factor ISGF-3: one gene encodes the 91-and 84-kDa ISGF-3 proteins that are activated by interferon alpha, *Proc. Natl. Acad. Sci. U.S.A.* **89**, 7836–7839.

Schindler, C., Shuai, K., Prezioso, V., Darnell, J.E. (1992b) Interferon-dependent tyrosine phosphorylation of a latent cytoplasmic transcription factor, *Science* **257**, 809–813.

Seidel, H.M., Milocco, L.H., Lamb, P., Darnell, J.E., Stein, R.B., Rosen, J.B. (1995) Spacing of palindromic half sites as a determinant of selective STAT (signal transducers and activators of transcription) DNA binding and transcriptional activity, *Proc. Natl. Acad. Sci. U.S.A.* **92**, 3041–3045.

Shimoda, K., van Deursen, J., Sangster, M.Y., Sarawar, S.R., Carson, R.T., Tripp, R.A., Chu, C., Quelle, F.W., Nosaka, T., Vignali, D.A., et al. (1996) Lack of IL-4 induce Th2 response in IgE class switching mice with disrupted Stat6 gene, *Nature* **380**, 630–633.

Shimoda, K., Kato, K., Aoki, K., Matsuda, T., Miyamoto, A., Shibamori, M., Yamashita, M., Numata, A., Takase, K., Kobayashi, S., et al. (2000) Tyk2 plays a restricted role in IFN alpha signaling, although it is required for IL-12-mediated T cell function, *Immunity* **13**, 561–571.

Shuai, K., Schindler, C., Prezioso, V., Darnell, J.E. (1992) Activation of transcription by IFN-$\gamma$: tyrosine phosphorylation of a 91-kDa DNA binding protein, *Science* **258**, 1808–1812.

Shuai, K., Stark, G.R., Kerr, I.M., Darnell, J.E. (1993) A single phosphotyrosine residue of Stat91 required for gene activation by interferon-$\gamma$, *Science* **261**, 1744–1743.

Silvennoinen, O., Schindler, C., Schlessinger, J., Levy, D.E. (1993) Ras-independent signal transduction in response to growth factors and cytokines by tyrosine phosphorylation of a common transcription factor, *Science* **261**, 1737–1739.

Stahl, N., Boulton, T.G., Farruggella, T., Ip, N.Y., Davis, S., Witthuhn, B.A., Quelle, F.W., Silvennoinen, O., Barbieri, G., Pellegrini, S., et al. (1994) Association and activation of Jak-Tyk kinases by CNTF-LIF-OSM-IL-6 beta receptor components, *Science* **263**, 92–95.

Starr, R., Willson, T.A., Viney, E.M., Murray, L.J., Rayner, J.R., Jenkins, B.J., Gonda, T.J., Alexander, W.S., Metcalf, D., Nicola, N.A., Hilton, D.J. (1997) A family of cytokine-inducible inhibitors of signaling, *Nature* **387**, 917–921.

Takeda, K., Clausen, B.E., Kaisho, T., Tsujimura, T., Terada, N., Forster, I., Akira, S. (1999) Enhanced Th1 activity and development of chronic enterocolitis in mice devoid of Stat3 in macrophages and neutrophils, *Immunity* **10**, 39–49.

Takeda, K., Noguchi, K., Shi, W., Tanaka, T., Matsumoto, M., Yoshida, N., Kishimoto, T., Akira, S. (1997) Targeted disruption of the mouse Stat3 gene leads to early embryonic lethality, *Proc. Natl. Acad. Sci. U.S.A.* **94**, 3801–3804.

Takeda, K., Tanaka, T., Shi, W., Matsumoto, M., Minami, M., Kashiwamura, S., Nakanishi, K., Yoshida, N., Kishimoto, T., Akira, S. (1996) Essential role of Stat6 in IL-4 signaling, *Nature* **380**, 627–630.

Teglund, S., McKay, C., Schuetz, E., Van Deursen, J.M., Stravopodis, D., Wang, D., Brown, M., Bodner, S., Grosveld, G., Ihle, J.N. (1998) Stat5a and Stat5b proteins have essential roles and nonessential, or redundant, roles in cytokine responses, *Cell* **93**, 841–850.

Thierfelder, W.E., van Deursen, J.M., Yamamoto, K., Tripp, R.A., Sarawar, S.R., Carson, R.T., Sangster, M.Y., Vignali, D.A., Doherty, P.C., Grosveld, G., Ihle, J.N. (1996) Requirement for Stat4 in interleukin-12 mediated response of natural killer and T cells, *Nature* **382**, 171–174.

Thomis, D.C., Gurniak, C.B., Tivol, E., Sharpe, A.H., Berg, L.J. (1995) Defects in B lymphocyte maturation and T lymphocyte activation in mice lacking Jak3, *Science* **270**, 794–797.

Varinou, L., Ramsauer, K., Karaghiosoff, M., Kolbe, T., Pfeffer, K., Muller, M., Decker, T. (2003) Phosphorylation of the Stat1 transactivation domain is required for full-fledged IFN-gamma-dependent innate immunity, *Immunity* **19**, 793–802.

Velazquez, L., Fellous, M., Stark, G.R., Pellegrini, S. (1992) A protein tyrosine kinase in the Interferon $\alpha/\beta$ signaling pathway, *Cell* **70**, 313–322.

Vinkemeier, U., Moarefi, I., Darnell, J.E. Jr., Kuriyan, J. (1998) Structure of the amino-terminal protein interaction domain of STAT-4, *Science* **279**, 1048–1052.

Wen, Z., Zhong, Z., Darnell, J.E. (1995) Maximal activation of transcription by Stat1 and Stat3 requires both tyrosine and serine phosphorylation, *Cell* **82**, 241–250.

Yasukawa, H., Ohishi, M., Mori, H., Murakami, M., Chinen, T., Aki, D., Hanada, T., Takeda, K., Akira, S., Hoshijima, M., et al. (2003) IL-6 induces an anti-inflammatory response in the absence of SOCS3 in macrophages, *Nat. Immunol.* **4**, 551–556.

# Part III
# Techniques

*Immunology. From Cell Biology to Disease.* Edited by Robert A. Meyers.

ISBN: 978-3-527-31770-7

# 10
# Flow Cytometry

*Michael G. Ormerod*
*34 Wray Park Road, Reigate, UK*

**1 Introduction 273**

**2 The Instrument 273**
2.1 The Flow Chamber 273
2.2 Fluidics 274
2.3 Light Source 274
2.4 Optics 275
2.5 Signal Processing 275

**3 Measurement 276**
3.1 Light Scatter 276
3.2 Fluorescence 276
3.3 Data Processing 276
3.4 Cell Sorting 277

**4 General Applications of Flow Cytometry 279**
4.1 Immunofluorescence 279
4.2 Measurement of DNA 280
4.3 Other Flow Cytometric Methods 281
4.3.1 Measuring Enzyme Kinetics 281
4.3.2 Following Changes in Membrane Potential 281
4.3.3 Intracellular Calcium Ions 282
4.3.4 Intracellular pH 282
4.3.5 Intracellular Glutathione 282
4.3.6 Measuring Oxidative Species 282

**5 Specific Fields of Application 283**
5.1 Clinical 283
5.2 Long Term Labels for Cells 283

*Immunology. From Cell Biology to Disease.* Edited by Robert A. Meyers.

ISBN: 978-3-527-31770-7

5.3 Measurement of Cell Proliferation 283
5.3.1 The DNA Histogram 284
5.3.2 Using BrdUrd/Anti-BrdUrd 284
5.3.3 The BrdUrd-Hoechst/PI Method 286
5.3.4 Measuring Cell Proliferation through the Division of Label between Daughter Cells 287
5.4 Studying Cell Death 287
5.4.1 Necrosis 287
5.4.2 Apoptosis 288
5.5 Analyzing and Sorting Chromosomes 290
5.6 Microbead Technology 291

**6 Applications Other Than Mammalian Cell Biology 291**
6.1 Microbiology 291
6.2 Marine Biology 291

**Bibliography 291**
Books and Reviews 291
Primary Literature 292
Websites 294
Some Companies Who Supply Flow Cytometers 294
Free Software 294
Other Useful Sites 294

## Keywords

**Flow Cytometry**

**Cell Sorting**

A flow cytometer measures between one and thirteen or more parameters on single cells. The flow system is used to deliver cells to the point of measurement. In most instruments, the cells are interrogated by laser light. Measurements are made of scattered light and of fluorescence, either from the cells or, more commonly, from labelled probes. Sub-populations of cells may also be physically sorted for further study. The article briefly describes the main attributes of the technology and its applications in clinical areas and in mammalian biology.

## 1 Introduction

Flow cytometry is widely used in fields as diverse as immunology, hematology, cell biology, marine biology, and microbiology. As its name implies, it is a technique for measuring cells in a flow system. Its most important features are that it makes measurements on single cells, it records between 2 and 13 parameters per cell depending on the instrument, and it can measure thousands of cells a second. These features enable the analysis of subpopulations, in particular, subsets with a frequency as low as 1 in $10^4$ can be enumerated. Additionally, some instruments are capable of sorting subpopulations of cells with a purity of 99%. In addition to cells, other particles, such as nuclei, chromosomes or small beads, can also be studied.

The disadvantage of the technique is that it requires a suspension of single cells, which can be difficult to obtain from solid tissues, and, while the flow cytometer can measure the amount of an entity in a cell, it cannot measure its intracellular distribution.

I have previously published a succinct description of flow cytometry and its applications. Other, longer, introductory books as well as books that describe protocols for many applications of flow cytometry are also available.

## 2 The Instrument

### 2.1 The Flow Chamber

The flow chamber lies at the heart of the instrument. In it, cells are injected into a stream of either water or saline solution (called the *sheath fluid*). The sample stream is thereby hydrodynamically focused so that single cells are delivered to a defined point at which a light beam is focused (Fig. 1). The light source is usually a laser, although some, mainly older instruments, use an arc lamp. As cells pass through the laser beam, they will scatter light and may also fluoresce. While strong autofluorescence may be observed in some cells, particularly marine algae, cells are usually labeled with one or more of a variety of fluorescent markers.

Most instruments will collect light scattered by the cells in a narrow angle in a forward direction (forward scattered light, FS) and also light scattered at right angles to the laser beam (side scatter, SS). Fluorescences are collected and separated by optical filters into different wavelengths. Typically, four fluorescent parameters can be acquired, although in some instruments, this can be as many as eleven.

There are two basic designs of flow chamber in general use; the cuvette flow chamber and the "jet-in-air" or "stream-in-air." The latter is only used in instruments that can physically sort cells (see Sect. 3.4 below).

In the cuvette chamber, the interrogation point, at which the laser beam is focused on the sample, is inside the chamber (Fig. 1). In an analyzer, the sheath and sample streams are taken to waste in a fully enclosed system; in a cell sorter, the stream emerges into the open air. In a "stream-in-air" chamber, the interrogation point is immediately outside the chamber. While this system has advantages for cell sorting, it has inferior optical properties. The airstream interface scatters light, which has to be partially blocked by an additional obscuration bar fitted in front of the lens used

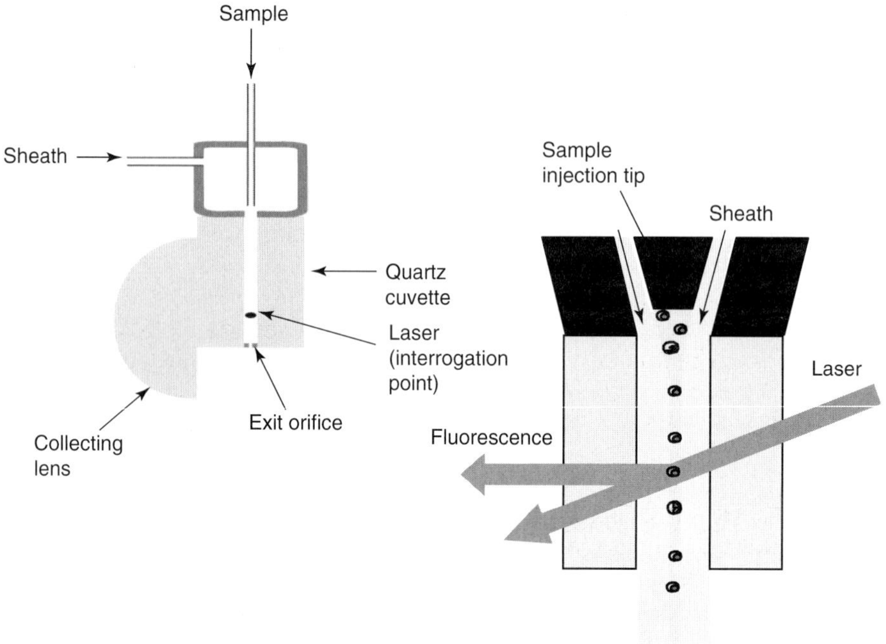


**Fig. 1** Diagrammatic representation of a flow chamber. (See color plate p. xxvii.)

to collect fluorescent light at right angles to the laser beam. Because it is less efficient, it requires a laser of higher power.

### 2.2 Fluidics

The sheath fluid is normally forced through the flow chamber under pressure. The sample is usually injected into the sheath stream by applying a difference in pressure between sample and sheath fluid containers. The difference will control the volume flow rate of the sample. Alternatively, it may be injected into the stream using a motor connected to the piston of a syringe. The flow rate of the sample is then controlled by changing the speed of the motor. With this arrangement, a predefined volume of sample can be used, which enables absolute cell counting.

### 2.3 Light Source

Lasers are the preferred light source because they produce a single wavelength of high intensity. Nearly all instruments are fitted with an argon-ion laser, which gives blue light at 488 nm – a convenient wavelength for exciting the commonly used fluorochrome, fluorescein. Bench-top instruments are usually fitted with an air-cooled laser producing about 15-mW power. "Stream-in-air" sorters need more powerful lasers, frequently water-cooled, which have the advantage that they can, if required, be tuned to other wavelengths.

A second or third laser may be fitted so that cells can be excited at more than one wavelength. The choice depends on the applications; often a He–Ne laser giving red light at 630 nm is installed. A third laser will be selected to give UV, for example, a He–Cd laser with an output at 325 nm. A wide range of solid state lasers are now available giving light at, for example, 405, 488, 530, or 635 nm.

## 2.4 Optics

The optical layout of typical flow cytometer is shown in Fig. 2.

Laser light is focused onto the point of measurement by a lens or lenses generally designed to give an elliptical laser spot. A blocker, or an obscuration bar is fitted behind the flow cell to block the laser beam. A light detector (such as a PIN diode) is positioned behind the obscuration bar to detect light scattered over a narrow angle in the forward direction.

Fluorescent and scattered light is collected at right angles to the laser beam and focused onto a series of photomultipliers (PMTs). Light of different wavelengths is selected using a series of dichroic and barrier (bandpass) optical filters. In the arrangement shown in Fig. 2, the dichroic filters are long pass filters, that is, they reflect light below a certain wavelength and pass light above that wavelength.

## 2.5 Signal Processing

Until recently, the signals from the PMTs were amplified and processed electronically before passing through analog-to-digital (ADC) converters, the digital signals being passed to a computer for data analysis. The trend now is to convert at an early stage and to manipulate the signals digitally. Using analog electronics, the signal is amplified in either a linear or logarithmic amplifier; using digital electronics, the amplified signal is presented on either a linear or a logarithmic scale. The signal may also be processed to output the width of the signal generated on the PMT, its height or the total integrated signal (area).

A threshold is set on the output voltage of one (or possibly two) of the measured parameters so that only signals above a certain limit are processed; this is sometimes referred to as the discriminator setting. The threshold, which is set by

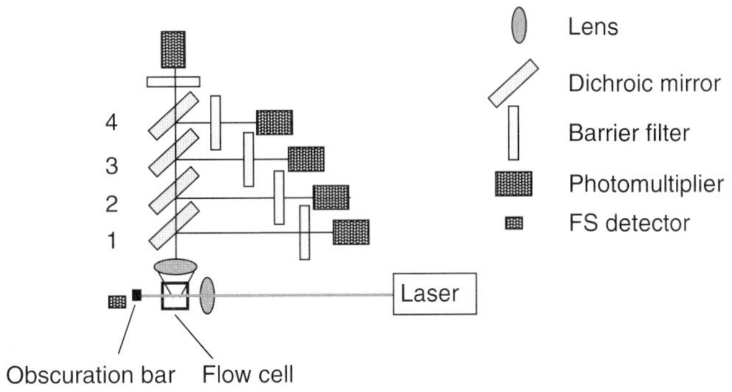


**Fig. 2** The optical layout of a typical flow cytometer. Using an argon-ion laser emitting light at 488 nm, the optical components could be as follows: (1) Dichroic selecting light <500 nm; blue bandpass (488 nm). (2) Dichroic selecting light <540 nm; green bandpass (520 nm). (3) Dichroic selecting light <595 nm; orange bandpass (575 nm). (4) Dichroic selecting light <640 nm; red bandpass (620 nm) Final filter: long pass >650 nm. (See color plate p. xxvii.)

the operator, ensures that the instrument processes signals from cells (or other particles under study) and not from small debris or noise in the system.

## 3 Measurement

### 3.1 Light Scatter

Light scattered by particles as they pass through the laser beam is measured over a narrow angle in a forward direction (forward angle light scatter, FS) and also at right angles to the laser beam (side scatter, SS). The intensity of scatter is proportional to the size, shape and optical homogeneity of cells (or other particles); it is strongly dependent on the angle over which it is measured. Forward scatter may show variation between different flow cytometers reflecting the different geometry of the instruments.

Forward scatter tends to be more sensitive to the size and surface properties and can be used to distinguish live from dead cells while side scatter tends to be more sensitive to inclusions within cells and can be used to distinguish granulated from nongranulated cells.

### 3.2 Fluorescence

In instruments fitted with a laser giving blue light at 488 nm, the filter set will usually be selected to observe fluorescence from fluorescein (512 nm), phycoerythrin (PE) (560 nm) and either phycoerythrin-cyanine5 (PC5, 670 nm) or peridinin-chlorophyll (PerCP, 675 nm), these being the fluorochromes most frequently used for immunofluorescence work. In instruments with more than one laser and more detectors, up to 11 different fluorescences can be measured. There is a wide range of other fluorescent probes available (see, for example, the Molecular Probes catalogue). They are usually selected for use with one of the standard optical filters; for certain probes, specific filter sets may be required.

Most fluorochromes emit light over a wide spectral range. Consequently, light from one fluorochrome will be detected on more than one PMT. For example, light from fluorescein is selected using a green filter and from PE using an orange filter. However, the emission from fluorescein has an orange component, which will give a signal on the PE detector. This spectral overlap can confuse the interpretation of the data and has to be corrected. The compensation parameters can either be set up in the instrument, so that compensated data is recorded, or the data can be compensated post-analysis using software. While electronic compensation is adequate if up to four fluorescent parameters are being measured, compensation for larger numbers of fluorescent parameters is best handled digitally, either in real time or off-line.

### 3.3 Data Processing

Because of the large number of parameters collected on, perhaps, 20 000 cells, it is difficult to visualize correlations between the parameters. The strategy adopted for the analysis of multiparameter data is to use "gating." Two parameters are displayed in a correlated dot plot (often called a *cytogram*). A region is generated around a cluster of cells of interest; this region acts as a "gate." Another cytogram is generated from two other parameters and the computer is instructed to display only those events that fall in the gate set on the first cytogram (see Fig. 3).

### 3.4 Cell Sorting

Some flow cytometers are able to sort selected subsets of cells. There are two types of cell sorters, piezo-electric and charged droplet.

The piezo-electric sorters separate cells mechanically. When a desired cell is detected, voltage is applied to a piezo-electric device, which deflects the cell into a collecting chamber. These sorters are relatively inexpensive, simple to operate

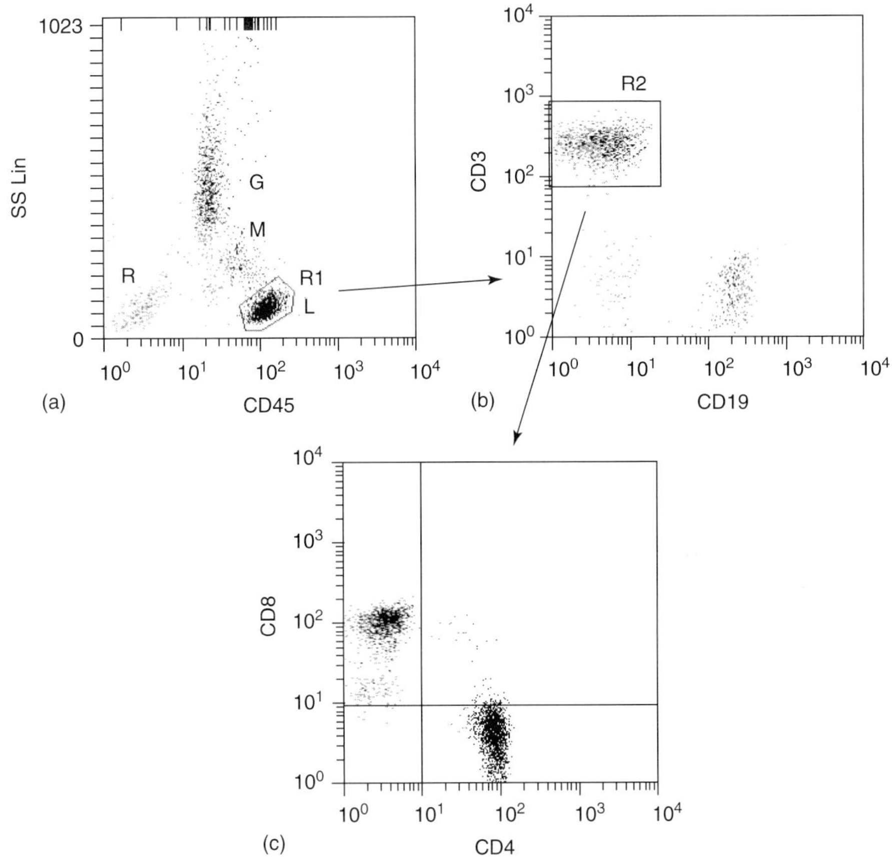


**Fig. 3** Display of typical flow cytometric data. Peripheral blood leucocytes were labeled with antibodies to the cell surface antigens, CD3 (labeled with PE-cyanine5), CD4 (PE), CD8 (PE-Texas Red) and CD45 (fluorescein). (a) Initially, a cytogram of side scatter versus the pan leucocyte marker, CD45, was generated. Four clusters of cells are evident, cluster G is from granulocytes, M from monocytes, L from lymphocytes and R from residual red cells. A region (R1) was drawn around the lymphocytes. This was used as a gate set on subsequent cytograms. (b) Shows a cytogram of CD3 versus CD19 fluorescence of the lymphocytes only, identifying the T cells and B cells respectively. (c) Cytogram C has been gated on R1 and R2 and shows the T-cell subsets identified by CD4 and CD8. Data recorded on a Beckman Coulter XL using an argon-ion laser at 488 nm.

and slow (maximum sort rate is 300 cells $s^{-1}$). The sorted sample is always highly diluted, being mixed with a large amount of sheath fluid.

In charged droplet cell sorting, the stream from the flow cell emerges into the open air. The sheath fluid contains saline so that it is conductive. The flow cell is vibrated by a piezo-electric device, typically at 30 Hz, breaking the stream into well-defined droplets. When a cell to be sorted is in the next droplet to

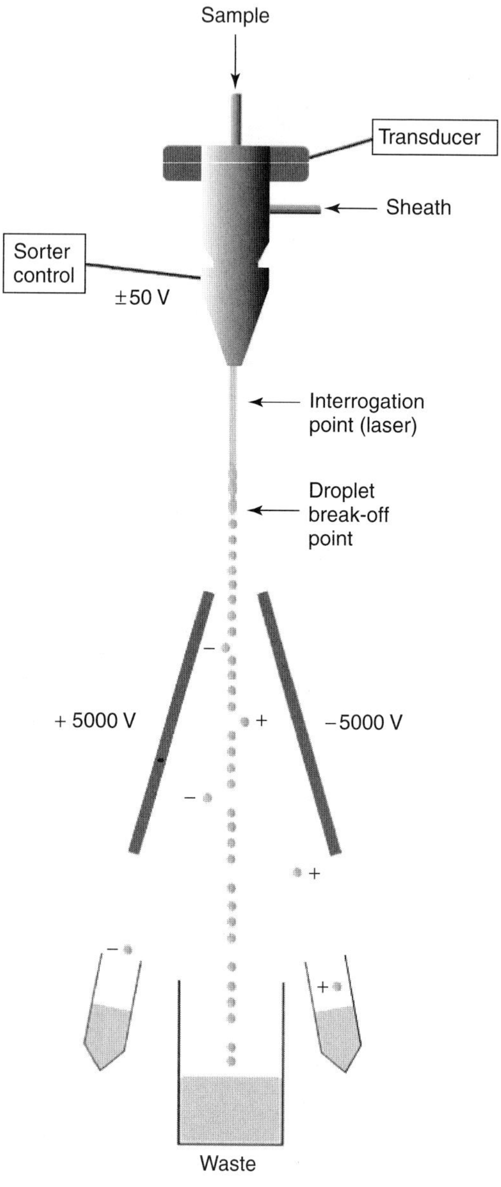


**Fig. 4** A droplet deflection cell sorter.

be formed, the flow cell and stream is momentarily charged, thereby charging that droplet. The droplets pass through a pair of high voltage plates, charged droplets being deflected into the collecting vessels (Fig. 4).

The sorter has to be calibrated so that delay time between the cell passing through the laser beam (and being analyzed) and a droplet being formed is known to the computer, which will trigger the charge pulse at the appropriate moment. The conditions must be stable throughout the sort so that droplet delay time does not change. On modern machines, the instrument can monitor the position of the droplet break-off and automatically correct any minor variation.

The original charged droplet cell sorters could sort cells at speeds up to 5000 cells $s^{-1}$. The modern instruments sort at, at least, 15 000 cells $s^{-1}$; some machines can run at 50 000 cells $s^{-1}$. Despite the increase in speed, flow cytometric cell sorters give a low yield compared to bulk methods, such as immunomagnetic separation. If the concentration of the subpopulation to be sorted is 10%, at a flow rate of 10 000 cells $s^{-1}$ and allowing for some losses through rejection of coincidences, the maximum number of cells collected will be about $3.10^6$ $h^{-1}$. The major advantage is that the purity is high (>98%) and the sort decision can be based on a multiparametric analysis so that small subpopulations can be purified.

## 4 General Applications of Flow Cytometry

There are a wide variety of applications of flow cytometry in most aspects of cell biology. Table 1 lists the most important.

**Tab. 1** Applications of flow cytometry.

| |
|---|
| ***Basic analysis*** |
| Immunofluorescence analysis |
| Cell surface markers |
| Intracellular and nuclear proteins |
| DNA content |
| Cell cycle analysis |
| Ploidy analysis |
| RNA content |
| Protein content |
| ***Measurement of functional parameters*** |
| Intracellular calcium ions |
| Intracellular pH |
| Intracellular glutathione |
| Membrane permeability |
| Membrane potential |
| including mitochondrial membrane potential |
| Measuring oxidative burst |
| ***Specific topics*** |
| Cell sorting |
| Immunophenotyping |
| Analysis of cell proliferation |
| Analysis of cell death |
| Monitoring fusion of cells |
| Measurement of drug uptake |
| Analyzing and sorting chromosomes |
| Kinetic analysis of intracellular enzymes |
| Tracking cells *in vivo* |
| Observing binding and endocytosis of ligands |
| Monitoring electropermeabilization |
| Gene tracking using fluorescent proteins |
| Microbead assays |

### 4.1 Immunofluorescence

This is the most widespread application of the technology. Any protein associated with a cell can be measured if a suitable antibody is available and if there are sufficient molecules present. The most commonly measured proteins, particularly in clinical laboratories, are those associated with leucocyte differentiation, the so-called, CD antigens, most of which are found on the surface of cells.

After fixing or permeabilizing cells, intracellular proteins can also be measured.

A variety of companies supply antibodies to most of the antigens of interest in routine clinical applications and for clinical and biological research. These antibodies are supplied labeled with different fluorochromes so that combinations can be selected to allow the simultaneous measurement of several proteins. Multicolor immunophenotypic analysis can reveal new, previously unrecognized, subsets of cells giving new insights into immunology and into the pathogenesis of diseases of the immune system, including cancer.

Fluorescence resonance energy transfer (FRET) is used in several applications in flow cytometry. If two fluorophores are sufficiently close to one another, energy may be transferred from one to the other. New fluorochromes for immunofluorescence have been made utilizing this principle. For example, if a conjugate of cyanine 5 and phycoerythrin (PE) is exposed to blue light, the PE will absorb the light, pass its energy to the cyanine, which fluoresces. The fluorescence is shifted from orange (PE alone) to red (cyanine 5). The principle can be applied to observe the interaction of two proteins on the surface of cells. Antibody to one protein is labeled with a donor dye, antibody to the other with an acceptor dye. When the proteins come into close contact, the fluorescence of the donor decreases while that of the acceptor increases.

### 4.2
### Measurement of DNA

Measurement of the DNA content of cells gives information about the DNA ploidy and the cell cycle. There are several compounds whose fluorescence increases on binding to DNA and which bind stoichiometrically. The most commonly used of these is propidium iodide (PI), which is excited by blue light and fluoresces red.

Examples of DNA histograms are shown in Fig. 5. In order to estimate the percentage of cells in the G0/G1, S, and G2/M phases of the cell cycle, the overlap between G1 and early S and between G2 and late S has to be resolved. This is generally accomplished by a computer program that attempts to model the DNA histogram.

DNA measurement can be combined with immunofluorescence measurements (Fig. 6). If several immunofluorescent parameters are to be measured, PI may cause a problem because its fluorescence

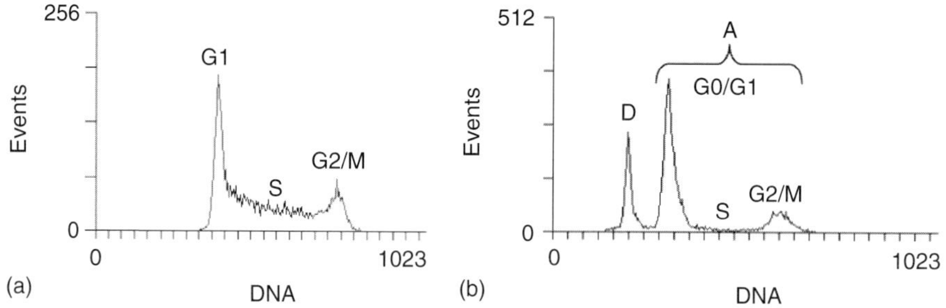


**Fig. 5** DNA histograms. (a) Murine leukemic cell line. Cells were fixed in 70% ethanol, suspended in phosphate-buffered saline and stained with PI. The phases of the cell cycle are labeled. (b) Nuclei extracted from a formalin-fixed, paraffin-embedded biopsy of a breast carcinoma. D, diploid cells; A, aneuploid tumor cells, with the phases of the cell cycle labeled. Data recorded on a Beckman Coulter Elite using an argon-ion laser at 488 nm.

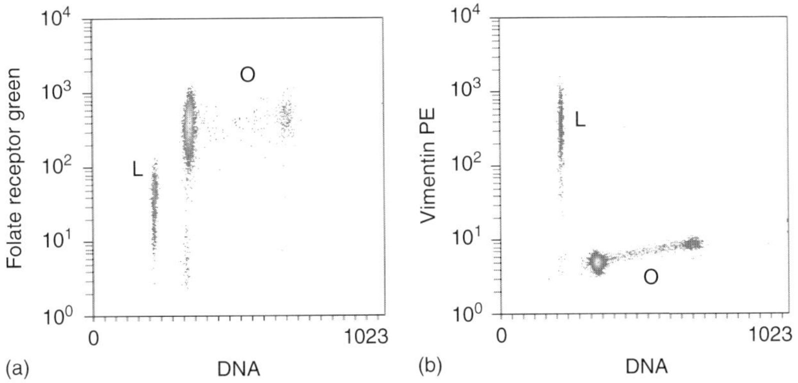


**Fig. 6** Immunofluorescence and DNA measurement combined. Ascitic cells from a patient with ovarian carcinoma. The cells were labeled with anti-folate receptor-FITC, anti-vimentin-PE and PI. Note that there are both diploid and aneuploid cells present. (a) The diploid cells are leucocytes, which express vimentin (b) while the vimentin-negative ovarian carcinoma cells express folate receptor. Data, supplied by Willem Corver, University of Leiden, was recorded on a BD FACScan using an argon-ion laser at 488 nm.

spectrum overlaps strongly with commonly used fluorochromes, such as phycoerythrin and phycoerythrin-cyanine 5. Depending on which lasers are available, alternative dyes are often used; for example, a *bis*-benzimadazole, such as Hoechst 33342, which is excited by UV to give blue fluorescence, or the anthraquinone, DRAQ5, which fluoresces a deep red (700 nm).

## 4.3 Other Flow Cytometric Methods

### 4.3.1 Measuring Enzyme Kinetics

An enzyme substrate can be linked to a fluorescent molecule to give a nonfluorescent compound. The action of the enzyme will release the substrate leaving a fluorescent compound. Fluorophores, which have been used in this application, include fluorescein and rhodamine 110. Generally, this method will only give a crude estimate of enzyme activity since the kinetics of the reaction can be affected by the rate of diffusion of the complex into the cell, the rate of diffusion of the fluorophore out the cell and the released substrate out of the cell, and compartmentalization within the cell.

This approach has been adapted to enable charged compounds to be loaded into cells. For example, the calcium indicator, indo-1, is loaded into the cell by incubation with the acetoxymethyl ester. In the cell, esterases remove the acetoxymethyl groups leaving the fluorescent indo-1, which, being charged, remains trapped in the cell.

### 4.3.2 Following Changes in Membrane Potential

There are several lipophilic dyes that partition between the cell and the surrounding medium according to the plasma membrane potential. Of particular use are a series of oxonol dyes which, depending on their substituents, have different wavelengths of excitation and emission.

Many cyanine dyes, such as dicyanine, 3′,3′-dihexyloxacarbocyanine have a positive charge and are taken up by the mitochondria and the strength of their fluorescence within the cell reflects the mitochondrial membrane potential. One of the cyanine dyes, abbreviated to JC-1, exists as a green fluorescent monomer at low concentrations and at low membrane potential. At high concentrations or higher membrane potential, it forms red-fluorescent aggregates. The dye can be excited at 488 nm and changes in mitochondrial membrane potential can be followed by measuring the ratio between green and red fluorescence.

#### 4.3.3 Intracellular Calcium Ions

$Ca^{2+}$ has an important role in cell signaling. The calcium indicator, indo-1, which is excited by UV, has a fluorescence emission maximum of about 490 nm in the absence of $Ca^{2+}$; in the presence of 1 mM $Ca^{2+}$, the emission maximum is about 410 nm. Changes in the concentration of intracellular $Ca^{2+}$ are recorded by observing fluorescence at 400 and 520 nm and measuring the ratio between the two fluorescences.

Other dyes, which are sensitive to $Ca^{2+}$ and can be excited by blue light, are fluo-3 and Fura Red. $Ca^{2+}$ concentration does not affect the emission wavelength, just the intensity of emission so that these indicators cannot be used to measure changes in $Ca^{2+}$ on a cell-by-cell basis; only the overall changes in a population of cells.

#### 4.3.4 Intracellular pH

There are several compounds that can be used to measure intracellular pH; the most useful is SemiNaphthoFluorescein (SNARF-1), usually used as the carboxy-derivative. Under acidic conditions, carboxy-SNARF-1 fluoresces yellow-orange (580 nm), under alkaline conditions, it fluoresces red (640 nm). The $pK_a$ is about 7.5 at room temperature and the pH is measured by recording the ratio between the fluorescences at 580 and 640 nm.

#### 4.3.5 Intracellular Glutathione

Glutathione has an important role in the cell in reacting with potential damaging oxidative species. Increased levels of glutathione may also be associated with resistance to some cytotoxic drugs.

One assay for glutathione is based on the activity of the enzyme, glutathione-*S*-transferase. Cells are incubated with the dye, monochlorobimane (MClB), which crosses the plasma membrane. In the cytoplasm, the enzyme catalyzes the reaction with glutathione yielding a fluorescent conjugate, which is trapped in the cell. MClB-glutathione fluoresces blue when excited by UV. MClB gives excellent results with rodent cells. It is less effective in human cells, which lack the necessary variant of glutathione-*S*-transferase. As an alternative, monobromobimane, which reacts directly with sulphydryl groups, can be used.

#### 4.3.6 Measuring Oxidative Species

There are several compounds which, in their reduced form, are nonfluorescent but which generate a fluorescent compound upon oxidation. Such compounds include dihydrorhodamine 123, dihydroethidium, and 2′,7′-dihydrodichlorofluorescein (dichlorofluorescin, DCFH). Dihydrorhodamine 123 and dihydroethidium diffuse freely into cells. Upon oxidation, rhodamine 123 is concentrated in the mitochondria and fluoresces green, while ethidium binds to DNA and fluoresces red.

DCFH can be loaded into cells as an ester. After conversion to DCFH by intracellular esterases, oxidation will yield a fluorescent product, 2′,7′-dichlorofluorescein.

These compounds have been used to observe oxidative burst in neutrophils, changes in the concentration of oxidative products during apoptosis and the production of intracellular oxidative species by ionizing radiation.

## 5
## Specific Fields of Application

### 5.1
### Clinical

Flow cytometry is universally used to classify leukemias. Immunophenotyping in combination with flow cytometry is also used to measure minimal residual disease in leukemias (Fig. 7), for stem cell enumeration in bone marrow transplant patients, monitoring HIV infection, quantifying feto-maternal hemorrhage, characterizing immunodeficiency diseases, analyzing paroxysmal nocturnal hemoglobinuria, in transplantation medicine, and for counting platelets and measuring platelet activation. Other routine clinical applications include the detection of auto-antibodies and the enumeration of reticulocytes in blood (Fig. 8).

Applications in the field of clinical research include the measurement of intracellular cytokines, the study of phagocyte biology and function (including phagocytosis and oxidative burst during activation), measurement of cell–cell interactions, the measurement of cell-mediated cytotoxicity and the study of viral infection.

### 5.2
### Long Term Labels for Cells

There are some fluorescent labels that will survive in cells for days, if not weeks. If the cells divide, the label will be carried over into the daughter cells. One of the several compounds used is carboxyfluorescein diacetate, succinimidyl ester (CFDA SE). The compound will diffuse into cells where the diacetate groups are cleaved, generating a fluorescent compound; the succinimidyl groups react with thiols in the cytoplasm, retaining the dye in the cell.

Applications include studying cell fusion or clustering, tracking cells *in vivo*, identifying one partner in the interaction between two types of cell (for, example, a cytotoxicity reaction), and following cell division.

Genes for naturally fluorescent proteins (such as green fluorescent protein, GFP) have been introduced onto the DNA of cultured cells, thereby rendering the cells fluorescent. The GFP gene can be introduced into a plasmid for use as a reported molecule after transfection. Transfected cells can be purified by cell sorting on the flow cytometer.

### 5.3
### Measurement of Cell Proliferation

The study of cell proliferation is an important and powerful application of flow cytometry. One can, for example, derive information about the proliferative status of tumors, observe the effects of cytotoxic drugs on the cell cycle or study the proliferation of subsets of lymphocytes undergoing immune stimulation.

There are several different approaches that can be used. One can measure the cell cycle phases, the rate of movement of cells through the cell cycle, the division of

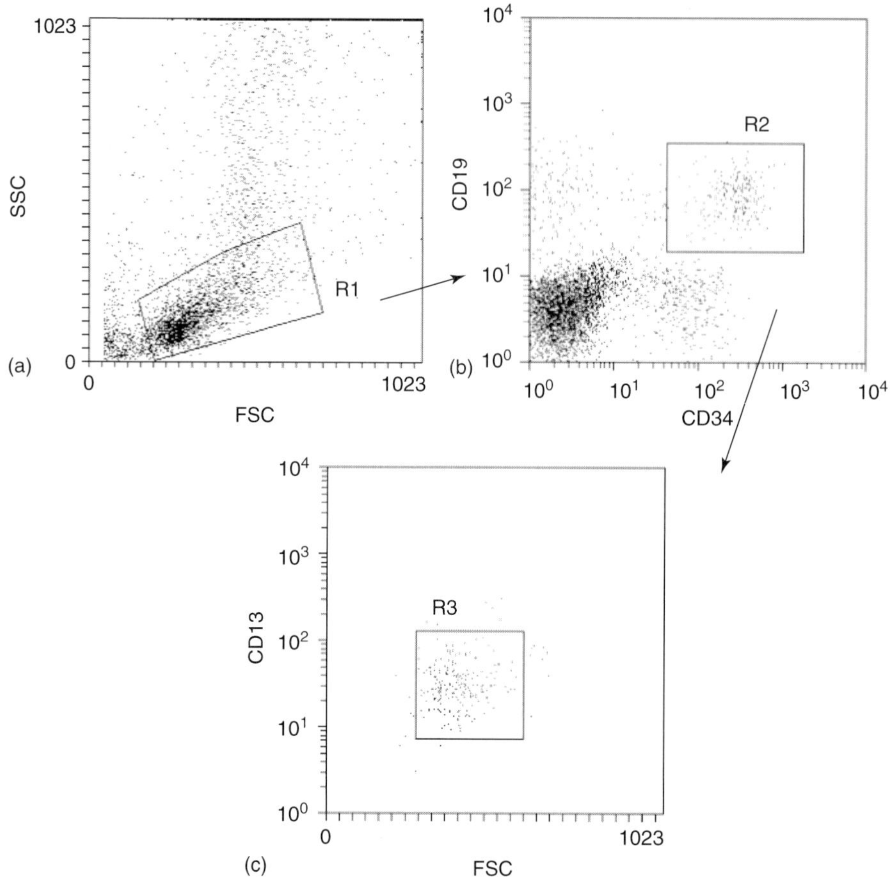


**Fig. 7** Bone marrow sample obtained from a child with acute lymphoblastic leukemia undergoing treatment and considered to be in clinical and morphological remission. Mononucleated cells were separated on a density gradient and labeled with three antibodies associated with the phenotype of the leukemia - CD19/CD34/CD13. (a) Cytogram of light scatter with a region, R1, set on the lymphocytes and lymphoblastic cells. (b) Cytogram of CD19 versus CD34 expression, gated on R1. A region has been drawn around the positive cells. (c) Cytogram of CD13 versus forward scatter. Region, R3, has been drawn around the CD13 positive, tumor cells. There were 199 events in R3 out of a total of 10 000 events recorded. Data, supplied by Dario Campana, St. Jude's Hospital, Memphis, was recorded on a BD FACScan using an argon-ion laser at 488 nm.

label between daughter cells or observe the distribution of cell cycle–related proteins.

### 5.3.1 The DNA Histogram

A measurement of the DNA histogram should always form the basis of any study of changes in the proliferative state of cells. Some effects of chemotherapeutic drugs on the cell cycle are shown in Fig. 9.

### 5.3.2 Using BrdUrd/Anti-BrdUrd

If cells are pulse-labeled with 5-bromodeoxyuridine (BrdUrd), only cells in S-phase

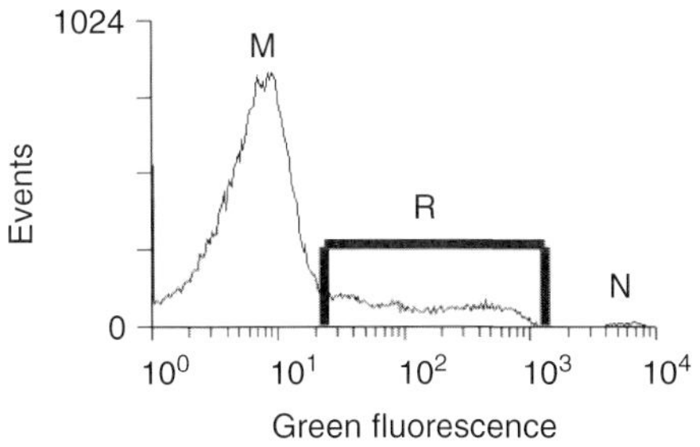


**Fig. 8** Reticulocyte analysis. Histogram of green fluorescence from blood incubated with thiazole orange, which stains RNA and DNA. M, mature erythrocytes; R, reticulocytes; N, nucleated red cells. Data, supplied by Terry Hoy, University of Wales, Cardiff, was recorded on a BD FACScan using an argon-ion laser at 488 nm.

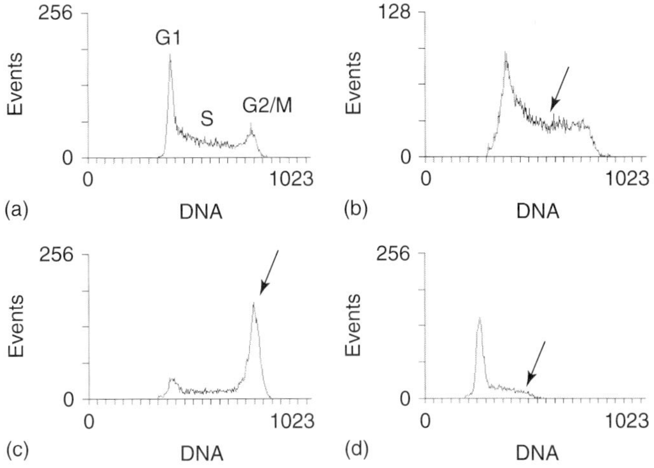


**Fig. 9** DNA histograms showing the effects of drugs on the cell cycle. (a) L1210 murine leukemic cell line, no drug; the stages of the cell cycle are marked. (b) 5 h after a 2-h incubation with a Pt(IV) dicarboxylate at $3 \times IC_{50}$. There was an accumulation of cells in S-phase (arrowed) caused by a slowdown in the movement of cells through S-phase. (c) As (b) but recorded after 24 h. The drug has caused a block in G2 and cells have accumulated in that phase of the cell cycle (arrowed). (d) HL60 cells 4 h after the addition of 10 μM camptothecin. S-phase was completely blocked; cells originally in G2 have divided and, as there was no movement of cells from S to G2, there is an absence of cells in G2 (arrowed). Other details as in Fig. 5.

will contain BrdUrd. During subsequent incubation of the cells, the BrdUrd-labeled cells will progress around the cell cycle. The progression can be followed by fixing the cells at different time intervals and visualizing the BrdUrd with an anti-BrdUrd antibody; PI is added to show the phases of the cell cycle (Fig. 10). If cells are incubated with a drug, irradiated or a growth factor is withdrawn immediately after labeling, the effects of the treatment on progression around the cell cycle can be followed. Alternatively, pulse-labeling after treatment of the cells would reveal the state of the S-phase cells (see Fig. 10).

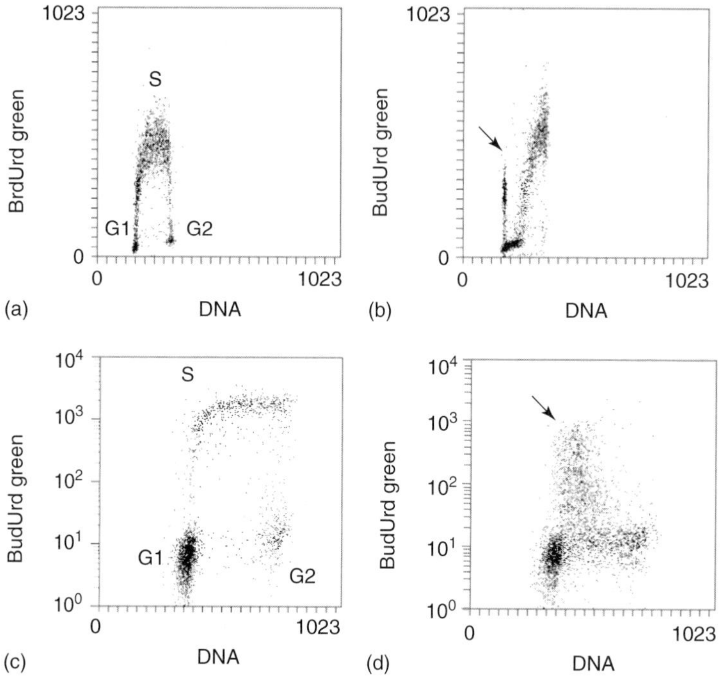


**Fig. 10** Cytograms of green versus red (PI/DNA) fluorescence from cultured cells pulse-labeled with BrdUrd, fixed and labeled with anti-BrdUrd-FITC and PI. (a) & (b) Chinese hamster V79 cells, 0 (a) and 5 h (b) after pulse-labeling. At 5 h, the labeled cells had moved about halfway through the cell cycle. Cells in late S-phase at time 0 had divided and were in G1 (arrowed). Data supplied by George Wilson, Karmanos Cancer Institute, Detroit. (c) & (d) Neuroblastoma cell line 0 h after pulse-labeling. (c): no drug; (d): 24 h after a 2 h incubation with cisplatin. Note that in (d), there was a population of S-phase cells that were not synthesizing DNA. A further population were accumulated in early/mid S-phase and were synthesizing less DNA than the control (arrowed). All data recorded on BD FACScans using an argon-ion laser at 488 nm. Noted that a linear amplifier was used in (a) & (b) to record green fluorescence and a logarithmic amplifier for (c) & (d).

This method can be applied *in vivo*. BrdUrd has a half-life of about 20 min in an animal; a single incubation of BrdUrd is equivalent to a pulse label. Figure 11 shows nuclei from a rat mammary tumor. The rat was killed and the tumor removed 4 h after the injection of the BrdUrd. From the observed movement of the cells through S-phase during this time, the S-phase transit time can be calculated.

### 5.3.3 The BrdUrd-Hoechst/PI Method

The fluorescence of *bis*-benzimidazole dyes (Hoechst 33258 and Hoechst 33342) bound to DNA is quenched by BrdUrd. Consequently, continuous labeling with BrdUrd and subsequent staining of DNA with Hoechst 33258 separates cells according to the number of replications they underwent during the period of labeling. Addition of a DNA label unaffected by BrdUrd, such as PI, resolves the cell

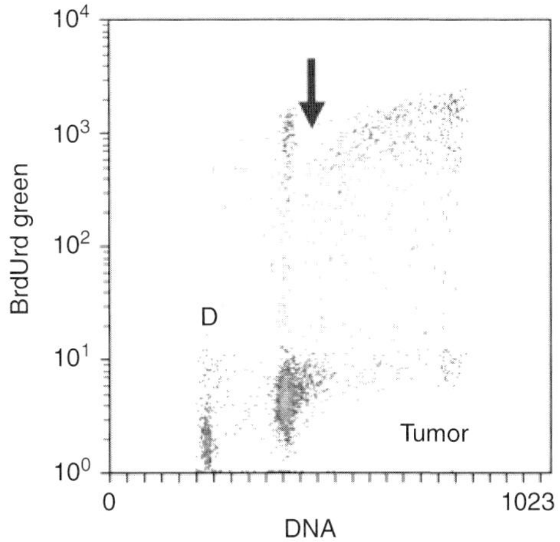


**Fig. 11** Cytogram of green (anti-BrdUrd) fluorescence versus PI/DNA of a rat mammary tumor. The rat was injected with BrdUrd 4 h before the tumor was excised. The diploid cells are marked, D; the tumor was aneuploid. From the length of the "window" (arrowed) between G1 and the S-phase label, the time of S-phase can be calculated. Cells prepared by Tim Smith, Royal Marsden NHS Trust, Sutton, and the sample analyzed on a Coulter Elite by Jenny Titley with an argon-ion laser giving 200 mW at 488 nm.

cycle into the G0/G1, S and G2/M compartments. Hoechst 333258 is excited by UV, and the small number of machines equipped with a UV laser has limited the use of this technique. Now that two and three laser instruments with UV lasers are becoming more common, this method may find more widespread use.

#### 5.3.4 Measuring Cell Proliferation through the Division of Label between Daughter Cells

Some labels will persist in cells for many days, if not weeks. Two compounds have been used for this application: carboxyfluorescein diacetate, succinimidyl ester (CFDA SE), and PKH26, an analog of acridine orange with an N-linked 26 carbon alkyl chain. CFDA SE diffuses into cells wherein it is converted to carboxyfluorescein, succinimidyl ester (CFSE); its reaction with amines stabilizes the fluorescent product in the cell. Because of its long alkyl side chain, PKH26 becomes anchored in the plasma membrane of cells.

To measure proliferation, the cells are labeled and their fluorescence recorded. After incubation under the appropriate experimental conditions, the fluorescence is rerecorded. The fluorescence from the cells that have divided once, twice, and so on, will have half, quarter the fluorescence of the undivided cells. Figure 12 shows data from murine T-cells in culture. After two days in the stimulated cultures, there are six peaks in the histogram of CFSE fluorescence; some of the cells had divided five times in 48 h, while other cells had not divided once.

### 5.4 Studying Cell Death

#### 5.4.1 Necrosis

Assays for necrosis measure the integrity of the plasma membrane. There are several dyes that fluoresce on binding to DNA but are excluded from the cell by the plasma membrane, the most commonly used being PI. On addition of the dye, those cells with a damaged membrane show positive fluorescence.

An alternative approach is to load the cells with a fluorescent dye, for example, by incubation with carboxyfluorescein diacetate. When the plasma

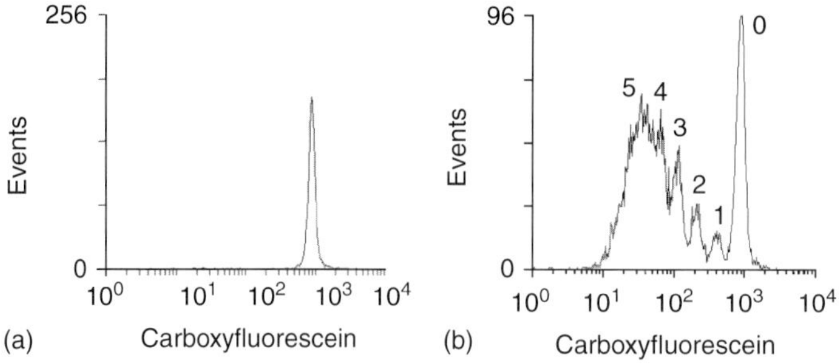


**Fig. 12** Murine CD4+ cells in culture labeled with CFSE. The unstimulated cells (a) gave a single peak on the fluorescence histogram while the histograms from stimulated cells (b) exhibited multiple peaks due to dilution of the label during cell division. The numbers of cell divisions are marked on histogram (b). The cells were stimulated by incubating with antigen-presenting cells and an appropriate peptide. Data supplied by Maria Daly, GlaxoSmith Kline, and recorded on a Beckman Coulter XL with an argon-ion laser.

membrane is damaged, the dye is lost from the cell. The two assays can be combined.

### 5.4.2 Apoptosis

The application of flow cytometry in following the apoptotic cascade has been reviewed. The features that can be measured using flow cytometry include

- expression of proteins involved in the apoptotic cascade, including proteins generated during the cascade, such as activated caspases and cleaved poly (ADP-ribose) polymerase (PARP);
- activation of caspases;
- changes in the mitochondrial membrane potential;
- changes in other functional parameters, such as intracellular pH, calcium ions, glutathione, and oxidative species;
- changes in the plasma membrane;
- cell shrinkage;
- DNA degradation.

5.4.2.1 **Activation of Caspases** There are antibodies available, which are specific for the activated form of many caspases. Caspase activity may be observed directly by incubating cells with a compound made of a peptide, specific for one of the caspases, connected to a potentially fluorescent group. An active enzyme will release the fluorescent group. Typically, a compound such as DEVD-rhodamine 110 is used. An alternative form of the assay uses a fluorescently labeled enzyme inhibitor, which will be concentrated in those cells containing active enzyme.

5.4.2.2 **Mitochondrial Membrane Potential (MMP)** The collapse of the MMP is a central feature of the apoptotic cascade and flow cytometry is the main method used for monitoring the MMP. Several dyes, including rhodamine 123, 3,3′-dihexyloxacarbocyanine iodide and 5, 5′,6,6′-tetrachloro-1,1′,3,3′-tetraethylbenzimidazolylcarbocyanine iodide (JC-1), are concentrated in mitochondria due to difference in the membrane potential of the mitochondria in relation to the cytoplasm. When the mitochondrial membrane potential collapses, the overall fluorescence is reduced. In the case of JC-1, this is accompanied by a shift in the fluorescence from red to green.

5.4.2.3 **Changes in the Plasma Membrane** During apoptosis, the distribution and packing of the lipids in the plasma membrane changes. The permeability of the plasma membrane is also increased. Phosphatidyl serine (PS), which is normally exposed on the internal face of the plasma membrane, moves to the exterior of the cell during apoptosis. This change can be detected using the recombinant protein, annexin V, which binds to PS. Annexin V is labeled with one of the common fluorochromes, such as fluorescein or PE. Annexin V staining can be combined with one or more immunofluorescence stains to determine the phenotype of the apoptotic cells. Frequently, PI is added to distinguish cells in secondary necrosis.

There are several DNA binding dyes that are excluded by normal cells but are taken up by apoptotic cells; these include the *bis*-benzimidazole, Hoechst 33342, ethidium bromide, 7-aminoactinomycin D (7-AAD) and the dicyanine, YO-PRO-1. Figure 13 shows an example using 7-AAD.

5.4.2.4 **DNA Degradation** Cleavage of DNA at the linkers between the nucleosomes is a late event in apoptosis, resulting in DNA oligomers of about 200 bp. There are two methods by which this effect may be used to visualize apoptotic cells.

If apoptotic cells are fixed in ethanol and then resuspended in buffer, the small fragments of DNA are extracted from the cell. Consequently, the apoptotic cells will have a lower DNA content than normal cells. This manifests itself as a "sub-G1" peak in the DNA histogram (Fig. 14).

For the Tdt-mediated dUTP nick end-labeling (TUNEL) assay, the cells are fixed in paraformaldehyde followed by ethanol. All the fragments of DNA are retained within the cell and are available for labeling by the enzyme, terminal deoxynucleotidyl transferase (Tdt). A variety of labels may be incorporated into the DNA at the sites of the strand breaks, such as FITC-dUrd, biotin-dUrd, BrdUrd, by adding the triphosphate derivative to the incubation mix. Biotin-dUrd is visualized by the addition of streptavidin-FITC and BrdUrd by anti-BrdUrd-FITC. PI is finally added to

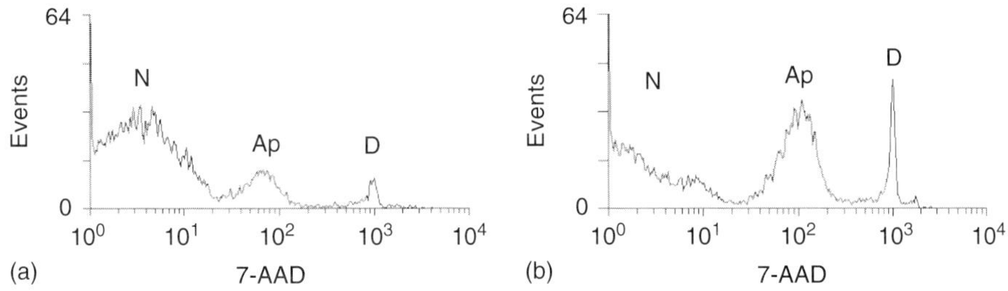


**Fig. 13** Immature human thymocytes incubated with (a) dexamethasone for 0 h and (b) 24 h and then incubated with 7-AAD. The normal (N), apoptotic (Ap) and cells that had undergone secondary necrosis (D) are marked. Data, supplied by Ingrid Schmid, UCLA School of Medicine, was recorded on a BD FACScan using an argon-ion laser.

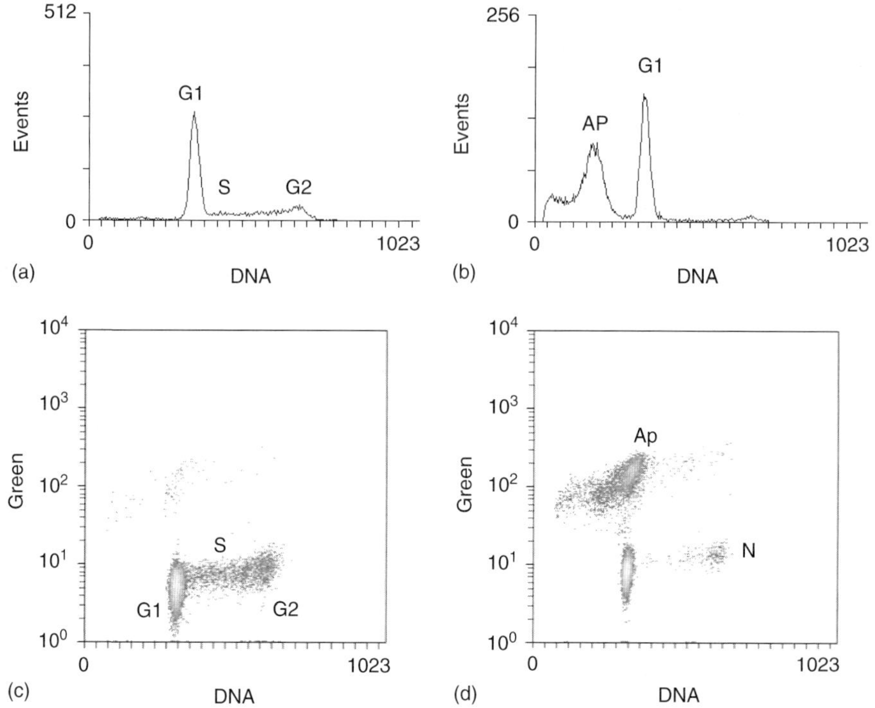


**Fig. 14** Murine hemopoietic cell line that requires interleukin-3 (IL-3) for growth. (a) & (c): grown in the presence of IL-3; (b) & (d): after 16 h in the absence of IL-3. (a) & (b): DNA histograms. The sub-G1 peak from apoptotic cells is marked "Ap". (c) & (d): Cells labeled using the TUNEL assay. The cytograms of green fluorescence versus red (DNA/PI) fluorescence show the phases of the cell cycle and the negative, normal (N) and the positive, apoptotic cells (Ap). Simone Detre, Royal Marsden NHS Trust, London labeled cells for the TUNEL assay, Data recorded on a Beckman Coulter Elite using an argon-ion laser at 488 nm.

label the DNA. The position in the cell cycle from which apoptosis occurred is revealed as well as the percentage of apoptotic cells (Fig. 14).

### 5.5
### Analyzing and Sorting Chromosomes

Sorting chromosomes requires a flow cytometer equipped with two large (5 W) argon-ion lasers and considerable expertise. A DNA stain of a suspension of chromosomes with a single dye (for example, PI) does not give enough resolution between the individual chromosomes. Two DNA stains are used – a *bis*-benzimidazole (Hoechst 33258) and chromomycin A3, each dye being excited by a separate laser. Hoechst 33258, excited by UV giving blue fluorescence, binds preferentially to AT-rich regions on the DNA; chromomycin A3, excited at 457 nm giving green fluorescence, binds preferentially to the GC-rich regions. The chromosomes can then be resolved on the basis of their total DNA content and their AT/GC ratio. All the individual chromosomes can be resolved with the exception of chromosomes 9–12, whose size and AT/GC ratios are not sufficiently separated.

### 5.6 Microbead Technology

Fluorescent microspheres (beads) offer a new emerging technology involving flow cytometry. A biomolecule can be attached to a bead and its interaction with other molecules studied; for example, the binding of two proteins; hybridization of DNA molecules; or the interaction of an enzyme with its substrate. Fluorescently labeled reporter molecules, such as antibodies, antigens, or nucleic acid probes can be used.

A particularly powerful application is the use of multiplexed arrays, for example, using two fluorescent dyes at eight different concentrations to give a set of 64 different beads. Each bead could carry a specific capture antibody; if the beads were fluorescent in the orange/red range, the detector antibodies might be labeled with fluorescein. Sixty-four assays could be carried out in the same tube simultaneously. The molecules measured might include hormones, cardiac markers, therapeutic drugs, drugs of abuse, and blood-borne viruses.

The same technology is being used for hybridization-based analyses for the detection of specific nucleic acid sequences.

Specific machines are being produced for these assays (for example, by Luminex Corporation).

## 6 Applications Other Than Mammalian Cell Biology

### 6.1 Microbiology

Generally, viruses are too small to be detected directly by flow cytometry but bacteria and other microorganizms, such as bacteria and yeasts, can be measured. Flow cytometric assays have been developed to determine features such as size, DNA content, surface receptors, membrane potential and intracellular pH.

One of the major applications has been in the study of antibiotic susceptibility. Flow cytometry is also being used to monitor water quality.

### 6.2 Marine Biology

The use of flow cytometry in monitoring the organisms in seawater is rapidly increasing. Many marine organizms are fluorescent and a combined measurement of light scatter and fluorescence at several wavelengths can yield a unique fingerprint for many organisms, such as the large variety of phytoplankton.

Instruments have been built to monitor seawater onboard a marine vessel and at fixed positions below sea level. Continuous monitoring has been achieved by mounting a small cytometer on a buoy and sending data to the mainland base by radio link (for example, Cytobuoy).

## Bibliography

### Books and Reviews

Darzynkiewicz, Z., Robinson, J.P., Crissman, H.A. (Eds.) (1999) Flow Cytometry, *Methods in Cell Biology*, 2nd edition, Vol. 41, Academic Press, San Diego.

Macey, M.G. (Ed.) (1994) *Flow Cytometry: Clinical Applications*, Blackwell Scientific, Oxford.

McCarthy, D.A., Macey, M.G. (Eds.) (2001) *Cytometric Analysis of Cell Phenotype and Function*, Cambridge University Press, Cambridge.

Ormerod, M.G. (1999) Flow Cytometry, *Royal Microscopical Society Handbook*, No. 44, Bios Scientific Publishers, Oxford.

Ormerod, M.G. (Ed.) (2000) *Flow Cytometry: A Practical Approach*, 3rd edition, IRL Press, Oxford University Press, Oxford.

Shapiro, H.M. (2003) Practical Flow Cytometry, 4th edition, Wiley-Liss, New York.

## Primary Literature

Bedner, E., Smolewski, P., Amstad, P., Darzynkiewicz, Z. (2000) Activation of caspases measured in situ by binding of fluorochrome-labeled inhibitors of caspases (FLICA): correlation with DNA fragmentation, *Exp. Cell Res.* **259**, 308–313.

Begg, A.C., McNally, N.J., Shrieve, D.C., Karcher, H. (1985) A method to measure the duration of DNA synthesis and potential doubling time from a single sample, *Cytometry* **6**, 620–626.

Borowitz, M.J., Bray, R., Gascoyne, R., Melnick, S., Parker, J.W., Picker, L., Stetler-Stevenson, M. (1997) US-Canadian consensus recommendations on immunophenotypic analysis of hematologic neoplasia by flow cytometry: data analysis and interpretation, *Cytometry* **30**, 236–444.

Bright, G.R., Whitaker, J.E., Haugland, R.P., Taylor, D.L. (1989) Heterogeneity of the changes in cytoplasmic pH upon serum stimulation of quiescent fibroblasts, *J. Cell. Physiol.* **141**, 410–419.

Bromilow, I.M., Duguid, J.K. (1997) Measurement of feto-maternal haemorrhage: a comparative study of three Kleihauer techniques and tow flow cytometry methods, *Clin. Lab. Haematol.* **19**, 137–142.

Campana, D. (1997) Immunophenotypic analysis in the monitoring of minimal residual leukemia, *Rev. Clin. Exp. Hematol.* **1**, 42–56.

Carrano, A.V., Gray, J.W., Langlois, R.G., Burkhart-Schultz, K.J., Van Dilla, M.A. (1979) Measurement and purification of human chromosomes by flow cytometry and sorting, *Proc. Natl. Acad. Sci. U.S.A.* **76**, 1382–1384.

Davey, H.M. (2002) Flow cytometric techniques for the detection of microorganisms, *Methods Cell Sci.* **24**, 91–97.

Davis, B.H., Bigelow, N.C. (1994) Automated reticulocyte analysis. Clinical practice and associated new parameters, *Hematol. Oncol. Clin. North Am.* **8**, 617–630.

Darzynkiewicz, Z., Bruno, S., Del Bino, G., Gorczyca, W., Hotz, M.A., Lassota, P., Traganos, F. (1992) Features of apoptotic cells measured by flow cytometry, *Cytometry* **13**, 795–808.

Darzynkiewicz, Z., Gong, J., Juan, G., Ardelt, B., Traganos, F. (1996) Cytometry of cyclin proteins, *Cytometry* **25**, 1–13.

Darzynkiewicz, Z., Juan, G., Li, X., Gorczyca, W., Murakami, T., Traganos, F. (1997) Cytometry in cell necrobiology: analysis of apoptosis and accidental cell death (necrosis), *Cytometry* **27**, 1–20.

De Rosa, S.C., Brenchley, J.M., Roederer, M. (2003) Beyond six colours: a new era in flow cytometry, *Nat. Med.* **9**, 112–117.

van Eeden, S.F., Klut, M.E., Walker, B.A., Hogg, J.C. (1999) The use of flow cytometry to measure neutrophil function, *J. Immunol. Methods* **232**, 23–43.

van Engeland, M., Nieland, L.J.W., Ramaekers, F.C.S., Schutte, B., Reutlingsperger, C.P.M. (1998) Annexin V-affinity assay: a review on an apoptosis detection system based on phosphatidyl serine exposure, *Cytometry* **31**, 1–9.

Givan, A.L. (2001) Flow Cytometry, *First Principles*, 2nd edition, Wiley-Liss, New York.

Hampton, M.B., Winterbourn, C.C. (1999) Methods for quantifying phagocytosis and bacterial killing by human neutrophils, *J. Immunol. Methods* **232**, 15–22.

Harrison, P. (2000) Progress in the assessment of platelet function, *Br. J. Haematol.* **111**, 733–744.

Hedley, D.W., Chow, S. (1994) Evaluation of methods for measuring cellular glutathione content using flow cytometry, *Cytometry* **15**, 349–358.

Hug, H., Los, M., Hirt, W., Debatin, K.-M. (1999) Rhodamine 110-linked amino acids and peptides as substrates to measure caspase activity upon apoptosis induction in intact cells, *Biochemistry* **38**, 13906–13911.

Jacobberger, J.W. (2000) Flow cytometric analysis of intracellular protein epitopes, in: Stewart, C.C., Nicholson, J.K.A. (Eds.) *Immunophenotyping, Cytometric Cellular Analysis*, John Wiley & Sons, New York, pp. 361–406.

Jennings, C.D., Foon, K.A. (1997) Recent advances in flow cytometry: applications to the diagnosis of hematologic malignancy, *Blood* **90**, 2863–2892.

Jepras, R.I., Carter, J.H., Pearson, S.C., Paul, F.E., Wilkinson, M.J. (1995) Development of a robust flow cytometric assay for determining viable numbers of bacteria, *Appl. Environ. Microbiol.* **61**, 2696–2701.

June, C.H., Rabinovitch, P.S. (1988) Flow cytometric measurement of cellular ionized calcium concentration, *Pathol. Immunopathol. Res.* **7**, 409–432.

Kallioniemi, O.P., Visakorpi, T., Holli, K., Isola, J.J., Rabinovitch, P.S. (1994) Automated peak detection and cell cycle analysis of flow cytometric DNA histograms, *Cytometry* **16**, 250–255.

Kellar, K.L., Iannone, M.A. (2002) Multiplexed microsphere-based flow cytometric assays, *Exp. Hematol.* **30**, 1227–1237.

Keren, D.F., McCoy, J.P., Carey, J.L. (Eds.) (2001) *Flow Cytometry in Clinical Diagnosis*, 3rd edition, American Society for Clinical Pathology Press, Chicago.

King, M.A. (2000) Detection of dead cells and measurement of cell killing by flow cytometry, *J. Immunol. Methods* **243**, 155–166.

Legendre, L., Courties, C., Troussellier, M. (2001) Flow cytometry in oceanography 1989–1999: environmental challenges and research trends, *Cytometry* **44**, 164–172.

Li, N., Goodall, A.H., Hjemdahl, P. (1999) Efficient flow cytometric assay for platelet-leucocyte aggregates in whole blood using fluorescence signal triggering, *Cytometry* **35**, 145–161.

Lyons, A.B. (2000) Analysing cell division in vivo and in vitro using flow cytometric measurement of CFSE dye dilution, *J. Immunol. Methods* **243**, 147–154.

McSharry, J.J. (2000) Analysis of virus-infected cells by flow cytometry, *Methods* **21**, 249–257.

Maino, V.C., Picker, L.J. (1998) Identification of functional subsets by flow cytometry: intracellular detection of cytokine expression, *Cytometry* **34**, 207–215.

Marchetti, P., Castedo, M., Susin, S.A., Zamzami, N., Hirsch, T., Macho, A., Haeffner, A., Hirsch, F., Geuskens, M., Kroemer, G. (1996) Mitochondrial permeability membrane transition is a central co-ordinating event in apoptosis, *J. Exp. Med.* **184**, 1155–1160.

Misteli, T., Spector, D.L. (1997) Applications of the green fluorescent protein in cell biology and biotechnology, *Nat. Biotechnol.* **15**, 961–964.

Nebe-von-Caron, G., Stephens, P.J., Hewitt, C.J., Powell, J.R., Badley, R.A. (2000) Analysis of bacterial function by multi-colour fluorescence flow cytometry and single cell sorting, *J. Microbiol. Methods* **42**, 97–114.

Nicholson, J.K.A. (1994) Immunophenotyping specimens from HIV-infected persons: laboratory guidelines from the Center for Disease Control and Prevention, *Cytometry* **18**, 55–59.

Nolan, J.P., Mandy, F.F. (2001) Suspension array technology: new tools for gene and protein analysis, *Cell Mol. Biol.* **47**, 1241–1256.

Ormerod, M.G. (1998) The study of apoptotic cells by flow cytometry, *Leukemia* **12**, 1013–1025.

Ormerod, M.G. (2001) Using flow cytometry to follow the apoptotic cascade, *Redox Rep.* **6**, 275–287.

Ormerod, M.G., Kubbies, M. (1992) Cell cycle analysis of asynchronous populations by flow cytometry using bromodeoxyuridine label and Hoechst-propidium iodide stain, *Cytometry* **13**, 678–685.

Petit, P.X., LeCoeur, H., Zorn, E., Dauguet, C., Mignotte, B., Gougeon, M.L. (1995) Alterations of mitochondrial structure and function are early events of dexamethasone-induced thymocyte apoptosis, *J. Cell Biol.* **130**, 157–165.

Prussin, K. (1997) Cytokine flow cytometry: understanding cytokine biology at the single-cell level, *J. Clin. Immunol.* **17**, 195–204.

Richards, S.J., Rawstron, A.C., Hillmen, P. (2000) Application of flow cytometry to the diagnosis of paroxysmal nocturnal hemoglobinuria, *Cytometry* **42**, 223–233.

Roederer, M. (2001) Spectral compensation for flow cytometry: visualization artifacts, limitations and caveats, *Cytometry* **45**, 194–205.

Rosenfeld, C.S., Bodensteiner, D.C. (1986) Detection of platelet alloantibodies by flow cytometry. Characterization and clinical significance, *Am. J. Clin. Pathol.* **85**, 207–212.

Shanahan, T. (1997) Application of flow cytometry in transplantation medicine, *Immunol. Invest.* **26**, 91–101.

Sintnicolaas, K., de Vries W., van der Linden, R., Gratama, J.W., Bolhuis, R.L. (1991) Simultaneous flow cytometric detection of antibodies against platelets, granulocytes and lymphocytes, *J. Immunol. Methods* **142**, 215–222.

Steen, H.B. (2000) Flow cytometry of bacteria: glimpses from the past with a view to the future, *J. Microbiol. Methods* **42**, 65–74.

Sutherland, D.R., Anderson, L., Keeney, M., Nayar, R., Chin-Yee, I. (1996) The ISHAGE guidelines for CD34+ cell determination by flow cytometry, *J. Hematother.* **3**, 213–226.

Szöllösi, J., Damjanovich, S., Matyus, L. (1998) Applications of fluorescence resonance energy transfer in the clinical laboratory: routine and research, *Cytometry* **34**, 159–179.

de Vries, E., Noordzij, J.G., Kuijpers, T.W., van Dongen, J.J. (2001) Flow cytometric immunophenotyping in the diagnosis and follow-up of immunodeficient children, *Eur. J. Pediatr.* **160**, 583–591.

Zamzami, N., Marchetti, P., Castedo, M., Decaudin, D., Macho, A., Hirsch, T., Susin, S.A., Petit, P.X., Mignotte, B., Kroemer, G. (1995) Sequential reduction of mitochondrial membrane potential and generation of reactive oxygen species in early programmed cell death, *J. Exp. Med.* **182**, 367–377.

## Websites

### Some Companies Who Supply Flow Cytometers

BD www.bd.com
Beckman Coulter www.beckmancoulter.com
Dakocytomation www.dakocytomation.com
Partec www.partec.de
Guava Technologies www.guavatechnologies.com
Luminex www.luminexcorp.com
Cytobuoy www.cytobuoy.com

### Free Software

WinMDI http://facs.scripps.edu/software.html
Cylchred www.uwcm.ac.uk/uwcm/hg/hoy/software.html

### Other Useful Sites

Molecular Probes www.probes.com
Information on flow cytometry and microbiology www.cyto.purdue.edu/flowcyt/research/micrflow/index.htm

# 11
# Immunoassays

*James P. Gosling*
*Department of Biochemistry, National University of Ireland, Galway, Ireland*

**1** **Antibodies for Immunoassays 299**
1.1 Immunoglobulin G Structure 299
1.2 Monoclonal, Polyclonal, and Anti-hapten Antibodies 301

**2** **Antibody–antigen Binding 301**
2.1 Structural Complementarity 301
2.2 Molecular Mobility Influences Binding 302
2.3 Epitopes 302

**3** **Labeling Substances 303**

**4** **Classification of Immunoassays 303**
4.1 Nomenclature 303
4.2 The Variety of Immunoassay Formats 305
4.3 Representing Assay Complexes in Text 305
4.4 The Four Main Groups of Immunoassays 306
4.4.1 Label-free Assays 307
4.4.2 Reagent-excess Assays 307
4.4.3 Reagent-limited Assays 308
4.4.4 Ambient Analyte Assays 308

**5** **Established Immunoassay Systems 309**
5.1 Automated Systems 309
5.2 General Laboratory Assays 309
5.3 Simple to use Test Devices 310

**6** **Immunoassays of the Future 311**
6.1 Multiplexed Immunoassays 311
6.2 New Assay Principles and Systems 311

*Immunology. From Cell Biology to Disease.* Edited by Robert A. Meyers.

ISBN: 978-3-527-31770-7

**7 The Limitations of Immunoassays 312**

**Bibliography 314**
Books and Reviews 314
Primary Literature 314

## Keywords

**Automated**
Automated immunoassays require complex, computer-controlled instrumentation that takes care of all repetitive steps such as sample handling, pipetting, washing, end point determination and the calculation of results. New larger machines integrate the functions of immunoassay analyzers and clinical chemistry analyzers into one system.

**Binding Affinity**
Binding affinity is the probability of association of a ligand (antigen) with a binding molecule (antibody). High affinity depends positively on the rate of association of the two molecules and negatively on the rate of dissociation of the complex formed.

**Classification**
In this context, "classification" refers to any system that may be used to explain how the very large numbers of different immunoassays are related to each other.

**Epitope**
An epitope is that portion of an antigen with which the binding site of an antibody (paratope) interacts specifically.

**Hapten**
A hapten is a low molecular weight antigen that is not inherently immunogenic.

**Hybridoma**
Monoclonal hybridoma cells are "immortal" cells that (in the present context) are capable of synthesizing antibodies with desirable properties. They are generated by fusing lymphocytes and myeloma cells, followed by dilution, culture, and selection steps to ensure enduring viability, homogeneity, and ability to secrete desirable monoclonal antibodies.

**Immunoglobulin Structure**
All immunoglobulin molecules are glycoproteins with similar characteristic complex structures that are rugged and flexible. They consist of multiple, paired domains that are all derived from "the immunoglobulin fold," a protein structural motif also found in many other proteins.

**Interference**

By interference is meant the phenomenon whereby an incorrect result is obtained in an immunoassay because of the presence, in the sample being tested, of a substance or substances (other than the analyte) capable of interacting with assay reagents so as to influence their normal roles.

**Label**

The term "label" is usually applied to the labeled antigen, labeled hapten, or labeled antibody employed as a reagent in immunoassay.

**Microarray**

In the context of immunoassay methodology, a microarray is usually an array or matrix of microdots each coated with a different specific antibody or antigen.

**Monoclonal Antibody**

A homogeneous population of antibody molecules, each with identical binding affinity and specificity, synthesized by a monoclonal cell line.

**Multiplexed**

Multiplexed immunoassays are capable of detecting or measuring two or more analytes following the execution of a single analytical procedure that usually has no additional steps as compared to an ordinary assay. Multiplex is a synonym of multiple.

**Paratope**

A paratope consists of complementarity determining regions (CDR) of the variable domains of an immunoglobulin and is that portion of an antibody that constitutes an antigen binding site.

**Polyclonal Antibody**

A heterogeneous population of antibody molecules raised by immunizing a suitable animal with a specific antigen.

**Rapid Test Device**

Rapid test devices (RTD) are small constructs usually consisting of one or more components, including moulded plastic structures, membranes, adsorbent strips, absorbent pads, and immobilized (or otherwise incorporated) assay reagents. An immunoassay formulated as a RTD enables the detection of tiny quantities of specific molecules at virtually any location by untrained persons.

**Reagent-excess**

In reagent-excess immunoassays, the principal reagents, such as immobilized antibody and labeled antibody or labeled antigen, are used in excess. Most reagent-excess immunoassays may also be described as "sandwich" immunoassays.

**Reagent-limited**
For reagent-limited immunoassays, the principal reagents, such as immobilized antibody and labeled antibody or labeled antigen, are used at limited concentrations, and a plot of "label bound" against analyte concentration displays an inverse relationship. Reagent-limited immunoassays are commonly referred to as "competitive" immunoassays. Separation-free immunoassays are usually "reagent-limited."

**Separation-free**
A separation-free (homogeneous) immunoassay has no separation step. Separation-free immunoassays are usually "reagent-limited." In a normal immunoassay, after incubation of reagents with analyte, bound label must be separated from free label to enable measurement of the bound fraction. In separation-free immunoassay, the binding of label to antibody modulates the activity of label such that analyte concentration determines *total* label activity.

**Solid Phase**
In immunoassay, a solid-phase component may be the internal wall of a tube or well, a bead or microparticle, a membrane or a thread. Linking a basic assay reagent such as antibody or antigen to a solid-phase component allows the easy separation of bound from free label, thereby facilitating the accurate measurement of bound (or free) label.

**Standardization**
By standardization is meant the development and validation of assays in such a way that all assays for the same analyte give the same (numeric) results if applied to the same samples. The same results should be obtained whether the assays were developed and manufactured by different companies, used at different locations, operated by different users, or used at different times.

**Structural Complementarity**
Structural complementarity is the term used to explain the high affinity binding of the paratope of a specific antibody to the epitope of an antigen. The paratope and epitope have complementary shapes, complementary distributions of electric charges, and opposing groups that form good noncovalent bonds. Structural complementarity, in all its variety and complexity, explains all strong and spontaneous molecular associations and is therefore an essential part of the ultimate explanation for all biological processes.

Specific, biologically related, intermolecular interaction is the basis of all "binding assays," recombinant DNA procedures, and, accompanied by catalysis, of enzymatic analysis. "Binding assays," also referred to by the term *saturation analysis*, depend on the "bio-recognition" of analytes by high affinity, specific binding proteins. In the last forty years "protein binding assays" have found so many applications in diverse areas of biochemical and chemical analysis that they now represent a standard analytical principle, like colorimetry or chromatography.

While there are many kinds of high affinity, specific binding proteins in nature (hormone receptors, for example), the special properties of antibodies have made them the most popular choice by far, for such methods. The term "immunoassay" applies to measurement procedures that use antibodies as specific binding reagents, as well as to assays that use antigens for the detection or quantification of specific antibodies. It is because of the extraordinary variety, specificity, and affinity of antibody–antigen binding reactions, that immunoassays are used for routine analyses and for research purposes throughout the biological and medical sciences.

Often an immunoassay procedure is chosen to measure a particular analyte because no other type of assay is technically feasible, which is true for specific antibodies, for most proteins and for many other complex biomolecules. But frequently, immunoassays are used because they match or exceed analytical and practical requirements and are more convenient and cost-effective than any alternative method.

When a broad perspective is adopted, it is clear that the history of immunotechnology as applied to the detection and quantitation of antigens did not begin with the invention of radioimmunoassay in the late 1950s. Rather, experimental immunologists have always used specific antibodies and "purified" antigens in their investigations, and major developments with respect to "immunoassay methodology" occurred when these were applied clinically and/or formulated to maximize their utility to clinical microbiologists, clinical chemists, and others. Table 1 lists some of the significant advances in this long history of the development of immunoassay methodology.

## 1 Antibodies for Immunoassays

### 1.1 Immunoglobulin G Structure

Antibodies, also defined as immunoglobulins and found in blood and other body fluids of vertebrates, are essential components of the adaptive, humoral immune system. Immunoglobulin G (IgG), the class of antibody used predominantly in immunoassays, is a 150 000 Da glycoprotein containing two identical heavy (H ~420 residues) and two identical light (L ~215 residues) polypeptide chains. The two identical antigen-binding sites (also called paratopes) of IgG are composed of the atoms of about fifty amino acid residues found in groups (complementarity determining regions, CDR) scattered along the variable segments of the H and L chains. In the 3-D structure of all immunoglobulins the six CDR occur as loops and it is CDR variability that accounts for almost all of the immense range of antibody specificity.

For analytical purposes, IgG molecules can be linked covalently to labeling substances and to solid-phase materials by any of a range of standard chemical procedures. They also readily bind noncovalently to plastics and glass to give "antibody coated" tubes, wells, or beads.

**Tab. 1** Milestones in the development of immunoassay methodology.

| ***Development*** | ***References*** |
|---|---|
| Detection of bacilli by antibody agglutination | Freund (1925) |
| Titration of antisera by antigen precipitation | Libby (1938) |
| Immunofluorescence staining | Coons (1942) |
| Albumin measurement by antibody precipitation | Chow (1947) |
| RIA for antigen | Berson and Yalow (1959); Ekins (1960) |
| Immunoenzymatic staining | Nakane and Pierce (1966); Avrameas and Uriel (1966) |
| Labeled antibody RIA | Miles and Hales (1968) |
| IRMA for antigen | Addison and Hales (1971), Wide (1971) |
| EIA for antigen | VanWeeman and Schuurs (1971) |
| EMIT® | Rubinstein et al. (1972) |
| FPIA® | Dandiker et al. (1973) |
| Microtitre-plate IEMA for antibody | Voller (1974) |
| IEMA for antigen | Maiolini and Masseyeff (1975) |
| Acridinium ester label | Simpson et al. (1979) |
| Theoretical basis for immunoassays | Jackson and Ekins (1986) |
| Chemiluminescence enzyme assay in EIA | Arakawa (1981); Thorpe et al. (1985) |
| Europium chelates in IFMA | Pettersson et al. (1983) |
| Fab'-enzyme conjugates in IEMA | Ishikawa et al. (1983) |
| ELISPOT for specific antibody-secreting cells | Czerkinsky et al. (1983) |
| Solid-phase separation with hapten anchor | Rattle et al. (1984) |
| Amplification enzyme assay in IEMA | Self (1985) |
| Immunoconcentration IEMA | Valkirs (1985) |
| CEDIA® | Henderson et al. (1986) |
| Microspot, multianalyte immunoassay | Ekins et al. (1990) |

[a]Abbreviations: RIA, radioimmunoassay; EIA, enzyme immunoassay; IRMA, immunoradiometric assay; IEMA, immunoenzymometric assay; IFMA, immunofluorometric assay.

Immunoglobulins are flexible molecules, and this facilitates their binding simultaneously to two (or more) adjacent identical antigens on the surface of a pathogen and enables the aggregation of soluble antigens. While all interdomain regions can flex and bend, the greatest flexibility is associated with the "hinge regions" of the H chains, near the C-termini of the L chains. These regions are highly susceptible to proteolytic cleavage, and pepsin specifically hydrolyzes a site located below the inter-H chain disulphydryl links to give $F(ab')_2$ dimers that are easily reduced to give Fab' fragments ($M_r \sim 50\,000$). Fab' fragments have at least one sulphydryl group (−SH) distally located from the binding site, which can be used to establish a covalent linkage to labeling molecules or solid-phase materials. Also because of their relatively small size and reduced tendency to bind "nonspecifically," Fab' fragments are highly suitable for the preparation of soluble reagents for use in immunoassays.

Fv fragments ($M_r \sim 25\,000$), which include only the variable segments of the

H and L chains, are minimal antibodies. Recombinant DNA methods enable the preparation of stable, single-chain Fv (sFv) fragments for any monoclonal antibody, and of chimeric proteins incorporating antibody-binding sites. (Phage display techniques allow the creation of vast repertoires of genes coding for heavy and light chains, *in vitro* selection of recombinant clones producing antibody of a desired specificity, and *in vitro* "affinity maturation.")

### 1.2
### Monoclonal, Polyclonal, and Anti-hapten Antibodies

Although during the 1980s the use of monoclonal antibodies in new immunoassays increased dramatically, even now in the twenty-first century, polyclonal antibodies are still preferred as secondary reagents in all commercial immunoassays, and is a primary reagent in a large minority of immunoassay kits. Obviously, differences with respect to cost, ease of preparation and availability are still sufficient to outweigh the advantages (homogeneity and continuity of supply) associated with monoclonal (or recombinant) antibodies.

Immunoassay antibodies come from a range of species including mice, rabbits, goats, and sheep. Avian IgY class antibodies, which may have much lower "nonspecific binding properties" are also used, and camel antibodies, which lack light chains, may become significant in the future. Usually, monoclonal antibodies are raised in mice but all the prerequisites are available for the use of rats.

The power and versatility of immunoassays is largely dependent on the availability of specific antibodies for most biologically important molecules with a distinctive chemical structure and of molecular weight $>150\ \mathrm{g\ mol^{-1}}$, whether or not they are inherently immunogenic. Immunologically, haptens are defined as molecules of limited molecular weight that are not naturally antigenic, but when conjugated to a carrier (bovine serum albumin or BSA for example, with 20–30 hapten molecules per molecule of BSA) give an immunogen that can be used to generate antibodies that specifically bind free hapten with high affinity. This approach is also highly effective for larger molecules that are weakly immunogenic. Therefore, antibodies for molecules that are newly identified or synthesized for the first time are readily prepared, in most cases.

## 2
## Antibody–antigen Binding

### 2.1
### Structural Complementarity

Antigen–antibody interactions are reversible and for binding to be defined as high affinity there must be a combination of easy association and reluctant dissociation. Association can be regarded as occurring in two phases, with the first requiring the overcoming of repulsive forces and the second involving the establishment of multiple hydrogen, ionic, and van der Waals attractive interactions between atoms on the antibody-binding site (paratope) and the antigenic determinant region (epitope).

When the epitope of a large protein antigen is bound in a high affinity complex with a paratope specific for it, the apparent area of close contact is about 700 to 900 $\mathrm{Å^2}$, with about 20 amino acid residues on both reactants apparently participating.

An appropriate pH, which is usually near pH 7, ensures that acidic and basic groups on both species are suitably dissociated or associated. However, the contribution to binding of any close sets of opposing amino acid residues may be either negligible or crucial and this may account for the fact that a genetically determined variation in an antigen may render it virtually undetectable by an immunoassay. For example, Pettersson and coworkers discovered that, in a significant subpopulation of normal women with a genetic variant of luteinizing hormone (LH), LH was almost undetectable when measured with an immunoassay employing a particular kind of monoclonal antibody.

When the high affinity complex is between a small antigen (hapten) and its specific paratope, the hapten is found to be located in a pocket, cavity, or groove on the antibody surface. Indeed, this may be an absolute requirement for *high* affinity. When bound, the hapten is surrounded by functional groups of the peptide residues and backbone that are structurally complementary to adjacent portions of the hapten molecule. Frequently, to give the number of weak bonds to the hapten that are necessary for tight binding, the hapten is buried to the extent of 80 to 90%, or even greater. Nevertheless, as established by Arevalo and coworkers, other compounds with significant structural differences from the original specific hapten, may be capable of binding tightly to the same paratope, provided they also are sufficiently complementary with respect to size, shape, and charge distribution.

As binding of epitope to paratope occurs, there may be significant structural adjustments of the paratope (and of the epitope) and this is termed *induced fit*. However, frequently there is still sufficient interstitial space to accommodate water molecules or ions, some of which may contribute significantly to binding. Therefore, the ionic composition of buffers may be relevant to the selection and performance of an antibody, and optimal assay conditions can only be estimated experimentally.

## 2.2 Molecular Mobility Influences Binding

Any interaction between antibody and antigen must be preceded by a collision. The rate at which collisions occur is dependent on rates of diffusion, which, in turn, are influenced by molecular mass and shape, temperature, and, of course, on whether both reactants are mobile or not. Therefore, since smaller molecules diffuse more rapidly, haptens tend to associate more readily with their specific antibodies than with large antigens, and hapten analytes associate more readily than the large hapten-enzyme conjugates used as labels in enzymeimmunoassays. In addition, antigen–antibody interactions reach equilibrium much more slowly when the antibody or antigen is immobilized to a solid phase as opposed to when both are free to diffuse. Consequentially, many automated immunoassays, which require short incubation times, involve the use of microparticulate solid-phase media and regular mixing, or are designed such that the binding reaction occurs in liquid phase and separation is effected later.

## 2.3 Epitopes

Each large antigen has a theoretically unlimited number of epitopes, a number that is limited in practice by immune tolerance in the immunized animal toward surface regions that share structural features with self-proteins. It is also difficult to differentiate between overlapping epitopes. Structurally, protein epitopes are often very similar to paratopes, being made up of a number of amino acid residues scattered, singly or in small groups, along the amino acid sequence, but close together on the surface of the intact antigen. Thus, many epitopes are "discontinuous." However, a single short peptide may represent a large fraction of an epitope and may, in isolation, be specifically bound by an antibody raised against the whole protein that it represents. Correspondingly, antibodies raised against such a peptide may bind to the region of the cognate protein containing the same sequence. Such epitopes may be referred to as "continuous." Not unusually, continuous epitopes correspond to exposed termini.

The chemical nature of epitopes is highly relevant to the development of sophisticated vaccines, and is also pertinent to immunoassays. For example, assays designed to detect specific antibodies against well-characterized viral pathogens (especially HIV), may employ single peptides (or more often sets of carefully chosen peptides), immobilized and/or linked to a labeling substance, instead of whole viral antigens. This allows precisely defined epitope specificity.

## 3 Labeling Substances

Although, as is explained further below, many immunoassays employ no discrete labeling "compound," immunoassays only came "of age" with the invention of radioimmunoassay by Rosaleen Yalow and Roger Ekins in the late 1950s, and the gradual development of reagent excess, sandwich assays in the 1970s. During this period also, data obtained with the aid of high-throughput immunoassays with lower detection limits of $10^{-9}$ to $10^{-15}$ M helped to revolutionize many branches of biological and medical science.

The sophistication of current immunoassay methodology allows the assay developer a wide choice of labeling substances and assay endpoints, but a small number of label types dominate. Crucial attributes of labels include ease of use and the availability of simple procedures and suitable equipment, and the ability to provide very high specific activity, that is, a very strong signal per unit mass of label. These attributes are provided to a greater or lesser extent by enzymes, chemiluminescent compounds, fluorescent compounds, latex particles, microspheres, and radioisotopes. Of particular interest to the future are the range of "rare earth" ions that can be used, each with a distinct peak wavelength of fluorescent light, facilitates dual and multiassays that simultaneously measure two or more analytes. Similarly, a set of populations of microspheres with fixed intensities of one or two fluorescent dyes, allows multiplexed (up to "100-plexed") immunoassays when measured in flow-cytometry systems with two lasers.

Ligands, such as biotin, are also used as primary labels in combination with another label type that provides the endpoint. Table 2 lists a selection of general-purpose labeling substances (including atoms, ions, small and large molecules, and chelates) and some of the methods used for detecting and measuring them.

## 4 Classification of Immunoassays

### 4.1 Nomenclature

The names of immunoassays can be confusing and make it difficult to understand

**Tab. 2** Some biomolecules, compound, ions, and other substances used for labeling and methods used to measure them.

| *Type* | *Substance* | *Endpoint detection* |
|---|---|---|
| Enzyme | Alkaline phosphatase | Choice of endpoint including:<br>amplified colorimetric<br>colorimetric<br>electrochemical<br>luminometric<br>visual assessment |
| | $\beta$-D-galactosidase | Choice of endpoint |
| | Glucose-6-P dehydrogenae | Colorimetric |
| | Horseradish peroxidase | Choice of endpoint |
| | Malate dehydrogenase | Colorimetric |
| Fluorescent | Fluorescein and other fluorescent dyes chemically coupled to antibodies or antigens, or/and in microspheres | Fluorimetric |
| | | Polarimetric–fluorimetric |
| | | Fluorescence flow cytometric |
| | Rare earth ions/chelates $Dy^{3+}$, $Eu^{3+}$, $Sm^{3+}$ or $Tb^{3+}$ | Time-resolved fluorimetric |
| Ligand | Biotin derivative | Avidin-acridinium ester, luminometric |
| | | Avidin-enzyme, choice of endpoint |
| | | Avidin-$Eu^{3+}$ chelator, TR fluorimetric |
| | FITC | AntiFITC-enzyme, choice of endpoint |
| | oligonucleotide | Polymerase chain reaction or other amplification, choice of endpoint |
| Luminescent | Acridinium ester | Luminometric |
| | Isoluminol derivatives | Luminometric |
| | Ruthenium(II) chelate, trispropylamine | Electroluminometric |
| Microparticle | Colloidal gold | Visual assessment |
| | Colored latex | Visual assessment |
| | Latex etc. | Particle counting |
| | | Visual assessment |
| Radioisotopic | $^{3}H$ | Liquid scintillation counting |
| | $^{125}I$, | Solid scintillation counting |
| | $^{125}I$-Bolton & Hunter reagent | Solid scintillation counting |

similarities and differences, and the diversity of immunoassay designs. In general, most assay names contain "immuno," the combining form of the adjective "immune," and another combining word indicating the type of label employed, for example, radioimmunoassay (RIA) or enzyme immunoassay (EIA). If an alternative biological binding protein to an antibody is employed, such as a hormone receptor, the corresponding name is radioreceptor assay.

The use of the terms RIA and EIA is usually (but not always) restricted to reagent-limited, competitive assays (see below), and reagent-excess assays (see below) are commonly distinguished by reversing the order of the combining forms, as in immunoradiometric assay (IRMA), immunofluorometric assay (IFMA), or immunoenzymometric assay (IEMA). The term *immunometric assay* is used to refer to reagent-excess assays in general. However, similar names have occasionally been used for competitive assays with labeled antibody.

The acronym ELISA (enzyme-linked immunosorbent assay) is generally applied to all kinds of microtiter plate assays with enzymatic labels and, when encountered, cannot be relied upon to indicate a particular assay mechanism. For clarity, EIA is to be preferred over ELISA for competitive assays equivalent to RIA and IEMA for reagent-excess assays equivalent to IRMA.

Commercially developed immunoassays with patented features are often named with "acronyms" that may or may not be defined. Three separation-free assays are EMIT® (enzyme modulated immunotest), FPIA® (fluorescence polarization immunoassay) and CEDIA® (combined enzyme donor immunoassay).

## 4.2 The Variety of Immunoassay Formats

The versatility of immunoassays is made possible largely by the wide variety of formats that can be implemented for assays of similar basic design. For example a basic, two-site, reagent excess, sandwich assay design for an antigen analyte may be used for high performance, high throughput, laboratory-based applications; or for simple, test devices to be used by untrained persons and sold over the counter in normal retail outlets. Much of the variety in format is dependent on the different solid-phase materials and configurations (e.g. microtiter wells, microparticles, membranes or simple test devices), on the different labels used (e.g. radioactive, fluorescent, enzymatic or colored particles), and on how the endpoint is measured (with a complex electronic instrument or visually).

## 4.3 Representing Assay Complexes in Text

In this article, the primary reagents used and the complexes of antibodies, antigens, and other components that form during the operation of immunoassays will be represented in the text by means of a standard notation system (not all aspects of which are exploited here).

According to this system, the analyte is shown in bold print, for example, "**Ag**," "**Ha**" or "**Ab**" for general antigen, hapten and antibody analytes, or more specifically, human chorionic gonadotropin (**hCG**) for the human pregnancy hormone. Where any one of a number of labeling substances could be used, it is represented by "L," otherwise an abbreviation such as "enz," or "$^{125}$I" is used.

Standard abbreviations are used when the class (IgG, IgM) or type of antibody (monoclonal, mAb, mAb1, mAb2), antibody fragment (Fab') or single-chain recombinant fragment (scFv) is to be specified. The origins of antibodies can be shown with single letter abbreviations (e.g. goat, G; human, H; mouse, M; rabbit, R; sheep, S) if this is relevant. The names of unusual components, or components that must be more exactly specified, can be abbreviated or spelled out.

The hyphen ("-") sign is used to indicate associations established before commencement of the assay such as an antibody-label conjugate or an antibody-enzyme conjugate (Ab-L, Ab-enz, respectively) or an antibody immobilized on a solid phase (sp-Ab). In contrast, the en dash ("–") represents associations formed during the course of the assay procedure. For example, the final complex of a solid-phase ELISA for an antigen is described as: sp-Ab–**Ag**–Ab-enzyme. For a reagent-limited competitive assay for a hapten analyte two complexes are formed, sp-Ab–**Ha** and sp-Ab–Ha-L, and, as explained below, only the complex that does *not* contain the analyte is measured.

## 4.4 The Four Main Groups of Immunoassays

Table 3 represents a system whereby most, if not all, immunoassays can be classified into four groups. In this context, it should also be borne in mind (as mentioned above) that binding proteins other than antibodies can be used as specific binding reagents in assays that are the exact equivalents to those mentioned below. For example, intrinsic factor and $\beta$-lactoglobin are used as specific binding reagents in group 2-type assays for vitamin $B_{12}$ and folate, respectively. Also, a combination of an antibody and a particular lectin may be used to measure specifically glycoproteins with particular carbohydrate side chains, by means of a reagent-excess, sandwich assay.

Most quantitative immunoassays developed for research purposes or for applications of limited scale are simple immunoprecipitation assays (Group 1), reagent excess, sandwich assays with either direct or indirect labels (Group 2) or reagent limited, competitive assays with a separation step (Group 3). The first three groups have long been very well represented among commercially developed immunoassays sold as

**Tab. 3** Classification of immunoassays into four groups.

| *Group* | *Type* | *Examples* |
|---|---|---|
| 1. Label free | Agglutination | Hemagglutination<br>Latex agglutination |
| | Precipitation | Ouchterlony gel diffusion<br>Rocket immunoelectrophoresis<br>Immunoturbidimetry/nephelometry |
| | Immunosensors | Resonant mirror waveguide (IAsys™)<br>Surface plasmon resonance (BIAcore™) |
| 2. Reagent excess | One-site | Immunostaining<br>Western blotting<br>Fluorescence activated flow cytometry |
| | Two-site | Immunoenzymometric assay, (IEMA, ELISA)<br>Immunofluorometric assay<br>Immunoradiometric assay |
| 3. Reagent limited | Labeled antigen or antibody | Radioimmunoassay<br>Enzymeimmunoassay<br>Fluoroimmunoassay |
| | Separation free | EMIT, CEDIA, FPIA |
| 4. Ambient analyte | Microarray/microchip | Microspot® |

stand-alone kits or run on automated analyzers.

### 4.4.1 Label-free Assays

Provided that the concentrations are sufficient, the molecular complexes ([Ab–Ag]*n*) generated by antibody–antigen interaction are visible to the naked eye, but smaller amounts may also be detected and measured due to their ability to scatter a beam of light. The formation of complexes indicates that both reactants are present, and in immunoprecipitation assays a constant concentration of a reagent antibody is used to measure specific antigen ([Ab–**Ag**]*n*), and reagent antigens are used to detect specific antibody ([**Ab**–Ag]*n*). If the reagent species is previously coated onto cells (as in hemagglutination assay) or very small particles (as in latex agglutination assay), "clumping" of the coated particles is visible at much lower concentrations.

A variety of assays based on these elementary principles are in common use, including Ouchterlony immunodiffusion assay, rocket immunoelectrophoresis, and immunoturbidometric and nephelometric assays. The main limitations of such assays are restricted sensitivity (lower detection limits) in comparison to assays employing labels and, in some cases, the fact that very high concentrations of analyte can actually inhibit complex formation, necessitating safeguards that make the procedures more complex. Some of these Group 1 assays date right back to the discovery of antibodies (Table 1) and none of them have an actual "label" (e.g. Ag-enz).

Other kinds of immunoassays that are label free depend on immunosensors, and a variety of instruments that can directly detect antibody–antigen interactions are now commercially available. Most depend on generating an evanescent wave on a sensor surface with immobilized ligand, which allows continuous monitoring of binding to the ligand. Immunosensors allow the easy investigation of kinetic interactions and, with the advent of lower-cost specialized instruments, may in the future find wide application in immunoanalysis.

### 4.4.2 Reagent-excess Assays

Group 2 contains assays that are typified by the use of an excess concentration of labeled antibody or antigen for the detection specific antigen or antibody.

"One-site" reagent-excess assays include western blotting, whereby proteins absorbed to nitrocellulose filters are located by "probing" with labeled specific antibodies (nitrocellulose-**Ag**–Ab-L) and immunostaining, such as immunofluorescence assay whereby antigen in a tissue section is visualized with specific antibodies conjugated to fluorescein (cell matrix-**Ag**–Ab-fluorescein). Fluorescence-activated cell sorting (FACS) also depend on specific immunostaining of free-floating cells (**cell-Ag**–Ab-fluorescein).

However, most often, separate binding reactions specific for two different sites on the analyte are involved, giving rise to a tri-molecular, or larger, complex with the analyte in the middle, hence the terms "two-site assay" and "sandwich assay." For example, two monoclonal antibodies against different, distally located epitopes on the analyte may be used to detect a protein hormone such as hCG with high specificity (sp-mAb1–**hCG**–mAb2-L). Similarly, two-site assays can be used to detect specific antibodies of a particular immunoglobulin class or subclass. For example, a purified allergen and an antibody against human IgE may be used to detect specific antibodies of the IgE class that bind to the allergen in question (sp-allergen–**IgE**–Ab-L). Please note also that to detect viral (e.g. HIV) infection, the antigen employed may be derived from a crude viral lysate, be prepared by a recombinant method; or be a synthetic peptide corresponding to an appropriate viral epitope, or a defined set of such peptides (see Sect. 2.3).

Assays to detect specific antibodies that employ immobilized-antigen to capture the analyte antibody and labeled antigen to titrate the bound antibody via its free antigen-binding site (IgG) or sites (IgM or IgA) are, strictly speaking, two-site assays (sp-Ag–**Ab**–Ag-L). However, since all sites on a single antibody are identical, such "antibody bridge" assays may be regarded as resembling one-site assays with respect to their specificity. "Antibody bridge" assays are very suitable for the early detection of viral infections such as HIV because IgM class antibodies are prominent in the early immune response, and in such assays the captured antiviral IgM antibodies may each bind multiple copies of the labeled antigen, enhancing the lower detection limit.

#### 4.4.3 Reagent-limited Assays

Group 3 contains reagent-limited assays, typified by classical RIA. These are also commonly referred to as *competitive assays* and may have labeled antigen or labeled antibody. They are used to measure the concentrations of antigens or, more commonly, of smaller analytes such as steroids or drugs. In contrast to almost all other immunoassays, the immune complexes measured or detected (sp-Ab–Ag-L or sp-Ag–Ab-L) do not contain the analyte and, therefore, all such assays give inverse (i.e. falling as opposed to rising) standard curves when the concentration of bound label is plotted against increasing analyte concentration. Because of their "indirect" nature, competitive assays are inherently less sensitive than reagent-excess assays.

However, competitive assays can readily be designed to obviate the need for a separation step before label measurement, and such assays are referred to as *separation-free* or *homogeneous assays*. To avoid the need for a separation step, the binding reaction must influence the ability of the label to generate the endpoint signal. Ideally, such modulation should be 100% efficient and the limited lower detection limits generally achievable testify to the fact that, in general, this is not realized. Separation-free assays include the commercial systems EMIT, FPIA, and CEDIA which are widely employed in the manual and automated measurement of therapeutic, toxic, and illegal drugs. (See also Sect. 5.2.)

#### 4.4.4 Ambient Analyte Assays

Ambient analyte immunoassays (Group 4) were proposed by Roger Ekins and are performed under conditions in which the concentration of immobilized captured antibodies (or antigen molecules) is so low that, as analyte binds to them, the concentration of analyte in the reaction medium is unaffected. As with many receptors in endocrine systems, fractional binding site occupancy is independent of both the amount of capture agent and reaction volume. For such a system, very high specific activity, fluorescent labels are necessary, and the small amount of capture antibody (or antigen) may be coated at very high density within a "microspot," facilitating the development of microarray-based multianalyte systems. A range of multianalyte microassay immunoassay systems for clinical applications have been proposed and are under development by different groups, but many do not correspond exactly to the kinetic aspects of Ekins' model.

## 5 Established Immunoassay Systems

Many immunoassay systems used in human clinical laboratories were first introduced in the 1970s and 1980s and have had long lives. For example, the separation-free systems EMIT (based on glucose 6-phosphate dehydrogenase labels) and FPIA were first described in the early 1970s and are still in widespread use across the world.

### 5.1 Automated Systems

Early clinical immunoassay analyzer systems, which provided automation of repetitive steps such as pipetting and washing, soon led to fully automated "batch analyzers;" and more recent systems are "random access." Sample and reagent volumes have also decreased and incubation times have shortened. The panels of analytes are now more extensive, so that hormones, specific antibodies and tumor markers may be available on a single instrument. New larger machines integrate the functions of immunoassay analyzers and clinical chemistry analyzers into one system.

Another trend has been away from the colorimetric determination of endpoints to chemiluminescent-based systems, whether to measure the activity of an enzyme label (e.g. DPC Immulite®) or to directly measure a luminescent label incorporating acridinium ester (Bayer ADVIA Centaur®) or Ruthenium(II) chelate (Roche Diagnostics Elecsys®).

The precision of automated immunoassays has improved in parallel with the above developments. As reported by Wheeler in 2001, the results of the UK National External Quality Assessment Schemes indicate that levels of within and between sample assay precision of less than 5% are now commonly achievable over much of the concentration range of analytes such as luteinizing hormone.

### 5.2 General Laboratory Assays

Most immunoassays in use today are variations on reagent-excess two-site systems for the measurement of larger antigen analytes and specific antibodies. In addition, reagent-limited assays are employed for the measurement of smaller molecules such as steroids or common drugs, usually with labeled hapten, rather than with labeled antibody. For the measurement of low concentrations of analyte (nmol $L^{-1}$ and below), assays with a separation step (usually involving antibody immobilized on a solid phase) are used. All of these formats, adapted for use on 96 well microtiter plates, are operated manually in many laboratories for a huge range of applications covering all areas of routine and research biology. In some areas with high throughput, such as human clinical chemistry, they have been adapted to the automated systems described briefly above; in others they have been adapted to give the rapid, easy-to-use tests described below.

For small analytes such as steroids in the low pmol $L^{-1}$ range, an assay with an extraction step to minimize sample matrix effects is to be preferred. Radioactive labels are now rarely used outside large laboratories where the appropriate traditions, facilities, and expertise persist.

Separation-free, reagent-limited immunoassays dominate in some areas, particularly for drug screening when concentrations are often in the μmol $L^{-1}$ range. The most widely used systems have been EMIT®, FPIA® (Abbot Laboratories) and CEDIA®.

EMIT®, is a registered name of the Syva Corporation and is an acronym for "enzyme multiplied immunotest." EMIT® is almost as old as EIA and dates from the classic paper of Rubinstein, Schneider, and Ullman in 1972. The invention of EMIT is one of the most important milestones in the history of immunoassays. In one stroke, radioisotopes were avoided, procedures simplified, "turn-around-times" greatly reduced, and automation on standard clinical chemistry analyzers made feasible.

CEDIA® is a major competitor of EMIT that depends on a detailed understanding of enzymology and protein structure. To devise CEDIA, recombinant DNA technology was exploited to produce new strains of *Escherichia coli* synthesizing inactive variants of $\beta$-D-galactosidase. These were of two complementary kinds, large fragments (enzyme acceptors, EA) and small fragments (enzyme donors, ED). Some matched pairs of EA and ED fragments could associate readily to give fully active enzyme and these were exploited (with EA as label) to develop a new, versatile separation-free immunoassay system, that is suited to automation by means of many general-purpose clinical analyzers.

There are also variations of standard immunoassays that play important specific roles in basic research. One example is the ELISPOT assay that for the last 20 years has been used in immunology research laboratories to quantify T-cell immune responses in individuals infected with different pathogens. ELISPOT tests are crucial to the evaluation of vaccines and other therapies when they are under trial, and are contributing to the fights against diseases such as AIDS.

## 5.3 Simple to use Test Devices

Many of the most impressive inventions involve the simplification of a previously complex task, but a long road with a series of separate inventions or developments may be necessary in some cases. Searching a database for articles on pregnancy tests reveals that in the 1950s live toads or frogs were used and that days passed before results were obtained. Obviously, and to say the least, such tests required many steps, laboratory facilities, and a skilled technician. Now the "patient" just holds a simple plastic "rapid test device" (RTD) in her urine stream for a few seconds, and minutes later she has the result.

The adoption of immunoassays, such as latex agglutination tests, to detect the pregnancy hormone hCG in urine was an important first step on the road. Later came dipsticks with multiple steps and reagents, and liquid waste to dispose of. Still later came what is now the "Beckman Coulter Hybritech ICON® hCG test," which combines simplicity of operation and a low detection limit in a reagent-excess IEMA with two monoclonal antibodies, an alkaline phosphatase label, and a visible result. Two drops serum or 5 of urine, appropriately prediluted, are added to a simple small test "pot." After two further reagent additions and washes with a simple dropper and suitable short delays, one colored dot indicates a valid negative result and two dots indicate a pregnancy. All liquid waste is retained in the "pot."

One-step operation, like that described above, arrived with the development of the

Unipath ClearView® test device. There are no liquid reagents or wash solutions to add and no need to precollect the urine sample. The "mobile" antibody is linked to intensely colored microbeads, obviating the need for a color development step. The simple RTD hides a basic reagent-excess sandwich immunoassay and has all the reagents preloaded, ready to be set in motion by the addition of water, and all the water needed comes with the urine sample.

Like for the ICON® RTD, there are now many similar devices from many companies and these have been adapted for a range of tests including fertility monitoring by detecting the preovulatory surge of LH in women. Similar RTDs are also widely used to detect specific antibodies for infectious diseases such as HIV infection, Lyme disease, etc.

## 6 Immunoassays of the Future

### 6.1 Multiplexed Immunoassays

Immunoassays capable of detecting simultaneously two or up to 10 or more analytes have been available for many years. These include the extraordinary MAST® system with up to 35 different allergens coated on cellulose threads for detecting elevated levels of specific IgE class antibodies in human serum. However, many of the multiplexed immunoassay systems of the future will be fully quantitative, capable of detecting larger numbers of varied and distinct analytes, and automated.

Developments in multianalyte microarray immunoassay systems (see also Sect. 4.4.4) have benefited from parallel advances in the methodology of genomics and proteomics. Computer controlled equipment is available for the immobilization of just hundreds of capture molecules per microspot on heterogeneous arrays of hundreds of spots. There are also automated liquid pipettors that deliver submicroliter volumes of reagents and of samples to be analyzed, where and when required. In addition, charge-coupled device (CCD) detectors can scan and quantify luminescent or fluorescent light emission from each spot in a large array. Planar waveguide evanescent field technology enables microarray assays without separation or washing steps.

Multiplexed assays based on heterogeneous populations of microbeads have also been developed in a number of laboratories. These benefit greatly from the availability of diverse sets of identifiable microspheres. Each set or population of microspheres may incorporate one or two fluorescent dyes at set densities (enabling the creation of a large number of sets) and is easily and specifically quantifiable by a suitable flow cytometer. Recent flow cytometers, dedicated to suspension-array assays, have dual lasers and are capable of up to 100-plexed operations. Immunoassays for the measurement of specific antibodies and antigens developed for such systems usually correspond to Group 2 and 3 assays as described in Sect. 4.

### 6.2 New Assay Principles and Systems

The detailed history of immunoassay methodology is dotted by hundreds of reports of new and original systems, labels and detection methods, each presented as offering multiple advantages such as ease of use, low detection limit, accuracy, versatility, and so on. Most disappear from view immediately, or after a flurry of articles from a single laboratory. Of the successful ones, many have endured, and have been refined gradually, over decades. Here are some new systems that may (or may not) prosper.

"Open sandwich" (OS) immunoassay could not have been envisaged before recombinant DNA methods had been used to prepare $V_H$ and $V_L$ polypeptides (see Sect. 1.1), because OS assays depend on antigen stabilization of the triple complex $V_H$–Ag–$V_L$. When free $V_H$ and $V_L$ were first studied, their association was found to give unstable Fv fragments, and that is why, for applications requiring minimal antibodies, single-chain sFv fragments, in which the two variable polypeptides are joined by a linker peptide, are always used. In contrast, Hiroshi Ueda found, in experiments with a surface plasmon resonance biosensor, that the strength of the interaction between $V_H$ and $V_L$ increases greatly when specific antigen is bound also. From this realization, he went on to devise a range of potentially promising one-site "sandwich" immunoassays, and separation-free assays involving fluorescence resonance energy transfer (FRET) or bioluminescence resonance energy transfer (BRET).

Colin Self is best known as the inventor of amplified enzyme assays and their application to the development of ultrasensitive immunoassays, but he has been long aware of the basic limitations of competitive immunoassays, which are the only standard immunoassays suitable for small molecular weight ($<1000\ g\ mol^{-1}$) "hapten" analytes. As summarized recently by Ashton, he has devised and patented a number of sophisticated reagent-excess immunoassay systems for small analytes. The "selective antibody system," involves a blocking molecule (BM) for free sites on the immobilized primary antibody (sp-Ab, sp-Ab–BM) and a labeled anti-idiotypic antibody (aiAb-L) that is added later and binds specifically to the occupied binding site of the primary antibody (sp-Ab–**Ha**, sp-Ab–**Ha**–aiAb-L) (an anti-idiotypic antibody is an antibody that binds to the variable region of another monoclonal antibody.) In this case, because of the role of the "blocker," the anti-idiotypic antibody does not have to be specific for the *occupied* site. Note that, unlike a competitive assay, the complex measured at the end of the assay *does* contain the analyte.

Colin Self's later "apposition system" is similar to the above but does not require the difficult task of raising a suitable anti-idiotypic antibody for *every* single primary antibody used. Rather, the challenge is to link, to an appropriate residue near the binding site of the primary antibody, a ligand to which a labeled binding protein can bind. Again, a blocking molecule is used to block all empty binding sites. In principle, the same ligand can be linked to all primary antibodies to be used in a range of assays (perhaps via a suitably located cysteine sulphydryl group inserted in the peptide sequence by "site-selective mutation" of the gene coding for the $V_H$ or $V_L$), making the labeled binding protein a "universal reagent."

## 7
## The Limitations of Immunoassays

Absolute specificity that depends on a binding reaction alone is unattainable. Whether the binding protein is an antibody, a hormone receptor or a lectin, the existence of significant concentrations of cross-reacting molecules in (unrefined) samples is a continuous possibility. After all, the science of pharmacology depends largely on finding or devising substances that strongly cross-react with "specific" binding sites.

With hapten analytes, a structurally similar cross-reactant may be able to bind to the specific paratope in a slightly different orientation, or a molecule with a single different functional group may bind because of the existence of a sufficiently large "pocket" in the binding site. There is also the well-accepted fact that changes at, or near, the site via which the hapten was conjugated to the immunogen carrier molecule often have little effect on affinity. "Dissimilar" cross-reactants, perhaps defined as such by crude comparisons of chemical formulae, are also always possible.

Large protein, glycoprotein or complex carbohydrate antigens interact with a large area of the paratope, thereby making effective epitope uniqueness more feasible. Nevertheless, cross-reacting molecules do occur and, in addition, the achievement of adequate specificity may be exacerbated by analyte heterogeneity. Two-site assays require the recognition of two spatially distinct epitopes and can be exploited to improve specificity when analytes are large, but then there are *two* interactions that are susceptible to general interference instead of one. In contrast, absolute specificity may not always be desirable. For example, as protein analytes are subject to genetic variation, subpopulations of individuals, from whom samples to be analyzed are obtained, may exist that express a variant epitope that is not recognized by a reagent antibody (see Sect. 2.1).

When the analyte is a complex biomolecule, standardization may be especially difficult to achieve because suitable primary standards and "gold-standard" reference analytical methods are unavailable. Lately, it has become possible to prepare by recombinant DNA methods primary standards for some complex analytes such as hCG and human kallikrein 2. Reference methods for such analytes also require epitope mapping of the analyte, well-characterized antibodies, and international collaboration to ensure their acceptance and use.

When the analytes are specific antibodies, the required specificity may be quite different from one application to another. The purity and suitability of the antigen reagent is critical and, when class (or subclass) specificity is also required, a two-site design is normally used, so that the specificity of the class-specific antibody used is also vital. In all these designs, antibodies against other antigens or polyspecific antibodies may be detected, and may even give rise to false-positive diagnoses.

However, perhaps no analytical method can be guaranteed in advance to be absolutely specific 100% of the time when the samples to be analyzed are complex and of variable composition. For some types of analyte, isotope dilution mass spectrometry (ID-MS) is used as the "gold standard" and this performs as well as it does because it involves at least two stages, one to separate the analyte from the general constituents of its sample matrix and one to measure it. Therefore, when the analyte is presented eventually to the mass spectrometer, it has been extensively purified by means of a high-resolution chromatographic method. In contrast, most immunoassay methods are expected to be accurate and precise when presented with untreated samples. The only preliminary treatment that is accepted as practicable in most situations is dilution.

Therefore, any really effective answer to interference and cross-reactivity as they may affect immunoassays must also involve preliminary purification of samples. Otherwise the accuracy of every result is completely dependent on the care taken in developing and operating the test. It depends on the careful selection and purity of antibodies (and/or antigens) and other reagents used, on the careful optimization, thorough analytical and biological validation of the assay, and on thorough quality control procedures.

## Bibliography

### Books and Reviews

Crowther, J.R. (2000) *ELISA Handbook*, Humana Press, Totowa, NJ.

Diamandes, E.P., Chrisopoulos, T.K. (Eds.) (1996) *Immunoassay*, Academic Press, San Diego, CA.

Gosling, J.P. (1990) A decade of development in immunoassay methodology, *Clin. Chem.* **36**, 1408–1427.

Gosling, J.P. (Ed.) (2000) *Immunoassays: A Practical Approach*, IRL Press, Oxford, UK.

Gosling, J.P., Basso, L.V. (Eds.) (1994) *Immunoassay: Laboratory Analysis and Clinical Applications*, Butterworth Heineman, Boston, MA.

Joos, T.O., Stoll, D., Templin, M.F. (2001) Miniaturised multiplexed immunoassays, *Curr. Opin. Chem. Biol.* **6**, 76–80.

Levinson, S.S., Miller, J.J. (2002) Towards a better understanding of heterophile (and the like) antibody interference with modern immunoassays, *Clin. Chim. Acta* **325**, 1–15.

Mashishi, T., Gray, C.M. (2002) The ELISPOT assay: an easily transferable method for measuring cellular responses and identifying T cell epitopes, *Clin. Chem. Lab. Med.* **40**, 903–910.

Price, C.P., Newman, D.J. (Eds.) (1997) *Principles and Practice of Immunoassay*, 2nd edition, Macmillan, London, UK.

Wild, D. (Ed.) (2001). *The Immunoassay Handbook*, 2nd edition, Nature Publishing Group, UK.

Wheeler, M.J. (2001) Automated immunoassay analysers, *Ann. Clin. Biochem.* **38**, 217–229.

### Primary Literature

Abdul-Ahad, W.G., Gosling, J.P. (1987) An enzyme-linked immunosorbent assay (ELISA) for bovine LH capable of monitoring fluctuations in baseline concentrations, *J. Reprod. Fertil.* **80**, 653–61.

Addison, G.M., Hales, C.N., (1971) The Immunometric Assay [and discussion], in: Kirkham K.E., Hunter W.M. (Eds.) *Radioimmunoassay Methods*, Churchill Livingstone, Edinburgh, Scotland, pp. 447–461 [and 467–488].

Arakawa, H., Maeda, M., Tsuji, A., Kambegawa, A. (1981) Chemiluminescence enzyme immunoassay of dehydroepiandrosterone and its sulphate using peroxidase as label, *Steroids* **38**, 453–464.

Ashton, V. (2004) Measurements in miniature, *Chemistry World* **10**, 42–45.

Avrameas, S., Uriel, J. (1966) Méthode de marquage d'antigénes at d'anticorps avec enzymes et son application en immunodiffusion, *Comptes Rendus Hebdomedaires des Seances de l'Academie des Sciences: D: Sciences naturelles (Paris)* **262**, 2543–2545.

Berson, S.A., Yalow, R.S. (1959) Assay of plasma insulin in human subjects by immunological methods, *Nature* **184**, 1648–1649.

Blackburn, G.F., Shah, H.P., Kenten, J.H., Leland, J., Kamin, R.A., Link, J., Peterman, J., Powell, M.J., Shah, A., Talley, D.B., Tyagi, S.K., Wilkins, E., Wu, T.G., Massey, R.J. (1991) Electrochemiluminescence detection for development of immunoassays and DNA probe assays for clinical diagnostics, *Clin. Chem.* **37**, 1534–1539.

Brown, C.R., Higgins, K.W., Frazer, K., Schoelz, L.K., Dyminiski, J.W., Marinkovich, V.A., Miller, S.P., Burd, J.F. (1984) Simultaneous determination of total IgE and allergen-specific IgE in serum by the MAST chemiluminescent system, *Clin. Chem.* **31**, 1500–1505.

Carson, R.T., Vignali, D.A. (1999) Simultaneous quantification of 15 cytokines in a multiplexed flow cytometric assay, *J. Immunol. Methods* **227**, 41–52.

Chow, B.F. (1947) The determination of plasma or serum albumin by means of a precipitin reaction, *J. Biol. Chem.* **167**, 757–763.

Coons, A.H., Creech, H.J., Jones, R.N., Berliner, E. (1942) Demonstration of pneumococcal antigen in tissues by use of fluorescent antibody, *J. Immunol.* **45**, 159–170.

Czerkinsky, C.C., Nilsson, L.A., Nygren, H., Ouchterlony, O., Tarkpowshy, A. (1983) A solid-phase enzyme-linked immunospot (ELISPOT) assay for enumeration of specific antibody-secreting cells, *J. Immunol. Methods* **65**, 109–121.

Dandiker, W.B., Kelly, R.J., Dandiker, J., Farcuhar, J., Levin, J. (1973) Fluorescence polarisation immunoassay. Theory and experimental methods, *Immunochemistry* **10**, 219–227.

Ekins, R.P. (1960) The estimation of thyroxine in human plasma by an electrophoretic technique, *Clin. Chim. Acta* **5**, 453–459.

Ekins, R.P., Chu, F., Biggart, E. (1990) The development of microspot, multianalyte ratiometric immunoassay using dual fluorescent labelled antibodies, *Anal. Chim. Acta* **227**, 73–96.

Emerson, J.F., Ngo, G., Emerson, S.S. (2003) Screening for interference in immunoassays, *Clin. Chem.* **49**, 1163–1169. (See also Span et al. 2003, below.)

Engel, W.D., Khanna, P.L. (1992) CEDIA in vitro diagnostics with a novel homogeneous immunoassay technique: current status and future prospects, *J. Immunol. Methods* **150**, 99–102.

Finckh, P., Berger, H., Karl, J., Eichenlaub, U., Weindel, K., Hornauer, H., Lenz, H., Sluka, P., Ehrleich Weinreich, G., Chu, F., Ekins, R. (1998) Proceedings of the UK NEQAS Endocrinology Meeting, Vol. 4, Association of Clinical Biochemists, London, 155.

Freund, J. (1925) Agglutination of tubercule bacilli, *Am. Rev. Tuberc.* **12**, 124–141.

Gonzalez, C., Guevara, P., Alarcon, I., Hernando, M., Navajo, J.A., Gonzalez-Buitrago, J.M. (2002) Antinuclear antibodies (ANA) screening by enzyme immunoassay with nuclear HEp-2 cell extract and recombinant antigens: analytical and clinical evaluation, *Clin. Biochem.* **35**, 463–469.

Gosling, J.P. (1997) Enzyme-immunoassay: With and Without Separation, in: Price, C.P., Newman D.J. (Eds.) *Principles and Practice of Immunoassay*, 2nd edition, Academic Press, San Diego, CA, pp. 350–388.

Gosling, J.P., Middle, J., Siekmann, L., Read, G. (1993) Standardization of hapten immunoprocedures: total cortisol, *Scand. J. Clin. Lab. Invest.* **53**(Suppl. 216), 3–41.

Gillis, E.H., Gosling, J.P., Sreenan, J.M., Kane, M. (2002) Development and validation of a biosensor-based immunoassay for progesterone in bovine milk, *J. Immunol. Methods* **267**, 131–138.

Haese, A., Vaisanen, V., Finlay, J.A., Pettersson, K., Rittenhouse, H.G., Partin, A.W., Bruzek, D.J., Sokoll, L.J., Lilja, H., Chan, D.W. (2003) Standardisation of two immunoassays for human glandular kallikrein 2, *Clin. Chem.* **49**, 601–610.

Henderson, D.R., Friedman, S.B., Harris, J.B., Manning, W.B., Zoccoli, M.A. (1986) 'CEDIA', a new homogeneous immunoassay system, *Clin. Chem.* **32**, 1637–41.

Herold, D.A., Fitzgerald, R.L. (2003) Immunoassays for testosterone in women: Better than a guess? *Clin. Chem.* **49**, 1250–1251.

Huang, R.P. (2001) Simultaneous detection of multiple proteins with an array-based enzyme-linked immunosorbent assay (ELISA) and enhanced chemiluminescence, *Clin. Chem. Lab. Med.* **39**, 209–214.

Ishikawa, E. (1973) Enzyme immunoassay on insulin by fluorimetry of insulin-glucoamylase complex, *J. Biochem.* **73**, 1319–21.

Ishikawa, E., Imagawa, M., Hashida, S., Yoshitake, S., Hamaguchi, Y., Ueno, T. (1983) Enzyme-labelling of antibodies and their fragments for enzyme immunoassay and immunohistochemical staining, *J. Immunoassay* **4**, 209–327.

Jackson, T.M., Ekins, R.P. (1986) Theoretical limitations on immunoassay sensitivity. Current practice and potential advantages of fluorescent Eu3+ chelates as non-radioisotopic tracers, *J. Immunol. Methods* **87**, 13–20.

Jani, I.V., Janossy, G., Brown, D.W.G., Mandy, F. (2002) Multiplexed immunoassays by flow cytometry for diagnosis and surveillance of infectious diseases in resource-poor settings, *Lancet Infect. dis.* **2**, 243–250.

Libby, R.L. (1938) A new and rapid quantitative technique for the determination of potency of types I and II antipneumococcal serum, *J. Immunol.* **34**, 269–279.

Luzzi, V.I., Saunders, A.N., Koenig, J.W., Turk, J., Lo, S.F., Carg, U.C., Dietzen, D.J. (2004) Analytic performance of immunoassays for drugs of abuse below established cutoff values, *Clin. Chem.* **50**, 717–722.

Maiolini, R., Masseyeff, R. (1975) A sandwich method of enzymeimmunoassay. I. Application to rat and human alpha-fetoprotein, *J. Immunol. Methods* **8**, 223–234.

Miles, L.E.M., Hales, C.N. (1968) Labelled antibodies and immunological assay systems, *Nature* **219**, 186–169.

Nakane, P.K., Pierce, G.B. (1966) Enzyme-labelled antibodies: preparation and application to the localization of antigens, *J. Histochem. Cytochem.* **14**, 929–931.

Nayak, P.N. (1981) The kinetics of solid-phase immunoassays, *Ligand Q.* **4**, 34–42.

O'Connor, A., Gosling, J.P. (1997) The dependence of detection limit on antibody affinity in competitive radioimmunoassays for progesterone, *J. Immunol. Methods* **208**, 181–189.

Odell, W.D., Griffin, J. (1987) Two-monoclonal-antibody "sandwich"-type assay of human lutropin, with no cross reaction with choriogonadotrophin, *Clin. Chem.* **33**, 1603–1607.

O'Rorke, A., Kane, M.M., Gosling, J.P., Tallon, D.F., Fottrell, P.F. (1994) Development and validation of a monoclonal antibody enzymeimmunoassay for the measurement of progesterone in saliva, *Clin. Chem.* **40**, 400–410.

Pettersson, K., Ding, Y.Q., Huhtaniemi, I. (1992) An immunologically anomalous luteinizing hormone variant in a healthy woman, *J. Clin. Endocrinol. Metab.* **74**, 164–171.

Pettersson, K., Siitari, H., Hemmilä, I., Soini, E., Lövgren, T., Hänninen, V., Tanner, P., Stenman, U-H. (1983) Time-resolved fluoroimmunoassay of human choriogonadotrophin, *Clin. Chem.* **29**, 60–64.

Rattle, S.J., Purnell, D.R., Williams, P.I.M., Siddle, K., Forrest, G.C. (1984) New separation method for monoclonal immunoradiometric assays and its application to assays for thyrotropin and human choriogonadotrophin, *Clin. Chem.* **30**, 1457–1461.

Rowe-Taitt, C.A., Hazzard, J.W., Hoffman, K.E., Cras, J.J., Golden, J.P., Ligler, F.S. (2000) Simultaneous detection of six biohazardous agents using a planar waveguide array biosensor, *Biosens. Bioelectron.* **15**, 579–589.

Rubinstein, K.E., Schneider, R.S., Ullman, E.F. (1972) Homogeneous enzyme immunoassay: a new immunochemical technique, *Biochem. Biophys. Res. Commun.* **47**, 846–851.

Schneyer, A.L., Sluss, P.M., Whitcomb, R.W., Hall, J.E., Crowley, W.F. Jr., Freeman, R.G. (1991) Development of a radioligand receptor assay for measuring follitropin in serum: application to premature ovarian failure, *Clin. Chem.* **37**, 508–514.

Self, C.H. (1985) Enzyme amplification – a general method applied to provide an immunoassisted assay for placental alkaline phosphatase, *J. Immunol. Methods* **76**, 389–393.

Span, P.N., Grebenchtchikov, N., Sweep, C.G.J. (2003) Screening for interference in immunoassays, *Clin. Chem.* **49**, 1708–1709.

Stenman, U-H., Bidart, J-M., Birkin, S., Mann, K., Nisula, B., Oconnor, J. (1993) Standardization of immunoprocedures: chorionic gonadotrophin (CG), *Scand. J. Clin. Lab. Invest.* **53**(Suppl. 216), 42–78.

Taieb, J., Mathian, B., Millot, F., Patricot, M-C., Mathieu, E., Queyrel, N., Lacroix, I., Somma-Delpero, C., Boudou., P. (2003) Testosterone measured in 10 immunoassays and by isotope-dilution gas chromatography-mass spectrometry in sera from 116 men, women and children, *Clin. Chem.* **49**, 1381–1395. (See also Herold and Fitzgerald 2003, above.)

Tamate, K., Kane, M.M., Charleton, M., Egan, D., Gosling, J.P., Ishikawa, M., Fottrell, P.F. (1997) Direct colorimetric monoclonal antibody enzymeimmunoassay for estradiol-17b in saliva, *Clin. Chem.* **43**, 1159–1164.

Thorpe, G.H.G., Kricka, L.J., Mosely, S.B., Whitehead, T.P. (1985) Phenols as enhancers of the chemiluminescent horseradish peroxidase-luminol-hydrogen peroxide reaction: application in luminescence monitored enzyme immunoassays, *Clin. Chem.* **31**, 1335–1341.

Ueda, H. (2002) Open sandwich immunoassay: a novel immunoassay approach based on the interchain interaction of an antibody variable region, *J. Biosci. Bioeng.* **94**, 614–619.

Valkirs, G.E., Barton, R. (1985) ImmunoConcentration™ – a new format for solid phase immunoassays, *Clin. Chem.* **31**, 1427–1431.

Van Weeman, B.K., Schuurs, A.H.W.M. (1971) Immunoassay using antigen-enzyme conjugates, *FEBS Lett.* **15**, 232–236.

Voller, A., Bidwell, D.E., Huldt, G., Engvall, E. (1974) A microplate method of enzyme-linked immunosorbent assay and its application to malaria, *Bull. W. H. O.* **51**, 209–211.

Wide, L. (1971) Solid Phase Antigen-antibody Systems, in: Kirkham, K.E., Hunter, W.M. (Eds.) *Radioimmunoassay Methods*, Churchill Livingstone, Edinburgh, Scotland, pp. 405–412.

# 12
# Genetic Engineering of Antibody Molecules

*Manuel L. Penichet and Sherie L. Morrison*
*Department of Microbiology, Immunology, and Molecular Genetics, University of California, Los Angeles, CA, USA*

**1 Antibody Structure and Engineering 321**
1.1 The Basic Structure of Antibodies 321
1.2 Classes and Subclasses of Antibodies 322

**2 From Mouse to Human Antibodies 324**
2.1 Murine Monoclonal Antibodies 324
2.2 Chimeric Antibodies 326
2.3 Humanized Antibodies 327
2.4 Human Monoclonal Antibodies in Mice 327

**3 *In Vitro* Antibody Production by Phage Libraries 328**

**4 Further Genetic Modifications of Antibodies 330**
4.1 Engineering Antibody Fragments: Monovalent, Bivalent, and Multivalent scFvs 330
4.2 Bispecific Antibodies 332
4.3 Polymers of Monomeric Antibodies 333
4.4 Antibody Fusion Proteins 334

**5 Expression Systems 336**

**6 Conclusion 336**

**Bibliography 337**
Books and Reviews 337
Primary Literature 337

*Immunology. From Cell Biology to Disease.* Edited by Robert A. Meyers.

ISBN: 978-3-527-31770-7

## Keywords

**Antibody/Antigen**
Antibodies are proteins known as *immunoglobulins* (Igs), which are produced by the immune system in response to the presence of a foreign substance. Antigens are molecules (usually proteins) on the surface of cells, viruses, fungi, bacteria, and some nonliving substances such as toxins, chemicals, drugs, and foreign particles, which are recognized and specifically bound by antibodies. The immune system recognizes immunogenic antigens (also known as *immunogens*), and produces antibodies that destroy or neutralize substances or organisms containing antigens.

**Antibody-dependent Cellular Cytotoxicity (ADCC)**
Cell-killing reaction in which the Fc receptor bearing killer cells recognizes target cells via specific antibodies.

**Bacteriophage**
Bacteriophages or phages are viruses that infect bacteria. Bacteriophage $\lambda$ is a temperate phage, which can grow lytically lysing the bacteria and forming a clear plaque on a lawn of bacteria. Infection with filamentous phages such as M13 is not lethal and the host bacteria do not lyse. Instead, their rate of growth slows and they form turbid plaques on the bacterial lawn.

**Complement**
Group of serum proteins participating in the lysis of foreign cells and pathogens [a process known as *complement-dependent cytotoxicity* (CDC)]; they also play an important role in phagocytosis.

**Constant Region**
Portion of the antibody molecule exhibiting little variation and determining the isotype (class or subclass) of the antibody.

**Fab**
The Fab fragment is a monovalent antigen-binding fragment of an antibody that consists of one light chain and part of one heavy chain (the variable region and the first constant region domain). It can be obtained by digestion of intact antibody with papain or by genetic engineering techniques.

**Fc**
The Fc is a nonantigen binding fragment of an antibody that consists of the carboxy-terminal portion of both heavy chains. It can be obtained by papain digestion of an intact antibody. Two Fabs and one Fc fragment comprise a complete IgG antibody.

**Fv/scFv**
The Fv is a monovalent antigen-binding fragment of an antibody composed of the variable regions from the heavy and light chains. scFv fragments are composed of the

variable domains of heavy and light chains ($V_L$ and $V_H$) joined by a synthetic flexible linker peptide. Thus, the scFv provides a fully functional antigen binding domain that is encoded in a single gene and expressed as a single polypeptide.

**Hybridoma**
Cell derived by a fusion between a normal cell, usually a lymphocyte, and a tumor cell, usually a myeloma cell.

**Variable Region**
Variable portion of the antibody molecule that is responsible for antigen binding.

Antibodies have long been appreciated for their exquisite specificity. With the development of the hybridoma technology, it was possible to produce rodent (mouse or rat) monoclonal antibodies that are the product of a single clone of antibody-producing cells and have only one antigen-binding specificity. Advances in genetic engineering and expression systems have been applied to overcome problems of immunogenicity of rodent-produced antibodies and to improve their ability to trigger human immune effector mechanisms. The production of chimeric, humanized, and totally human antibodies as well as antibodies with novel structures and functional properties has resulted in improved monoclonal antibodies. As a consequence, recombinant antibody-based therapies are now used to treat a variety of diverse conditions that include infectious diseases, inflammatory disorders, and cancer. This article summarizes and compares different strategies for developing recombinant antibodies and their derivatives.

## 1 Antibody Structure and Engineering

### 1.1 The Basic Structure of Antibodies

Antibodies are molecules with multiple properties that make them a critical component of the immune system. These properties include the ability to recognize a vast array of different molecules known as *antigens* and to interact with and activate the host effector systems.

The basic structure of all antibodies, also known as *immunoglobulins* (Igs), is a unit consisting of two identical light polypeptide chains and two identical heavy polypeptide chains linked together by disulfide bonds (Fig. 1). Heavy and light chains are encoded by separate genes and are organized into discrete globular domains separated by short peptide segments. The amino-terminus end of both heavy and light chains is the antigen binding site and consists of one domain characterized by sequence variability (variable region or V) in both the heavy and light chains, called the $V_H$ and $V_L$ regions respectively. The rest of the molecule has a relatively constant (C) structure. The

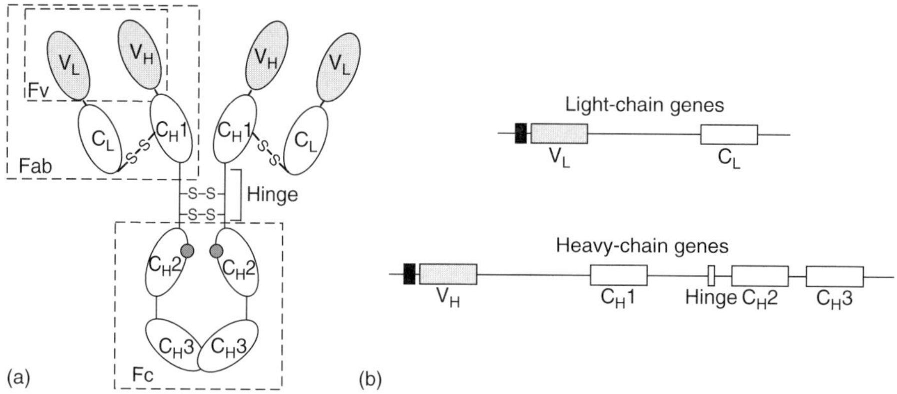


**Fig. 1** (a) Diagram of an immunoglobulin G (IgG) molecule (the most abundant antibody in serum) and the active fragments that can be derived from it. The antibody molecule is divided into discrete functional domains: two domains constitute the light chain ($V_L$ and $C_L$), while four domains make up the heavy chain ($V_H$, $C_H1$, $C_H2$, and $C_H3$). The variable region domains make the antibody binding site and are designated as the Fv region. The effector functions of the antibody are properties of the constant region domains. The carbohydrate units (black circles) present within the $C_H2$ domains contribute to the functional properties of the antibody. The hinge region provides flexibility to the antibody molecule, facilitating antigen binding and some effector functions. The enzyme papain cleaves the antibody into two Fab fragments containing the antigen binding sites and an Fc fragment responsible for the effector functions. (b) Genes that encode the heavy and light chains. In the genes, each domain is encoded by a discrete exon (indicated by boxes) separated by intervening sequences (introns) indicated by the line; the intervening sequences are present in the primary transcript but are removed from the mature mRNA by splicing. Both heavy and light chains contain hydrophobic leader sequences (indicated by the black exon) necessary for their secretion. This leader sequence is present in the newly synthesized heavy and light chains but is cleaved from them after they enter the endoplasmic reticulum and, therefore, is not present in the mature chains.

constant region of the light chain is termed the $C_L$ region. The constant region of the heavy chain is further divided into three structural domains stabilized by intrachain disulfide bonds: $C_H1$, $C_H2$, $C_H3$ (Fig. 1). The $C_H3$ domain of the heavy chain represents its carboxy-terminus. The domain structure of the antibodies is very important for genetic engineering because it facilitates protein engineering, allowing the exchange between molecules of functional domains carrying antigen-binding activities (Fabs or Fvs) or effector functions (Fc).

The hinge region, a segment of heavy chain between the $C_H1$ and $C_H2$ domains, provides flexibility in the molecule. Papain digestion of IgG yields two Fab fragments and one Fc fragment. The Fab region binds antigen, while the Fc region mediates effector functions such as complement activation, antibody-dependent cellular cytotoxicity (ADCC), and placental transmission. All antibodies are glycoproteins, and the carbohydrate present in the constant region has been shown to be essential for many of its effector functions.

## 1.2
## Classes and Subclasses of Antibodies

The constant region of the heavy chain determines the class or isotype of the

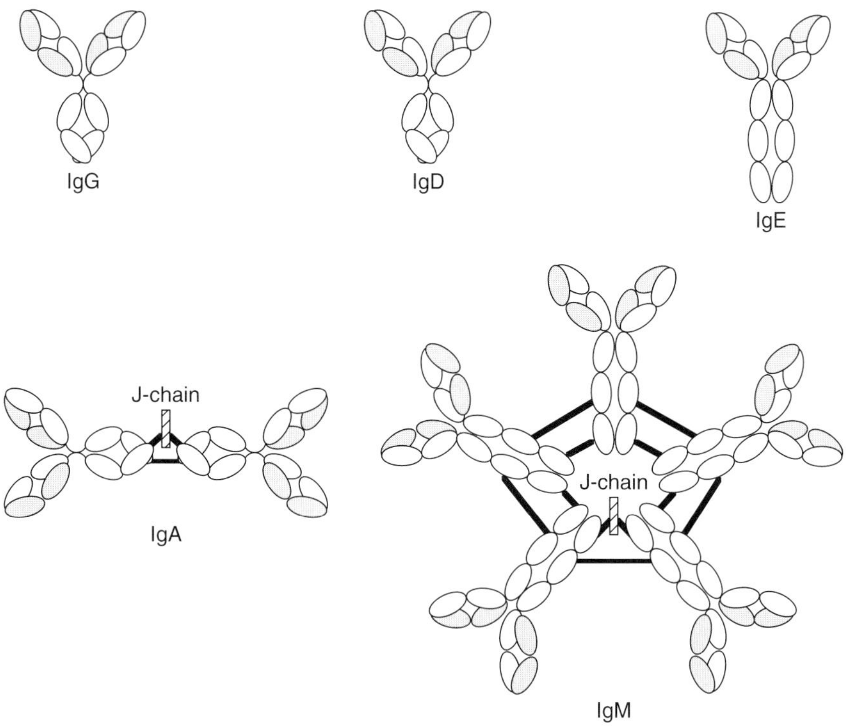


**Fig. 2** The five classes of antibodies: IgG, IgD, and IgE are monomeric antibodies; IgA and IgM are polymeric antibodies.

antibody. Figure 2 shows the five different classes of antibodies, which differ in the constant region of their respective heavy chains. The different heavy-chain isotypes that define these classes of antibodies are designated by lower-case Greek letters: $\gamma$ for IgG, $\delta$ for IgD, $\varepsilon$ for IgE, $\alpha$ for IgA, and $\mu$ for IgM. IgM and IgE heavy chains contain an extra $C_H$ domain ($C_H4$) and lack the hinge region found in IgG, IgD, and IgA. However, the absence of a hinge region does not imply that IgM and IgE lack flexibility; electron micrographs of IgM molecules binding to ligands have shown that the Fab arms can bend, relative to the Fc fragment. There are also two different light-chain isotypes, which are designated by the lower-case Greek letters $\kappa$ and $\lambda$. Light chains of both isotypes are found associated with all the heavy-chain isotypes.

In humans, there is only one class of IgD, IgE, and IgM with monomeric structures having a molecular weight (MW) of around 190 kDa. However, human IgA antibodies can be further subdivided in two subclasses (IgA1 and IgA2) with a molecular weight of around 160 kDa for each monomer, and human IgG antibodies can be subdivided into four subclasses: IgG1, IgG2, IgG4 (all three with a molecular weight of around 150 kDa) and IgG3 with a molecular weight of around 165 kDa. The higher molecular weight exhibited by human IgG3 is due to the presence of an extended hinge region, which provides extraordinary flexibility. In fact, IgG3 is the most flexible human IgG.

Although all antibody molecules are constructed from the basic unit of two heavy and two light chains ($H_2L_2$), both IgA and IgM form polymers (Fig. 2). IgA forms a dimeric structure in which two $H_2L_2$ units are joined by a J-chain [$(H_2L_2)_2J$], and IgM forms either a pentameric structure with five $H_2L_2$ units joined by a J-chain [$(H_2L_2)_5J$] or a hexameric structure [$(H_2L_2)_6$], which does not have the J-chain (not shown). IgM and IgA heavy-chain constant regions contain a "tailpiece" of 18 amino acids that contains a cysteine residue essential for polymerization. The J-chain is a 15 kDa polypeptide produced by B-lymphocytes and plasma cells (the same cells that produce the antibodies) that promotes polymerization by linking to the cysteine of the tailpiece (Fig. 2). Differences in the isotype of the heavy chain determine the number of carbohydrate units and the ability to engage in various effector functions such as complement activation, ADCC, and placental transmission. Differences in the isotype of the light chain do not appear to significantly influence the structure or the effector functions of the antibody molecule.

## 2 From Mouse to Human Antibodies

### 2.1 Murine Monoclonal Antibodies

During the normal immune response, a wide variety of antibodies is produced. These antibodies, known as *polyclonal antibodies* (i.e. they are the product of many different antibody-producing cells), include antibodies with different variable regions as well as antibodies with the same variable regions associated with different constant regions. Rarely do different individuals mount an identical immune response. This heterogeneity in the immune response plus ethical and safety concerns has made it difficult to use polyclonal antibodies for many applications.

A significant breakthrough was made when it became possible to produce stable cell lines that synthesize a single homogeneous antibody (Fig. 3). By fusing a normal B-cell from the splenocytes of an immunized animal (initially a mouse or a rat) with a myeloma cell, it is possible to generate a "hybridoma," which possesses the immortality of the myeloma cell and secretes the antibody characteristic of the normal B-cell. Antibodies produced by hybridomas are monoclonal (i.e. they are the product of a single antibody-producing cell) and therefore have a single variable region associated with only one constant region. The immortality of the hybridoma ensures the continued availability of a well-characterized antibody. Once a hybridoma cell line is developed, it can be grown *in vitro* or *in vivo* for large-scale production of the monoclonal antibody, or it can be used to clone the variable regions of the monoclonal antibody for genetic engineering purposes.

Owing to their high affinity and exquisite specificity, murine monoclonal antibodies seemed to be the ideal "magic bullets" for diagnosis or therapy of multiple diseases including cancer. However, the progression of "magic bullets" from dream to reality has been slow because mouse (murine) monoclonal antibodies are not the ideal agents to be administered into a human. Murine monoclonal antibodies compared to human antibodies require more frequent dosing to maintain a therapeutic level of monoclonal antibodies because of a shorter circulating half-life in humans than human antibodies. In addition, the

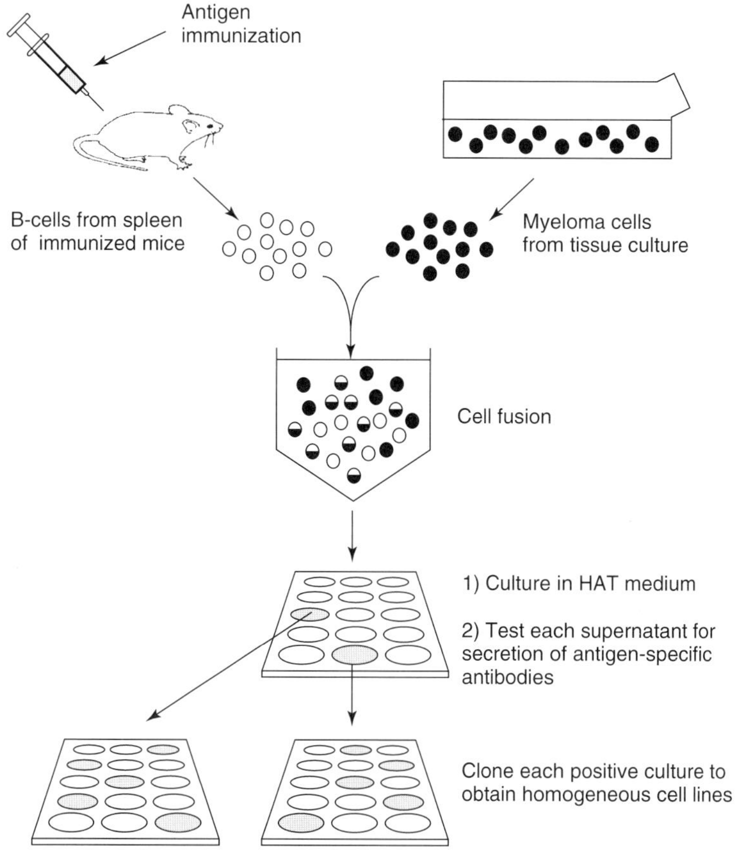


**Fig. 3** Production of monoclonal antibodies. Mice are immunized with the antigen of interest and spleen cells are isolated and then fused with myeloma cells. HAT (hypoxanthine, aminopterin, and thymidine) medium is used to separate unfused myeloma cells from fused hybridoma cells. It takes advantage of the fact that normal mammalian cells can synthesize nucleotides by both a *de novo* and a salvage pathway, while the myeloma cells used have a defect in the salvage pathway. When the *de novo* pathway is blocked by aminopterin, cells must then utilize the enzymes of the salvage pathway. Thus, in the presence of aminopterin, unfused myeloma cells will die. Normal spleen cells do not grow. Only cells that have acquired the enzymes of the salvage pathway from the normal cells and the capacity for continuous growth from the myeloma cell will survive. Supernatants of hybridomas are screened for the presence of the antibody with the desired specificity, and positive cell cultures are subcloned to obtain a homogeneous cell line.

human immune system recognizes the mouse protein as foreign, generating a human antimouse antibody (HAMA) response, which results in an even more rapid clearance of the murine antibody (rendering the therapeutic useless) and, in some cases, in a severe allergic reaction. Moreover, murine constant regions can be ineffective in interacting with the human immune effector system. These problems would be overcome by producing human monoclonal antibodies using human B-cells from immunized human donors. However, human monoclonal antibodies are difficult to produce using the hybridoma technology originally designed for the production of murine antibodies: the cell lines are unstable and frequently produce antibodies of the IgM isotype, which have low affinity for the antigen and are difficult to purify and handle. In addition, there are ethical and safety problems in obtaining humans immunized with certain antigens.

### 2.2 Chimeric Antibodies

To overcome the problems associated with the administration of murine monoclonal antibodies to humans, protein engineering has been used to convert murine monoclonal antibodies to mouse/human chimeric antibodies by genetically fusing the mouse variable regions to the human constant regions (Fig. 4), a procedure that is facilitated by the structure of antibodies.

Initially, variable region genes were obtained from genomic or cDNA libraries produced using DNA or mRNA from antibody-producing cells. More recently, cloning of genes encoding for the antibody variable region has been greatly facilitated by polymerase chain reaction (PCR)-based procedures. The variable region consists of hypervariable complementarity determining regions CDRs that are responsible for antibody specificity supported by relatively conserved framework regions. A limited number of different hydrophobic leader sequences are also found associated with the different variable regions. Therefore, it is possible to design sets of oligonucleotide primers on the basis of either the framework or the leader regions that will bind to virtually all mouse variable regions. A limited number of primers are required for all constant regions. Using upstream consensus primers for framework or leader regions in light- and heavy-chain variable regions and downstream primers for the constant regions, PCR can be used to amplify cDNAs generated by reverse-transcription (RT-PCR) directly from hybridoma mRNA.

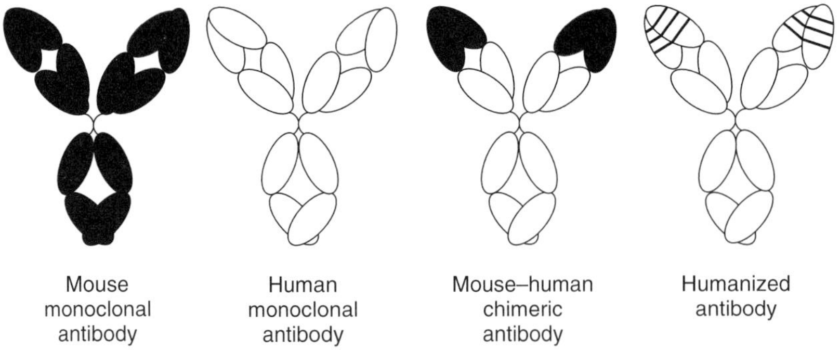


**Fig. 4** Schematic representation of a murine monoclonal antibody, a human monoclonal antibody, a mouse–human chimeric antibody, and a CDR-grafted or humanized antibody. Chimeric antibodies have murine derived variable regions and binding specificities joined to human constant regions with their corresponding effector functions. Humanized antibodies are composed mostly of human sequences except for the areas in contact with the antigen (CDR regions), which are derived from mouse sequences.

This product is cloned and sequenced, and then used for the construction of chimeric antibodies or for further genetic modifications, as described below.

For the most part, chimeric antibodies retain their target specificity and show reduced HAMA responses. An example of a successful mouse–human chimeric antibody approved for clinical use is Rituximab (Rituxan/Mabthera), which targets the CD20 antigen and is now widely used to treat non-Hodgkin's lymphoma.

## 2.3
## Humanized Antibodies

Although mouse–human chimeric antibodies are less immunogenic than mouse antibodies, in some cases they can still elicit a significant human antichimeric antibody (HACA) response. One approach to overcome this problem is to further manipulate the antibody variable region encoding for the antigen binding site resulting in humanized antibodies (Fig. 4). Each variable domain consists of a $\beta$-barrel with seven antiparallel $\beta$-strands connected by loops. Among the loops are the CDR regions. It is feasible to move the CDRs and their associated specificity from one scaffolding $\beta$-barrel to another, thereby creating "CDR-grafted" or "humanized" antibodies. However, it is rarely sufficient to move only the CDRs from a murine antibody onto a completely human framework because the resulting antibody frequently has reduced or no binding activity. In these cases, additional mouse residues near the CDRs are incorporated until binding is restored.

An example of a successful humanized monoclonal antibody that has been approved for clinical use is Trastuzumab (Herceptin), which has demonstrated significant antitumor activity in patients affected with breast cancer overexpressing the tumor-associated antigen HER2/*neu*.

## 2.4
## Human Monoclonal Antibodies in Mice

Recently, mice have been produced that make antigen-specific antibodies that are totally human. To accomplish this goal, transgenic mice carrying portions of the human IgH and Ig$\kappa$ loci [in germ line configuration using megabase-sized YACs (yeast artificial chromosome)] were obtained that included the majority of the variable region repertoire, the genes for C$\mu$, C$\delta$ and C$\gamma$1, C$\gamma$2, or C$\gamma$4, as well as the *cis* elements required for their function. The IgH and Ig$\kappa$ transgenic was then bred into a genetic background deficient in the production of murine immunoglobulin. Therefore, the resulting mouse model, named *XenoMouse*, has elements of the human heavy- and light-chain loci in a murine context in which endogenous mouse heavy and light chains were disrupted.

The large and complex human variable region repertoire encoded on the immunoglobulin transgenes in XenoMouse strains support the development of large peripheral B-cell compartments and the generation of a diverse primary immune repertoire similar to that of adult humans. The human genes are compatible with mouse enzymes mediating class switching from IgM to IgG as well as somatic hypermutation and affinity maturation. Importantly, the immune system of the XenoMouse recognizes human antigens as foreign, with a concomitant strong human humoral immune response. The use of XenoMouse mice in conjunction with well-established hybridoma procedures (see Fig. 3) reproducibly results in human IgG monoclonal antibodies with high affinity for human antigens and suitable for repeated administration to humans. To date, these engineered mice can produce only some of the human isotypes, but further genetic modification of these mice promises to expand their potential.

## 3 *In Vitro* Antibody Production by Phage Libraries

The production of monoclonal antibodies frequently requires several injections of antigen, which can take weeks to months. Hybridoma construction and screening require additional time. In addition, the immune response is often biased toward certain "immunodominant" epitopes of the antigen, making it difficult or impossible to produce monoclonal antibodies with the specificity desired for a particular use. Moreover, the production of antibodies against antigens conserved among species may be difficult or impossible. Phage display, a technology that allows the expression of immunoglobulin genes in bacteriophages (viruses that infect bacteria) without the need for developing hybridomas, provides an alternative approach that overcomes many of these problems.

In phage display, the heavy and light V-gene obtained from the spleen of an immunized mouse or from the peripheral blood of a naive or immunized donor are expressed as Fab or single-chain Fv (scFv) on the surface of filamentous phages (f1, M13, and fd). Phage libraries can be generated using variable antibody gene repertoires from any species including humans, or even synthetic sequences. Figure 5 shows the development of scFv libraries using a human donor.

Both scFv and Fab fragments can be expressed on the surface of filamentous bacteriophages (f1, M13, and fd) as either single or multiple copies of the antibody of interest depending on the phage protein used for fusion. Expression of multiple copies facilitates the identification and isolation of low-affinity antibodies. When functional antibody V domains are displayed on the surface of filamentous phages, the resulting phages bind specifically to antigen, and rare phages can be isolated on the basis of their ability to bind antigen. Multiple rounds of enrichment consisting of binding to immobilized antigen, expanding the bound phage, and further enriching by again binding to immobilized antigen can yield specific phages even if the desired specificity was present on less than 1 in $10^6$ of the original phages in the library. The selected variable regions will generally have affinities similar to monoclonal antibodies and can be expressed as antibody fragments in the bacterium *Escherichia coli* or they can be used to produce complete antibodies and can be expressed in mammalian hosts. In addition, specific variable regions can be mutagenized and phages that express variable regions with increased affinity can be selected. The bacteriophage expression systems are designed to allow the genes encoding for heavy and light chains to undergo random combinations, which are tested for their ability to bind the desired antigen.

The decision of whether to produce scFv or Fab libraries depends partly on the intended use. The single gene format of the scFv is an advantage for the construction of fusion proteins such as "immunotoxins" (antibodies or antibody fragments fused to a toxin) or for targeted gene therapy approaches where the *scFv* gene is fused to a viral envelope protein gene. The use of scFv also appears to

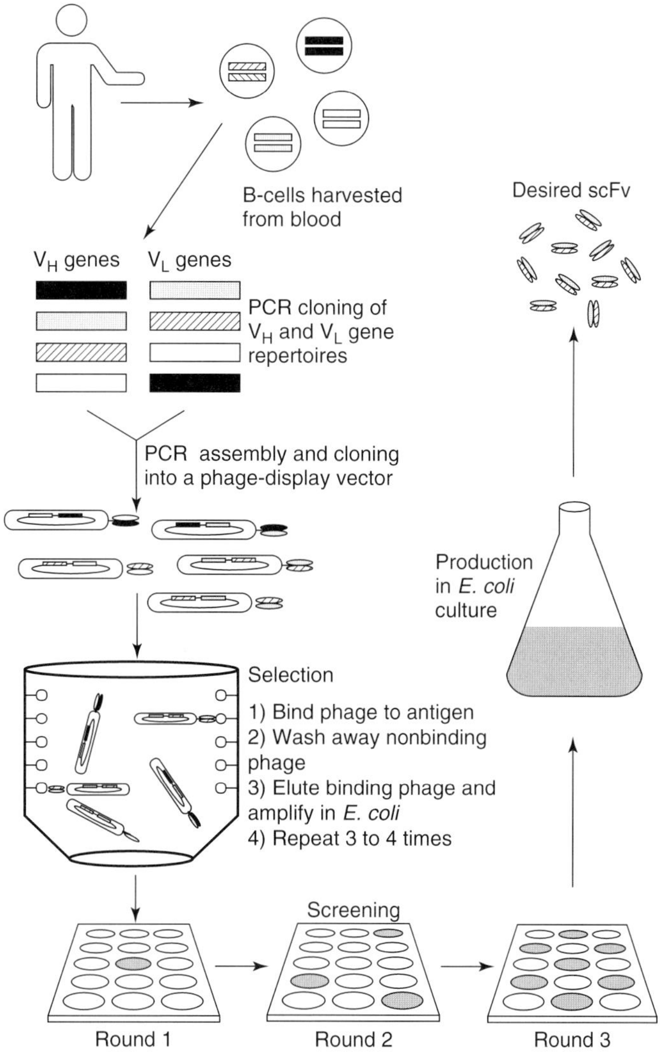


**Fig. 5** *In vitro* human antibody production using phage libraries (phage-display technology). Peripheral blood mononuclear cells are harvested from a nonimmunized or immunized human donor, and the heavy- and light-chain V-gene regions ($V_L$ and $V_H$) are amplified by PCR and assembled as a single-chain Fv region (scFv). Note that the original heavy- and light-chain pairings become scrambled during scFv assembly. The scFv genes are cloned into filamentous bacteriophages, where the encoded scFvs are displayed in a functional form on the phage surface. Multiple rounds of selection with a solid-phase antigen allow the isolation of even rare phages from the original library. The selected scFv can be expressed in *E. coli* (The illustration was adapted from Powers D.B. and Marks J.D. Monovalent Phage Display of Fab and scFv fusions. In *Antibody fusion proteins*. John Wiley & Son, Inc., New York, 1999).

be a better option for "intracellular immunization," a procedure in which the gene encoding the antibody binding site is delivered intracellularly to achieve phenotypic knockout. However, the use of Fab fragments offers the advantage that heavy- and light-chain libraries can be produced independently and reassorted. The use of two vectors also facilitates the construction of hierarchical libraries in which a fixed heavy or light chain is paired with a library of partners. In theory, Fab libraries may be preferred where the final product will be a complete antibody, since, in some instances, removal of the scFv linker might alter the antigen-binding properties. However, scFvs have been successfully converted into complete antibodies.

Phage display has proved to be very effective in generating a variety of antibodies that are difficult to obtain using the hybridoma technology. For preexisting mouse monoclonal antibodies, phage display can be used to obtain antibodies that are entirely human in sequence, but which bind to the same part of the antigen (epitope) as the mouse monoclonal antibodies. Antibodies to targets previously inaccessible using immunization approaches (for example, self-antigens, ubiquitous compounds, or toxic compounds) have been isolated by selection of phage on antigen *in vitro*. Building on the concept of phage display, alternative display strategies such as ribosome display and cell-surface display have been recently developed.

## 4 Further Genetic Modifications of Antibodies

### 4.1 Engineering Antibody Fragments: Monovalent, Bivalent, and Multivalent scFvs

As explained in Sect. 3, scFv fragments consist of the variable domains of heavy and light chains ($V_L$ and $V_H$) genetically fused by a flexile linker peptide. scFvs are much smaller than intact antibodies (Fig. 6) (25–27 kDa vs around 150 kDa for intact IgG), yet can retain the specificity and the affinity of the parental molecule.

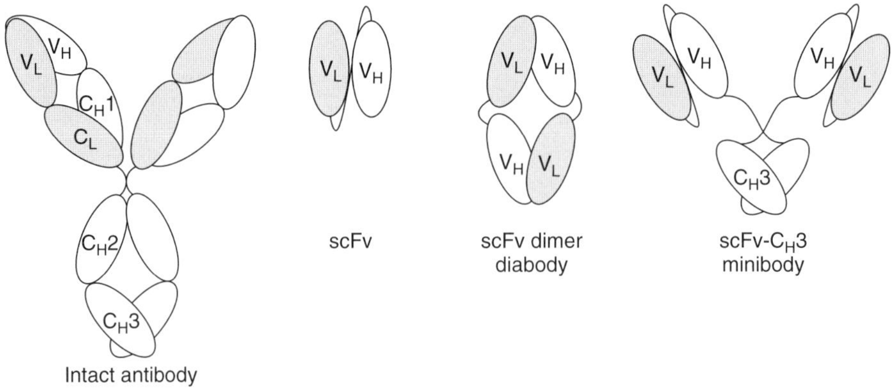


**Fig. 6** Schematic representation showing the relative size and domain relationships between intact IgG (around 150 kDa) and engineered single-chain Fv (scFv, 25–27 kDa), scFv dimer or diabody (55 kDa), and minibody (scFv-$C_H3$, 80 kDa) fragments.

The small size of scFvs is a tremendous advantage in certain applications such as tumor targeting for detection and/or therapeutic purposes. In fact, scFv fragments display a combination of rapid, high-level tumor targeting with concomitant clearance from normal tissues and circulation that have made radiolabeled scFvs important tools for detection and treatment of cancer metastasis in both preclinical models and patients. An example of a successful scFv in the clinic is MFE-23. This scFv is specific for carcinoembryonic antigen (CEA), a glycoprotein that is highly expressed in colorectal adenocarcinomas. MFE-23, expressed in bacteria, has been used in two clinical trials: a gamma camera imaging trial using $^{123}$I-MFE-23 and a radioimmunoguided surgery trial using $^{125}$I-MFE-23, in which tumor deposits are detected by a handheld probe during surgery. Both trials showed that MFE-23 is safe and effective in localizing tumor deposits in patients with cancer.

Despite their successful use in some tumor-targeting studies, a significant limitation of using scFvs for targeting *in vivo* is their monovalent binding to antigen. Intact antibodies have significant avidity as a result of the presence of two antigen binding sites. To address this problem, antibody engineers have used the monomeric scFv as a building block for larger engineered fragments. One approach consists of producing dimers of scFv by incorporating a carboxy-terminal cysteine residue so that a disulfide bridge forms, yielding $(scFv)_2$ fragments. Alternately, the single-chain concept has been extended by using an additional linker peptide to join the two scFv molecules in tandem. Another approach for the production of scFv dimers results from the observation that the use of a very short linker peptide to connect the antibody variable regions caused the formation of "cross-paired" dimers, in which the $V_L$ of one molecule associates with the $V_H$ of a second and the $V_L$ of the second molecule associates with the $V_H$ of the first (Fig. 6). These noncovalent dimers, also known as *diabodies*, are capable of bivalent binding to antigen. scFv dimers and diabodies have a molecular weight similar to that of the antibody Fab fragments (55–60 kDa) but contain two antigen binding sites. Diabodies show significant improvement in tumor targeting compared to monovalent scFv. scFv fragments can also be fused to an immunoglobulin $C_H3$ domain, resulting in a self-assembling bivalent "minibody" (Fig. 6).

To further increase the avidity of antibody fragments, several laboratories have generated fragments with an increased valence. One strategy to develop fragments with increased valence has been to extend the diabody approach by decreasing the length of the interdomain linker peptide, which may result in the formation of tribodies and tetrabodies. It is also possible to obtain larger bivalent or multivalent fragments by fusing scFvs to protein domains normally involved in protein association such as helix bundles or leucine zippers. An alternative approach for producing multivalent fragments has been the fusion of scFvs to the bacterial protein streptavidin. Since streptavidin is a tetramer composed of four noncovalently linked monomers, four scFvs assemble to form a tetrameric structure with four antigen binding sites. A practical example of this technology was the development of Rh-specific scFv fused to streptavidin. This fusion protein named scFv::strep was able to directly agglutinate antigen-positive red blood cells (a reaction that is impossible to achieve by using antigen-specific monomeric scFv or IgG), suggesting the potential use of scFv::strep as a blood-typing reagent.

### 4.2 Bispecific Antibodies

In contrast to normal antibodies, which are monospecific, bispecific antibodies contain binding sites with two different specificities. Antibodies with two different specificities have been prepared by chemical modification to combine univalent fragments of different pepsin-treated antibodies, by fusing two hybridomas secreting antibodies of different specificities yielding a quadroma and by joining two different single-chain antibodies (scFvs). However, chemical modification of antibodies is inefficient and can lead to side reactions that damage the combining site. In a quadroma, it is difficult to separate the desired bispecific antibodies from the mixed population of heavy and light chains produced by the two hybridomas, and scFvs lacking constant regions also lack the antibody effector functions, which may be critical in certain applications. To address these problems, novel bispecific antibodies have been produced in which an scFv of one specificity has been genetically fused after the hinge (hinge-scFv) or at the carboxy-terminus ($C_H3$-scFv) of an antibody with a different specificity. Both fusion proteins were expressed by gene transfection in the context of a murine variable region. Transfectomas secreted a homogeneous population of the recombinant antibody with the two different specificities, one at the amino-terminus (anti-dextran) and the other at the carboxy-terminus (anti-dansyl). The $C_H3$-scFv antibody, which maintains the constant region of human IgG3, has some of the associated effector functions such as long half-life and Fc receptor binding. As expected, hinge-scFv antibody, which lacks the $C_H2$ and $C_H3$ domains, has no known effector functions.

Bispecific antibodies provide potential tools for use in immunotherapy. They take advantage of the great specificity of variable regions for their antigens and can be envisioned as transporters of therapeutic drugs or molecules or even immune effector cells to the specific targets identified by one of their binding sites. In fact, bispecific antibodies have been shown to be beneficial in the recruitment of immune cells for the treatment of cancer. Depending on the effector/target interaction, bispecific antibodies enhance cytotoxicity and phagocytosis. Several of these antibodies are currently undergoing evaluation in phase I and II clinical trials. Among the most extensively studied bispecific antibodies are 2B1 and MDX-210. 2B1 produced from a hybrid hybridoma is specific for the tumor-associated antigens HER2/*neu* and Fc$\gamma$RIII (CD16, the Fc gamma receptor expressed by key cytotoxic effector cells such as natural killer cells, neutrophils, and activated mononuclear phagocytes), while MDX-210 is a chemically conjugated hetero F(ab)′2 fragment specific for HER2/*neu* and Fc$\gamma$RI (CD64, the Fc gamma receptor expressed by key cytotoxic effector cells such as monocytes, macrophages, and IFN-$\gamma$–activated granulocytes). Both antibodies showed therapeutic promise in late-stage cancer patients whose tumor was refractory to conventional therapy. Since conventionally prepared bispecific antibodies have already shown promise in clinical trials and results from preclinical studies of recombinant bispecific antibodies are encouraging, in the future recombinant bispecific antibodies will undoubtedly be used in multiple clinical trials.

### 4.3 Polymers of Monomeric Antibodies

Although IgM is a naturally occurring polymeric antibody (see Fig. 2), for several applications it is desirable to have other classes of antibodies in an IgM-like (polymeric) structure. For example, IgM does not bind the Fc receptors on phagocyte cells, whereas IgG effectively triggers effector functions mediated through gamma-specific Fc receptors. In addition, since in IgG both heavy and light variable regions are usually somatically mutated (because of isotype switching to IgG), it is expected that polymeric IgG may be of higher intrinsic and functional affinity than the currently available IgM, resulting in more sensitivity and/or specificity. In addition, many methods such as protein-A binding are available that facilitate the isolation of IgG.

Vectors have been developed for the construction and expression of human polymeric IgG. It was observed that the 18–amino acid carboxyl-terminal tailpiece from human $\mu$ heavy chain is sufficient for polymer assembly. This finding was exploited to produce IgM-like polymers of IgG by fusing the 18–amino acid carboxyl-terminal tailpiece from human $\mu$ to the carboxyl-terminal of $\gamma$ constant regions (Fig. 7). Using this technique, IgM-like polymers of IgG1, IgG2, IgG3, and IgG4 have been produced. IgGs obtained by this approach possess up to 6 Fcs and 12 antigen-combining sites, greatly increasing the avidity of their interactions with other molecules. These polymeric antibodies possess the Fc$\gamma$ receptor–binding properties of IgG. Not surprisingly, the complement activity of normally active IgG1 and IgG3 and somewhat less-active IgG2 antibodies is dramatically enhanced upon polymerization. An unexpected result is that IgG4, normally devoid of complement activity, when polymerized in the same fashion directs complement-mediated lysis of target cells almost as effectively as the other polymers. These experiments demonstrate that polymerization of monomeric antibodies such as IgG is an effective approach to obtain antibodies with broader and more powerful effector functions than their wild-type counterparts.

An alternative strategy to make polymers of IgG is to genetically fuse chicken

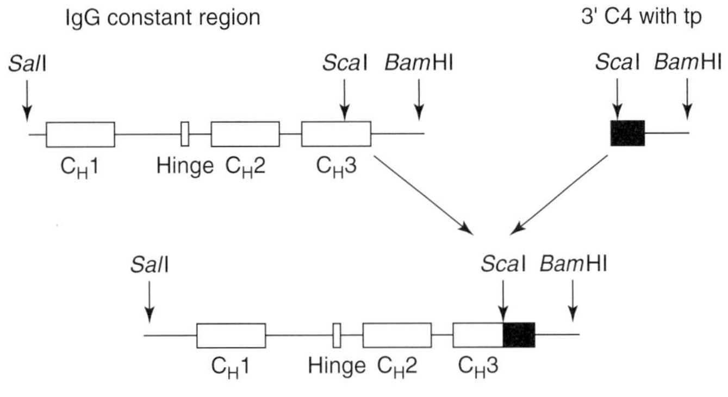


**Fig. 7** Strategy for the construction of polymeric IgG. Using appropriate restriction sites, the $\mu$ tailpiece ($\mu$tp) of human IgM is genetically fused to the end of the heavy chain of human IgG.

avidin to the carboxy-terminus of the heavy chain of IgG. This approach is similar to that described in Sect. 4.1 (production of tetramers of scFv by fusing it with streptavidin) based on the fact that both streptavidin and avidin are tetramers of four noncovalently linked monomers. Since each antibody-avidin protein contains two molecules of avidin (one genetically fused at the carboxy-terminus of each heavy chain), two independent antibody fusion proteins bind to each other through their respective avidins forming a dimeric structure. One example of this approach is a recently developed human IgG3-avidin fusion protein specific for the transferrin receptor (TfR). The anti-TfR IgG3-avidin was able to function as a universal vector to deliver different biotinylated compounds into cancer cells overexpressing the TfR. Furthermore, it was unexpectedly discovered that anti-TfR IgG3-avidin, but not a recombinant anti-TfR IgG3 or a nonspecific IgG3-avidin, possesses a strong antiproliferative/proapoptotic activity against hematopoietic malignant cell lines. Studies confirmed that anti-TfR IgG3-avidin exists as a dimer, suggesting that cross-linking of the surface transferrin receptor may be responsible for the cytotoxic activity. These findings demonstrate that it is possible to transform an antibody specific for a growth factor receptor that does not exhibit inhibitory activity into a novel drug with significant intrinsic cytotoxic activity against selected cells by fusing it with avidin. The antitumor activity may be enhanced by delivering biotinylated therapeutics into cancer cells. Further development of this technology may lead to effective therapeutics for *in vivo* eradication of hematological malignancies and *ex vivo* purging of cancer cells in autologous transplantation.

## 4.4 Antibody Fusion Proteins

Fusion proteins with nonantibody molecules fused to antibodies can be produced using different approaches. Antibody fusion proteins that contain an intact antigen binding site should retain the ability to bind antigen, while the attached nonantibody partner should be able to exert its function. Such molecules, which have been called *immunoligands*, can be produced in several different ways (Fig. 8 a–f). When the nonantibody partner is fused to the end of the $C_H3$ domain ($C_H3$-ligand) (Fig. 8a), the antibody-combining specificity can be used to deliver an associated biological activity as well as antibody-related effector functions. An example is the anti-TfR IgG3-avidin fusion protein described in Sect. 4.3. Other examples are antibodies targeting cancer cells fused with interleukin-2 and GM-CSF. The goal of this approach to cancer therapy is to concentrate the cytokine in the tumor microenvironment and, by so doing, enhance the tumoricidal effect of the antibody and/or the host immune response against the tumor, while limiting severe toxic side effects associated with a high dose of cytokine administration. Such antibody–cytokine fusion proteins have shown significant antitumor activity in mice bearing tumors, leading to clinical trials. Immunoligands with the nonantibody partner fused immediately after the hinge (H-ligand) (Fig. 8b) or to the $C_H1$ domain ($C_H1$-ligand) (Fig. 8c) may be useful when the antibody-related effector functions are unnecessary or harmful. In addition, for many applications such as tumor targeting, the small size of the H-ligand and $C_H1$-ligand may be an advantage over the larger $C_H3$-ligand.

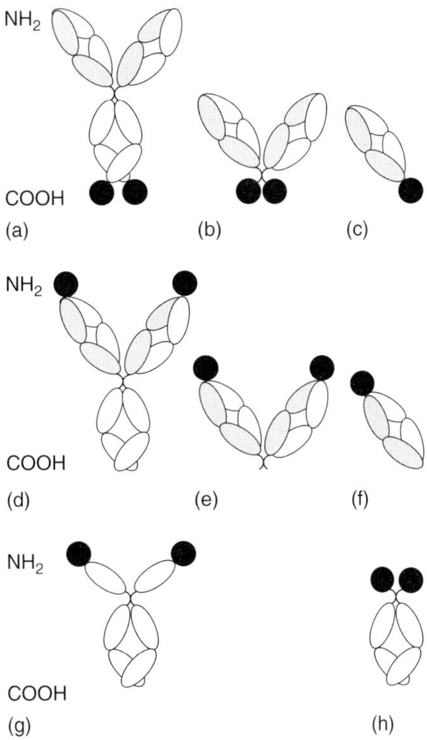


**Fig. 8** Schematic representation of antibody fusion proteins. (a) to (c) represent different antibody fusion proteins in which the nonantibody partner was fused at the carboxy-terminus after the $C_H3$ domain (a), immediately after the hinge (b), or after the $C_H1$ domain (c). (d) to (f) represent antibody fusion proteins in which the nonantibody partner has been joined to the amino-terminus of the full-length heavy chain (d) or the truncated heavy chain (e and f). (g) and (h) represent two fusion proteins with the nonantibody partner fused to the amino-terminus of the $C_H1$ domain (g) or immediately before the hinge (h).

An alternative approach is to construct antibody fusion proteins with the ligand fused to the amino-terminus of the heavy chain (Fig. 8 d–f). This may be necessary for proteins that require N-terminal processing or folding for activity such as nerve growth factor (NGF), the costimulatory molecule B7.1, and interleukin-12 (IL-12). In fact, antibody-NGF, antibody-(B7.1), and antibody-(IL-12) fusion proteins containing the ligand fused to the amino-terminus of the antibody retain both the ability to bind antigen and the activity of the nonantibody partner.

Nonantibody sequences can also be used to replace the $V_H$ domain or the $V_H-C_H1$ domains (Fig. 8 g and h). These molecules, which lack the ability to bind antigen, have been called *immunoadhesins* because they contain an adhesive molecule linked to the immunoglobulin Fc effector domains. In these proteins, the fused moiety acquires antibody-associated properties such as effector functions or improved pharmacokinetics. An example is the tumor necrosis factor (TNF) receptor IgG fusion protein, which binds to TNF (a mediator of inflammation) and neutralizes its activity. In fact, this molecule has been demonstrated to be efficacious for the treatment of rheumatoid arthritis.

Although Fig. 8 shows the nonantibody partner fused to the heavy chain, it should be appreciated that the nonantibody partner can also be fused to the light chain. It is also possible to construct antibody fusion proteins that combine more than one kind of nonantibody partner at the amino- or carboxy-termini of the heavy and/or light chains. In addition, a nonantibody molecule can be fused to an scFv molecule as described in Sect. 4.1. Moreover, two antibody Fc fragments can be genetically fused resulting in a molecule with novel biological properties. An example of this approach is the protein GE2, which is the product of the fusion between the Fc fragments of human IgG and human IgE ($Fc\gamma$–$Fc\varepsilon$ fusion protein). GE2 does not have an Fab and, as a consequence, is unable to target an antigen. However, GE2 was able to form complexes with both $Fc\gamma$RII and $Fc\varepsilon$RI, resulting in an inhibition of mast cell and basophil activation that results in the blocking of the anaphylaxis in transgenic mice expressing human $Fc\varepsilon RI\alpha$. This approach has therapeutic potential in IgE- and $Fc\varepsilon$RI-mediated diseases such as allergic asthma, allergic rhinitis, chronic urticaria, angioedema, and anaphylaxis.

## 5 Expression Systems

A large variety of expression systems have been used for the production of genetically engineered antibodies and antibody fragments. These expression systems include bacteria, yeast, plants, baculovirus, and mammalian cells.

Antibody fragments are commonly expressed in bacteria and yeast. The bacterium *E. coli* is a frequently used expression system owing to its rapid growth and easy genetic manipulation. However, proteins expressed in *E. coli* are frequently insoluble and/or inactive, and refolding may be required to obtain functional fragments. Secretion of fragments into the bacterial periplasm or culture supernatant provides an alternative means to obtain the desired functional fragments without the need for refolding. Although the results are highly antibody-dependent, there are many examples in which the bacterial expression system is successful.

Complete functional antibodies have been most successfully expressed in mammalian cells, as these cells possess the mechanisms required for correct immunoglobulin assembly, posttranslational modification (glycosylation), and secretion. Posttranslational modifications can influence the biologic properties and effector functions, important considerations especially when the antibody is to be used for therapy. Examples of mammalian cells that have been successfully used to express properly assembled and glycosylated antibodies and antibody fusion proteins are the mouse myeloma cell lines P3X63Ag8.653, Sp2/0-Ag14, and NS0/1. These three myeloma cell lines have lost the ability to produce endogenous H and L chains and are derived from the parent myeloma. Antibodies produced in nonlymphoid cell lines such as Chinese hamster ovary (CHO), HeLa, C6, and PC12 are also properly assembled and glycosylated. Owing to the slower growth of mammalian cells, mammalian expression requires a longer time frame and higher costs than bacterial or yeast expression, but it is preferred when complete functional antibodies with proper glycosylation and disulfide bonds are required. Other expression systems that have been extensively used to produce complete functional antibodies include insect cells and plants. There is no "universal" expression system – each system has its advantages and disadvantages.

## 6 Conclusion

Rapid progress has been made in producing genetically engineered antibodies. The ability to express foreign DNA in a variety of host cells has made it possible to produce chimeric, humanized, and human antibodies as well as antibodies with novel structures and functional properties in quantities sufficiently large for many applications including clinical therapy. The available experience suggests that antibody-based therapies can be successfully developed for use in clinical situations in which no alternative effective therapy is available. However, continued progress in the development of antibody-based therapies will require extensive research to further define the mechanism of antibody action and on how to optimally use the novel proteins with unique functional properties.

## Bibliography

### Books and Reviews


Carter, P. (2001) Improving the efficacy of antibody-based cancer therapies, *Nat. Rev. Cancer* **1**, 118–29.

Chamow, S.M., Ashkenazi, A. (Eds.) (1999) *Antibody fusion proteins*, John Wiley & Son, New York.

Helguera, G.F., Morrison, S.L., Penichet, M.L. (2002) Antibody-cytokine fusion proteins: harnessing the combined power of cytokines and antibodies to cancer therapy, *Clin. Immunol.* **105**, 233–246.

Janeway, C.A., Travers, P., Walport, M., Shlomchik, M. (Eds.) (2001) The Generation of Lymphocyte Antigen Receptors, *Immunobiology: The Immune System in Health and Disease*, 5th edition, Garland Publisher, New York, pp. 123–154.

Kriangkum, J., Xu, B., Nagata, L.P., Fulton, R.E., Suresh, M.R. (2001) Bispecific and bifunctional single chain recombinant antibodies, *Biomol. Eng.* **18**, 31–40.

Rader, C., Barbas, C.F. (1997) Phage display of combinatorial antibody libraries, *Curr. Opin. Biotechnol.* **8**, 503–508.

Sensel, M.G., Coloma, M.J., Harvill, E.T., Shin, S.U., Smith, R.I., Morrison, S.L. (1997) Engineering novel antibody molecules, *Chem. Immunol.* **65**, 129–158.

Verma, R., Boleti, E., George, A.J. (1998) Antibody engineering: comparison of bacterial, yeast, insect and mammalian expression systems, *J. Immunol. Methods* **216**, 165–181.

Wu, A.M., Yazaki, P.J. (2000) Designer genes: recombinant antibody fragments for biological imaging, *Q. J. Nucl. Med.* **44**, 268–283.

Yoo, E.M., Chintalacharuvu, K.R., Penichet, M.L., Morrison, S.L. (2002) Myeloma expression system, *J. Immunol. Methods* **261**(1–2), 1–20.


### Primary Literature


Abramowicz, D., Crusiaux, A., Goldman, M. (1992) Anaphylactic shock after retreatment with OKT3 monoclonal antibody, *N. Engl. J. Med.* **327**, 736.

Atwell, J.L., Breheney, K.A., Lawrence, L.J., McCoy, A.J., Kortt, A.A., Hudson, P.J. (1999) scFv multimers of the anti-neuraminidase antibody NC10: length of the linker between VH and VL domains dictates precisely the transition between diabodies and triabodies, *Protein Eng.* **12**, 597–604.

Bajorin, D.F., Chapman, P.B., Wong, G.Y., Cody, B.V., Cordon-Cardo, C., Dantes, L., Templeton, M.A., Scheinberg, D., Oettgen, H.F., Houghton, A.N. (1992) Treatment with high dose mouse monoclonal (anti-GD3) antibody R24 in patients with metastatic melanoma, *Melanoma Res.* **2**, 355–362.

Barbas, C.F., Bain, J.D., Hoekstra, D.M., Lerner, R.A. (1992) Semisynthetic combinatorial antibody libraries: a chemical solution to the diversity problem, *Proc. Natl. Acad. Sci. U.S.A.* **89**, 4457–4461.

Biocca, S., Pierandrei-Amaldi, P., Campioni, N., Cattaneo, A. (1994) Intracellular immunization with cytosolic recombinant antibodies, *Biotechnology (NY)* **12**, 396–399.

Boulianne, G.L., Hozumi, N., Shulman, M.J. (1984) Production of functional chimaeric mouse/human antibody, *Nature* **312**, 643–646.

Challita-Eid, P.M., Penichet, M.L., Shin, S.-U., Mosammaparast, N., Poles, T.M., Mahmood, K., Slamon, D.L., Morrison, S.L.,

Rosenblatt, J.D. (1997) A B7.1-Ab fusion protein retains antibody specificity and ability to activate via the T cell costimulatory pathway, *J. Immunol* **160**, 3419–3426.

Chester, K.A., Bhatia, J., Boxer, G., Cooke, S.P., Flynn, A.A., Huhalov, A., Mayer, A., Pedley, R.B., Robson, L., Sharma, S.K., Spencer, D.I., Regent, R.H. (2000) Clinical applications of phage-derived sFvs and sFv fusion proteins, *Dis. Markers* **16**, 53–62.

Chester, K.A., Mayer, A., Bhatia, J., Robson, L., Spencer, D.I., Cooke, S.P., Flynn, A.A., Sharma, S.K., Boxer, G., Pedley, R.B., Begent, R.H. (2000) Recombinant anti-carcinoembryonic antigen antibodies for targeting cancer, *Cancer Chemother. Pharmacol.* **46**, S8–S12.

Coloma, M.J., Hastings, A., Wims, L.A., Morrison, S.L. (1992) Novel vectors for the expression of antibody molecules using variable regions generated by polymerase chain reaction, *J. Immunol. Methods* **152**, 89–104.

Coloma, M.J., Morrison, S.L. (1997) Design and production of novel tetravalent bispecific antibodies [see comments], *Nat. Biotechnol.* **15**, 159–163.

Dangl, J.L., Wensel, T.G., Morrison, S.L., Stryer, L., Herzenberg, L.A., Oi, V.T. (1988) Segmental flexibility and complement fixation of genetically engineered chimeric human, rabbit and mouse antibodies, *EMBO J.* **7**, 1989–1994.

Davis, A.C., Roux, K.H., Pursey, J., Shulman, M.J. (1989) Intermolecular disulfide bonding in IgM: effects of replacing cysteine residues in the mu heavy chain, *EMBO J.* **8**, 2519–2526.

Davis, A.C., Roux, K.H., Shulman, M.J. (1988) On the structure of polymeric IgM., *Eur. J. Immunol.* **18**, 1001–1008.

Davis, A.C., Shulman, M.J. (1989) IgM–molecular requirements for its assembly and function, *Immunol. Today* **10**, 127, 128.

Dubel, S., Breitling, F., Fuchs, P., Zewe, M., Gotter, S., Welschof, M., Moldenhauer, G., Little, M. (1994) Isolation of IgG antibody Fv-DNA from various mouse and rat hybridoma cell lines using the polymerase chain reaction with a simple set of primers, *J. Immunol. Methods* **175**, 89–95.

Dubel, S., Breitling, F., Kontermann, R., Schmidt, T., Skerra, A., Little, M. (1995) Bifunctional and multimeric complexes of streptavidin fused to single chain antibodies (scFv), *J. Immunol. Methods* **178**, 201–209.

Ernst, M., Meier, D., Sonneborn, H.H. (1999) From IgG monoclonals to IgM-like molecules, *Hum. Antibodies* **9**, 165–170.

Figini, M., Marks, J.D., Winter, G., Griffiths, A.D. (1994) In vitro assembly of repertoires of antibody chains on the surface of phage by renaturation, *J. Mol. Biol.* **239**, 68–78.

George, A.J., Titus, J.A., Jost, C.R., Kurucz, I., Perez, P., Andrew, S.M., Nicholls, P.J., Huston, J.S., Segal, D.M. (1994) Redirection of T cell-mediated cytotoxicity by a recombinant single- chain Fv molecule, *J. Immunol.* **152**, 1802–1811.

Glennie, M.J., Brennand, D.M., Bryden, F., McBride, H.M., Stirpe, F., Worth, A.T., Stevenson, G.T. (1988) Bispecific F(ab' gamma)2 antibody for the delivery of saporin in the treatment of lymphoma, *J. Immunol.* **141**, 3662–3670.

Glennie, M.J., McBride, H.M., Worth, A.T., Stevenson, G.T. (1987) Preparation and performance of bispecific F(ab' gamma)2 antibody containing thioether-linked Fab' gamma fragments, *J. Immunol.* **139**, 2367–2375.

Goldenberg, M.M. (1999) Trastuzumab, a recombinant DNA-derived humanized monoclonal antibody, a novel agent for the treatment of metastatic breast cancer, *Clin. Ther.* **21**, 309–318.

Green, N.M. (1990) Avidin and streptavidin, *Methods Enzymol.* **184**, 51–67.

Green, L.L. (1999) Antibody engineering via genetic engineering of the mouse: XenoMouse strains are a vehicle for the facile generation of therapeutic human monoclonal antibodies, *J. Immunol. Methods* **231**, 11–23.

Griffiths, A.D., Malmqvist, M., Marks, J.D., Bye, J.M., Embleton, M.J., McCafferty, J., Baier, M., Holliger, K.P., Gorick, B.D., Hughes-Jones, N.C., Hoogboom, H., Winter, H. (1993) Human anti-self antibodies with high specificity from phage display libraries, *EMBO J.* **12**, 725–734.

Hanes, J., Jermutus, L., Weber-Bornhauser, S., Bosshard, H.R., Plückthun, A. (1998) Ribosome display efficiently selects and evolves high-affinity antibodies *in vitro* from immune libraries, *Proc. Natl. Acad. Sci. U.S.A.* **95**, 14130–14135.

He, M., Taussig, M.J. (1997) Antibody-ribosome-mRNA (ARM) complexes as efficient selection particles for *in vitro* display and evolution of antibody combining sites, *Nucleic Acids Res.* **25**, 5132–5134.

Higuchi, K., Araki, T., Matsuzaki, O., Sato, A., Kanno, K., Kitaguchi, N., Ito, H. (1997) Cell display library for gene cloning of variable regions of human antibodies to hepatitis B surface antigen, *J. Immunol. Methods* **202**, 193–204.

Holliger, P., Prospero, T., Winter, G. (1993) "Diabodies": small bivalent and bispecific antibody fragments, *Proc. Natl. Acad. Sci. U.S.A.* **90**, 6444–6448.

Hu, S., Shively, L., Raubitschek, A., Sherman, M., Williams, L.E., Wong, J.Y., Shively, J.E., Wu, A.M. (1996) Minibody: a novel engineered anti-carcinoembryonic antigen antibody fragment (single-chain Fv-CH3) which exhibits rapid, high-level targeting of xenografts, *Cancer Res.* **56**, 3055–3061.

Hudson, P.J., Kortt, A.A. (1999) High avidity scFv multimers; diabodies and triabodies, *J. Immunol. Methods* **231**, 177–189.

Iliades, P., Kortt, A.A., Hudson, P.J. (1997) Triabodies: single chain Fv fragments without a linker form trivalent trimers, *FEBS Lett.* **409**, 437–441.

Jaffers, G.J., Fuller, T.C., Cosimi, A.B., Russell, P.S., Winn, H.J., Colvin, R.B. (1986) Monoclonal antibody therapy. Anti-idiotypic and non-anti-idiotypic antibodies to OKT3 arising despite intense immunosuppression, *Transplantation* **41**, 572–578.

James, N.D., Atherton, P.J., Jones, J., Howie, A.J., Tchekmedyian, S., Curnow, R.T. (2001) A phase II study of the bispecific antibody MDX-H210 (anti-HER2 × CD64) with GM-CSF in HER2+ advanced prostate cancer, *Br. J. Cancer* **85**, 152–156.

Jespers, L.S., Roberts, A., Mahler, S.M., Winter, G., Hoogenboom, H.R. (1994) Guiding the selection of human antibodies from phage display repertoires to a single epitope of an antigen, *Biotechnology (NY)* **12**, 899–903.

Kohler, G., Milstein, C. (1975) Continuous cultures of fused cells secreting antibody of predefined specificity, *Nature* **256**, 495–497.

Kontsekova, E., Kolcunova, A., Kontsek, P. (1992) Quadroma-secreted bi(interferon alpha 2–peroxidase) specific antibody suitable for one-step immunoassay, *Hybridoma* **11**, 461–468.

Kortt, A.A., Lah, M., Oddie, G.W., Gruen, C.L., Burns, J.E., Pearce, L.A., Atwell, J.L., McCoy, A.J., Howlett, G.J., Metzger, D.W., Webster, R.G., Hudson, P.J. (1997) Single-chain Fv fragments of anti-neuraminidase antibody NC10 containing five- and ten-residue linkers form dimers and with zero- residue linker a trimer, *Protein Eng.* **10**, 423–433.

Kostelny, S.A., Cole, M.S., Tso, J.Y. (1992) Formation of a bispecific antibody by the use of leucine zippers, *J. Immunol.* **148**, 1547–1553.

Kuus-Reichel, K., Grauer, L.S., Karavodin, L.M., Knott, C., Krusemeier, M., Kay, N.E. (1994) Will immunogenicity limit the use, efficacy, and future development of therapeutic monoclonal antibodies? *Clin. Diagn. Lab. Immunol.* **1**, 365–372.

Leget, G.A., Czuczman, M.S. (1998) Use of rituximab, the new FDA-approved antibody, *Curr. Opin. Oncol.* **10**, 548–551.

Lewis, L.D., Cole, B.F., Wallace, P.K., Fisher, J.L., Waugh, M., Guyre, P.M., Fanger, M.W., Curnow, R.T., Kaufman, P.A., Ernstoff, M.S. (2001) Pharmacokinetic-pharmacodynamic relationships of the bispecific antibody MDX-H210 when administered in combination with interferon gamma: a multiple-dose phase-I study in patients with advanced cancer which overexpresses HER-2/neu, *J. Immunol. Methods* **248**, 149–165.

Liu, S.J., Sher, Y.P., Ting, C.C., Liao, K.W., Yu, C.P., Tao, M.H. (1998) Treatment of B-cell lymphoma with chimeric IgG and single-chain Fv antibody-interleukin-2 fusion proteins, *Blood* **92**, 2103–2112.

Lloyd, F. Jr., Goldrosen, M. (1991) The production of a bispecific anti-CEA, anti-hapten (4-amino-phthalate) hybrid-hybridoma, *J. Natl. Med. Assoc.* **83**, 901–904.

Mallender, W.D., Voss, E.W. Jr. (1994) Construction, expression, and activity of a bivalent bispecific single-chain antibody, *J. Biol. Chem.* **269**, 199–206.

Marasco, W.A. (1997) Intrabodies: turning the humoral immune system outside in for intracellular immunization, *Gene Ther.* **4**, 11–15.

Marks, J.D., Hoogenboom, H.R., Bonnert, T.P., McCafferty, J., Griffiths, A.D., Winter, G. (1991) By-passing immunization. Human antibodies from V-gene libraries displayed on phage, *J. Mol. Biol.* **222**, 581–597.

McGrath, J.P., Cao, X., Schutz, A., Lynch, P., Ebendal, T., Coloma, M.J., Morrison, S.L.,

Putney, S.D. (1997) Bifunctional fusion between nerve growth factor and a transferrin receptor antibody, *J. Neurosci. Res.* **47**, 123–133.

Milstein, C., Cuello, A.C. (1983) Hybrid hybridomas and their use in immunohistochemistry, *Nature* **305**, 537–540.

Moreland, L.W., Baumgartner, S.W., Schiff, M.H., Tindall, E.A., Fleischmann, R.M., Weaver, A.L., Ettlinger, R.E., Cohen, S., Koopman, W.J., Mohler, K., Widmer, M.B., Blosch, C.M. (1997) Treatment of rheumatoid arthritis with a recombinant human tumor necrosis factor receptor (p75)-Fc fusion protein [see comments], *N. Engl. J. Med.* **337**, 141–147.

Moreland, L.W., Schiff, M.H., Baumgartner, S.W., Tindall, E.A., Fleischmann, R.M., Bulpitt, K.J., Weaver, A.L., Keystone, E.C., Furst, D.E., Mease, P.J., Ruderman, E.M., Horwitz, D.A., Arkfeld, D.G., Garrison, L., Burge, D.J., Blosch, C.M., Lange, M.L., McDonnell, N.D., Weinblatt, M.E. (1999) Etanercept therapy in rheumatoid arthritis. A randomized, controlled trial, *Ann. Intern. Med.* **130**, 478–486.

Morrison, S.L., Johnson, M.J., Herzenberg, L.A., Oi, V.T. (1984) Chimeric human antibody molecules: mouse antigen-binding domains with human constant region domains, *Proc. Natl. Acad. Sci. U.S.A.* **81**, 6851–6855.

Ng, P.P., Dela Cruz, J.S., Sorour, D.N., Stinebaugh, J.M., Shin, S.U., Shin, D.S., Morrison, S.L., Penichet, M.L. (2002) An anti-transferrin receptor-avidin fusion protein exhibits both strong proapoptotic activity and the ability to deliver various molecules into cancer cells, *Proc. Natl. Acad. Sci. U.S.A.* **99**, 10706–10711.

Orlandi, R., Gussow, D.H., Jones, P.T., Winter, G. (1989) Cloning immunoglobulin variable domains for expression by the polymerase chain reaction, *Proc. Natl. Acad. Sci. U.S.A.* **86**, 3833–3837.

Pack, P., Plückthun, A. (1992) Miniantibodies: use of amphipathic helices to produce functional, flexibly linked dimeric FV fragments with high avidity in *Escherichia coli*, *Biochemistry* **31**, 1579–1584.

Parmley, S.F., Smith, G.P. (1988) Antibody-selectable filamentous fd phage vectors: affinity purification of target genes, *Gene* **73**, 305–318.

Peng, L.S., Penichet, M.L., Morrison, S.L. (1999) A single-chain IL-12 IgG3 antibody fusion protein retains antibody specificity and IL-12 bioactivity and demonstrates antitumor activity, *J. Immunol.* **163**, 250–258.

Penichet, M.L., Dela Cruz, J.S., Shin, S.U., Morrison, S.L. (2001) A recombinant IgG3-(IL-2) fusion protein for the treatment of human HER2/neu expressing tumors, *Hum. Antibodies* **10**, 43–49.

Penichet, M.L., Harvill, E.T., Morrison, S.L. (1998) An IgG3-IL-2 fusion protein recognizing a murine B cell lymphoma exhibits effective tumor imaging and antitumor activity, *J. Interferon Cytokine Res.* **18**, 597–607.

Plückthun, A., Pack, P. (1997) New protein engineering approaches to multivalent and bispecific antibody fragments, *Immunotechnology* **3**, 83–105.

Raso, V., Griffin, T. (1981) Hybrid antibodies with dual specificity for the delivery of ricin to immunoglobulin-bearing target cells, *Cancer Res.* **41**, 2073–2078.

Russell, S.J., Hawkins, R.E., Winter, G. (1993) Retroviral vectors displaying functional antibody fragments, *Nucleic Acids Res.* **21**, 1081–1085.

Shin, S.U., Wu, D., Ramanathan, R., Pardridge, W.M., Morrison, S.L. (1997) Functional and pharmacokinetic properties of antibody-avidin fusion proteins, *J. Immunol.* **158**, 4797–4804.

Shusta, E.V., Kieke, M.C., Parke, E., Kranz, D.M., Wittrup, K.D. (1999) Yeast polypeptide fusion surface display levels predict thermal stability and soluble secretion efficiency, *J. Mol. Biol.* **292**, 949–956.

Smith, G.P. (1985) Filamentous fusion phage: novel expression vectors that display cloned antigens on the virion surface, *Science* **228**, 1315–1317.

Smith, R.I., Coloma, M.J., Morrison, S.L. (1995) Addition of a mu-tailpiece to IgG results in polymeric antibodies with enhanced effector functions including complement-mediated cytolysis by IgG4, *J. Immunol.* **154**, 2226–2236.

Smith, R.I., Morrison, S.L. (1994) Recombinant polymeric IgG: an approach to engineering more potent antibodies, *Biotechnology (NY)* **12**, 683–688.

Somia, N.V., Zoppe, M., Verma, I.M. (1995) Generation of targeted retroviral vectors by using single-chain variable fragment: an approach to *in vivo* gene delivery, *Proc. Natl. Acad. Sci. U.S.A.* **92**, 7570–7574.

Tada, H., Kurokawa, T., Seita, T., Watanabe, T., Iwasa, S. (1994) Expression and characterization of a chimeric bispecific antibody against fibrin and against urokinase-type plasminogen activator, *J. Biotechnol.* **33**, 157–174.

Valone, F.H., Kaufman, P.A., Guyre, P.M., Lewis, L.D., Memoli, V., Deo, Y., Graziano, R., Fisher, J.L., Meyer, L., Mrozek-Orlowski, M., Wardwell, K., Guyre, V., Morley, T.L., Arvizu, C., Fanger, M.W. (1995) Phase Ia/Ib trial of bispecific antibody MDX-210 in patients with advanced breast or ovarian cancer that overexpresses the proto-oncogene HER-2/neu, *J. Clin. Oncol.* **13**, 2281–2292.

van Dijk, M.A., van de Winkel, J.G. (2001) Human antibodies as next generation therapeutics, *Curr. Opin. Chem. Biol.* **5**, 368–374.

Vaughan, T.J., Williams, A.J., Pritchard, K., Osbourn, J.K., Pope, A.R., Earnshaw, J.C., McCafferty, J., Hodits, R.A., Wilton, J., Johnson, K.S. (1996) Human antibodies with sub-nanomolar affinities isolated from a large non-immunized phage display library, *Nat. Biotechnol.* **14**, 309–314.

Verhoeyen, M., Milstein, C., Winter, G. (1988) Reshaping human antibodies: grafting an antilysozyme activity, *Science* **239**, 1534–1536.

Weiner, L.M., Clark, J.I., Ring, D.B., Alpaugh, R.K. (1995) Clinical development of 2B1, a bispecific murine monoclonal antibody targeting c-erbB-2 and Fc gamma RIII, *J. Hematother.* **4**, 453–456.

Weinstein, J.N., Eger, R.R., Covell, D.G., Black, C.D., Mulshine, J., Carrasquillo, J.A., Larson, S.M., Keenan, A.M. (1987) The pharmacology of monoclonal antibodies, *Ann. N.Y. Acad. Sci.* **507**, 199–210.

Zhu, D., Kepley, C.L., Zhang, M., Zhang, K., Saxon, A. (2002) A novel human immunoglobulin Fc gamma Fc epsilon bifunctional fusion protein inhibits Fc epsilon RI-mediated degranulation, *Nat. Med.* **8**, 518–521.

# Part IV
# Immunological Disorders

*Immunology. From Cell Biology to Disease.* Edited by Robert A. Meyers.

ISBN: 978-3-527-31770-7

# 13 Autoantibodies and Autoimmunity

*Kenneth Michael Pollard*
*The Scripps Research Institute, La Jolla, CA, USA*

1 **Autoantibodies and Autoimmunity** 346

2 **Autoantibodies as Diagnostic Markers** 350

3 **Autoantibodies as Molecular and Cellular Probes** 352

4 **Autoantibodies in Experimental Models of Autoimmunity** 358

5 **Perspectives** 364

**Bibliography** 364
Books and Reviews 364
Primary Literature 365

## Keywords

**Antibody**
A protein product of B cells that combines with a specific molecular target called an antigen.

**Antigen**
A substance that interacts with an antibody.

**Epitope**
The region of an antigen that is directly recognized by a specific antibody.

*Immunology. From Cell Biology to Disease.* Edited by Robert A. Meyers.

ISBN: 978-3-527-31770-7

**Immunogen**
A substance that elicits an antibody response.

**Indirect Immunofluorescence (IIF)**
A technique whereby an antibody is overlaid onto an antigen containing cellular substrate and the antigen–antibody formed complex is detected by a fluorescently labeled anti-antibody.

Autoimmunity is an immunological reaction against constituents of the organism that are normally tolerated by the immune system of that organism. Autoimmune reactions can be either cell- or antibody-mediated. Autoantibodies are therefore antibodies that recognize normally tolerated cell and tissue constituents (or autoantigens). The antigenic specificity of an autoantibody can be a useful aid in clinical diagnosis. Autoantibodies are either cell (or tissue) specific, as found in organ-specific autoimmune diseases such as autoimmune thyroiditis, or non–organ-specific and reactive with ubiquitous intracellular antigens, as found in multisystem autoimmune diseases such as the systemic rheumatic diseases. The latter group includes autoantibodies that recognize components of macromolecular complexes of nucleic acids and/or proteins such as small nuclear ribonucleoprotein (snRNP) particles, nucleosomal and subnucleosomal structures, and tRNA synthetases, which are intrinsic components of all cell types present in an organism. Autoantibodies also recognize components of subcellular structures, including mitochondria, ribosomes, Golgi apparatus, nuclear membrane, and substructures within the nucleus and nucleolus. The ability of autoantibodies to recognize components of the cellular machinery of replication, transcription, RNA processing, RNA translation, and protein processing has made them important reagents for isolating cDNA clones that code for proteins involved in these cellular processes and for probing the relationship between molecular and cellular structure and function. The evolutionarily conserved nature of many autoantigens allows the use of autoantibodies to identify their target antigens in diverse species, ranging in some cases from humans to lower eukaryotes such as yeast. Autoantibodies have been used to inhibit the biological function of autoantigens and/or to recognize autoantigens in a defined functional state.

## 1 Autoantibodies and Autoimmunity

An autoimmune response is an attack by the immune system on the host itself. In healthy individuals, the immune system is "tolerant" of its host ("self") but attacks foreign ("nonself") constituents such as bacteria and viruses. The ability to distinguish self from nonself is considered to be the determining factor in whether the immune system responds to a suspected challenge. Although it may appear obvious, there is actually considerable debate over what constitutes "self" and "nonself," and what cellular/molecular mechanisms are involved. Possible discriminators between "self" and "nonself" include recognition of infection or identification of danger signals. The outcome of the debate on self/nonself discrimination notwithstanding, autoimmunity represents an obvious disruption of the mechanism by which the immune system regulates its activities. Importantly, the responsible effector mechanisms appear to be no different from those used to combat exogenous infective reagents, and include soluble products such as antibodies (humoral immunity) as well as direct cell-to-cell contact resulting in specific cell lysis (cell-mediated immunity). No single mechanism has been described that can account for the diversity of autoimmune responses or the production of autoantibodies. Figure 1 outlines the common features of hypothetical models of autoantibody elicitation. Most models, particularly those relating to autoimmune disease in animals, include a genetic predisposition. Breeding experiments between inbred strains of mice have shown that the genetic control of autoantibody production is complex, involving multiple genes. Although most of the required genetic elements remain to be characterized, it appears that both acceleration and suppression of autoimmune responses are under genetic control. The most frequently observed genetic requirement involves the major histocompatibility complex (MHC) class II genes, which encode proteins responsible for the presentation of processed antigen to $CD4^+$ T cells via the T-cell receptor.

The most perplexing and challenging aspect of autoimmunity and autoantibody elicitation is the identification of the events involved in the initiation of the response. Although these early events are poorly understood for most autoimmune diseases, it is thought that an exogenous trigger can provide the first step in the initiation of some autoimmune responses. The best evidence for this comes from drug- and

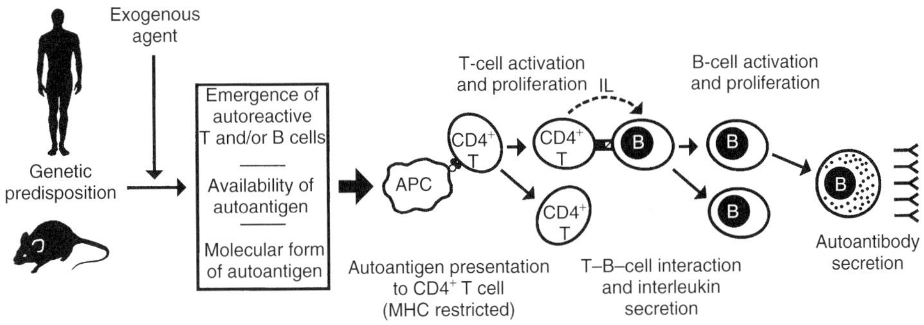


**Fig. 1** Hypothetical pathway of autoantibody elicitation in human disease and experimental animal models. This model combines features from the most commonly accepted postulated mechanisms for autoantibody production. Genetically predisposed individuals may be triggered to begin the response by an exogenous agent such as exposure to a drug, chemical toxin, or other environmental influence. The events that follow (listed in large box) are poorly understood but must involve the emergence of autoreactive lymphoid cells and the presence of autoantigen in a molecular form reactive with autoreactive cells. Once the presentation of autoantigen has activated autoreactive lymphoid cells, the production of autoantibody proceeds essentially as it would for a nonautoimmune antibody response.

chemical-induced autoimmunity, which has been described in both human disease and animal models of autoimmunity. However, even in exogenously induced autoimmunity, many of the events between the administration of a chemical or a drug and the appearance of autoantibodies remain to be unveiled. Induction of autoantibodies by exogenous agents can take from several weeks to many months. Drug-induced systemic autoimmunity in humans can take prolonged periods of time to develop and can be provoked by a large number of chemically unrelated drugs. The autoantibody response, however, appears quite restricted, targeting histones and histone–DNA complexes, the components of chromatin. Complexes of drug and autoantigen are not the immunogens responsible for the autoantibody response, since the drug is not required for autoantibody interaction with the autoantigen. Withdrawal of the drug often leads to cessation of clinical symptoms, clearly implicating the participation of the drug in some mechanism inciting the autoimmune response, although the autoantibody may persist for months in the absence of the drug. In several animal models, exposure to chemicals, particularly inorganic forms of heavy metals such as mercury, silver, or gold, can lead to autoantibody expression within weeks. In these murine models, the autoantibody response is again restricted, but here the predominant targets are nonchromatin components of the nucleolus. The development of restricted autoantibody specificities in humans given many different drugs or in mice given heavy metals suggests that it is not the parent molecule that is important but rather the metabolic products of these compounds that lead on the one hand to antichromatin autoantibodies and on the other to antinucleolar antibodies. In human drug-induced autoimmunity, a common pathway of oxidative metabolism via the ubiquitous neutrophil has been suggested as a means of producing reactive drug metabolites that may perturb immune regulation sufficiently to produce autoimmune disease. Another mechanism that has been proposed is disruption by drug metabolites of positive selection of T cells during their development in the thymus. This mechanism has been shown to result in mature $CD4^+$ T cells that are able to respond to self-antigen leading to T-cell proliferation as well as autoantibody production by B cells.

In Fig. 1, the large boxed area highlights several concepts that form pivotal points in many hypothetical postulates of autoantibody elicitation but about which little is known. How do B- and T cells, with receptors for autoantigen, emerge from and escape the regulatory mechanisms that normally keep them in check, then make their way to the secondary lymphoid tissues? Studies involving transgenic mice possessing neoautoantigens suggest that possible mechanisms include avoidance of apoptotic elimination, escape from tolerance induction, and reversal of an anergic state. Molecular identification of autoantigens, their presence in macromolecular complexes, the occurrence of autoantibodies in different components of the same complex, and the appearance of somatic mutations in the variable regions of autoantibodies have suggested that it is the autoantigen that drives the autoimmune response. It remains unclear how autoantigens, particularly intracellular autoantigens, are made available to autoreactive lymphoid cells, and what molecular forms of these complex macromolecular structures interact with autoreactive lymphoid cells. One mechanism that has been proposed as a means by which autoantigens might be made available to the immune system is apoptotic cell death. The impetus for this hypothesis is the finding that many autoantigens undergo proteolytic cleavage during apoptotic cell death and that apoptotic bodies (debris from dying cells) contain multiple autoantigens. Processing and presentation of such material by antigen-presenting cells (APC) has been suggested as a means of providing antigen to autoreactive T cells. However, uptake of apoptotic cellular material does not lead to the activation of APCs, which is necessary if APCs are to activate T cells. Inability of apoptotic material to activate APCs may stem from the observation that apoptosis is a descriptor for programmed cell death (PCD), which is a physiological process. This contrasts sharply with necrotic cell death, which is a nonphysiological process that produces cellular material that activates APCs. Also of note is that necrotic cell death induced by mercury leads to proteolytic cleavage of the autoantigen fibrillarin. Immunization with the N-terminal fragment of such cleavage leads to autoantibodies against fibrillarin that possess some of the characteristics of the antifibrillarin response elicited by mercury alone. In contrast, the antibody response elicited by immunization with full-length fibrillarin does not mimic the mercury-induced response, suggesting that processing and presentation of fragmented autoantigens may allow loss of self/nonself discrimination. Examination of the molecular forms of autoantigens during and after cell death and their roles in activating both APC and T cells will be fruitful areas of future research.

Roles in autoantibody production have been argued for pathways that either are or are not dependent on the presence of T cells. A T cell–dependent response is shown in Fig. 1 with an APC supplying processed antigen to $CD4^{+}$ T cells. An essential element in any model of autoantibody elicitation is the emergence of antibody-secreting B cells, which recognize material derived from the host. The antibody secreted by a B cell is directed against a single region (or epitope) on an antigen. An autoantibody response can target a number of epitopes on any one antigen, clearly showing that multiple autoreactive B-cell clones are activated during an autoimmune response. In the systemic autoimmune diseases, many autoantigens are complexes of nucleic acid and/or protein, and an autoimmune response may target several of the components of a complex. It is unknown whether the autoantibody responses to the components of a complex arise simultaneously, sequentially, independently, or through interrelated mechanisms.

In only a few diseases have autoantibodies been shown to be the causative agents of pathogenesis (e.g. antiacetylcholine receptor autoantibodies in myasthenia gravis, antithyroid stimulating hormone receptor autoantibodies in Graves' disease). It is noteworthy not only that these diseases are organ specific but also that their autoantigens are extracellular or on the surface of cell membranes and are therefore easily targeted by the immune system. In some individuals, the largest organ, the skin, can suffer insult from several blistering conditions now known to be autoimmune diseases characterized by autoantibodies against products of keratinocytes. The autoantigens involved are cell adhesion molecules that are important in maintaining the integrity of the skin by cell–cell contact between the various cell layers in the epidermis and at the dermal–epidermal junction. In the non–organ-specific autoimmune disease systemic lupus erythematosus (SLE), anti–double-stranded DNA (dsDNA) autoantibodies have been shown to participate in pathogenic events by way of complexing with their cognate antigen to cause immune complex–mediated inflammation. These examples show that in both the organ-specific and systemic autoimmune diseases, *in vivo* disposition of autoantibody in tissues and organs has clinical significance inasmuch as it indicates sites of inflammation, which may contribute to the pathological process. Moreover, detection of autoantibody deposits in the organ-specific autoimmune diseases has particular significance because some organ-specific autoantibodies have been found to be the direct mediators of pathological lesions. In most autoimmune diseases, however, it has not been determined whether autoantibodies cause or contribute to disease or are merely a secondary consequence of the underlying clinical condition.

## 2
## Autoantibodies as Diagnostic Markers

The diseases associated with autoantibodies can be divided into two broad groups: the organ-specific autoimmune diseases, in which autoantibodies have the ability to react with autoantigens from a particular organ or tissue, and the multisystem autoimmune diseases, in which autoantibodies react with common cellular components that appear to bear little relevance to the underlying clinical picture. In both cases, particular autoantibody specificities can serve as diagnostic markers (Table 1).

In the multisystem autoimmune diseases, there are several features of the relationship between autoantibody specificity and diagnostic significance that bear consideration. Autoantigens in these diseases are components of macromolecular structures such as the nucleosome of chromatin and the small nuclear ribonucleoprotein (snRNP) particles of the spliceosome, among others. Autoantibodies to different components of the same macromolecular complex can be diagnostic for different clinical disorders. Thus, the core proteins B, B′, D, and E, which are components of the U1, U2, and U4–U6 snRNPs and are antigenic targets in the anti-Smith antigen (Sm) response in SLE, are different from the U1 snRNP–specific proteins of 70 kDa, A and C, which are targets of the anti-nRNP response in mixed connective tissue disease (MCTD; see Table 1). It has also been observed that certain autoantibody responses are consistently associated with one another. The anti-Sm response, which is diagnostic of SLE, is commonly associated with the anti-nRNP response, but the anti-nRNP response can occur without the anti-Sm response, in which case it can be diagnostic of MCTD. These two observations suggest that the snRNP complexes responsible for the autoantibody response against the spliceosome in MCTD may differ from the snRNP complexes that produce the antispliceosome response in SLE. Other autoantibody responses demonstrate similar associations and restrictions. The anti-SS-A/Ro response (see Table 1) frequently occurs alone in SLE, but the anti-SS-B/La response in Sjögren's syndrome is almost always associated with the anti-SS-A/Ro response. Similarly, the antichromatin response occurs alone in drug-induced lupus but is usually associated with the anti-dsDNA response in idiopathic SLE.

**Tab. 1** Examples of clinical diagnostic specificity of autoantibodies.

| *Autoantibody specificity*[a] | *Molecular specificity* | *Clinical association* |
|---|---|---|
| *Organ-specific autoimmune diseases* | | |
| Antiacetylcholine receptor* | Acetylcholine receptor | Myasthenia gravis |
| Anti-TSH receptor* | TSH receptor | Graves' disease |
| Antithyroglobulin* | Thyroglobulin | Chronic thyroiditis |
| Antithyroid peroxidase* | Thyroid peroxidase | Chronic thyroiditis |
| Antimitochondria* | Pyruvate dehydrogenase complex | Primary biliary cirrhosis |
| Antikeratinoctye* | Desmoplakin I homologue | Bullous pemphigoid |
| Antikeratinoctye* | Desmoglien | Pemphigus foliaceus |
| *Multisystem autoimmune diseases* | | |
| Anti–double-stranded DNA* | B-form of DNA | SLE |
| Anti-Sm* | B, B′, D, and E proteins of U1, U2, U4–U6 snRNP | SLE |
| Anti-nRNP | 70 kDa, A and C proteins of U1-snRNP | MCTD, SLE |
| Anti-SS-A/Ro | 60 and 52 kDa proteins associated with hY1-Y5 RNP complex | SS, neonatal lupus, SLE |
| Anti-SS-B/La | 47 kDa phosphoprotein complexed with RNA polymerase III transcripts | SS, neonatal lupus, SLE |
| Anti-Jo-1* | Histidyl tRNA synthetase | Polymyositis |
| Antifibrillarin* | 34 kDa protein of box C/D containing snoRNP (U3, U8, etc.) | Scleroderma |
| Anti-RNA polymerase 1* | Subunits of RNA polymerase 1 complex | Scleroderma |
| Anti-DNA topoisomerase 1 (anti-Scl-70)* | 100 kDa DNA topoisomerase I | Scleroderma |
| Anti-centromere* | Centromeric proteins CENP-A, B, C | CREST (limited Scleroderma) |
| Canca | Serine proteinase (proteinase 3) | Wegener's vasculitis |

[a]Disease-specific diagnostic marker antibodies indicated by an asterisk.
*Notes*: SLE: systemic lupus erythematosus; MCTD: mixed connective tissue disease; SS: Sjögren's syndrome; cANCA: cytoplasmic antineutrophil cytoplasmic antibody; TSH: thyroid-stimulating hormone; CREST: calcinosis, Raynaud's phenomenon, esophageal dysmotility, sclerodactyly, telangiectasia.

Autoantibody specificities may occur at different frequencies in a variety of diseases, and the resultant profile consisting of distinct groups of autoantibodies in different diseases can have diagnostic use. In some cases, the grouping of autoantibody specificities, such as the preponderance of antinucleolar autoantibodies in scleroderma (Table 1), provides provocative but as yet little-understood relationships with clinical diagnosis. Unlike SLE, where a single patient may have multiple autoantibody specificities to a number of unrelated nuclear autoantigens (e.g. DNA, Sm, SS-A/Ro), scleroderma patients infrequently have multiple autoantibody specificities to nucleolar autoantigens that are unrelated at the macromolecular level (i.e. not part of the same macromolecular complex).

## 3
## Autoantibodies as Molecular and Cellular Probes

Autoantibodies can be used for the detection of their cognate antigens using immunoprecipitation, immunoblotting, enzyme-linked immunosorbent assay (ELISA), and a variety of microscopy techniques including immunoelectron microscopy. The most visually impressive demonstration of the usefulness of autoantibodies as biological probes is the indirect immunofluorescence (IIF) test. In this technique (Fig. 2), a cell or tissue source containing the autoantigen of interest is permeabilized, to allow entry of the antibody into the cell, and fixed, to ensure that the target antigen is not leached away during the procedure. Although some procedures are inappropriate for particular antigens, workable means of cell permeabilization and fixation have been developed. The cell substrate is incubated with the autoantibody to allow interaction with the antigen, and any excess is washed away. The location of the antigen/autoantibody complex within the cell is revealed by addition of an anti-antibody tagged with a fluorochrome. Fluorescence microscopy is then used to view the cells to determine the location of the antigen-/autoantibody-/fluorochrome-tagged anti-antibody complex. Using this technique, investigators are identifying an increasing number of autoantibody specificities that recognize cellular substructures and domains (Table 2, Figs. 3 and 4). The nucleus can be identified by a variety of autoantibodies such as those against chromatin and DNA or, as shown in Fig. 3(a), autoantibodies to the nuclear lamina, which underlies the nuclear envelope and produces a ringlike fluorescence around the nucleus. The nucleolus and its subdomains can also be identified by a variety of autoantibody specificities (Table 2). Autoantibodies against the 34 kDa protein fibrillarin, a component of the C/D box containing small nucleolar ribonucleoprotein (snoRNP) particles, label the nucleolus in a distinctive "clumpy" pattern (Fig. 3e). The list of autoantibodies that are able to distinguish subnuclear domains and compartments, some considerably smaller than the nucleolus, continues to grow. One example is the Cajal body, a small subnuclear structure described using light microscopy by the Spanish cytologist Santiago Ramon y Cajal in 1903 and subsequently named after him. This nuclear domain can now be easily identified using human autoimmune sera that react with p80 coilin (Fig. 3d), a protein highly enriched in the Cajal body. Using other autoantibodies, it has been found that Cajal bodies contain snRNP particles and fibrillarin (previously thought to be restricted to the nucleolus and prenucleolar bodies). Knowledge of the functional associations of these coiled-body constituents suggests that the Cajal body may play a role in RNA processing and/or in the accumulation of components involved in RNA processing.

Many features of subcellular structures such as size, shape, and distribution can be studied by IIF during the cell cycle, viral infection, mitogenesis, or any cellular response that may result in changes in the distribution of an antigen or a subcellular structure. As shown in Fig. 3(a) (arrowheads), antinuclear lamin autoantibodies can be used to reveal re-formation of the lamina during telophase. Autoantibodies

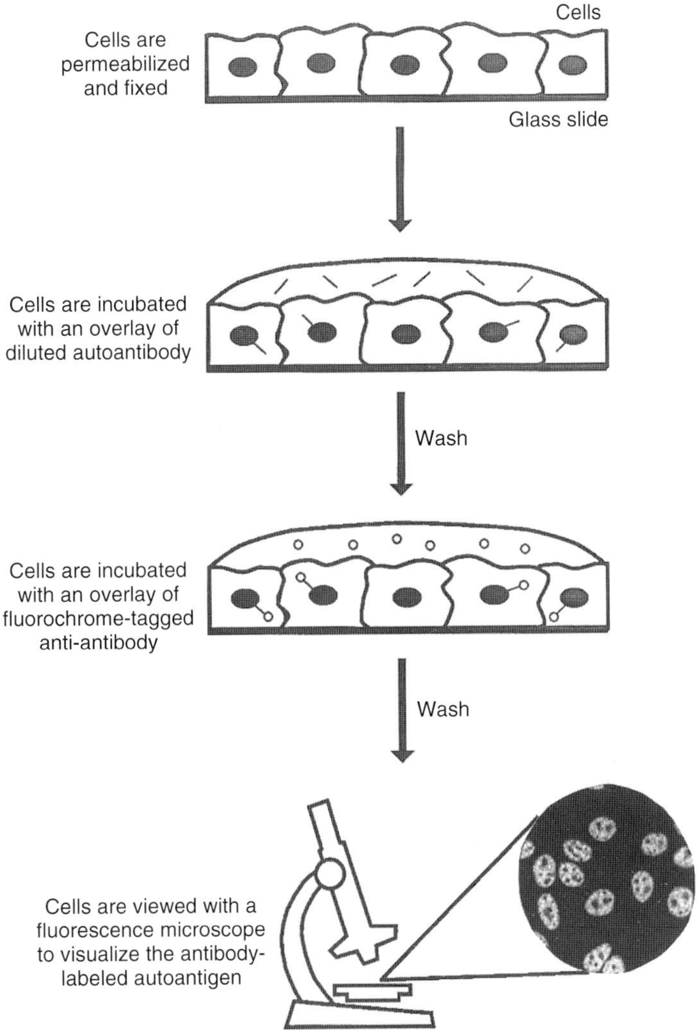
Cells
Cells are permeabilized and fixed
Glass slide
Cells are incubated with an overlay of diluted autoantibody
Wash
Cells are incubated with an overlay of fluorochrome-tagged anti-antibody
Wash
Cells are viewed with a fluorescence microscope to visualize the antibody-labeled autoantigen

**Fig. 2** Diagrammatic representation of the steps involved in the indirect immunofluorescence (IIF) test: see text for explanation.

have identified unexpected protein distributions such as the localization of the nucleolar protein fibrillarin to the outer surface of the chromosomes during cell division (Fig. 3e, arrowheads). The localization of some autoantigens during the cell cycle has aided in their identification. Detection of proliferating cell nuclear antigen (PCNA) in S-phase cells (Fig. 3c) suggested its involvement in DNA synthesis, while the distribution of speckles along the metaphase plate produced by other autoantibodies (Fig. 3f, arrowheads) was an important clue in their identification as autoantibodies to centromeric proteins A, B, and C.

The IIF test has also proved useful in the identification of autoantibodies

**Tab. 2** Examples of subcellular structures and domains recognized by autoantibodies.

| ***Autoantibody*** | ***Molecular specificity*** | ***Subcellular structure*** |
|---|---|---|
| *Nuclear components* | | |
| Antichromatin | Nucleosomal and subnucleosomal complexes of histones and DNA | Chromatin |
| Anti-nuclear pore | 210 kDA glycoprotein (gp210) | Nuclear pore |
| Antilamin | Nuclear lamins A, B, C | Nuclear lamina |
| Anticentromere | Centromere proteins (CENP) A, B, C, F | Centromere |
| Anti-p80 coilin | p80-coilin (80 kDa protein) | Coiled body |
| Anti-PIKA | p23–25 kDa proteins | Polymorphic interphase karyosomal association (PIKA) |
| Anti-NuMA | 238 kDa protein | Mitotic spindle apparatus |
| *Nucleolar components* | | |
| Antifibrillarin | 34 kDa fibrillarin | Dense fibrillar component of nucleolus |
| Anti-RNA polymerase 1 | RNA polymerase 1 | Fibrillar center of nucleolus |
| Anti-Pm-Scl | 75 and 100 kDa proteins of Pm-Scl complex | Granular component of nucleolus |
| Anti-NOR 90 | 90 kDa doublet of (human) upstream binding factor (hUBF) | Nucleolar organizer region (NOR) |
| *Cytosolic components* | | |
| Antimitochondria | Pyruvate dehydrogenase complex | Mitochondria |
| Antiribosome | Ribosomal P-proteins ($P_0$, $P_1$, $P_2$) | Ribosomes |
| Anti-Golgi | 95 and 160 kDa golgins | Golgi apparatus |
| Antiendosome | 180 kDa protein | Early endosomes |
| Antimicrosomal | Cytochrome P450 superfamily | Microsomes |
| cANCA | Serine proteinase (proteinase 3) | Lysosomes |
| Antimidbody | 38 kDa protein | Midbody |
| Anticentrosome/centriole | Pericentrin (48 kDa) | Centrosome/centriole |

*Notes*: NuMA: nuclear mitotic apparatus; Pm-Scl: polymyositis-scleroderma; cANCA: cytoplasmic antineutrophil cytoplasmic antibody.

that react with subcellular structures other than the nucleus. Figure 4 shows the IIF patterns produced by some of these autoantibodies. Prior knowledge of subcellular organelles and their relative cellular distribution was instrumental in identifying the structures recognized by these and other autoantibodies. In turn, autoantibodies, by virtue of their reactivity with individual autoantigens, have allowed cell and molecular biologists an insight into the molecular constituents of these subcellular organelles.

Comparative studies using human autoantibodies, and nonhuman autoantibodies raised by immunization against specific cellular proteins, can be useful in determining the cellular distribution of the specific protein. This is achieved by using antihuman antibodies labeled with one chromophore and antibodies specific for the nonhuman antibody that

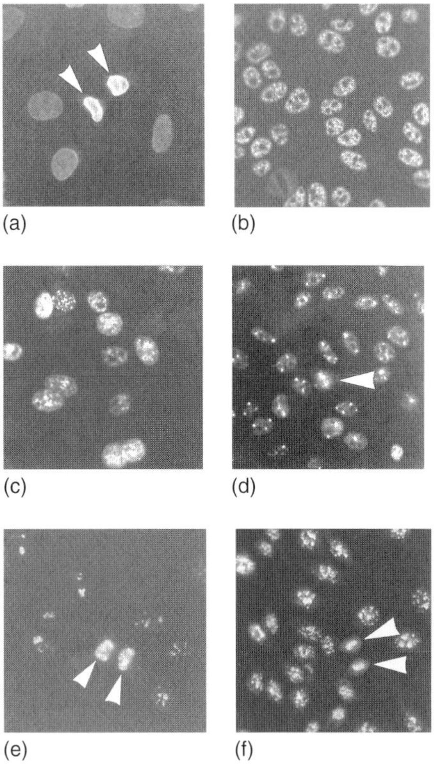


**Fig. 3** Immunofluorescence patterns produced by autoantibodies recognizing structural and functional domains within the cell nucleus (magnification, 350×). (a) Antinuclear lamin B1 antibodies identify the periphery of the nucleus; arrowheads show the re-formation of the nuclear envelope during late telophase. (b) Anti-Sm antibodies localize the U1, U2, and U4–U6 snRNP particles as a speckled nuclear pattern. (c) Anti-PCNA antibodies recognize the auxiliary protein of DNA polymerase delta during active DNA synthesis, producing different fluorescence patterns as cells progress through mitosis. (d) Anti-p80 coilin antibodies highlight subnuclear domains known as Cajal bodies, which disappear during metaphase (arrowhead). (e) Antifibrillarin antibodies target the nucleolus and produce a characteristic clumpy pattern in interphase cells, decorating the chromosomes from late metaphase until cell division (arrowhead). (f) Antibodies to centromeric proteins A, B, and C produce a discrete speckling of the interphase nucleus and identify the centromeric region of the dividing chromosomes during cell division (arrowheads).

are labeled with a different chromophore and by comparing the fluorescence patterns. Immunolocalization of the non-snRNP spliceosome component SC-35 was achieved in this way by comparison of anti-SC-35 antisera with the IIF pattern of autoimmune anti-Sm sera, which recognize protein components of the spliceosome.

One feature of autoantibodies that distinguishes them from antibodies raised by specific immunization, and underscores their uniqueness, is their ability to recognize their target antigen not only from the host but also from a variety of species. The extent of this species cross-reactivity is dependent on the evolutionary conservation of the autoantigen and is related to the conservation of protein sequence. One example is the snoRNP protein fibrillarin. Using autoantibodies in a variety of immunological techniques, including IIF, this protein can be recognized from species as diverse as humans and the unicellular yeast *Saccharomyces cerevisiae*. cDNA cloning of fibrillarin has confirmed the expected high degree of conservation of the protein sequence.

The reactivity of autoantibodies with conserved sequence and conformational protein elements has made them useful reagents in the cloning of cDNAs of expressed proteins from cDNA libraries from a variety of species. However, because of their reactivity with the human

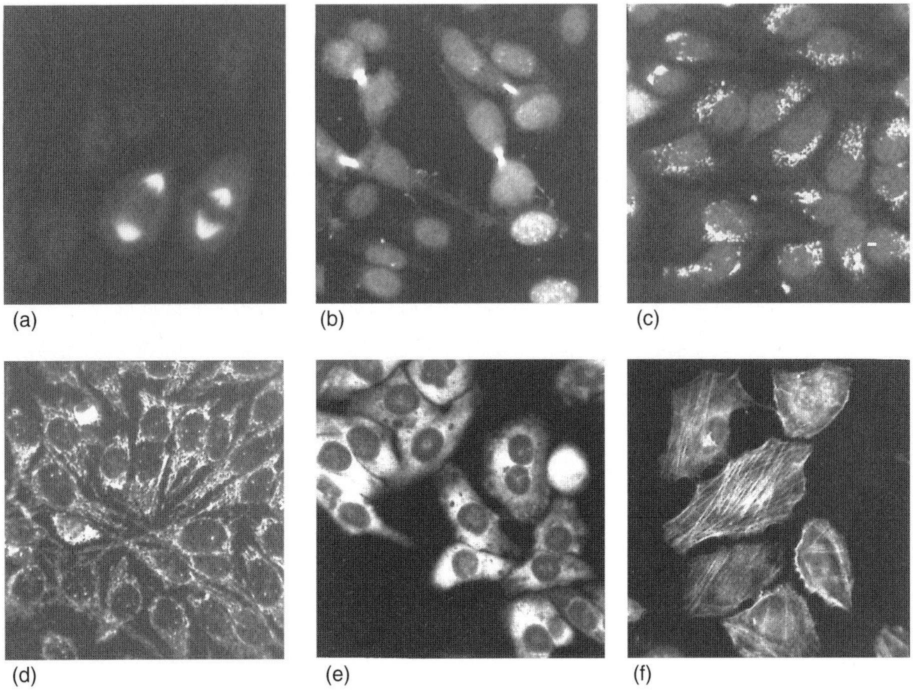

**Fig. 4** Immunofluorescence patterns produced by autoantibodies recognizing intracellular structures other than the nucleus. (a) Antimitotic spindle apparatus antibodies identify spindle poles and spindle fibers during cell division. (b) Antimidbody antibodies react with the bridgelike midbody that connects daughter cells following chromosome segregation but before cell separation. (c) Anti-Golgi complex antibodies decorate the Golgi apparatus, which in most cells is shown as an accumulation of fluorescence in a discrete cytoplasmic region. (d) Antimitochondrial antibodies demonstrate the presence of mitochondria throughout the cytoplasm; the discrete nuclear dots represent an additional autoantibody specificity in this serum unrelated to mitochondria. (e) Antiribosome antibodies produce a diffuse cytoplasmic staining pattern that spares the nucleus but may show some weak nucleolar fluorescence. (f) Anticytoskeletal antibodies react with a variety of cytoskeletal components; in this case, the antibody reacts with nonmuscle myosin. (magnification: a, 700×; b–f, 350×).

protein they have found most use in the cloning and characterization of the primary structure of numerous human cellular proteins. This diversity of targets that can be exploited by this approach is clearly illustrated in Tables 1 and 2.

Elucidation of the structure of the autoantigens that are the targets of autoantibodies from systemic autoimmune diseases has revealed that many are functional macromolecular complexes involved in nucleic acid and protein synthesis and processing (Table 3). A distinguishing feature of many of these complexes of nucleic acids and/or proteins is that autoantibodies do not recognize all the components of the complex. An extreme, but useful, example is the ribosome, which in eukaryotes may contain more than 70 proteins. However, only the P-proteins ($P_0$, $P_1$, and $P_2$), S10, and L12 are recognized by autoantibodies. Nonetheless,

**Tab. 3** Examples of the function of nuclear autoantigens and the effects of autoantibody on antigen function.

| *Autoantigen* | *Function* | *Autoantibody effect*[a] |
|---|---|---|
| | *Known function* | |
| Sm/nRNP (U1, 2, 4–6 snRNP) | Pre-mRNA splicing | Inhibition of pre-mRNA splicing |
| PCNA (DNA polymerase delta auxiliary protein) | DNA replication | Inhibition of DNA replication and repair |
| RNA polymerase I | Transcription of rRNA | Inhibition of rRNA transcription |
| tRNA polymerase | Aminoacylation of tRNA | Inhibition of charging of tRNA |
| Ribosomal RNP | mRNA translation | Inhibition of protein synthesis |
| Centromere/kinetochore | Microtubule-based chromosome movement during mitosis | Inhibition of centromere formation and function |
| | *Probable function* | |
| Fibrillarin (Box C/D snoRNP) | Processing and methylation of pre-rRNA | Blocks translocation of fibrillarin during the cell cycle, thereby influencing the ultrastructure of the nucleolus. |
| NOR-90 | Nucleolar transcription factor | Not tested |

[a] Inhibition of function has been demonstrated *in vitro* or following injection of autoantibody into living cells.

the use of autoantibodies that identify specific components of such complexes has aided in identifying other subunits of these complexes, with profound consequences. Thus, the initial identification of anti-Sm and anti-nRNP autoantibodies in SLE led to the observation that they recognize some of the protein components of the snRNP particles, fueling subsequent studies that showed the snRNPs as components of the spliceosome complex that functions in pre-mRNA splicing.

As the functional associations of autoantigens have become known, attempts to uncover the role of the autoantigen itself have revealed that autoantibodies can directly inhibit the function of their target autoantigen (Table 3). Although it remains to be determined, it seems likely that such inhibition reflects the involvement of a conserved protein sequence or structure in functional activity. An increasing number of autoantibodies, many of unknown molecular specificity, recognize their autoantigen only in a particular functional state or phase of the cell cycle. Of the several examples known, the best characterized is PCNA, which is the auxiliary protein of DNA polymerase delta and is recognized by autoantibodies only during mitosis, even though PCNA is present throughout the cell cycle. When a population of cells at different stages of the cell cycle is used as substrate in the IIF test, anti-PCNA autoantibodies produce varying degrees of fluorescence intensity, being negative for $G_0$ cells and highly positive for S-phase cells (Fig. 3c). These intriguing features of some autoantibodies have added new dimensions to the biological usefulness of these proteins and have suggested that functionally active macromolecular complexes may play a role in the elicitation of the autoantibody response.

The presence of multiple autoantibody specificities in the blood of individual human patients with autoimmune diseases poses a limitation on their use in studies involving a single autoantigen. Only infrequently are patients found whose autoantibody response is so restricted that they express autoreactivity to a single autoantigen or autoantigenic complex; such autoantibodies are termed *monospecific*. For some autoantibody specificities, this condition has been overcome by the production of hybridomas secreting a monoclonal autoantibody. Some hybridomas have been produced by fusion of B cells from human patients, but most have come from fusion of lymphoid cells from animal models of autoimmunity, particularly inbred murine strains. Monoclonal antibody specificities include reactivity against the nucleic acids DNA and RNA, subunits of chromatin, protein components of snRNP particles, fibrillarin, and immunoglobulin.

## 4 Autoantibodies in Experimental Models of Autoimmunity

Research into the mechanisms of autoimmunity and the antigenic specificity, and possible pathogenic role, of autoantibodies has been significantly advanced by the availability of animal models. Four different types of models have been used (Table 4). Specific antigen immunization models are produced by direct injection of purified antigen into animals to elicit autoantibody. Direct immunization has proven most useful when the autoantigen is extracellular or is on the cell membrane. In such examples, the elicited autoantibody response can produce pathological consequences such as the myasthenia gravis–like disease produced in rodents following immunization with purified acetylcholine receptor. The animals used in this type of model are most often healthy, normal individuals with fully functional immune systems and are able to downregulate the autoimmune response produced by the immunization of autoantigen. As a result, direct immunization models often produce transient autoimmune responses and the animals return to a healthy state.

Comparison of the autoantigenic reactivities of antibodies raised by immunization, especially to intracellular autoantigens, has revealed distinct differences in comparison to autoantibodies found in human autoimmune disease. Direct immunization requires a purified antigen, which means subjecting the antigen to rigorous biophysical, biochemical, and sometimes immunological separation techniques. The resulting preparation may therefore be partially or totally denatured and no longer in association with other cellular components that constitute its *in vivo* molecular form. Even if the native *in vivo* macromolecular complex can be purified, direct immunization experiments are "best guess" attempts to mimic the natural autoimmunization process because the molecular structure of the putative autoimmunogen that contains the autoantigen of interest is unknown. A further complication is the use of adjuvants to boost the immune response. As a result, direct immunization produces antibodies that, although they react with the autoantigen, usually recognize a denatured form rather than

**Tab. 4** Examples of animal models of autoantibody production.

| Model | Animal | Human disease | Autoantibody specificity |
|---|---|---|---|
| *Spontaneous* | | | |
| NZB | Mouse | Hemolytic anemia | Erythrocyte |
| (NZB × NZW) F1 | Mouse | SLE | Chromatin, DNA |
| MRL-*lpr/lpr* | Mouse | SLE | Chromatin, DNA, Sm, ribosome |
| MRL-+/+ | Mouse | SLE | Chromatin, DNA, Sm |
| BXSB | Mouse | SLE | Chromatin, DNA |
| Obese strain (white leghorn chicken) | Bird | Thyroiditis | Thyroglobulin |
| *Induced by exogenous agents* | | | |
| Chronic GVHD | Mouse | SLE | DNA, chromatin, snRNP, ribosome |
| Mercuric chloride | Mouse (H-$2^s$) | Scleroderma[a] (immune-complex nephritis) | Fibrillarin (Box C/D snoRNP) |
| Mercuric chloride | Mouse (non-H-2 restricted) | SLE | Chromatin |
| Mercuric chloride | Rat ($RT1^n$) | Immune-complex nephritis | GBM |
| Pristane | Mouse | SLE | DNA, Sm, RNP, Su |
| *Direct immunization (antigen)* | | | |
| EAT (thyroglobulin) | Rabbit, mouse (H-$2^{k, s, \text{ or } q}$) | Thyroiditis | Thyroglobulin |
| GMB nephritis (GBM) | Sheep, mouse | Immune-complex nephritis | GBM |
| EMG (acetylcholine receptor) | Lewis rat | Myasthenia gravis | Acetylcholine receptor |
| *Gene mutation* | | | |
| C1q knockout | Mouse (MRL-+/+) | SLE | DNA, rheumatoid factor |
| Dnase1 knockout | Mouse | SLE | Chromatin, DNA |
| SAP knockout | Mouse | SLE | Chromatin, DNA |
| c-mer knockout | Mouse | SLE | Chromatin |
| IFN-γ transgenic | Mouse | SLE | DNA, histones |

[a]Autoantibody specificity is specific for scleroderma, but a scleroderma-like disease has not been described in mice treated with mercuric chloride.
*Notes*: GVHD: graft-versus-host disease; GMB: glomerular basement membrane; EAT: experimental autoimmune thyroiditis; EMG: experimental myasthenia gravis; C1q: component of serum complement; Dnase1: deoxyribonuclease 1; SAP: serum amyloid P component; c-mer: tyrosine kinase.

the autoantigen in its native state; only rarely do they react with the autoantigen when it is in association with other cellular subunits that make up its *in vivo* molecular form. Although direct-immunization antibodies can recognize conformational epitopes, they do not appear to recognize conserved epitopes and therefore cannot exhibit the same lack of species specificity that allows, for example, antifibrillarin autoantibodies to recognize fibrillarin in all species that contain this protein. Lack of reaction against conserved epitopes means that direct immunization antibodies are less efficient at inhibiting the functional activity of their target autoantigen than autoantibodies from patients with autoimmune diseases. Animal models of other types, described next, can produce autoantibodies with reactivities that are extremely difficult to differentiate from those of human patients. As a result, such models more closely approximate their human counterparts.

The second type of model also involves the manipulation of normal, nonautoimmune animals to produce an autoimmune response. In these cases, the triggering event is the introduction of exogenous material into the animal, which, unlike the case of direct immunization, may appear to bear little relationship to the ensuing autoimmune response. An excellent example of this type of model is the autoimmunity induced by heavy metals. Administration of mercury by several different routes and in several different forms, most notably subcutaneous injection of mercuric chloride, produces in mice an autoantibody response that targets the nucleolus. The principal autoantigen involved is the 34 kDa protein fibrillarin (Fig. 5), a protein component of the box C/D snoRNP particles. Mercury induces this autoantibody response in a restricted number of histocompatibility genotypes, most commonly H-$2^s$. Although offspring of crosses between the autoimmune-sensitive H-$2^s$ strains and the autoimmune-resistant strains such as C57BL/6 (H-$2^b$) or DBA/2 (H-$2^d$) are sensitive to antifibrillarin induction following $HgCl_2$ treatment, the response does not appear to be solely due to the product of a dominant *H-2* gene but involves multiple loci as well. This is supported by back-crossing of hybrids onto the autoimmune-sensitive H-$2^s$ background, where the $HgCl_2$-induced antifibrillarin response is even less frequent, even though 50% of the mice would be expected to be homozygous for H-$2^s$. Although antifibrillarin autoantibodies are a marker for human scleroderma, mercury does not appear to produce a scleroderma-like disease in mice; the importance of the model lies instead in the similarity of this toxin-induced murine autoantibody's response to the spontaneous antifibrillarin autoantibody in human scleroderma.

Another example of the exogenous factor-type model is murine graft-versus-host disease (GVHD). In this model, the offspring of the mating of two inbred nonautoimmune mouse strains are injected (grafted) with lymphocytes from one of the parental strains. The injected lymphocytes recognize genetic differences in the host strain that are inherited from the other parental strain and are stimulated to mount a variety of immune responses against the host animal; hence the name "graft versus host." Unlike the case of direct immunization models, the autoimmunity produced in this type of model can lead to severe pathological consequences including lethal immune-complex disease. The immunological sequelae that occur during a GVHD response depend on the murine inbred strains used. Injection of

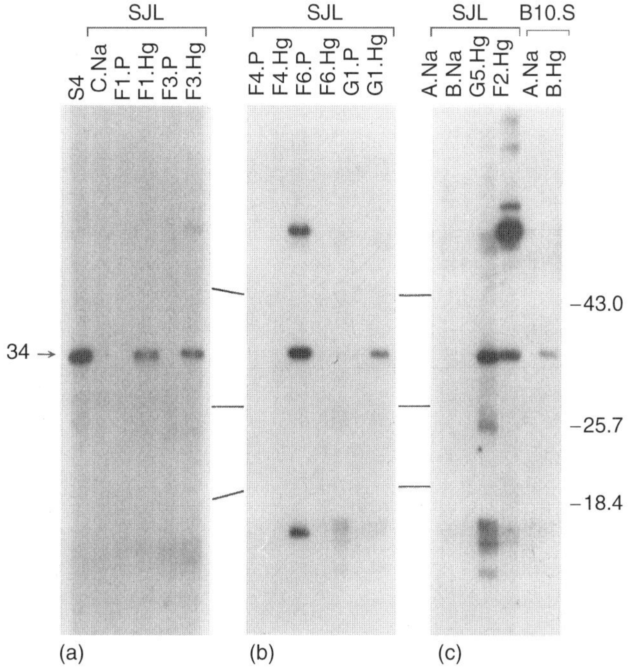


**Fig. 5** Antifibrillarin autoantibodies in H-$2^s$ murine strains (SJL and B10.S) after treatment with mercuric chloride. Autoantibodies were detected by immunoblotting using mouse liver nucleoli resolved by sodium dodecyl sulfate/polyacrylamide gel electrophoresis (SDS-PAGE). (a) Prototype human antifibrillarin serum (S4), serum from a NaCl-treated SJL mouse (C.Na), and serum from mice F1 and F3 before (P) and after (Hg) mercuric chloride treatment. (b) Immunoblotting using sera from three additional SJL mice before and after mercuric chloride treatment. (c) Immunoblotting using sera from two SJL mice treated with NaCl (A.Na; B.Na); two SJL mice treated with mercuric chloride (G5.Hg; F2,Hg); and two B10.S mice, one treated with NaCl (A.Na) and the other treated with mercuric chloride (B.Hg). (Reproduced from Hultman et al., *Clin. Exp. Immunol.* 78 (1989), 470–477 with permission from the publishers.)

DBA/2 lymphocytes into a cross between the DBA/2 and C57B16 mice produces a chronic GVHD, which results in an autoimmune response similar to human SLE including the presence of autoantibodies against chromatin and DNA. Injection of lymphocytes from the A/J strain into Balb/c X A/J hybrids also produces a chronic GVHD in which autoantibodies to snRNP particles including the U3 snoRNP are found. The relationship of different autoantibody specificities to the use of different strains of inbred mice in the GVHD model again highlights the influence of genotype on autoimmunity and autoantibodies.

The third type of model does not require any manipulation of the animal at all; the disease develops spontaneously. The best described of these are the murine strains

BXSB, (NZB × NZW) F1, NZM, MRL-+/+, and MRL-*lpr*/*lpr*, which develop forms of SLE that serve as excellent models of the autoantibody specificities and pathology of the human disease. While the variety of autoantibodies developed by these different strains continues to be investigated, the common autoantibody response, like human SLE, is against chromatin and its subcomponents including DNA. In the (NZB × NZW) F1 strain, autoimmune disease and autoantibodies occur earlier and more frequently in female mice, a finding that has been found to be associated with the presence of female sex hormones. Because of this and other features, the (NZB × NZW) F1 strain is considered the best animal model of human SLE. As noted above, it is the genetic makeup of these inbred strains that has significant potential to address the genesis of the autoimmune response. A much studied aspect of several of the spontaneous models of systemic autoimmune disease is the presence of single-gene defects that accelerate or exacerbate autoimmunity in these already susceptible mouse strains. In the MRL substrains, the *lpr* phenotype is responsible for massive lymphoproliferation of $CD4^-$, $CD8^-$, and $B200^+$ T cells and an accelerated occurrence of autoimmune phenomena compared to the MRL-+/+. Recent studies have indicated that the *lpr* defect is due to a mutation in the *fas* gene that leads to defective expression of *fas* on T- and B cells, which allows them to escape apoptotic elimination and reach the peripheral circulation. Breeding experiments to impart the *lpr* gene to nonautoimmune genetic backgrounds have shown that the *fas* defect does influence the development of autoimmunity and the expression of autoantibodies. A dominant role for *fas* in the initiation of autoimmunity and autoantibodies is questionable, however, because the MRL-+/+, which does not have the *fas* defect, does develop a autoantibody profile and immunopathological disease that is similar to the MRL-*lpr*/*lpr*, albeit at a much later age. Other genes that appear to play a role in acceleration of autoimmunity include *gld*, the ligand for *fas*, and *Yaa*, a sex-linked gene that produces a defect in B cells and is the accelerator gene of autoimmunity in the male BXSB mouse. Exposure of lupus-prone strains to exogenous agents known to elicit autoimmunity in normal mice can result in accelerated appearance of disease features including autoantibodies. In some cases, the exogenous agent accelerates the appearance of idiopathic disease, while in others the elicited disease has features of xenobiotic-induced disease. Thus, mercury exposure accelerates idiopathic disease in BXSB mice including antichromatin autoantibodies of the IgG2a subclass, while pristane injection into (NZB × NZW) F1 mice elicits anti-Sm/RNP and Su autoantibodies that are not part of the idiopathic disease of the (NZB × NZW) F1 but are found in pristane-induced autoimmunity. These observations suggest that not only idiopathic and induced autoimmunity may arise through different mechanisms but also exogenous triggers can influence disease expression.

The fourth type of model involves genetic manipulation in which a gene is deleted ("knockout") or added ("transgenic") in order to influence the expression of autoimmunity. Both types of genetic modification can be used to study the influence of single genes on the animal models described above. Perhaps not unexpectedly, many gene deletions have little or no effect on the expression of autoimmunity and autoantibodies. Such negative effects need to be interpreted carefully as they may indicate a genetically redundant process rather than an unimportant gene. Other gene deletions have been reported to influence differing aspects of autoimmunity in a gene-specific manner, although the extent of the effect may vary between experimental models. Some gene deletions exhibit highly consistent responses. Thus, deletion of the gene for the pleiotropic cytokine interferon-$\gamma$ (IFN-$\gamma$) abrogates autoantibody production and immunopathology in mercury-induced autoimmunity of B10.S mice and spontaneous autoimmunity in MRL-*lpr*/*lpr* mice. The significance of IFN-$\gamma$ in systemic autoimmunity has been demonstrated in nonautoimmune-prone mice made transgenic for IFN-$\gamma$ expression in the epidermis. The increased expression of IFN-$\gamma$ leads to a lupus-like disease characterized by the production of autoantibodies and immune complex–mediated tissue damage. Further evidence for the importance of IFN-$\gamma$ has come from an examination of gene expression in the Nba2 locus of chromosome 1 of the mouse. Nba2 is a genetic interval identified as a locus of genetic susceptibility for lupus in the NZB strain. The offspring of Nba2 interval–specific congenic C57BL/6 mice mated with NZW mice develop autoimmunity similar to the SLE-prone (NZB × NZW) F1 mouse. Examination of gene expression by DNA array revealed a relationship between increased expression of interferon inducible gene (*Ifi*) 202 and features of systemic autoimmunity. Importantly, the gene for Ifi202 lies within the Nba2 interval. Confirmation that increased expression of Ifi202 occurs in other models of systemic lupus would significantly enhance its stature as a lupus susceptibility gene. However, as susceptibility for SLE maps to multiple genetic loci, it is highly likely that additional genes contribute to full disease expression in the (NZB × NZW) F1 mouse.

A number of other gene deletions are associated with expression of autoimmunity and autoantibodies. Some of these such as deficiency of C1q, a component of the complement system, have particular relevance as complement deficiencies in humans and can lead to development of systemic lupus. Significantly, lack of C1q is not sufficient for the development of murine lupus; this gene deletion must occur on genetic backgrounds carrying additional susceptibility genes for autoimmunity to occur. It must also be noted that although many knockout and transgenic models exhibit features of autoimmunity, they may also exhibit other features that are not consistent with the known spectrum of clinical and immunological facets of autoimmune diseases. More telling is the finding that many of the genetic mutations that lead to autoimmunity in mice are not necessary for the development of human systemic autoimmune disease. As described above, mutation in the *fas* gene contributes significantly to the severity of murine SLE. However, mutations in the *fas* gene are not associated with human SLE but rather with autoimmune lymphoproliferative syndrome (ALPS). ALPS is characterized by lymphoproliferation of double-negative T cells and autoantibodies against DNA and cardiolipin, features found in mice with a *fas* mutation but without other lupus susceptibility genes. Similarly, Dnase1-deficient mice develop a lupus-like disease with antichromatin autoantibodies, but deficiency of Dnase1 is not common in human systemic lupus. A nonsense mutation in exon 2 of the *DNASE1* gene has been reported in two apparently unrelated young Japanese patients but not in SLE patients in Spain or the United States.

Although care must be taken in their interpretation, identification and characterization of genes that are associated with autoimmunity and autoantibody production constitute fertile ground for the molecular biologist. Elucidation of the roles of the many genes that appear to contribute to the development of autoimmunity will help define the critical molecular events in the disease process. The murine strains described above have proven valuable model systems for studies on a variety of facets of autoimmunity, and will play significant roles in future genetic studies. It will be important to focus attention not only on the genetic loci that impart susceptibility to autoimmunity but also on those that may allow an individual to resist the development of autoimmune phenomena.

## 5
## Perspectives

Initially used as aids in clinical diagnosis, autoantibodies have become increasingly useful "reporter" molecules in the identification of structure–function relationships. New autoantigens continue to be discovered, while many described autoantigens remain to be characterized both structurally and functionally. Autoantibodies will figure prominently in these characterization studies. As the molecular structures of the interaction between autoantigen and autoantibody become known, it should be possible to design peptide configurations capable of perturbing the functional activity of numerous cellular processes.

Understanding the influence of genes and their products not only on susceptibility but also on resistance to autoimmunity and autoantibody expression is in its infancy. However, the tools to mature this field (inbred animal models of spontaneous and induced autoimmunity, and the molecular techniques of transgenics and gene knockout) are already available. They await the complex but potentially fruitful identification and functional analysis of candidate genes.

## Bibliography

### Books and Reviews

Earnshaw, W.C., Rattner, J.B. (1991) The Use of Autoantibodies in the Study of Nuclear and Chromosomal Organization, in: Hamkalo, B.A., Elgin, S.C.R. (Eds.) *Methods in Cell Biology*, Vol. 35, Academic Press, New York.

Kono, D.H., Theofilopoulos, A.N. (2001) Genetics of murine models of systemic lupus erythematosus, in: Wallace, D.J., Hahn, B.H. (Eds.) *Dubois' Systemic Lupus Erythematosus*, 6th edition, Williams & Wilkins, Baltimore.

Matzinger, P. (2002) The danger model: a renewed sense of self, *Science* **296**, 301–305.

Medzhitov, R., Janeway, C.A. Jr. (2002) Decoding the patterns of self and nonself by the innate immune system, *Science* **296**, 298–300.

Peter, J.B., Shoenfeld, Y. (Eds.) (1996) *Autoantibodies*, Elsevier, Amsterdam.

Pollard, K.M., Hultman, P. (1997) Effects of mercury on the immune system, *Met. Ions Biol. Syst.* **34**, 421–440.

Rose, N.R., Mackay, I.R. (Eds.) (1998) *The Autoimmune Diseases*, 3rd edition, Academic Press, New York.

Rubin, R.L., Pollard, K.M. (1997) Chemical-Induced Autoimmunity, in: Hertzenberg, L.A., Weir, D.M., Hertzenberg, L.A., Blackwell, C. (Eds.) *Handbook of Experimental Immunology*, 5th edition, Blackwell Scientific Publications, London.

Tan, E.M. (1989) Antinuclear antibodies: diagnostic markers for autoimmune diseases and probes for cell biology, *Adv. Immunol.* **44**, 93–151.

## Primary Literature

Balomenos, D., Rumold, R., Theofilopoulos, A.N. (1998) Interferon-gamma is required for lupus-like disease and lymphoaccumulation in MRL-lpr mice, *J. Clin. Invest.* **101**, 364–371.

Bickerstaff, M.C., Botto, M., Hutchinson, W.L., Herbert, J., Tennent, G.A., Bybee, A. et al. (1999) Serum amyloid P component controls chromatin degradation and prevents antinuclear autoimmunity, *Nat. Med.* **5**, 694–697.

Chatenoud, L., Salomon, B., Bluestone, J.A. (2001) Suppressor T cells – they're back and critical for regulation of autoimmunity! *Immunol. Rev.* **182**, 149–163.

Gall, J.G. (2000) Cajal bodies: the first 100 years, *Annu. Rev. Cell. Dev. Biol.* **16**, 273–300.

Kono, D.H., Theofilopoulos, A.N. (2000) Genetics of systemic autoimmunity in mouse models of lupus, *Int. Rev. Immunol.* **19**, 367–387.

Kono, D.H., Balomenos, D., Pearson, D.L., Park, M.S., Hildebrandt, B., Hultman, P., Pollard, K.M. (1998) The prototypic Th2 autoimmunity induced by mercury is dependent on IFN-gamma and not Th1/Th2 imbalance, *J. Immunol.* **161**, 234–240.

Kono, D.H., Park, M.S., Szydlik, A., Haraldsson, K.M., Kuan, J.D., Pearson, D.L., Hultman, P., Pollard, K.M. (2001) Resistance to xenobiotic-induced autoimmunity maps to chromosome 1, *J. Immunol.* **167**, 2396–2403.

Matzinger, P. (2002) The danger model: a renewed sense of self, *Science* **296**, 301–305.

Medzhitov, R., Janeway, C.A. Jr. (2002) Decoding the patterns of self and nonself by the innate immune system, *Science* **296**, 298–300.

Mitchell, D.A., Pickering, M.C., Warren, J., Fossati-Jimack, L., Cortes-Hernandez, J., Cook, H.T., Botto, M., Walport, M.J. (2002) C1q deficiency and autoimmunity: the effects of genetic background on disease expression, *J. Immunol.* **168**, 2538–2543.

Napirei, M., Karsunky, H., Zevnik, B., Stephan, H., Mannherz, H.G., Moroy, T. (2000) Features of systemic lupus erythematosus in Dnase1-deficient mice, *Nat. Genet.* **25**, 177–181.

Pollard, K.M. (2002) Cell death, autoantigen cleavage, and autoimmunity, *Arthritis Rheum.* **46**, 1699–1702.

Pollard, K.M., Hultman, P. (1997) Effects of mercury on the immune system, *Met. Ions Biol. Syst.* **34**, 421–440.

Pollard, K.M., Pearson, D.L., Bluthner, M., Tan, E.M. (2000) Proteolytic cleavage of a self-antigen following xenobiotic-induced cell death produces a fragment with novel immunogenic properties, *J. Immunol.* **165**, 2263–2270.

Pollard, K.M., Pearson, D.L., Hultman, P., Deane, T.N., Lindh, U., Kono, D.H. (2001) Xenobiotic acceleration of idiopathic systemic autoimmunity in lupus-prone bxsb mice, *Environ. Health Perspect.* **109**, 27–33.

Pollard, K.M., Pearson, D.L., Hultman, P., Hildebrandt, B., Kono, D.H. (1999) Lupus-prone mice as models to study xenobiotic-induced acceleration of systemic autoimmunity, *Environ. Health Perspect.* **107**(Suppl. 5), 729–735.

Rozzo, S.J., Allard, J.D., Choubey, D., Vyse, T.J., Izui, S., Peltz, G., Kotzin, B.L. (2001) Evidence for an interferon-inducible gene, Ifi202, in the susceptibility to systemic lupus, *Immunity* **15**, 435–443.

Rubin, R.L., Kretz-Rommel, A. (2001) A nondeletional mechanism for central T-cell tolerance, *Crit. Rev. Immunol.* **21**, 29–40.

Salomon, B., Bluestone, J.A. (2001) Complexities of CD28/B7: CTLA-4 costimulatory pathways in autoimmunity and transplantation, *Annu. Rev. Immunol.* **19**, 225–252.

Satoh, M., Richards, H.B., Shaheen, V.M., Yoshida, H., Shaw, M., Naim, J.O., Wooley, P.H., Reeves, W.H. (2000) Widespread susceptibility among inbred mouse strains to the induction of lupus autoantibodies by pristane, *Clin. Exp. Immunol.* **121**, 399–405.

Seery, J.P., Carroll, J.M., Cattell, V., Watt, F.M. (1997) Antinuclear autoantibodies and lupus nephritis in transgenic mice expressing interferon gamma in the epidermis, *J. Exp. Med.* **186**, 1451–1459.

Steinman, R.M., Nussenzweig, M.C. (2002) Avoiding horror autotoxicus: the importance of dendritic cells in peripheral T cell tolerance, *Proc. Natl. Acad. Sci. USA* **99**, 351–358.

von Muhlen, C.A., Tan, E.M. (1995) Autoantibodies in the diagnosis of systemic rheumatic diseases, *Semin. Arthritis Rheum.* **24**, 323–358.

Vyse, T.J., Kotzin, B.L. (1998) Genetic susceptibility to systemic lupus erythematosus, *Annu. Rev. Immunol.* **16**, 261–292.

# 14
# Synovial Mast Cells in Inflammatory Arthritis

*Theoharis C. Theoharides*
*Tufts University School of Medicine, Boston, MA, USA*


**1 Osteoarthritis Versus Rheumatoid Arthritis 369**
1.1 Specific Triggers Versus Lack of Inhibitory Molecules 369

**2 Mast Cells in the Joints 370**
2.1 Inflammatory Mediators 370

**3 Chemoattractants 371**
3.1 Animal Models 371

**4 Pathophysiology of Mast Cells 372**
4.1 Triggers 372
4.2 Mediators 373
4.3 Selective Secretion 375

**5 Neurohormonal Stimulation of Mast Cells and Inflammatory Disorders 377**
5.1 Anatomical Observations 377
5.2 Stress and the HPA Axis 377
5.3 Proinflammatory Effects of Stress 377
5.4 CRH Effects on Mast Cells 378

**6 Therapy 378**
6.1 Proteoglycans 378
6.2 Flavonoids 379
6.3 Synergistic Effects 380

**7 Perspectives 380**

**Acknowledgment 381**


*Immunology. From Cell Biology to Disease.* Edited by Robert A. Meyers.

ISBN: 978-3-527-31770-7

**Bibliography 382**
Books and Reviews 382
Primary Literature 383

## Keywords

**Activation**
Stimulation of cells to undergo partial, selective, or complete synthesis and secretion of mediators, or generation of other molecules or responses in response to a variety of triggers; such processes may be associated with subtle morphologic changes discernible only by appropriate ultrastructural studies.

**Corticotropin-releasing Hormone**
A peptide originally shown to be secreted from the hypothalamus in response to stress and primarily stimulating the anterior pituitary cells to secrete ACTH. It has since been shown to affect numerous systems including the immune system, and we proposed that it be renamed "stress response hormone (SRH)" instead of CRH.

**Degranulation**
The release of the contents of secretory granules, from mast cells or other secretory cells such as polymorphonuclear leukocytes, platelets, pituitary, or $\beta$-pancreatic cells in response to specific receptor-mediated triggers through a process dependent on energy and extracellular calcium ions.

**Inflammation**
An orchestrated, often excessive response of the organism, involving the immune, endocrine, and neural systems against real or perceived triggers.

**Protease-activated Receptors (PARs)**
Unique receptors on cell surfaces, stimulation of which leads to widespread inflammation.

**Stress**
The response of cells and tissues to real or perceived noxious stimuli with anti-inflammatory effects through brain secretion of CRH, but proinflammatory effects through peripheral secretion of CRH and related peptides.

**Tryptase**
One of many unique proteases secreted from mast cells that have multiple actions, such as microvascular leakage, stimulation of protease-activated receptors (PARs) and induction of inflammation, generation of active peptides, degradation of cytokines.

Inflammatory arthritis is a highly prevalent condition with substantial morbidity and degree of disability. The pathogenesis is still poorly understood limiting effective therapy. Increasing evidence has implicated synovial mast cells in the initiation of inflammatory arthritis. Mast cells are commonly known for their involvement in allergic reactions through degranulation and release of histamine. However, these cells are capable of releasing over 50 potent vasoactive, proinflammatory and neurosensitizing molecules in response to nonallergic triggers without degranulation. Such selective release of cytokines, especially in response to stress hormones, provides a unique explanation and possible therapeutic targets for inflammatory arthritis.

## 1 Osteoarthritis Versus Rheumatoid Arthritis

A 2002 report from the US Center for Disease Control (CDC) indicated that one-third of all adults in the United States (almost 70 million) suffer from arthritis or chronic joint pain, up from one-fifth in 1993, with an estimated annual cost of $82 billion. The impact of arthritis is comparable in Australia, Canada, Europe, and the United Kingdom, with an estimated total of another 60 million affected individuals.

Osteoarthritis (OA) is a degenerative joint disease and is characterized by the breakdown of the joint's cartilage and loss of joint space especially in hands and weight-bearing joints, such as knees, hips, feet, and the back. Cartilage breakdown allows bones to rub against each other, causing inflammation, pain, and loss of movement. Although age is a risk factor, obesity, and joint injuries due to sports, work-related activity, or accidents increase the risk of developing OA.

Rheumatoid arthritis (RA) involves chronic inflammation and destruction of the joints. RA is a systemic disease that affects the entire body and is characterized by inflammation and reactive hyperplasia of the synovium, which causes redness, swelling, stiffness, and pain. RA typically affects many different joints and is noted for remissions and flares, often precipitated or exacerbated by stress. Articular damage in RA involves cartilage erosion, inflammatory cell accumulation and finally to bone destruction and reactive bone spur formation. The involved joints lose their shape and alignment, resulting in pain and loss of movement.

### 1.1 Specific Triggers Versus Lack of Inhibitory Molecules

As in common practice in medicine, there have been numerous investigations in the potential trigger(s) implicated in the pathophysiology of inflammatory and especially of rheumatoid arthritis. In certain cases, as in crystal (gout) arthritis or Lyme arthritis, the etiologic agents are known. However, such triggers are not known for RA. There is some evidence from animal models that inflammatory arthritis may involve an inappropriate autoimmune response against structural (colagen) or infection (bacterial, viral) antigens. However, little attention has been paid to the potential lack of innate inhibitory molecules, such as $a_1$-antitrypsin or $a_2$-macroglobulin in other tissues.

## 2 Mast Cells in the Joints

The inflammatory and subsequent lytic processes in inflammatory arthritis are not well understood, especially the initial triggering events. Until recently, discussion of the cellular elements involved did not even mention mast cells, even though a number of papers had reported the presence of mast cells in joints. Over the last 10 years, histochemical observations were correlated to clinical findings. Increased numbers of mast cells have been routinely noted and their mediators could initiate and promote inflammation while mast cells constitute a small fraction of the cells in a normal human synovium, they increase to almost 10% in RA. Mast cells, cytokines, and metalloproteinases were colocalized in the RA lesion and activation of synovial mast cells was shown to be autoregulated by their own mediators' histamine and tryptase (Fig. 1).

It was recently convincingly argued that mast cells may be involved in inflammatory arthritis; however, this presentation was still limited primarily to the role of histamine and tumor necrosis factor, TNF-$\alpha$, and did not discuss potential triggers such as C3a and C5a. One of the reasons why mast cells have been neglected comes from the persistence of associating them only with allergies in spite of mounting evidence to the contrary. Another reason is the fact that mast cells are not commonly seen to degranulate in tissues such as joints that do not get allergic reactions; they were, therefore, considered an "innocent bystander's" renewed evidence that mast cells can release mediators selectively without degranulation, especially in response to nonallergic triggers (discussed in the following) has expanded the potential importance of mast cells.

### 2.1 Inflammatory Mediators

Mast cell-derived TNF-$\alpha$ mediates inflammation and induces hypertrophy of draining lymph nodes during infection.

Mast cells are known to secrete IL-6 and recent studies showed that mast cell–deficient mice could not increase their serum IL-6 in response to acute stress. This is of interest in view of the fact that IL-6 is elevated, especially in juvenile RA. IL-6 is

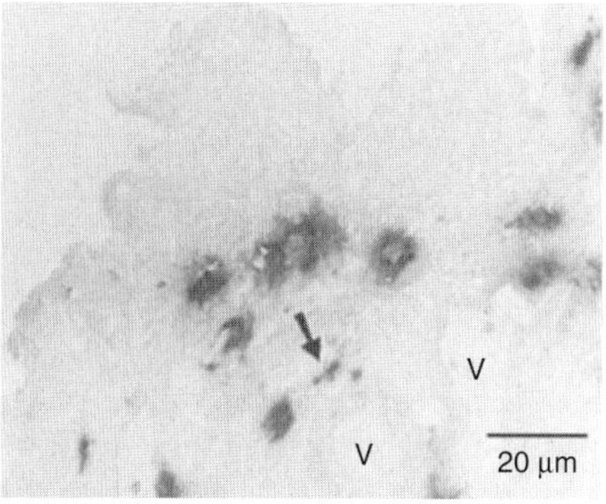


**Fig. 1** Tryptase-positive mast cells (brown color) adjacent to blood vessels (V) from the joint of a patient with rheumatoid arthritis (See color plate p. xxviii).

also important in collagen-induced and antigen-induced arthritis.

A number of chemokine receptors have also been identified on mast cells, especially CXCR3 protein, which was preferentially expressed on mast cells in the synovium of RA patients.

## 3 Chemoattractants

Tryptase can cause microvascular leakage and stimulate protease-activated receptors (PAR) leading to widespread inflammation, including autocrine stimulation of mast cells. Protease also acts as a tissue metalloproteinase, degrading matrix, as well as gelatinase, thus contributing to tissue remodeling. Tryptase also induces hyperexcitability of submucosal neurons, while histamine directly stimulates SP and calcitonin gene-related peptide (CGRP)-containing neurons of the trigeminal system.

Cells obtained from fluid aspirated from joints of patients with arthrosynovitis express RANTES and MCP-1, both of which are potent mast cell chemoattractants. In addition, stem cell factor (SCF) and nerve growth factor (NGF) are also known to induce mast cell migration, proliferation, and secretion. Other mast cell chemoattractants include vascular endothelial growth factor (VEGF), epidermal growth factor (EGF), and platelet-derived growth factor (PDGF).

### 3.1 Animal Models

In spite of the fact that TNF blockers are used clinically with considerable success in adult RA, increasing evidence indicates that IL-6 may be equally or more important. In fact, we recently showed that experimental inflammatory arthritis in mice was unaffected in TNF −/− mice. Instead, the IL-6 −/− mice were resistant to inflammatory arthritis, and humanized antibody to human IL-6 receptor inhibited collagen-induced arthritis. These findings are of particular significance as we recently showed that mast cells are the sole source of serum IL-6 elevations induced by acute stress in mice. Most recently, mast cell deficient mice were shown to be resistant to autoimmune and inflammatory arthritis.

In particular, inflammatory arthritis induced by the injection of carrageenan in the hind knee joints, increased the joint size by $1.94 \pm 0.41$ mm in C57BL mice in 4 days, by which time the mice were obviously limping (Fig. 2). These effects were inhibited in $W/W^v$ mast cell deficient and corticotropin-releasing hormone (CRH) −/− mice, which were clinically indistinguishable from mice injected with normal saline (Fig. 2). These results are of interest because CRH is increased in the joints of RA patients and CRH receptors (CRHR) are present on articular mast cells, implying that CRH could trigger or regulate mast cell activation. In fact, we showed that CRH could activate mast cells in the skin and increase vascular permeability in rodents. CRH was recently shown to also induce mast cell–dependent vascular permeability in humans. The pathophysiological implication of such finding is that CRH could be released locally under stress and could exacerbate inflammatory diseases (Fig. 3).

It is quite interesting that stress increased serum IL-6 levels in care givers of chronically ill patients, but it was decreased in those who went to church regularly, indicating that reduction of stress could lead to decrease in a key proinflammatory cytokine producer.

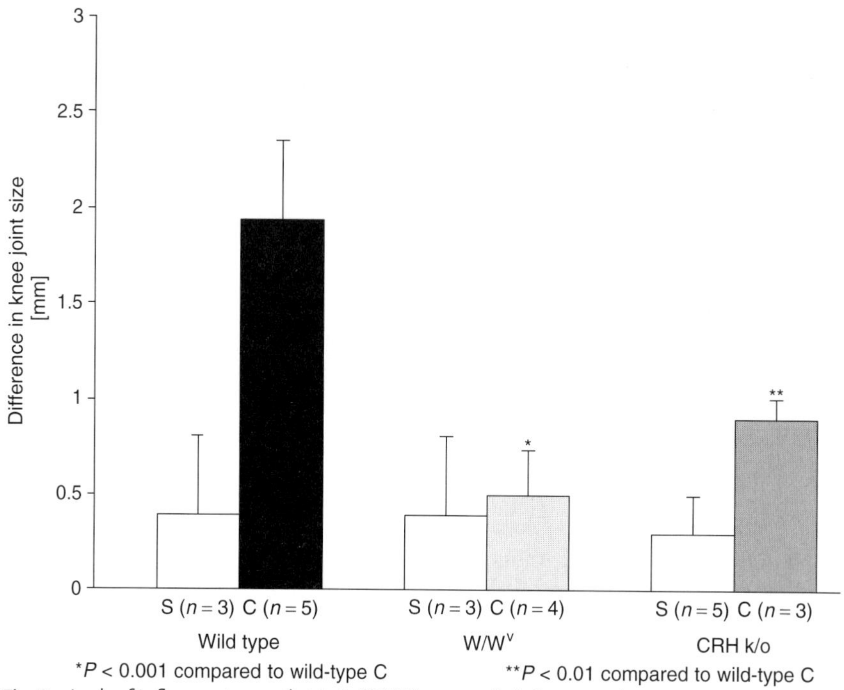


**Fig. 2** Lack of inflammatory arthritis in W/W$^V$ mast cell deficient and CRH −/− mice. W/W$^V$ = mast cell deficient mice; k/o = knockout mice; S = normal saline; C = carrageenan.

The K/BxN mouse has been used as a model for human inflammatory arthritis and it was recently shown that the presence of autoantibodies against glucose-G phosphate isomerase is dependent on mast cells.

## 4 Pathophysiology of Mast Cells

Mast cells derive from a bone marrow progenitor and mature in tissues depending on microenvironmental conditions; they are characterized by the surface expression of the stem cell factor (CSF) receptor c-kit and the high affinity receptor for immunoglobulin E (FcεRI). Mast cells are important not only for allergic reactions but also in inflammation, autoimmunity, and inflammatory arthritis. However, though mast cells do not always express their c-kit receptor or FceRJ, they do so during inflammation. They are ubiquitous in the body, including the brain and joints that are not known to develop allergic reactions. Mature mast cells vary considerably in their cytokine and proteolytic enzyme content. However, the phenotypic expression of mast cells does not appear to be fixed.

In fact, mast cells cultured together with pheochromocytoma cells were shown to develop contacts with the latter and acquire the ability to respond to neuropeptides.

### 4.1 Triggers

In addition to IgE and antigen, anaphylatoxins, C3a, C5a antibody light chains, bacterial or viral superantigens aggregated

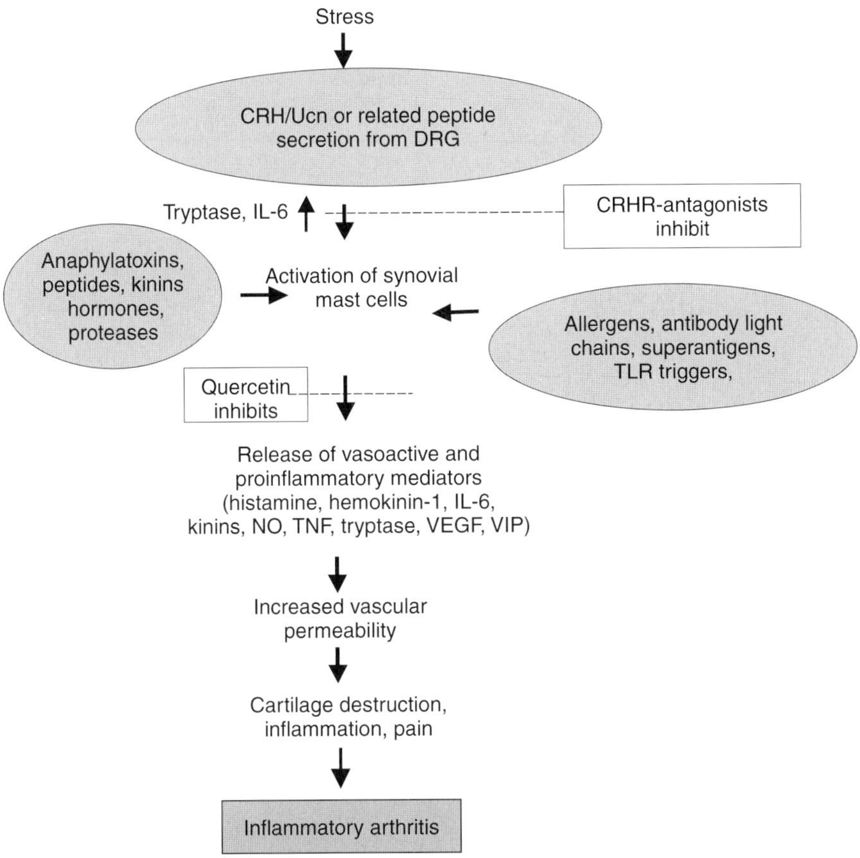


**Fig. 3** Schematic representation of the possible role of mast cells in inflammatory arthritis. DRG = dorsal root ganglia; NO = nitric oxide; VIP = vasoactive intestinal peptide; VEGF = vascular endothelial growth factor (see color plate p. xxviii).

IgG cytokines, histamine-releasing factors, hormones, and neuropeptides can trigger mast cell secretion (Table 1). The latter include SP, somatostatin, neurotensin (NT), parathyroid hormone, and NGF, which is also released under stress.

Mast cells can also be activated by endothelin, and blockade of the endothelin receptor was recently shown to block *in vivo* effects attributed to mast cells in mice. Proteases released from mast cells or other inflammatory cells can generate biologically active peptides, like histamine-releasing peptide (HRP) from plasma albumin, and HRP is found in synovial fluids obtained from patients with RA. Adherent *Escherichia coli* can activate mast cells that express Toll-like (Toll) receptors 2, activated by *E. coli* lipopolysaccharide (LPS); such receptors are critical in innate and acquired immunity.

### 4.2 Mediators

Mast cells synthesize and secrete over 50 biologically potent molecules that include most of the known cytokines. They

**Tab. 1** Triggers of mast cell activation.

| | |
|---|---|
| ***Antigens*** | ***Hormones*** |
| Allergens/IgE | ACTH |
| Aggregated IgE | CRH |
| Antibody light chains | Estradiol |
| Chemokine receptor ligands (CXCR3) | PTH |
| Superantigens (bacterial, viral) | Urocortin |
| Toll-like receptor (TLR 2, 4, 9) triggers (LPS, PDG, dd DNA) | |
| | *Peptides* |
| *Anaphylatoxins* | Albumin-derived histamine-releasing molecules |
| C3a, C5a | Bradykinin |
| | CGRP |
| *Cytokines* | Endorphins |
| IL-1 | Endothelin |
| TNF-$\alpha$ | Histamine-releasing factor |
| | NT |
| *Drugs* | SRIF |
| Contrast media | SP |
| Curare | VIP |
| Morphine | |
| | *Neurotransmitters* |
| *Enzymes* | Acetylcholine |
| Chymase | Purines |
| PARs | |
| Tryptase | *Free Radicals* |
| | Oxygen |
| *Growth Factors* | Hydroxy |
| NGF | |
| SCF | |

Notes: dd: double stranded; LPS: lipopolysaccharides; PDG: peptidoglycan; TLR: Toll-like receptor.

can, therefore, easily induce vasodilation, vascular endothelial molecule expression, and chemoattraction, which is necessary for the initiation of inflammation.

These vasodilatory, angiogenic, proinflammatory, and neurosensitizing mediators (Table 2) include histamine, heparin, kinins, neuropeptides such as vasoactive intestinal peptide (VIP) and proteases, such as chymase and tryptase (preformed) as well as chemoattractants, cytokines, growth factors, leukotrienes, prostaglandins, nitric oxide (NO), SCF, and VEGF (newly synthesized) (Table 2).

The most abundant mediators are proteases that can cause microvascular leakage, activation of PAR receptors and inflammation, as well as activation of neuropeptide precursors and direct proteolytic damage to synovium. Mast cells are also particularly rich in TNF-$\alpha$ and IL-6 both of which are implicated in arthritis, as discussed earlier. Mast cells are also the richest source of CRH outside the brain and could help propagate local inflammatory processes in response to chemical, mechanical, or oxidative stress.

**Tab. 2** Mast cell mediators and their biologic effects.

| ***Molecules*** | ***Major actions*** |
|---|---|
| ***Prestored*** | |
| *Enzymes* | |
| Arylsulfatases | Lipid/proteoglycan hydrolysis |
| Carboxypeptidase A | Peptide processing |
| Cholinesterase | Angiotensin II synthesis |
| Chymase | Kinins synthesis vasodilation, *pain* |
| Kinogenases | Arachidonic acid generation |
| Phospholipases | Angiotensin I generation |
| Renin | Tissue damage, microvascular leakage, gelatin and |
| Tryptase (7 such proteases) | matrix degradation inflammation, *pain* |
| *Biogenic Amines* | |
| Histamine | Vasodilation, angiogenesis, mitogenesis, *pain* |
| 5-HT (serotonin) | Vasoconstriction, *pain* |
| *Peptides* | |
| Bradykinin | Vasodilation, *pain* |
| CRH | Vasodilation, inflammation |
| Endorphins | Analgesia |
| SP | Anti-inflammatory (?) |
| SRIF | Inflammation, *pain* |
| Ucn | Vasodilation, inflammation |
| VIP | Vasodilation |
| *Proteoglycans* | |
| Chondroitin sulfate | Cartilage synthesis, anti-inflammatory |
| Heparin | Angiogenesis, NGF stabilization |
| Hyaluronic acid | Cell surface recognition |
| ***De novo synthesized*** | |
| $LTB_4$ | Leukocyte chemotaxis |
| $LTC_4$ | Vasoconstriction, *pain* |
| NO | Vasodilation |
| PAF | Platelet activation and serotonin release |
| PGD2 | Vasodilation, *pain* |
| *Chemokines* | |
| IL-8, MCP-1, MCP-3, MCP-4, RANTES | Chemoattraction, leukocyte infiltration |
| *Cytokines* | |
| IL-1,2,3,4,5,6,9,10,13,16 | Inflammation, leukocyte migration, *pain* |
| INF-$\alpha$; MIF; TNF-$\alpha$ | Inflammation, leukocyte proliferation/activation |
| *Growth Factors* | |
| CSF, GM-CSF, $\beta$-FGF, NGF, VEGF | Endothelial, immune and neuronal cell growth |

### 4.3 Selective Secretion

One of the reasons mast cells have remained fairly obscure in inflammatory arthritis has been the fact that they appear "intact," giving the impression that they are quiescent.

Unlike degranulation, where most of the 500 secretory granules, each 1000 nm in diameter release their contents in an explosive manner, mast cells can undergo

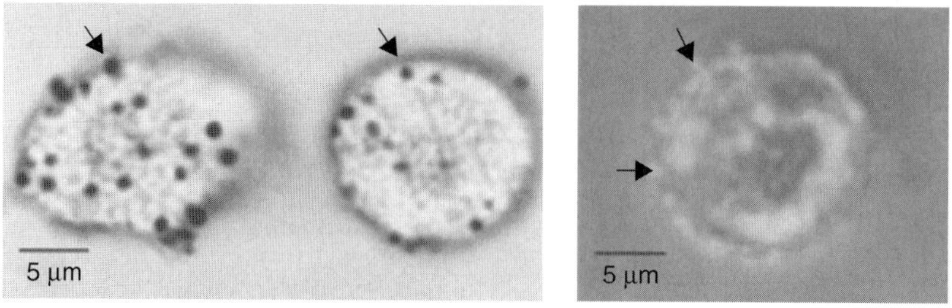


(a) (b)

**Fig. 4** Punctate release of contents of individual secretory granules from purified rat peritoneal mast cells incubated with calcium containing liposomes. (a) Content of individual granules (arrow) visualized by ruthenium red staining (with a complementary filter to show black) using Nomarski optics; (b) individual granules in the process of exocytosis from a mast cell visualized by ruthenium red staining (see color plate p. xxix).

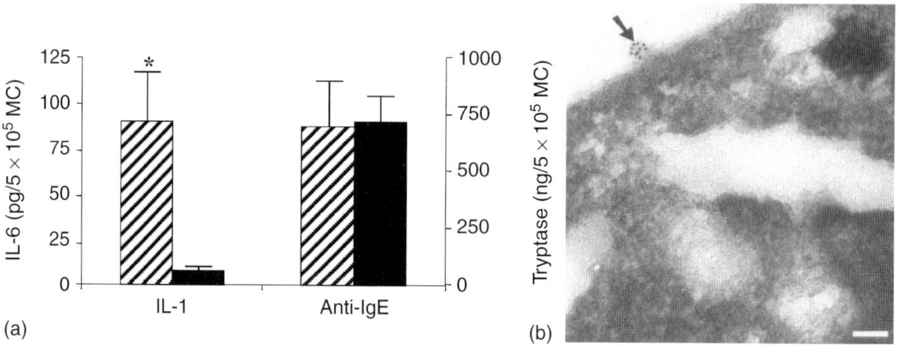


**Fig. 5** Selective vesicular IL-6 release induced by IL-1 from normal human cultured mast cells. (a) IL-6 and tryptase release in response to either IL-1 or anti-IgE ($n = 8$ $^*p < 0.05$); (b) a vesicle (arrow) budding from a mast cell observed by cryo-immuno-ultrastructural microscopy "white" areas represent secretory granules. Scale bar = 70 nm.

secretion of the content of individual granules (Fig. 4).

Mast cells also have the ability to release some mediators selectively without degranulation. This unique phenomenon has been shown for biogenic amines, eicosanoids, and IL-6. In fact, IL-1 was shown to induce selective release of IL-6 without histamine or tryptase, through a process that involves small vesicles (50–70 nm each in diameter) (Fig. 5). Mast cells *in vivo* also often have ultrastructural alterations of their electron dense granular core indicative of secretion, but without degranulation, a process termed *activation*, or *intragranular activation*, or *piecemeal* degranulation. This appearance was prominent in brain mast cells of nonhuman primates with experimental allergic encephalomyelitis (EAE), and in human inflammatory conditions such as inflammatory bowel disease (IBD) or interstitial cystitis (IC) of the urinary bladder. Moreover, in certain diseases such as scleroderma and IC, mast cells appear totally depleted of their granule content and they could not be recognized by light microscopy (phantom mast cells).

Unpublished evidence indicates that synovial mast cells in RA undergo mostly selective release of mediators that is not obvious by routine histopathology.

## 5 Neurohormonal Stimulation of Mast Cells and Inflammatory Disorders

Mast cells could also be activated by antidromic nerve stimulation as shown for trigeminal or cervical ganglion stimulation that may result in neuropeptide secretion of reactive fibers when localized close to mast cells, release of SP from sensory afferents could stimulate mast cell secretion *in vivo*. NT also could be released from dorsal root ganglia (DRG) alone or together with CRH and further stimulate mast cells.

### 5.1 Anatomical Observations

Mast cells are located perivascularly in close proximity to neurons. A functional association between mast cells and neurons has been reported in 1999, and the potential pathophysiological role of brain mast cells has been reviewed by Theoharides in 1996.

### 5.2 Stress and the HPA Axis

Chronic stress affects illness and suppresses the immune system. Acute stress, however, can exacerbate inflammatory syndromes. In fact, there is evidence that acute stress can stimulate the immune system. Stress has been shown to worsen arthritis and activate mast cells.

Corticotropin-releasing hormone or CRH is a 41-amino acid peptide that regulates the hypothalamic-pituitary adrenal (HPA) axis and coordinates the stress response through activation of the sympathetic nervous system. CRH acts through specific receptors, which include CRHR-1 and CRHR-2. Both receptor types are located on brain neurons, but CRHR-2 has also been identified on cerebral arterioles that could be stimulated by CRH and urocortin (Ucn) directly. CRHR-2 has been further subdivided to CRHR-2$\alpha$ and CRHR-2$\beta$, which are best activated by Ucn, a peptide with about 50% structural similarity to CRH; Ucn II and Ucn III have also been identified and are potent CRHR-2 agonists.

### 5.3 Proinflammatory Effects of Stress

Both CRH and CRH mRNA have been demonstrated in rodent spleen and thymus, while human peripheral blood leukocytes and enterochromaffin cells express mRNA for Ucn. CRH and CRHR mRNA is expressed in rodent and human skin, CRH-like immunoreactivity is present in the dorsal horn of the spinal cord and in DRG, as well as in sympathetic ganglia. We recently showed that acute stress could increase the skin content of CRH, most likely from DRG. We also showed that human mast cells contain both CRH and Ucn that could be released in response to immunologic stimulation.

CRH can stimulate leukocytes to produce $\beta$-endorphin, adrenocorticotropic hormone (ACTH), and $\alpha$-melanocyte stimulating hormone ($\alpha$-MSH), as well as monocytes to secrete interleukin-1 (IL-1), and lymphocytes to produce IL-2.

CRH also stimulated lymphocyte proliferation, increased IL-2 receptor expression on T lymphocytes, was chemotactic for mononuclear leukocytes and activated CRHR-1 on spleen cells.

Relevant examples of possible proinflammatory actions of CRH include carrageenan-induced aseptic inflammation and arthritis, where both CRH and Ucn have been identified in the joints. Moreover, inflammatory arthritis shown to be absent in $W/W^v$ mast cell deficient mice, was blocked by the preferential CRHR-1 antagonist Antalarmin and was greatly attenuated in CRH knockout mice. In fact, mast cells in the joints of RA patients were reported to express CRH receptors.

### 5.4 CRH Effects on Mast Cells

Mast cells are localized close to CRH positive neurons in the rat median eminence. CRH administration in humans causes peripheral vasodilation and flushing reminiscent of mast cell activation. Iontophoresis of CRH increased human skin vasodilation, detected by laser Doppler, which was dependent on CRHR-1 and mast cells. Intradermal CRH administration leads to histamine-dependent swelling; in rodents, intradermal CRH leads to activation of mast cells and Evans blue extravasation was also induced. Moreover, CRHR-1 is involved in stress-induced exacerbation of chronic contact dermatitis in rats.

In addition to CRH stimulating mast cells, mast cell mediators could influence CRH release. For instance, human mast cells synthesize and secrete large amounts of CRH. Histamine can increase CRH mRNA expression in the hypothalamus, and mast cells could stimulate the HPA axis. This could occur possibly through IL-6 and IL-1, both of which are released from mast cells; conversely, CRH stimulates IL-6 release. Acute stress also increased BBB permeability in rats and mice only in brain areas containing mast cells.

CRH Ucn can directly activate mast cells that express CRHRs; in fact, mast cells were recently shown to express five CRHR-1 isoforms that could have different functions.

## 6 Therapy

Typical therapy for osteoarthritis involves weight loss, exercise, and use of nonsteroidal anti-inflammatory drugs (NSAIDs). In RA, active articular inflammation often requires immunosuppressant and immune modifying drugs. Severe RA requires multiple agents that include steroids, methotrexate, penicillamine, gold salts, and most recently TNF blockers, such as etanercept or infliximab. Unfortunately, recent evidence that cyclooxygenase 2 (COX-2) inhibitors increase cardiovascular disease has forced the withdrawal or limited use of such drugs.

Except for glucocorticoids, there are no clinically effective mast cell inhibitors.

### 6.1 Proteoglycans

Glucosamine sulfate presumably acts as a building block for new cartilage, while chondroitin sulfate acts as a "ready-made" component.

D-Glucosamine, *N*-acetylglucosamine, glucosamine sulfate, and chondroitin are commonly used for arthritis. A meta analysis of clinical trials using glucosamine and/or chondroitin originally had indicated potential usefulness in OA. The validation of WOMAC, a self-administered health status instrument for OA has helped with the design of double-blind studies; similarly the Cedars–Sinai Health Related Quality of Life Instrument (CSHQ-RA) can be used for RA research. One of the first double-blind studies compared 500-mg glucosamine sulfate tid to 400-mg ibuprofen tid for 4 weeks. From the second week onwards, the clinical improvement was the same (~50%) in both groups, but 35% of those on ibuprofen reported adverse effects as compared to 6% on glucosamine. At least two studies have shown that 1500 mg of glucosamine per day for 3 years can delay progression of OA and knee space reduction. In another randomized, double-blind placebo-controlled trial, 212 patients with OA of the knee were assigned either 1500 mg of oral glucosamine sulfate or a placebo for 3 years. Patients on active treatment had clinical improvement of symptoms and no significant joint space loss, compared to progressive narrowing of those on placebo. In a subsequent study, 202 patients with OA of the knee were randomized to either 1500 mg of glucosamine sulfate or placebo for 3 years; symptoms on the active group improved by about 25% and joint space narrowing (>0.5 mm) occurred in 5% of those on the active arm as compared to 15% on the placebo. However, more recent studies with glucosamine, especially when administered for shorter duration, have failed to show consistent benefit. Moreover, use of 1500 mg of glucosamine per day could interfere with insulin sensitivity.

The proteoglycan chondroitin sulfate has also been added to glucosamine, but there are a number of problems with such preparations. Very little oral chondroitin sulfate is absorbed in powder form due to its high molecular weight (150 000–1 000 000 Da) and degree of sulfation. In particular, less than 5% of chondroitin sulfate is absorbed intact when administered orally. In another study, after oral intravenous administration of chondroitin sulfate (16 000 Da), the absolute bioavailability was 13%, but when chondroitin sulfate of about double the size (26 000 Da) was used, less than 4% of the oral dose administered in rats was intact chondroitin sulfate in the blood. However, the smaller size (16 000 Da) has almost no protective value in laboratory experiments. Other papers found no oral absorption of chondroitin sulfate and concluded that any protective effect in the joints after oral administration was unfounded.

There are no good studies on the use of proteoglycans in RA, presumably because the aim has traditionally been the synovial inflammatory response. However, chondroitin sulfate was shown to inhibit activation of connective tissue mast cells. There is also some evidence that bacterial adhesion and invasion depends on sulfated surface polysaccharides and use of chondroitin sulfate as "decoy" may prevent bacteria from adhering to the cell surface and causing infection. Aloe vera has been reported to reduce mast cell secretion and mast cell infiltration in an inflamed synovial pouch model.

## 6.2 Flavonoids

Few clinically available drugs can effectively inhibit human mast cell activation. For instance, even though disodium cromoglycate (cromolyn) had been known to inhibit activation of rodent mast cells, it was unable to inhibit human mast cell activation. The most well documented evidence published to date on the inhibitory

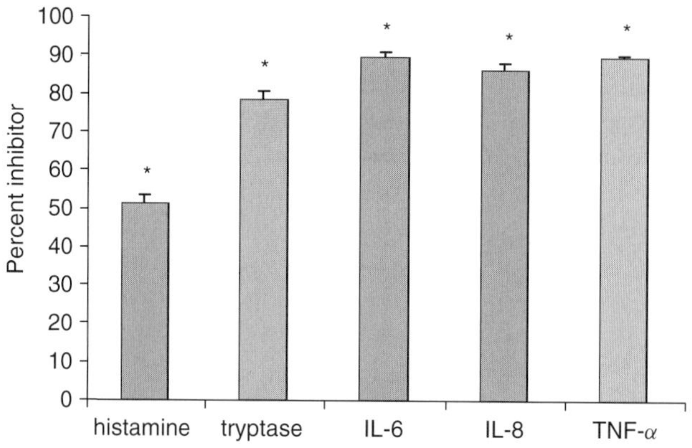


**Fig. 6** Inhibitory effect of quercetin on normal human mast cell proinflammatory mediator secretion. Mast cells were sensitized with human-IgE and was pretreated for 15 min prior to stimulation with anti-IgE (10 μg $mL^{-1}$) for 30 min (histamine, $n = 6–31$; and tryptase, $n = 6–16$) and for 6 h (IL-6, IL-8, and TNF-$\alpha$, $n = 4–7$) at 37 °C. Quercetin significantly inhibited all these mediators when compared to anti-IgE alone treated mast cells. $* = p < 0.05$.

action of mast cells has focused on the naturally occurring flavonoids. For instance, quercetin can inhibit not only the prestored mediators histamine and tryptase from normal human mast cells but also the synthesis of the cytokines IL-6, IL-8, and TNF-$\alpha$ (Fig. 6).

Some flavonoids like morin have weak inhibitory activity, even though it differs from quercetin on the addition of one hydroxyl group in the $\beta$-phenolic ring.

### 6.3 Synergistic Effects

However, chondroitin sulfate appears to block mast cell activation and histamine release; quercetin blocks mast cell secretion and particularly cytokine secretion. The two molecules together could have synergistic actions. Patients do not have an allergic reaction to glucosamine sulfate or chondroitin sulfate if they are allergic to sulfonamide antibiotics. Rutin, the glycoside form of quercetin, also has antiarthritis properties.

The possibility is rather unlikely, because there was absence of cross-reactivity between sulfonamide antibiotics and sulfonamide nonantibiotics.

## 7 Perspectives

Evidence dating back over 20 years that mast cells may participate in inflammation through selective release of proinflammatory molecules had been ignored until recently (see Figure 7). It now appears that mast cells are critical for the initiation of inflammation, especially in arthritis. Unfortunately, the triggers of synovial mast cell activation are not known and there are no clinically available mast cell inhibitors.

A combination of proteoglycans and select flavonoids, such as quercetin (e.g. ArthroSoft®) together with low dose buffered salicylates, (e.g. Trilisate) may

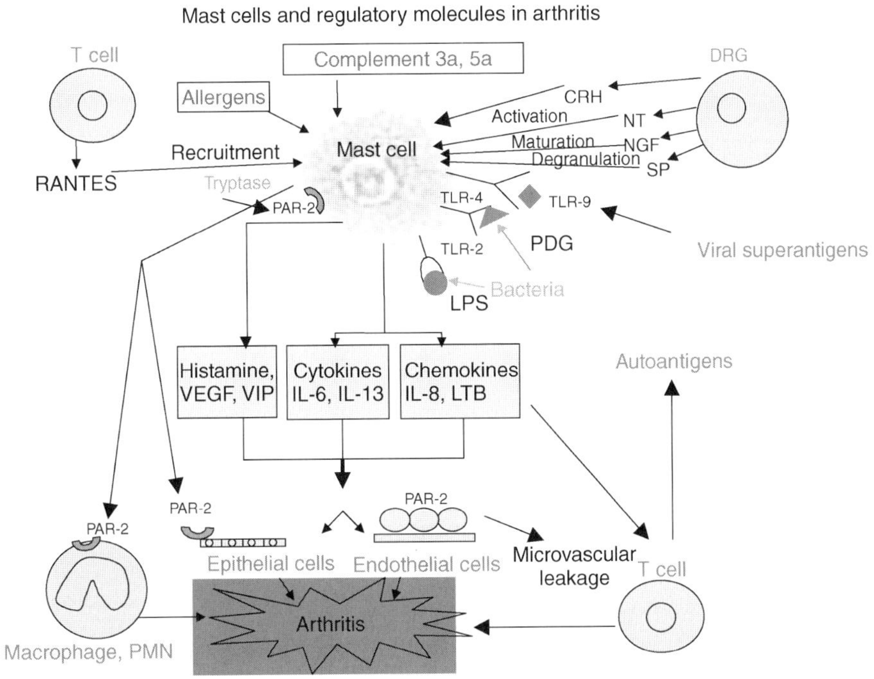


**Fig. 7** Diagrammatic representation of synovial mast cells interaction with other cells in the initiation of synovial inflammation and arthritis (see color plate p. xxix).

be the best initial approach for RA. The additional benefit of quercetin is its unique ability to inhibit secretion of many proinflammatory molecules, especially IL-6 that has been increasingly implicated in juvenile RA. Such molecules could be combined with CRH receptor antagonists when they become clinically available.

## Acknowledgment

Aspects of our own research have been funded by NIH grants No. DK 42409, DK 44816, DK 62861, NS38326, and AR47652. Use of CRH receptor antagonists, as well as proteoglycans and flavonoids, for inflammatory conditions is covered by US patents # 6,020,305; 6,624,148; # 6,635,625; 6,641,806; 6,645,482; 6,689,748 and 09/771,669 (allowed Aug 10, 2004) awarded to TCT and assigned to Theta Biomedical Consulting and Development, Co., Inc. (Brookline, MA). Thanks are due to Dr D. Kempuraj and Dr N. Papadopoulou, as well as to Mrs. Jill Donelan and Mrs. Jessica Christian for some of the graphics, and to the latter also for her patience and word processing skills.

## Bibliography

### Books and Reviews

Akira, S., Takeda, K., Kaisho, T. (2001) Toll-like receptors: critical proteins linking innate and acquired immunity, *Nat. Immunol.* **2**, 675–680.

Chalmers, D.T., Lovenberg, T.W., Grigoriadis, D.E., Behan, D.P., DeSouza, E.B. (1996) Corticotropin-releasing factor receptors: from molecular biology to drug design, *Trends Pharmacol. Sci.* **17**, 166–172.

Chrousos, G.P. (1995) The hypothalamic-pituitary-adrenal axis and immune-mediated inflammation, *N. Engl. J. Med.* **332**, 1351–1362.

Dautzenberg, F.M., Hauger, R.L. (2002) The CRF peptide family and their receptors: yet more partners discovered, *Trends Pharmacol. Sci.* **23**, 71–77.

Dieterich, K.D., Lehnert, H., De Souza, E.B. (1997) Corticotropin-releasing factor receptors: an overview, *Exp. Clin. Endocrinol. Diabetes* **105**, 65–82.

Fitzgerald, G.A. (2005) Coxibs and cardiovascular disease, *N. Engl. J. Med.* **351**, 1709–1711.

Foreman, J.C. (1987) Peptides and neurogenic inflammation, *Brain Res. Bull.* **43**, 386–398.

Galli, S.J. (1993) New concepts about the mast cell, *N. Engl. J. Med.* **328**, 257–265.

Goetzl, E.J., Cheng, P.P.J., Hassner, A., Adelman, D.C., Frick, O.L., Speedharan, S.P. (1990) Neuropeptides, mast cells and allergy: novel mechanisms and therapeutic possibilities, *Clin. Exp. Allergy* **20**, 3–7.

Grammatopoulos, D.K., Chrousos, G.P. (2002) Functional characteristics of CRH receptors and potential clinical applications of CRH-receptor antagonists, *Trends Endocrinol. Metab.* **13**, 436–444.

Karalis, K., Louis, J.M., Bae, D., Hilderbrand, H., Majzoub, J.A. (1997) CRH and the immune system, *J. Neuroimmunol.* **72**, 131–136.

Khansari, D.N., Murgo, A.J., Faith, R.E. (1990) Effects of stress on the immune system, *Immunol. Today* **11**, 170–175.

Lindahl, U., Hook, M. (1978) Glycosaminoglycans and their binding to biological macromolecules, *Annu. Rev. Biochem.* **47**, 385–417.

Manek, N.J. (2001) Medical management of osteoarthritis, *Mayo Clin. Proc.* **76**, 533–539.

McAlindon, T. (2003) Why are clinical trials of glucosamine no longer uniformly positive? *Rheum. Dis. Clin. North Am.* **29**, 789–801.

Rabkin, J.G., Struening, E.L. (1976) Life events, stress and illness, *Science* **194**, 1013–1020.

Redegeld, F.A., Nijkamp, F.P. (2005) Immunoglobulin free light chains and mast cells: pivotal role in T-cell-mediated immune reactions? *Trends Immunol.* **24**, 181–185.

Rosenwasser, L.J., Boyce, J.A. (2003) Mast cells: beyond IgE, *J. Allergy Clin. Immunol.* **111**, 24–32.

Rostand, K.S., Esko, J.D. (1997) Minireview: microbial adherence to and invasion through proteoglycans, *Infect. Immunol.* **65**, 1–8.

Schmidlin, F., Bunnett, N.W. (2001) Protease-activated receptors: how proteases signal to cells, *Curr. Opin. Pharmacol.* **1**, 575–582.

Schwartz, L.B. (1987) Mediators of human mast cells and human mast cell subsets, *Ann. Allergy* **58**, 226–235.

Serafin, W.E., Austen, K.F. (1987) Mediators of immediate hypersensitivity reactions, *N. Engl. J. Med.* **317**, 30–34.

Stead, R.H., Bienenstock, J. (1990) Cellular Interactions Between the Immune and Peripheral Nervous System. A Normal Role for Mast Cells? in: Burger, M.M., Sordat, B., Zinkernagel, R.M. (Eds.) *Cell to Cell Interaction*, Karger, Basel, pp. 170–187.

Taylor, P.C. (2003) Anti-TNFalpha therapy for rheumatoid arthritis: an update, *Intern. Med.* **42**, 15–20.

Theoharides, T.C. (1996) Mast cell: a neuroimmunoendocrine master player, *Int. J. Tissue React.* **18**, 1–21.

Theoharides, T.C. (2002) Mast cells and stress – a psychoneuroimmunological perspective, *J. Clin. Psychopharmacol.* **22**, 103–108.

Theoharides, T.C. (2003) Dietary supplements for arthritis and other inflammatory conditions: key role of mast cells and benefit of combining anti-inflammatory and proteoglycan products, *Eur. J. Inflamm.* **1**, 1–8.

Theoharides, T.C., Bielory, L. (2004) Mast cells and mast cell mediators as targets of dietary supplements, *Ann. Allergy Asthma Immunol.* **93**, S24–S34.

Theoharides, T.C., Cochrane, D.E. (2004) Critical role of mast cells in inflammatory diseases and the effect of acute stress, *J. Neuroimmunol.* **146**, 1–12.

Theoharides, T.C., Sant, G.R. (2003) Neuroimmune Connections and Regulation of Function in the Urinary Bladder, in: Bienenstock, J., Goetzl, E., Blennerhassett, M. (Eds.) *Autonomic Neuroimmunology*, Hardwood Academic Publishers, Lausanne, Switzerland, pp. 345–369.

Theoharides, T.C., Alexandrakis, M., Kempuraj, D., Lytinas, M. (2001) Anti-inflammatory actions of flavonoids and structural requirements for new design, *Int. J. Immunopathol. Pharmacol.* **14**, 119–127.

Theoharides, T.C., Donelan, J.M., Papadopoulou, N., Cao, J., Kempuraj, D., Conti, P. (2004) Mast cells as targets of corticotropin-releasing factor and related peptides, *Trends Pharmacol. Sci.* **25**, 563–568.

Topol, E.J. (2005) Failing the public health – rofecoxib, Merck, and the FDA, *N. Engl. J. Med.* **351**, 1707–1709.

Walker-Bone, K., Javaid, K., Arden, N., Cooper, C. (2000) Regular review: medical management of osteoarthritis, *BMJ* **321**, 936–940.

Wasserman, S.I. (1984) The mast cell and synovial inflammation, *Arthritis Rheum.* **27**, 841–844.

Woolley, D.E. (2003) The mast cell in inflammatory arthritis, *N. Engl. J. Med.* **348**, 1709–1711.

### Primary Literature

Akira, S., Takeda, K., Kaisho, T. (2001). Toll-like receptors: critical proteins linking innate and acquired immunity, *Nat. Immunol.* **2**, 675–680.

Alkaabi, J.K., Ho, M., Levison, R., Pullar, T., Belch, J.J. (2003) Rheumatoid arthritis and macrovascular disease, *Rheumatology (Oxford)* **42**, 292–297.

Aloe, L., Levi-Montalcini, R. (1977) Mast cells increase in tissues of neonatal rats injected with the nerve growth factor, *Brain Res.* **133**, 358–366.

Alonzi, T., Fattori, E., Lazzaro, D., Costa, P., Probert, L., Kollias, G., De Benedetti, F., Poli, V., Ciliberto, G. (1998) Interleukin 6 is required for the development of collagen-induced arthritis, *J. Exp. Med.* **187**, 461–468.

Ando, T., Rivier, J., Yanaihara, H., Arimura, A. (1998) Peripheral corticotropin-releasing factor mediates the elevation of plasma IL-6 by immobilization stress in rats, *Am. J. Physiol.* **275**, R1461–R1467.

Angioni, S., Petraglia, F., Gallinelli, A., Cossarizza, A., Franceschi, C., Muscettola, M., Genazzani, A.D., Surico, N., Genazzani, A.R. (1993) Corticotropin-releasing hormone modulates cytokines release in cultured human peripheral blood mononuclear cells, *Life Sci.* **53**, 1735–1742.

Askenase, P.W. (2005) Mast cells and the mediation of T-cell recruitment in arthritis. *N. Engl. J. Med.* **349**, 1294.

Baici, A., Horler, D., Moser, B., Hofer, H.O., Fehr, K., Wagenhauser, F.J. (1992) Analysis of glycosaminoglycans in human serum after oral administration of chondroitin sulfate, *Rheumatol. Int.* **12**, 81–88.

Baigent, S.M. (2001) Peripheral corticotropin-releasing hormone and urocortin in the control of the immune response, *Peptides* **22**, 809–820.

Bamberger, C.M., Wald, M., Bamberger, A.-M., Ergun, S., Beil, F.U., Schulte, H.M. (1998) Human lymphocytes produce urocortin, but not corticotropin-releasing hormone, *J. Clin. Endocrinol. Metab.* **83**, 708–711.

Batler, R.A., Sengupta, S., Forrestal, S.G., Schaeffer, A.J., Klumpp, D.J. (2002) Mast cell activation triggers a urothelial inflammatory response mediated by tumor necrosis factor-alpha, *J. Urol.* **168**, 819–825.

Benyon, R., Robinson, C., Church, M.K. (1989) Differential release of histamine and eicosanoids from human skin mast cells activated by IgE-dependent and non-immunological stimuli, *Br. J. Pharmacol.* **97**, 898–904.

Bethin, K.E., Vogt, S.K., Muglia, L.J. (2000) Interleukin-6 is an essential, corticotropin-releasing hormone-independent stimulator of the adrenal axis during immune system activation, *Proc. Natl. Acad. Sci. U.S.A.* **97**, 9317–9322.

Bienenstock, J., Tomioka, M., Matsuda, H., Stead, R.H., Quinonez, G., Simon, G.T., Coughlin, M.D., Denburg, J.A. (1987) The role of mast cells in inflammatory processes: evidence for nerve mast cell interactions, *Int. Arch. Allergy Appl. Immunol.* **82**, 238–243.

Bischoff, S.C., Sellge, G., Lorentz, A., Sebald, W., Raab, R., Manns, M.P. (1999) IL-4 enhances proliferation and mediator release in mature human mast cells, *Proc. Natl. Acad. Sci. U.S.A.* **96**, 8080–8085.

Boe, A., Baiocchi, M., Carbonatto, M., Papoian, R., Serlupi-Crescenzi, O. (1999) Interleukin 6 knock-out mice are resistant to antigen-induced experimental arthritis, *Cytokine* **11**, 1057–1064.

Bradding, P., Okayama, Y., Howarth, P.H., Church, M.K., Holgate, S.T. (1995) Heterogeneity of human mast cells based on cytokine content, *J. Immunol.* **155**, 297–307.

Bridges, A.J., Malone, D.G., Jicinsky, J., Chen, M., Ory, P., Engber, W., Graziano, F.M. (1991) Human synovial mast cell involvement in rheumatoid arthritis and osteoarthritis. Relationship to disease type, clinical activity, and antirheumatic therapy, *Arthritis Rheum.* **34**, 1116–1124.

Buchanan, W.W., Goldsmith, C.H., Campbell, J., Stitt, L.W., Bellamy, N. (1988) Validation study of WOMAC: a health status instrument for measuring clinically important patient relevant outcomes to antirheumatic drug therapy in patients with osteoarthritis of the hip or knee, *J. Rheumatol.* **15**, 1833–1840.

Bugajski, A.J., Chlap, Z., Gadek-Michalska, A., Borycz, J., Bugajski, J. (1995) Degranulation and decrease in histamine levels of thalamic mast cells coincides with corticosterone secretion induced by compound 48/80, *Inflamm. Res.* **44**(Supp.1), S50–S51.

Cao, J., Papadopoulou, N., Kempuraj, D., Sugimoto, K., Cetrulo, C.L., Theoharides, T.C. (2005) Human mast cells express corticotropin-releasing hormone (CRH) receptors and CRH leads to selective secretion of vascular endothelial growth factor (VEGF), *J. Immunol.*, in press.

Carraway, R., Cochrane, D.E., Lansman, J.B., Leeman, S.E., Paterson, B.M., Welch, H.J. (1982) Neurotensin stimulates exocytotic histamine secretion from rat mast cells and elevates plasma histamine levels, *J. Physiol.* **323**, 403–414.

Ceponis, A., Konttinen, Y.T., Takagi, M., Xu, J.W., Sorsa, T., Matucci-Cerinic, M., Santavirta, S., Bankl, H.C., Valent, P. (1998) Expression of stem cell factor (SCF) and SCF receptor (c-kit) in synovial membrane in arthritis: correlation with synovial mast cell hyperplasia and inflammation, *J. Rheumatol.* **25**, 2304–2314.

Chalmers, D.T., Lovenberg, T.W., DeSouza, E.B. (1995) Localization of novel corticotropin-releasing factor receptor (CRF2) mRNA expression to specific subcortical nuclei in rat brain: comparison with CRF1 receptor mRNA expression, *J. Neurosci.* **10**, 6340–6350.

Chalmers, D.T., Lovenberg, T.W., Grigoriadis, D.E., Behan, D.P., DeSouza, E.B. (1996) Corticotropin-releasing factor receptors: from molecular biology to drug design, *Trends Pharmacol. Sci.* **17**, 166–172.

Chen, R., Lewis, K.A., Perrin, M.H., Vale, W.W. (1993) Expression cloning of a human corticotropin-releasing factor receptor, *Proc. Natl. Acad. Sci. U.S.A.* **90**, 8967–8971.

Chrousos, G.P. (1995) The hypothalamic-pituitary-adrenal axis and immune-mediated inflammation, *N. Engl. J. Med.* **332**, 1351–1362.

Claman, H.N., Choi, K.L., Sujansky, W., Vatter, A.E. (1986) Mast cell "disappearance" in chronic murine graft-vs-host disease (GVHD)-ultrastructural demonstration of "phantom mast cells", *J. Immunol.* **137**, 2009–2013.

Clifton, V.L., Crompton, R., Smith, R., Wright, I.M. (2002) Microvascular effects of CRH in human skin vary in relation to gender, *J. Clin. Endocrinol. Metab.* **87**, 267–270.

Cochrane, D.E., Carrawan, R.E., Miller, L.A., Feldberg, R.S., Bernheim, H. (2003) Histamine releasing peptide (HRP) has proinflammatory effects and is present at sites of inflammation, *Biochem. Pharmacol.* **66**, 331–342.

Conte, A., de Bernardi, M., Palmieri, L., Lualdi, P., Mautone, G., Ronca, G. (1991) Metabolic fate of exogenous chondroitin sulfate in man, *Arzneimittelforschung* **41**, 768–772.

Conte, A., Volpi, N., Palmieri, L., Bahous, I., Ronca, G. (1995) Biochemical and pharmacokinetic aspects of oral treatment with chondroitin sulfate, *Arzneimittelforschung* **45**, 918–925.

Conti, P., Pang, X., Boucher, W., Letourneau, R., Reale, M., Barbacane, R.C., Thibault, J., Theoharides, T.C. (1997) Impact of Rantes and MCP-1 chemokines on *in vivo* basophilic mast cell recruitment in rat skin injection model and their role in modifying the protein and mRNA levels for histidine decarboxylase, *Blood* **89**, 4120–4127.

Conti, P., Reale, M., Barbacane, R.C., Castellani, M.L., Orso, C. (2002) Differential production of RANTES and MCP-1 in synovial fluid from the inflamed human knee, *Immunol. Lett.* **80**, 105–111.

Conti, P., Reale, M., Barbacane, R.C., Letourneau, R., Theoharides, T.C. (1998) Intramuscular injection of hrRANTES causes mast cell recruitment and increased transcription of histidine decarboxylase: lack of effects in

genetically mast cell-deficient W/W$^{v}$ mice, *FASEB J.* **12**, 1693–1700.
Correa, S.G., Riera, C.M., Spiess, J., Bianco, I.D. (1997) Modulation of the inflammatory response by corticotropin-releasing factor, *Eur. J. Pharmacol.* **319**, 85–90.
Crisp, A.J., Champan, C.M., Kirkham, S.E., Schiller, A.L., Keane, S.M. (1984) Articular mastocytosis in rheumatoid arthritis, *Arthritis Rheum.* **27**, 845–851.
Crofford, L.J., Sano, H., Karalis, K., Webster, E.A., Friedman, T.C., Chrousos, G.P., Wilder, R.L. (1995) Local expression of corticotropin-releasing hormone in inflammatory arthritis, *Ann. N.Y. Acad. Sci.* **771**, 459–471.
Crofford, L.J., Sano, H., Karalis, K., Friedman, T.C., Epps, H.R., Remmers, E.F., Mathern, P., Chrousos, G.P., Wilder, R.L. (1993) Corticotropin-releasing hormone in synovial fluids and tissues of patients with rheumatoid arthritis and osteoarthritis, *J. Immunol.* **151**, 1–10.
Crompton, R., Clifton, V.L., Bisits, A.T., Read, M.A., Smith, R., Wright, I.M. (2003) Corticotropin-releasing hormone causes vasodilation in human skin via mast cell-dependent pathways, *J. Clin. Endocrinol. Metab.* **88**, 5427–5432.
Dautzenberg, F.M., Hauger, R.L. (2002) The CRF peptide family and their receptors: yet more partners discovered, *Trends Pharmacol. Sci.* **23**, 71–77.
Davis, R.H., Stewart, G.J., Bregman, P.J. (1992) Aloe vera and the inflamed synovial pouch model, *J. Am. Podiatr. Med. Assoc.* **82**, 140–148.
De Benedetti, F., Martini, A. (1998) Is systemic juvenile rheumatoid arthritis an interleukin 6 mediated disease? *J. Rheumatol.* **25**, 203–207.
De Benedetti, F., Massa, M., Pignatti, P., Albani, S., Novick, D., Martini, A. (1994) Serum soluble interleukin 6 (IL-6) receptor and IL-6/soluble IL-6 receptor complex in systemic juvenile rheumatoid arthritis, *J. Clin. Invest.* **93**, 2114–2119.
de Hooge, A.S., van De Loo, F.A., Arntz, O.J., van den Berg, W.B. (2000) Involvement of IL-6, apart from its role in immunity, in mediating a chronic response during experimental arthritis, *Am. J. Pathol.* **157**, 2081–2091.
de Paulis, A., Ciccarelli, A., Marinò, I., de Crescenzo, G., Marinò, D., Marone, G. (1997) Human synovial mast cells. II. Heterogeneity of the pharmacologic effects of antiinflammatory and immunosuppressive drugs, *Arthritis Rheum.* **40**, 469–478.
de Paulis, A., Marino, I., Ciccarelli, A., de Crescenzo, G., Concardi, M., Verga, L., Arbustini, E., Marone, G. (1996) Human synovial mast cells. I. Ultrastructural in situ and in vitro immunologic characterization, *Arthritis Rheum.* **39**, 1222–1233.
De Simone, R., Alleva, E., Tirassa, P., Aloe, L. (1990) Nerve growth factor released into the bloodstream following intraspecific fighting induces mast cell degranulation in adult male mice, *Brain Behav. Immun.* **4**, 74–81.
Dhabhar, F., McEwen, B.S. (1996) Stress-induced enhancement of antigen-specific cell-mediated immunity, *J. Immunol.* **156**, 2608–2615.
Dhabhar, F.S., McEwen, B.S. (1999) Enhancing versus suppressive effects of stress hormones on skin immune function, *Proc. Natl. Acad. Sci. U.S.A.* **96**, 1059–1064.
Dhabhar, F.S., Miller, A.H., McEwen, B.S., Spencer, R.L. (1995) Effects of stress on immune cell distribution: dynamics and hormonal mechanisms, *J. Immunol.* **154**, 5511–5527.
Dieterich, K.D., Lehnert, H., De Souza, E.B. (1997) Corticotropin-releasing factor receptors: an overview, *Exp. Clin. Endocrinol. Diabetes* **105**, 65–82.
Dimitriadou, V., Pang, X., Theoharides, T.C. (2000) Hydroxyzine inhibits experimental allergic encephalomyelitis (EAE) and associated brain mast cell activation, *Int. J. Immunopharmacol.* **22**, 673–684.
Dimitriadou, V., Aubineau, P., Taxi, J., Seylaz, J. (1987) Ultrastructural evidence for a functional unit between nerve fibers and type II cerebral mast cells in the cerebral vascular wall, *Neuroscience* **22**, 621–630.
Dimitriadou, V., Lambracht-Hall, M., Reichler, J., Theoharides, T.C. (1990) Histochemical and ultrastructural characteristics of rat brain perivascular mast cells stimulated with compound 48/80 and carbachol, *Neuroscience* **39**, 209–224.
Dimitriadou, V., Buzzi, M.G., Moskowitz, M.A., Theoharides, T.C. (1991) Trigeminal sensory fiber stimulation induces morphologic changes reflecting secretion in rat dura mast cells, *Neuroscience* **44**, 97–112.
Dimitriadou, V., Rouleau, A., Trung Tuong, M.D., Newlands, G.J.F., Miller, H.R.P.,

Luffau, G., Schwartz, J.-C., Garbarg, M. (1997) Functional relationships between sensory nerve fibers and mast cells of dura mater in normal and inflammatory conditions, *Neuroscience* **77**, 829–839.

Ditzel, H.J. (2005) The K/BxN mouse: a model of human inflammatory arthritis, *Trends Mol. Med.* **10**, 40–45.

Douglas, W.W. (1974) Involvement of calcium in exocytosis and the exocytosis-vesiculation sequence, *Biochem. Soc. Symp.* **39**, 1–28.

Duensing, T.D., Wing, J.S., vanPutten, J.P.M. (1999) Sulfated polysaccharide-directed recruitment of mammalian host proteins: a novel strategy in microbial pathogenesis, *Infect. Immunol.* **67**, 4463–4468.

Dunlop, D.D., Manheim, L.M., Yelin, E.H., Song, J., Chang, R.W. (2003) The costs of arthritis, *Arthritis Rheum.* **49**, 101–113.

Dvorak, A.M., Macglashan, D.W., Morgan, E.S., Lichtenstein, L.M. Jr. (1996) Vesicular transport of histamine in stimulated human basophils, *Blood* **88**, 4090–4101.

Dvorak, A.M., McLeod, R.S., Onderdonk, A.B., Monahan-Earley, R.A., Cullen, J.B., Antonioli, D.A., Morgan, E., Blair, J.E., Estrella, P., Cisneros, R.L., Cohen, Z., Silen, W. (1992) Human gut mucosal mast cells: ultrastructural observations and anatomic variation in mast cell-nerve associations *in vivo*, *Int. Arch. Allergy Immunol.* **98**, 158–168.

Esposito, P., Basu, S., Letourneau, R., Jacobson, S., Theoharides, T.C. (2003) Corticotropin-releasing factor (CRF) can directly affect brain microvessel endothelial cells, *Brain Res.* **968**, 192–198.

Esposito, P., Gheorghe, D., Kandere, K., Pang, X., Conally, R., Jacobson, S., Theoharides, T.C. (2001) Acute stress increases permeability of the blood-brain-barrier through activation of brain mast cells, *Brain Res.* **888**, 117–127.

Fajardo, I., Pejler, G. (2003) Human mast cell beta-tryptase is a gelatinase, *J. Immunol.* **171**, 1493–1499.

Fitzgerald, G.A. (2005) Coxibs and cardiovascular disease, *N. Engl. J. Med.* **351**, 1709–1711.

Foreman, J.C. (1987) Peptides and neurogenic inflammation, *Brain Res. Bull.* **43**, 386–398.

Gadek-Michalska, A., Chlap, Z., Turon, M., Bugajski, J., Fogel, W.A. (1991) The intracerebroventicularly administered mast cells degranulator compound 48/80 increases the pituitary-adrenocortical activity in rats, *Agents Actions* **32**, 203–208.

Gagari, E., Tsai, M., Lantz, C.S., Fox, L.G., Galli, S.J. (1997) Differential release of mast cell interleukin-6 via c-kit, *Blood* **89**, 2654–2663.

Galli, S.J. (1993) New concepts about the mast cell, *N. Engl. J. Med.* **328**, 257–265.

Genedani, S., Bernardi, M., Baldini, M.G., Bertolini, A. (1992) Influence of CRH and α-MSH on the migration of human monocytes *in vitro*, *Neuropeptides* **23**, 99–102.

Genovese, A., Borgia, G., Bouvet, J.P., Detoraki, A., de Paulis, A., Piazza, M., Marone, G. (2005) Protein Fv produced during viral hepatitis is an endogenous immunoglobulin superantigen activating human heart mast cells, *Int. Arch. Allergy Immunol.* **132**, 336–345.

Goetzl, E.J., Cheng, P.P.J., Hassner, A., Adelman, D.C., Frick, O.L., Speedharan, S.P. (1990) Neuropeptides, mast cells and allergy: novel mechanisms and therapeutic possibilities, *Clin. Exp. Allergy* **20**, 3–7.

Gotis-Graham, I., Smith, M.D., Parker, A., McNeil, H.P. (1998) Synovial mast cell responses during clinical improvement in early rheumatoid arthritis, *Ann. Rheum. Dis.* **57**, 664–671.

Grabbe, J., Welker, P., Möller, A., Dippel, E., Ashman, L.K., Czarnetzki, B.M. (1994) Comparative cytokine release from human monocytes, monocyte-derived immature mast cells and a human mast cell line (HMC-1), *J. Invest. Dermatol.* **103**, 504–508.

Grammatopoulos, D.K., Chrousos, G.P. (2002) Functional characteristics of CRH receptors and potential clinical applications of CRH-receptor antagonists, *Trends Endocrinol. Metab.* **13**, 436–444.

Hagen, L.E., Schneider, R., Stephens, D., Modrusan, D., Feldman, B.M. (2003) Use of complementary and alternative medicine by pediatric rheumatology patients, *Arthritis Rheum.* **49**, 3–6.

Hardingham, T.E. (1986) Structure and biosynthesis of proteoglycans, *J. Rheumatol.* **10**, 143–183.

He, S., Walls, A.F. (1997) Human mast cell tryptase: a stimulus of microvascular leakage and mast cell activation, *Eur. J. Pharmacol.* **328**, 89–97.

He, S., Gaca, M.D., Walls, A.F. (1998) A role for tryptase in the activation of human mast cells:

modulation of histamine release by tryptase and inhibitors of tryptase, *J. Pharmacol. Exp. Ther.* **286**, 289–298.

He, S., Gaca, M.D., Walls, A.F. (2001) The activation of synovial mast cells: modulation of histamine release by tryptase and chymase and their inhibitors, *Eur. J. Pharmacol.* **412**, 223–229.

Herrmann, M., Scholmerich, J., Straub, R.H. (2000) Stress and rheumatic diseases, *Rheum. Dis. Clin. North Am.* **26**, 737–763.

Hiromatsu, Y., Toda, S. (2003) Mast cells and angiogenesis, *Microsc. Res. Tech.* **60**, 64–69.

Huang, M., Berry, J., Kandere, K., Lytinas, M., Karalis, K., Theoharides, T.C. (2002) Mast cell deficient $W/W^v$ mice lack stress-induced increase in serum IL-6 levels, as well as in peripheral CRH and vascular permeability, a model of rheumatoid arthritis, *Int. J. Immunopathol. Pharmacol.* **15**, 249–254.

Hughes, R., Carr, A. (2002) A randomized, double-blind, placebo-controlled trial of glucosamine sulphate as an analgesic in osteoarthritis on the knee, *Rheumatology (Oxford)* **41**, 279–284.

Jain, S., Stevenson, J.R. (1991) Enhancement by restraint stress of natural killer cell activity and splenocyte responsiveness to concanavalin A in Fischer 344 rats, *Immunol. Invest.* **20**, 365–376.

Janiszewski, J., Bienenstock, J., Blennerhassett, M.G. (1994) Picomolar doses of substance P trigger electrical responses in mast cells without degranulation, *Am. J. Physiol.* **267**, C138–C145.

Johnson, E.O., Moutsopoulos, M. (1992) Neuroimmunological axis and rheumatic diseases, *Eur. J. Clin. Invest.* **22**, S2–S5.

Kandere-Grzybowska, K., Letourneau, R., Boucher, W., Bery, J., Kempuraj, D., Poplawski, S., Athanassiou, A., Theoharides, T.C. (2003) IL-1 induces vesicular secretion of IL-6 without degranulation from human mast cells, *J. Immunol.* **171**, 4830–4836.

Kaneko, K., Kawana, S., Arai, K., Shibasaki, T. (2003) Corticotropin-releasing factor receptor type 1 is involved in the stress-induced exacerbation of chronic contact dermatitis in rats, *Exp. Dermatol.* **12**, 47–52.

Kaneko, S., Satoh, T., Chiba, J., Ju, C., Inoue, K., Kagawa, J. (2000) Interleukin-6 and interleukin-8 levels in serum and synovial fluid of patients with osteoarthritis, *Cytokines Cell. Mol. Ther.* **6**, 71–79.

Karalis, K., Louis, J.M., Bae, D., Hilderbrand, H., Majzoub, J.A. (1997) CRH and the immune system, *J. Neuroimmunol.* **72**, 131–136.

Karalis, K., Sano, H., Redwine, J., Listwak, S., Wilder, R.L., Chrousos, G.P. (1991) Autocrine or paracrine inflammatory actions of corticotropin-releasing hormone *in vivo*, *Science* **254**, 421–423.

Kempuraj, D., Huang, M., Kandere-Grzybowska, K., Basu, S., Boucher, W., Letourneau, R., Athanasiou, A., Theoharides, T.C. (2003) Azelastine inhibits secretion of IL-6, TNF-$\alpha$ and IL-8 as well as NF-$\kappa$B activation and intracellular calcium ion levels in normal human mast cells, *Int. Arch. Allergy Immunol.* **132**, 231–239.

Kempuraj, D., Madhappan, B., Christodoulou, S., Boucher, W., Cao, J., Papadopoulou, N., Cetrulo, C.L., Theoharides, T.C. (2005) Flavonols inhibit pro-inflammatory mediator release, intracellular calcium ion levels and protein kinase C theta phosphorylation in human mast cells, *Br. J. Pharmacol.*, in press.

Kempuraj, D., Papadopoulou, N.G., Lytinas, M., Huang, M., Kandere-Grzybowska, K., Madhappan, B., Boucher, W., Christodoulou, S., Athanasiou, A., Theoharides, T.C. (2004) Corticotropin-releasing hormone and its structurally related urocortin are synthesized and secreted by human mast cells, *Endocrinology* **145**, 43–48.

Keul, R., Heinrich, P.C., Muller-Newen, G., Muller, K., Woo, P. (1998) A possible role for soluble IL-6 receptor in the pathogenesis of systemic onset juvenile chronic arthritis, *Cytokine* **10**, 729–734.

Khansari, D.N., Murgo, A.J., Faith, R.E. (1990) Effects of stress on the immune system, *Immunol. Today* **11**, 170–175.

Kiecolt-Glaser, J.K., Preacher, K.J., MacCallum, R.C., Atkinson, C., Malarkey, W.B., Glaser, R. (2003) Chronic stress and age-related increases in the proinflammatory cytokine IL-6, *Proc. Natl. Acad. Sci. U.S.A.* **100**, 9090–9095.

Kiener, H.P., Baghestanian, M., Dominkus, M., Walchshofer, S., Ghannadan, M., Willheim, M., Sillaber, C., Graninger, W.B., Smolen, J.S., Valent, P. (1998) Expression of the C5a receptor (CD88) on synovial mast cells in patients with rheumatoid arthritis, *Arthritis Rheum.* **41**, 233–245.

Kobayashi, H., Ishizuka, T., Okayama, Y. (2000) Human mast cells and basophils as sources of cytokines, *Clin. Exp. Allergy* **30**, 1205–1212.

Kohno, M., Kawahito, Y., Tsubouchi, Y., Hashiramoto, A., Yamada, R., Inoue, K.I., Kusaka, Y., Kubo, T., Elenkov, I.J., Chrousos, G.P., Kondo, M., Sano, H. (2001) Urocortin expression in synovium of patients with rheumatoid arthritis and osteoarthritis: relation to inflammatory activity, *J. Clin. Endocrinol. Metab.* **86**, 4344–4352.

Kops, S.K., Theoharides, T.C., Cronin, C.T., Kashgarian, M.G., Askenase, P.W. (1990) Ultrastructural characteristics of rat peritoneal mast cells undergoing differential release of serotonin without histamine and without degranulation, *Cell Tissue Res.* **262**, 415–424.

Kops, S.K., Van Loveren, H., Rosenstein, R.W., Ptak, W., Askenase, P.W. (1984) Mast cell activation and vascular alterations in immediate hypersensitivity-like reactions induced by a T cell derived antigen-binding factor, *Lab. Invest.* **50**, 421–434.

Krüger-Krasagakes, S., Möller, A.M., Kolde, G., Lippert, U., Weber, M., Henz, B.M. (1996) Production of inteuleukin-6 by human mast cells and basophilic cells, *J. Invest. Dermatol.* **106**, 75–79.

Kubler, A., Rothacher, G., Knappertz, V.A., Kramer, G., Nink, M., Beyer, J., Lehnert, H. (1994) Intra and extracerebral blood flow changes and flushing after intravenous injection of human corticotropin-releasing hormone, *Clin. Invest.* **72**, 331–336.

Leal-Berumen, I., Conlon, P., Marshall, J.S. (1994) IL-6 production by rat peritoneal mast cells is not necessarily preceded by histamine release and can be induced by bacterial lipopolysaccharide, *J. Immunol.* **152**, 5468–5476.

Lee, D.M., Friend, D.S., Gurish, M.F., Benoist, C., Mathis, D., Brenner, M.B. (2002) Mast cells: a cellular link between autoantibodies and inflammatory arthritis, *Science* **297**, 1689–1692.

Letourneau, R., Pang, X., Sant, G.R., Theoharides, T.C. (1996) Intragranular activation of bladder mast cells and their association with nerve processes in interstitial cystitis, *Br. J. Urol.* **77**, 41–54.

Letourneau, R., Rozniecki, J.J., Dimitriadou, V., Theoharides, T.C. (2003) Ultrastructural evidence of brain mast cell activation without degranulation in monkey experimental allergic encephalomyelitis, *J. Neuroimmunol.* **145**, 18–26.

Leu, S.J., Singh, V.K. (1992) Stimulation of interleukin-6 production by corticotropin-releasing factor, *Cell. Immunol.* **143**, 220–227.

Levi-Schaffer, F., Shalit, M. (1989) Differential release of histamine and prostaglandin D2 in rat peritoneal mast cells activated with peptides, *Int. Arch. Allergy Appl. Immunol.* **90**, 352–357.

Levi-Schaffer, F., Austen, K.F., Gravallese, P.M., Stevens, R.L. (1986) Co-culture of interleukin 3-dependent mouse mast cells with fibroblasts results in a phenotypic change of the mast cells, *Proc. Natl. Acad. Sci. U.S.A.* **83**, 6485–6488.

Lewis, K., Li, C., Perrin, M.H., Blount, A., Kunitake, K., Donaldson, C., Vaughan, J., Reyes, T.M., Gulyas, J., Fischer, W., Bilezikjian, L., Rivier, J., Sawchenko, P.E., Vale, W.W. (2001) Identification of urocortin III, an additional member of the corticotropin-releasing factor (CRF) family with high affinity for the CRF 2 receptor, *Proc. Natl. Acad. Sci. U.S.A.* **98**, 7570–7575.

Lindahl, U., Hook, M. (1978) Glycosaminoglycans and their binding to biological macromolecules, *Annu. Rev. Biochem.* **47**, 385–417.

Lovenberg, T.W., Chalmers, D.T., Liu, C., DeSouza, E.B. (1995a) CRF2$\alpha$ and CRF2$\beta$ receptor mRNAs are differentially distributed between the rat central nervous system and peripheral tissues, *Endocrinology* **136**, 4139–4142.

Lovenberg, T.W., Liaw, C.W., Grigoriadis, D.E., Clevenger, W., Charmers, D.T., DeSouza, E.B., Oltersdorf, T. (1995b) Cloning and characterization of a functionally distinct corticotropin-releasing factor receptor subtype from rat brain, *Proc. Natl. Acad. Sci. U.S.A.* **92**, 836–840.

Lowry, P.J., Woods, R.J., Baigent, S. (1996) Corticotropin releasing factor and its binding protein, *Pharmacol. Biochem. Behav.* **54**, 305–308.

Lutgendorf, S.K., Russell, D., Ullrich, P., Harris, T.B., Wallace, R. (2004) Religious participation, interleukin-6, and mortality in older adults, *Health Psychol.* **23**, 465–475.

Lytinas, M., Kempuraj, D., Huang, M., Boucher, W., Esposito, P., Theoharides, T.C. (2003) Acute stress results in skin corticotropin-releasing hormone secretion, mast cell activation and vascular permeability, an effect mimicked by intradermal corticotropin-releasing hormone and inhibited by histamine-1 receptor antagonists, *Int. Arch. Allergy Immunol.* **130**, 224–231.

Malaviya, R., Ross, E., Jakschik, B.A., Abraham, S.N. (1994) Mast cell degranulation induced by type 1 fimbriated *Escherichia coli* in mice, *J. Clin. Invest.* **93**, 1645–1653.

Malone, E.D. (2002) Managing chronic arthritis, *Vet. Clin. North Am. Equine Pract.* **18**, 411–437.

Manek, N.J. (2001) Medical management of osteoarthritis, *Mayo Clin. Proc.* **76**, 533–539.

Mastorakos, G., Chrousos, G.P., Weber, J.S. (1993) Recombinant interleukin-6 activates the hypothalamic-pituitary-adrenal axis in humans, *J. Clin. Endocrinol. Metab.* **77**, 1690–1694.

Matsuda, H., Kawakita, K., Kiso, Y., Nakano, T., Kitamura, Y. (1989) Substance P induces granulocyte infiltration through degranulation of mast cells, *J. Immunol.* **142**, 927–931.

Matsuda, H., Kannan, Y., Ushio, H., Kiso, Y., Kanemoto, T., Suzuki, H., Kitamura, Y. (1991) Nerve growth factor induces development of connective tissue-type mast cells *in vitro* from murine bone marrow cells, *J. Exp. Med.* **174**, 7–14.

Matsumoto, I., Inoue, Y., Shimada, T., Aikawa, T. (2001) Brain mast cells act as an immune gate to the hypothalamic-pituitary-adrenal axis in dogs, *J. Exp. Med.* **194**, 71–78.

Matsushima, H., Yamada, N., Matsue, H., Shimada, S. (2004) The effects of endothelin-1 on degranulation, cytokine, and growth factor production by skin-derived mast cells, *Eur. J. Immunol.* **34**, 1910–1019.

Mattheos, S., Christodoulou, S., Kempuraj, D., Kempuraj, B., Karalis, K., Theoharides, T.C. (2003) Mast cells and corticotropin-releasing hormone (CRH) are required for experimental inflammatory arthritis, *FASEB J.* **17**, C44.

Maurer, M., Wedemeyer, J., Metz, M., Piliponsky, A.M., Weller, K., Chatterjea, D., Clouthier, D.E., Yanagisawa, M.M., Tsai, M., Galli, S.J. (2004) Mast cells promote homeostasis by limiting endothelin-1-induced toxicity, *Nature* **432**, 512–516.

McAlindon, T. (2003) Why are clinical trials of glucosamine no longer uniformly positive? *Rheum. Dis. Clin. North Am.* **29**, 789–801.

McAlindon, T.E., LaValley, M.P., Gulin, J.P., Felson, D.T. (2000) Glucosamine and chondroitin for treatment of osteoarthritis, *JAMA* **283**, 1469–1475.

McCurdy, J.D., Olynych, T.J., Maher, L.H., Marshall, J.S. (2003) Cutting edge: distinct Toll-like receptor 2 activators selectively induce different classes of mediator production from human mast cells, *J. Immunol.* **170**, 1625–1629.

McEvoy, A.N., Bresnihan, B., FitzGerald, O., Murphy, E.P. (2001) Corticotropin-releasing hormone signaling in synovial tissue from patients with early inflammatory arthritis is mediated by the type 1$\alpha$ corticotropin-releasing hormone receptor, *Arthritis Rheum.* **44**, 1761–1767.

McLachlan, J.B., Hart, J.P., Pizzo, S.V., Shelburne, C.P., Staats, H.F., Gunn, M.D., Abraham, S.N. (2003) Mast cell-derived tumor necrosis factor induces hypertrophy of draining lymph nodes during infection, *Nat. Immunol.* **4**, 1199–1205.

Merchenthaler, I., Hynes, M.A., Vingh, S., Schally, A.V., Petrusz, P. (1983) Immunocytochemical localization of corticotropin-releasing factor (CRF) in the rat spinal cord, *Brain Res.* **275**, 373–377.

Middleton, E., Kandaswami, C., Theoharides, T.C. Jr. (2000) The effects of plant flavonoids on mammalian cells:implications for inflammation, heart disease and cancer, *Pharmacol. Rev.* **52**, 673–751.

Mihara, M., Kotoh, M., Nishimoto, N., Oda, Y., Kumagai, E., Takagi, N., Tsunemi, K., Ohsugi, Y., Kishimoto, T., Yoshizaki, K., Takeda, Y. (2001) Humanized antibody to human interleukin-6 receptor inhibits the development of collagen arthritis in cynomolgus monkeys, *Clin. Immunol.* **98**, 319–326.

Million, M., Maillot, C., Saunders, P., Rivier, J., Vale, W., Taché, Y. (2002) Human urocortin II, a new CRF related peptide, displays selective CRF2-mediated action on gastric transit in rats, *Am. J. Physiol. Gastrointest. Liver Physiol.* **282**, G34–G40.

Muller-Fassbender, H., Bach, G.L., Haase, W., Rovati, L.C., Setnikar, I. (1994) Glucosamine sulfate compared to ibuprofen in osteoarthritis of the knee, *Osteoarthritis Cartil.* **2**, 61–69.

Navarra, P., Tsagarakis, S., Faria, M.S., Rees, L.H., Besser, G.M., Grossman, A.B. (1991) Interleukins-1 and -6 stimulate the release of corticotropin-releasing hormone-41 from rat hypothalamus *in vitro* via the eicosanoid cyclooxygenase pathway, *Endocrinology* **128**, 37–44.

Nigrovic, P.A., Lee, D.M. (2005) Mast cells in inflammatory arthritis, *Arthritis Res. Ther.* **7**, 1–11.

Nishioka, T., Kurokawa, H., Takao, R., Kumon, Y., Nishiya, K., Hashimoto, K. (1996) Differential changes of corticotropin releasing hormone (CRH) concentrations in plasma and synovial fluids of patients with rheumatoid arthritis (RA), *Endocr. J.* **43**, 241–247.

Ohshima, S., Saeki, Y., Mima, T., Sasai, M., Nishioka, K., Nomura, S., Kopf, M., Katada, Y., Tanaka, T., Suemura, M., Kishimoto, T. (1998) Interleukin 6 plays a key role in the development of antigen-induced arthritis, *Proc. Natl. Acad. Sci. U.S.A.* **95**, 8222–8226.

Olsson, N., Ulfgre, A.K., Nilsson, G. (2001) Demonstration of mast cell chemotactic activity in synovial fluid from rheumatoid patients, *Ann. Rheum. Dis.* **60**, 187–193.

Ostrakhovitch, E.A., Afanas'ev, I.B. (2001) Oxidative stress in rheumatoid arthritis leukocytes: suppression by rutin and other antioxidants and chelators, *Biochem. Pharmacol.* **62**, 743–746.

Palmieri, L., Conte, A., Giovannini, L., Lualdi, P., Ronca, G. (1990) Metabolic fate of exogenous chondroitin sulfate in the experimental animal, *Arzneimittelforschung* **40**, 319–323.

Pavelka, K., Gatterova, J., Olejarova, M., Machacek, S., Giacovelli, G., Rovati, L.C. (2002) Glucosamine sulfate use and delay of progression of knee osteoarthritis: a 3-year, randomized, placebo-controlled, double-blind study, *Arch. Intern. Med.* **162**, 2113–2123.

Pendleton, A., Arden, N., Dougados, M., Doherty, M., Bannwarth, B., Bijlsma, J.W., Cluzeau, F., Cooper, C., Dieppe, P.A., Gunther, K.P., Hauselmann, H.J., Herrero-Beaumont, G., Kaklamanis, P.M., Lequesne, M., Lohmander, S., Mazieres, B., Mola, E.M., Pavelka, K., Serni, U., Swoboda, B., Verbruggen, A.A., Weseloh, G., Zimmermann-Gorska, I. (2000) EULAR recommendations for the management of knee osteoarthritis: report of a task force of the Standing Committee for International Clinical Studies Including Therapeutic Trials (ESCISIT), *Ann. Rheum. Dis.* **59**, 936–944.

Pignatti, P., Vivarelli, M., Meazza, C., Rizzolo, M.G., Martini, A., De Benedetti, F. (2001) Abnormal regulation of interleukin 6 in systemic juvenile idiopathic arthritis, *J. Rheumatol.* **28**, 1670–1676.

Rabkin, J.G., Struening, E.L. (1976) Life events, stress and illness, *Science* **194**, 1013–1020.

Radulovic, M., Dautzenberg, F.M., Sydow, S., Radulovic, J., Spiess, J. (1999) Corticotropin-releasing factor receptor 1 in mouse spleen: expression after immune stimulation and identification of receptor-bearing cells, *J. Immunol.* **162**, 3013–3021.

Redegeld, F.A., Nijkamp, F.P. (2005) Immunoglobulin free light chains and mast cells: pivotal role in T-cell-mediated immune reactions? *Trends Immunol.* **24**, 181–185.

Reginster, J.Y., Deroisy, R., Rovati, L.C., Lee, R.L., Lejeune, E., Bruyere, O., Giacovelli, G., Henrotin, Y., Dacre, J.E., Gossett, C. (2001) Long-term effects of glucosamine sulphate on osteoarthritis progression: a randomized, placebo-controlled clinical trial, *Lancet* **357**, 251–256.

Rinner, I., Schauenstein, K., Mangge, H., Porta, S., Kvetnansky, R. (1992) Opposite effects of mild and severe stress on in vitro activation of rat peripheral blood lymphocytes, *Brain. Behav. Immunol.* **6**, 130–140.

Rodewald, H.-R., Dessing, M., Dvorak, A.M., Galli, S.J. (1996) Identification of a committed precursor for the mast cell lineage, *Science* **271**, 818–822.

Rosenwasser, L.J., Boyce, J.A. (2003) Mast cells: beyond IgE, *J. Allergy Clin. Immunol.* **111**, 24–32.

Rostand, K.S., Esko, J.D. (1997) Minireview: microbial adherence to and invasion through proteoglycans, *Infect. Immunol.* **65**, 1–8.

Roupe van der Voort, C., Heijnen, C.J., Wulffraat, N., Kuis, W., Kavelaars, A. (2000) Stress induces increases in IL-6 production by leucocytes of patients with the chronic inflammatory disease juvenile rheumatoid arthritis: a putative role for alpha(1)-adrenergic receptors, *J. Neuroimmunol.* **110**, 223–229.

Rozniecki, J.J., Dimitriadou, V., Lambracht-Hall, M., Pang, X., Theoharides, T.C. (1999) Morphological and functional demonstration of rat dura mast cell-neuron interactions *in vitro* and *in vivo*, *Brain Res.* **849**, 1–15.

Ruschpler, P., Lorenz, P., Eichler, W., Koczan, D., Hanel, C., Scholz, R., Melzer, C., Thiesen, H.J., Stiehl, P. (2005) High CXCR3 expression in synovial mast cells associated with CXCL9 and CXCL10 expression in inflammatory synovial tissues of patients with rheumatoid arthritis, *Arthritis Res. Ther.* **5**, R241–R252.

Sakai, K., Clemmons, D.R. (2005) Glucosamine induces resistance to insulin-like growth factor I (IGF-I) and insulin in Hep G2 cell cultures: biological significance of IGF-I/insulin hybrid receptors, *Endocrinology* **144**, 2388–2395.

Schmidlin, F., Bunnett, N.W. (2001) Protease-activated receptors: how proteases signal to cells, *Curr. Opin. Pharmacol.* **1**, 575–582.

Schwartz, L.B. (1987) Mediators of human mast cells and human mast cell subsets, *Ann. Allergy* **58**, 226–235.

Serafin, W.E., Austen, K.F. (1987) Mediators of immediate hypersensitivity reactions, *N. Engl. J. Med.* **317**, 30–34.

Shanas, U., Bhasin, R., Sutherland, A.K., Silverman, A.-J., Silver, R. (1998) Brain mast cells lack the c-kit receptor: immunocytochemical evidence, *J. Neuroimmunol.* **90**, 207–211.

Singh, L.K., Boucher, W., Pang, X., Letourneau, R., Seretakis, D., Green, M., Theoharides, T.C. (1999) Potent mast cell degranulation and vascular permeability triggered by urocortin through activation of CRH receptors, *J. Pharmacol. Exp. Ther.* **288**, 1349–1356.

Singh, V.K. (1989) Stimulatory effect of corticotropin-releasing neurohormone on human lymphocyte proliferation and interleukin-2 receptor expression, *J. Neuroimmunol.* **23**, 256–262.

Singh, V.K., Leu, C.S.J. (1990) Enhancing effect of corticotropin-releasing neurohormone on the production of interleukin-1 and interleukin-2, *Neurosci. Lett.* **120**, 151–154.

Skofitsch, G., Savitt, J.M., Jacobowitz, D.M. (1985a) Suggestive evidence for a functional unit between mast cells and substance P fibers in the rat diaphragm and mesentery, *Histochemistry* **82**, 5–8.

Skofitsch, G., Zamir, N., Helke, C.J., Savitt, J.M., Jacobowitz, D.M. (1985b) Corticotropin-releasing factor-like immunoreactivity in sensory ganglia and capsaicin sensitive neurons of the rat central nervous system: colocalization with other neuropeptides, *Peptides* **6**, 307–318.

Slominski, A., Pisarchik, A., Tobin, D.J., Mazurkiewicz, J., Wortsman, J. (2004) Differential expression of a cutaneous corticotropin-releasing hormone system, *Endocrinology* **145**, 941–950.

Slominski, A., Wortsman, J., Luger, T., Paus, R., Solomon, S. (2000) Corticotropin releasing hormone and proopiomelanocortin involvement in the cutaneous response to stress, *Physiol. Rev.* **80**, 979–1020.

Spinedi, E., Hadid, R., Daneva, T., Gaillard, R.C. (1992) Cytokines stimulate the CRH but not the vasopressin neuronal system: evidence for a median eminence site of interleukin-6 action, *Neuroendocrinology* **56**, 46–53.

Stead, R.H., Bienenstock, J. (1990) Cellular Interactions Between the Immune and Peripheral Nervous System. A Normal Role for Mast Cells? in: Burger, M.M., Sordat, B., Zinkernagel, R.M. (Eds.) *Cell to Cell Interaction*, Karger, Basel, pp. 170–187.

Stead, R.H., Tomioka, M., Quinonez, G., Simon, G.T., Felten, S.Y., Bienenstock, J. (1987) Intestinal mucosal mast cells in normal and nematode-infected rat intestines are in intimate contact with peptidergic nerves, *Proc. Natl. Acad. Sci. U.S.A.* **84**, 2975–2979.

Strom, B.L., Schinzel, R., Apter, A.J., Margolis, D.J., Lautenbach, E., Hennessy, S., Bilker, W.B., Pettitt, D. (2003) Absence of cross-reactivity between sulfonamide antibiotics and sulfonamide nonantibiotics, *N. Engl. J. Med.* **349**, 1628–1635.

Suda, T., Tomori, N., Tozawa, F., Mouri, T., Demura, H., Shizume, K. (1984) Distribution and characterization of immunoreactive corticotropin-releasing factor in human tissues, *J. Clin. Endocrinol. Metab.* **59**, 861–866.

Takagi, M., Nakahata, T., Kubo, T., Shiohara, M., Koike, K., Miyajima, A., Arai, K.-I., Nishikawa, S.-I., Zsebo, K.M., Komiyama, A. (1992) Stimulation of mouse connective tissue-type mast cells by hemopoietic stem cell factor, a ligand for the *c-kit* receptor, *J. Immunol.* **148**, 3446–3453.

Tal, M., Liberman, R. (1997) Local injection of nerve growth factor (NGF) triggers degranulation of mast cells in rat paw, *Neurosci. Lett.* **221**, 129–132.

Taylor, P.C. (2003) Anti-TNFalpha therapy for rheumatoid arthritis: an update, *Intern. Med.* **42**, 15–20.

Taylor, A.M., Galli, S.J., Coleman, J.W. (1995) Stem-cell factor, the kit ligand, induces direct degranulation of rat peritoneal mast cells in vitro and in vivo: dependence of the in vitro effect on period of culture and comparisons of stem- cell factor with other mast cell-activating agents, *Immunology* **86**, 427–433.

Tetlow, L.C., Woolley, D.E. (1995b) Distribution, activation and tryptase/chymase phenotype of mast cells in the rheumatoid lesion, *Ann. Rheum. Dis.* **54**, 549–555.

Tetlow, L.C., Woolley, D.E. (1995a) Mast cells, cytokines, and metalloproteinases at the rheumatoid lesion: dual immunolocalization studies, *Ann. Rheum. Dis.* **54**, 896–903.

Theoharides, T.C. (1996) Mast cell: a neuroimmunoendocrine master player, *Int. J. Tissue React.* **18**, 1–21.

Theoharides, T.C. (2002) Mast cells and stress – a psychoneuroimmunological perspective, *J. Clin. Psychopharmacol.* **22**, 103–108.

Theoharides, T.C. (2003) Dietary supplements for arthritis and other inflammatory conditions: key role of mast cells and benefit of combining anti-inflammatory and proteoglycan products, *Eur. J. Inflamm.* **1**, 1–8.

Theoharides, T.C., Bielory, L. (2004) Mast cells and mast cell mediators as targets of dietary supplements, *Ann. Allergy Asthma Immunol.* **93**, S24–S34.

Theoharides, T.C., Cochrane, D.E. (2004) Critical role of mast cells in inflammatory diseases and the effect of acute stress, *J. Neuroimmunol.* **146**, 1–12.

Theoharides, T.C., Douglas, W.W. (1978a) Secretion in mast cells induced by calcium entrapped within phospholipid vesicles, *Science* **201**, 1143–1145.

Theoharides, T.C., Douglas, W.W. (1978b) Somatostatin induces histamine secretion from rat peritoneal mast cells, *Endocrinology* **102**, 1637–1640.

Theoharides, T.C., Sant, G.R. (2003) Neuroimmune Connections and Regulation of Function in the Urinary Bladder, in: Bienenstock, J., Goetzl, E., Blennerhassett, M. (Eds.) *Autonomic Neuroimmunology*, Hardwood Academic Publishers, Lausanne, Switzerland, pp. 345–369.

Theoharides, T.C., Alexandrakis, M., Kempuraj, D., Lytinas, M. (2001) Anti-inflammatory actions of flavonoids and structural requirements for new design, *Int. J. Immunopathol. Pharmacol.* **14**, 119–127.

Theoharides, T.C., Bondy, P.K., Tsakalos, N.D., Askenase, P.W. (1982) Differential release of serotonin and histamine from mast cells, *Nature* **297**, 229–231.

Theoharides, T.C., Sieghart, W., Greengard, P., Douglas, W.W. (1980) Antiallergic drug cromolyn may inhibit histamine secretion by regulating phosphorylation of a mast cell protein, *Science* **207**, 80–82.

Theoharides, T.C., Donelan, J.M., Papadopoulou, N., Cao, J., Kempuraj, D., Conti, P. (2004) Mast cells as targets of corticotropin-releasing factor and related peptides, *Trends Pharmacol. Sci.* **25**, 563–568.

Theoharides, T.C., Sant, G.R., El-Mansoury, M., Letourneau, R.J., Ucci, A.A., Jr., Meares, E.M., Jr. (1995a) Activation of bladder mast cells in interstitial cystitis: a light and electron microscopic study, *J. Urol.*, **153**, 629–636.

Theoharides, T.C., Singh, L.K., Boucher, W., Pang, X., Letourneau, R., Webster, E., Chrousos, G. (1998) Corticotropin-releasing hormone induces skin mast cell degranulation and increased vascular permeability, a possible explanation for its pro-inflammatory effects, *Endocrinology* **139**, 403–413.

Theoharides, T.C., Patra, P., Boucher, W., Letourneau, R., Kempuraj, D., Chiang, G., Jeudy, S., Hesse, L., Athanasiou, A. (2000) Chondroitin sulfate inhibits connective tissue mast cells, *Br. J. Pharmacol.* **131**, 1039–1049.

Theoharides, T.C., Spanos, C.P., Pang, X., Alferes, L., Ligris, K., Letourneau, R., Rozniecki, J.J., Webster, E., Chrousos, G. (1995b) Stress-induced intracranial mast cell degranulation. A corticotropin releasing hormone-mediated effect, *Endocrinology* **136**, 5745–5750.

Thomason, B.T., Brantley, P.J., Jones, G.N., Dyer, H.R., Morris, J.L. (1992) The relation between stress and disease activity in rheumatoid arthritis, *J. Behav. Med.* **15**, 215–220.

Topol, E.J. (2005) Failing the public health – rofecoxib, Merck, and the FDA, *N. Engl. J. Med.* **351**, 1707–1709.

Tsai, M., Takeishi, T., Thompson, H., Langley, K.E., Zsebo, K.M., Metcalfe, D.D., Geissler, E.N., Galli, S.J. (1991) Induction of mast cell proliferation, maturation and heparin synthesis by the rat c-*kit* ligand, stem cell factor, *Proc. Natl. Acad. Sci. U.S.A.* **88**, 6382–6386.

Tsakalos, N.D., Theoharides, T.C., Kops, S.K., Askenase, P.W. (1983) Induction of mast cell secretion by parathormone, *Biochem. Pharmacol.* **32**, 355–360.

Uzuki, M., Sasano, H., Muramatsu, Y., Totsune, K., Takahashi, K., Oki, Y., Iino, K.,

Sawai, T. (2001) Urocortin in the synovial tissue of patients with rheumatoid arthritis, *Clin. Sci.* **100**, 577–589.

Vaughan, J., Donaldson, C., Bittencourt, J., Perrin, M.H., Lewis, K., Sutton, S., Chan, R., Turnbull, A.V., Lovejoy, D., Rivier, C., Rivier, J., Sawchenko, P.E., Vale, W. (1995) Urocortin, a mammalian neuropeptide related to fish urotensin I and to corticotropin-releasing factor, *Nature* **378**, 287–292.

Venihaki, M., Dikkes, P., Carrigan, A., Karalis, K.P. (2001) Corticotropin-releasing hormone regulates IL-6 expression during inflammation, *J. Clin. Invest.* **108**, 1159–1166.

Walker-Bone, K., Javaid, K., Arden, N., Cooper, C. (2000) Regular review: medical management of osteoarthritis, *BMJ* **321**, 936–940.

Wasserman, S.I. (1984) The mast cell and synovial inflammation, *Arthritis Rheum.* **27**, 841–844.

Webster, E.L., Barrientos, R.M., Contoreggi, C., Isaac, M.G., Ligier, S., Gabry, K.E., Chrousos, G.P., McCarthy, E.F., Rice, K.C., Gold, P.W., Sternberg, E.M. (2002) Corticotropin releasing hormone (CRH) antagonist attenuates adjuvant induced arthritis: role of CRH in peripheral inflammation, *J. Rheumatol.* **29**, 1252–1261.

Weisman, M.H., Paulus, H.D., Russak, S.M., Lubeck, D.P., Chiou, C.F., Sengupta, N., Ofman, J.J., Borenstein, J., Moadel, A.B., Sherbourne, C.D. (2003) Development of a new instrument for rheumatoid arthritis: the Cedars-Sinai Health-Related Quality of Life instrument (CSHQ-RA), *Arthritis Rheum.* **49**, 78–84.

Wipke, B.T., Wang, Z., Nagengast, W., Reichert, D.E., Allen, P.M. (2005) Staging the initiation of autoantibody-induced arthritis: a critical role for immune complexes, *J. Immunol.* **172**, 7694–7702.

Wood, P.G., Karol, M.H., Kusnecov, A.W., Rabin, B.S. (1993) Enhancement of antigen-specific humoral and cell-mediated immunity by electric footshock stress in rats, *Brain Behav. Immun.* **7**, 121–134.

Woolley, D.E. (2003) The mast cell in inflammatory arthritis, *N. Engl. J. Med.* **348**, 1709–1711.

Woolley, D.E., Tetlow, L.C. (2000) Mast cell activation and its relation to proinflammatory cytokine production in the rheumatoid lesion, *Arthritis Res.* **2**, 65–74.

Yu, J.G., Boies, S.M., Olefsky, J.M. (2005) The effect of oral glucosamine sulfate on insulin sensitivity in human subjects, *Diabetes Care* **26**, 1941–1942.

# 15
# Molecular and Cell Biology of AIDS/HIV

*Andrew M. L. Lever*
*University of Cambridge, Cambridge, UK*

**1 Origins of HIV 396**

**2 The Molecular Biology of HIV 398**
2.1 Virus Structure 398
2.2 Life cycle 399
2.3 Additional Accessory Proteins 404
2.3.1 Virion Associated Proteins 404
2.3.2 Nonvirion Associated Proteins 404
2.4 Envelope Variants 405

**3 Routes of Transmission 406**

**4 Natural History of Infection 406**

**5 Immune Response to HIV 408**

**6 Drug Therapy of HIV 410**

**7 Viral Escape 411**

**8 Vaccines 412**

**9 Summary 413**

**Acknowledgments 413**

**Bibliography 413**
Books and Reviews 413
Primary Literature 414

*Immunology. From Cell Biology to Disease.* Edited by Robert A. Meyers.

ISBN: 978-3-527-31770-7

## Keywords

**HIV (Human Immunodeficiency Virus)**
A retrovirus of the lentivirus family responsible for AIDS.

**AIDS (Acquired Immune Deficiency Syndrome)**
A clinical state of profound susceptibility to infection with low pathogenicity (opportunistic) infectious agent and unusual malignancies.

**Retrovirus**
A virus characterized by a diploid RNA genome converted to a DNA provirus by its reverse transcriptase enzyme and integrated into the target cell DNA.

**Provirus**
The integrated DNA form of the virus.

**Antiretroviral Drugs**
Pharmacological agents targeting virus-specific processes used in the treatment of AIDS.

This article describes the basic molecular biology of the viruses causing AIDS and the nature of the disease in causes, summarizing the immune response, antiviral therapy and varrine prospects.

## 1 Origins of HIV

HIV is a retrovirus of the lentivirus family. Closely related viruses have been discovered in many groups of African primates and sequence comparison would suggest that there has been more than one transspecies transmission of these viruses from monkeys into humans within the last 80 years. In the simian population, these viruses are transmitted by blood contact during biting and fighting. The butchering of monkeys for sale as bush meat is a highly plausible route of transmission of these viruses into humans through monkey blood contamination of cuts and scratches on human hands. Hypotheses surrounding contaminated polio vaccine have been comprehensively disproved.

There are two major divisions of HIV that have infected humans, HIV-1 and HIV-2. HIV-1 is phylogenetically closely related to $SIV_{cpz}$, a lentivirus of chimpanzees. HIV-2 is different in sequence from both of these and more closely related to the sooty mangabey lentivirus, $SIV_{smm}$. HIV-1 is divided into a number of different groups, the main (M) and the out (O) and new (N) groups within each of which are clades. The M group is estimated to have entered the human

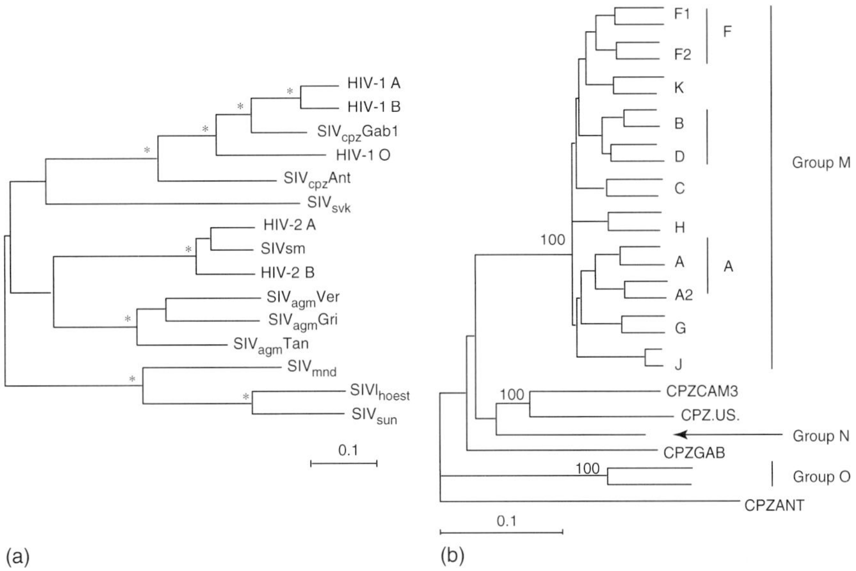


**Fig. 1** (a) Relationships of HIV and SIV based on *pol* sequence comparison and (b) clades of HIV.

race in the 1930s. Within the M group, nine clades are recognized (Fig. 1a and b). In addition to clades, many recombinant or "mosaic" subtypes have also been identified. HIV-1 and HIV-2 have a very similar genetic structure except as detailed below. The sequence of the virus is, however, very variable indeed and two clades within the M group may differ in certain regions of the genome at the amino acid level by 20 to 40%. This is an astonishing level of variation and contributes significantly to the virus' success in establishing infection in the human population.

In their native hosts, simian lentiviruses appear to cause little disease. SIV from the African green monkey, $SIV_{agm}$, is asymptomatic in its natural host but produces profound immunodeficiency when introduced into rhesus macaques or cynomologus macaques. This may give us clues to the reason why HIV is so catastrophic in humans given the short time (<100 years) that it has been infecting humans and the consequent lack of host or pathogen adaptation.

The name lentivirus (lenti = slow) is derived from the archetypal member of this family, Maedi-Visna virus, which infects sheep and causes a very slow progressive disease involving pneumonic change and wasting, sometimes associated with central nervous system defects. In HIV, there may be symptoms associated with the initial viremia but thereafter, the speed of onset of clinical disease following infection is variable and infected individuals may remain asymptomatic for a number of years before the immunodeficiency becomes severe enough to cause illness.

Both HIV-1 and HIV-2 infection lead to a progressive decline in immune competence to the point where it becomes incapable of protecting the individual from otherwise very low pathogenicity infections, which are easily dealt with by a normal individual. This progressive loss of immune competence has given the disease its name, *acquired immune deficiency syndrome* (*AIDS*). In general, the tempo of the disease is slower with HIV-2, possibly related to its replicating to a lower titer. This may also explain why it is less easily transmitted vertically from mother to child than HIV-1.

## 2 The Molecular Biology of HIV

Understanding the interactions of the virus with the immune system, its ability to cause disease, and potential therapies requires an understanding of the retroviral life cycle. The genetic structure of the virus is shown in Fig. 2. It has the conventional *gag*, *pol*, and *env* open reading frames of all retroviruses together with at least seven additional accessory and regulatory genes.

### 2.1 Virus Structure

HIV is a conventionally structured retrovirus (Fig. 3) with a C-type assembly mechanism. On the exterior, it has a lipid envelope derived from the surface plasma membrane of the cell from which it budded and in this are the envelope glycoproteins consisting of trimers of a transmembrane (TM) protein, each of which is noncovalently linked to an external surface (SU) glycoprotein. TM and SU are often referred to by their molecular weights as gp41 and gp120 respectively (in HIV-1).

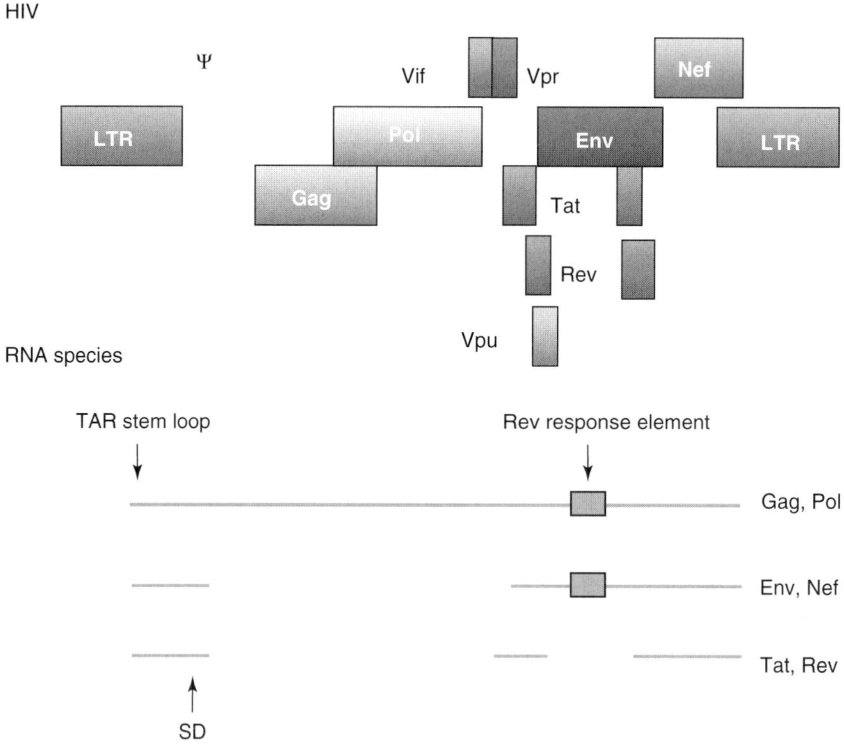


**Fig. 2** Genetic structure of lentivirus DNA provirus and families of RNAs transcribed.

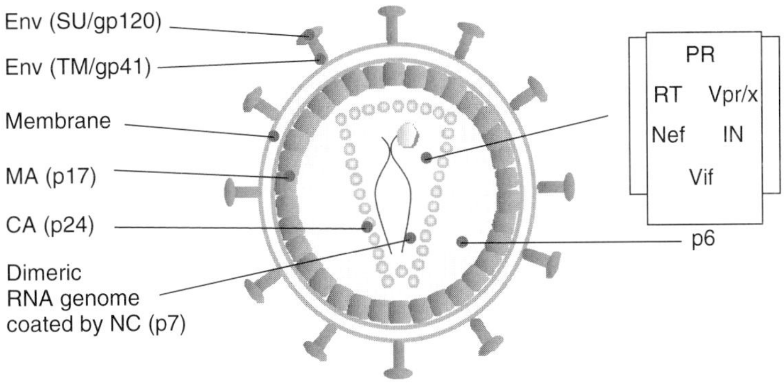


**Fig. 3** Diagram of HIV 1 virus particle.

Within the mature virus particle, the inner surface of the envelope is lined by the matrix protein (MA), a cleavage derivative of the major core polyprotein, Gag. Within this is a pyramidal core structure made up of capsid (CA) protein subunits. This encloses the two copies of the RNA genome that are coated with the nucleocapsid (NC) protein. The two RNAs are tightly linked at a site known as the *dimer linkage site* (DLS). Each has a cell-derived $tRNA_{lys}$ annealed to a complementary sequence at the 5′ end of the genome, the primer binding site. Within the particle are a number of viral accessory proteins including Vif and Vpr. The virus also captures a number of cellular proteins including glycoproteins from the plasma membrane of the cell in its envelope and a number of cytoplasmic proteins including cyclophilin that binds specifically to the capsid protein in the core. The viral gene products of the *pol* gene are also incorporated into the virus. These include the protease (Pro) that cleaves the Gag and Pol precursor polyproteins to generate Gag products (p6, p7, p17, and p24) and Pol products including the reverse transcriptase enzyme (RT), which is responsible for generating a DNA copy from the RNA genome and the integrase enzyme (IN) that inserts the viral DNA into the chromosomes of the target cell.

## 2.2 Life cycle

Infection of a cell by HIV occurs following binding of the virus through its SU glycoproteins to the cell surface molecule CD4 and coreceptors (Fig. 4). The CD4 glycoprotein is found on a variety of cells, predominantly in those of the immune system. HIV tropism is broadly divided into viruses that can infect T lymphocytes (T-cell tropic) and those that infect cells of the macrophage/monocyte lineage (M tropic) (see Sect. 2.4).

Binding to the cell is followed by a conformational change in TM in which its hydrophobic terminus penetrates the target cell membrane and approximates it to the viral envelope lipids, fusing the two membranes together. Cellular lipid rafts appear to play a role in virus entry. The capsid and its contents enter the cell cytoplasm and the processes of disassembly and reverse transcription begin in which the viral RNA is copied into a DNA form. Reverse transcription is shown in Fig. 5. More detailed accounts can be found in the literature. Two important features of the

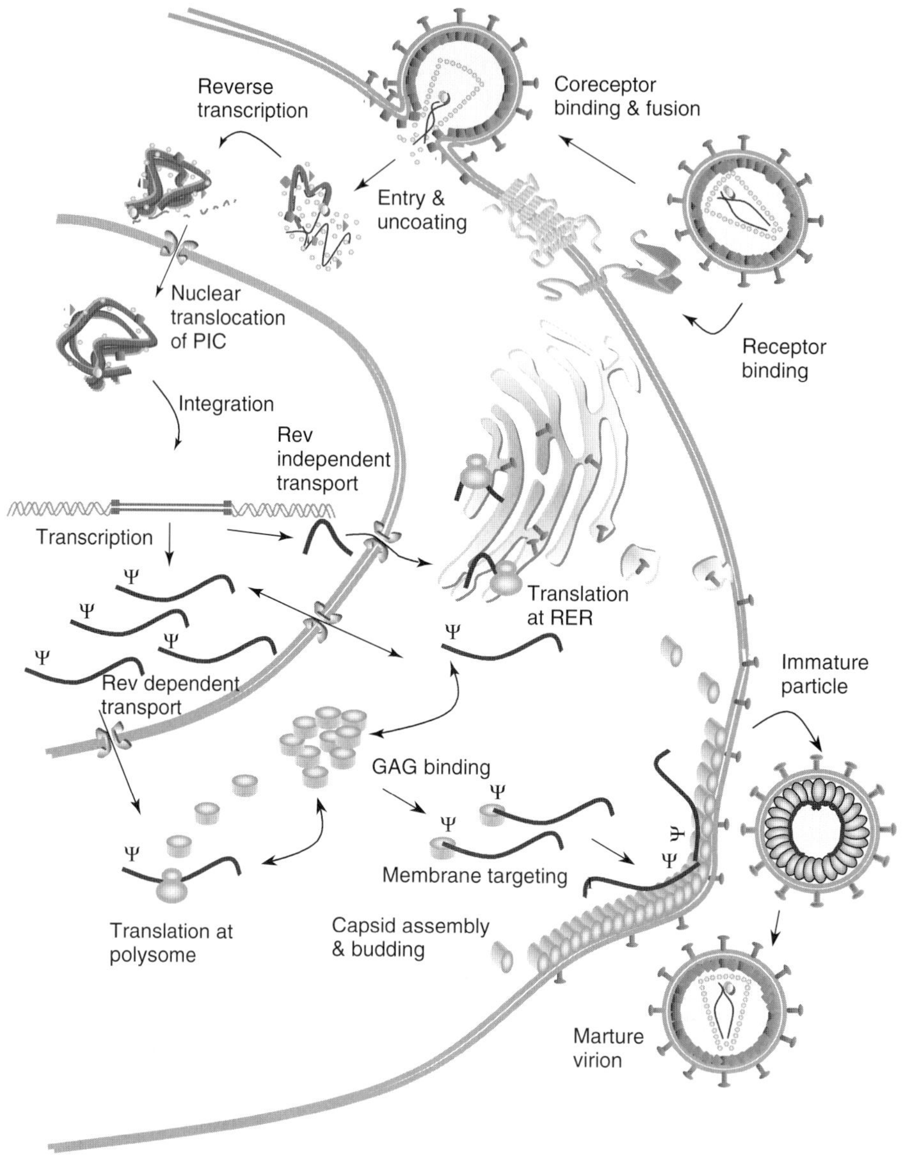


**Fig. 4** Retroviral life cycle.

reverse transcription process are that both the RNA genomes are used as a template to produce a DNA copy in which template switching can occur such that the resultant DNA may be a combination of copies of segments from each of the RNA strands. Secondly, the resulting double-stranded DNA terminal sequences have both been templated by the 3′ long terminal repeat. It is during this reverse transcription step that the combination of template switching and nucleotide misincorporation, due

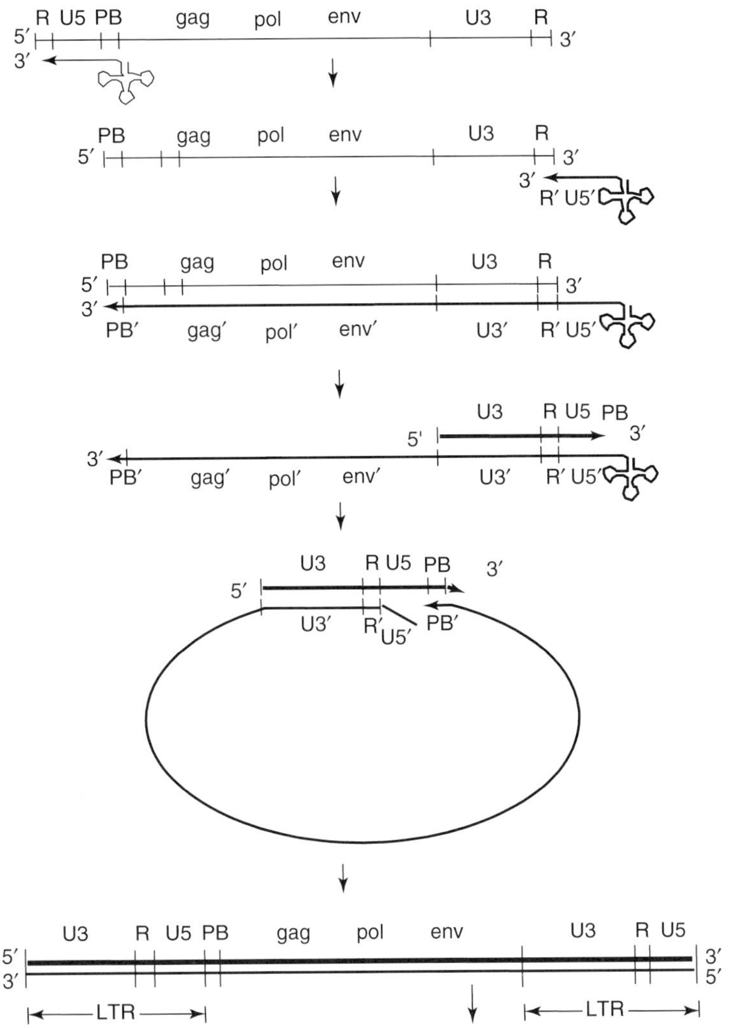


**Fig. 5** Reverse transcription.

to the lack of RT proof-reading, produces the remarkable sequence diversity seen in HIV.

The DNA provirus is transported into the nucleus as a preintegration complex (PIC) using karyophilic signals found on proteins with the complex including the matrix protein, the integrase, and the Vpr accessory protein. There is some functional redundancy between these signals. In lentiviruses, the PIC can cross an intact nuclear membrane, meaning that these viruses have the unique capability among retroviruses of being able to integrate into cells that are not undergoing mitosis. The integration step involves the integrase enzyme cleaving the cellular DNA and making a 3′ to 5′ recess in the two ends of the proviral DNA. The chromosomal DNA undergoes a similar staggered

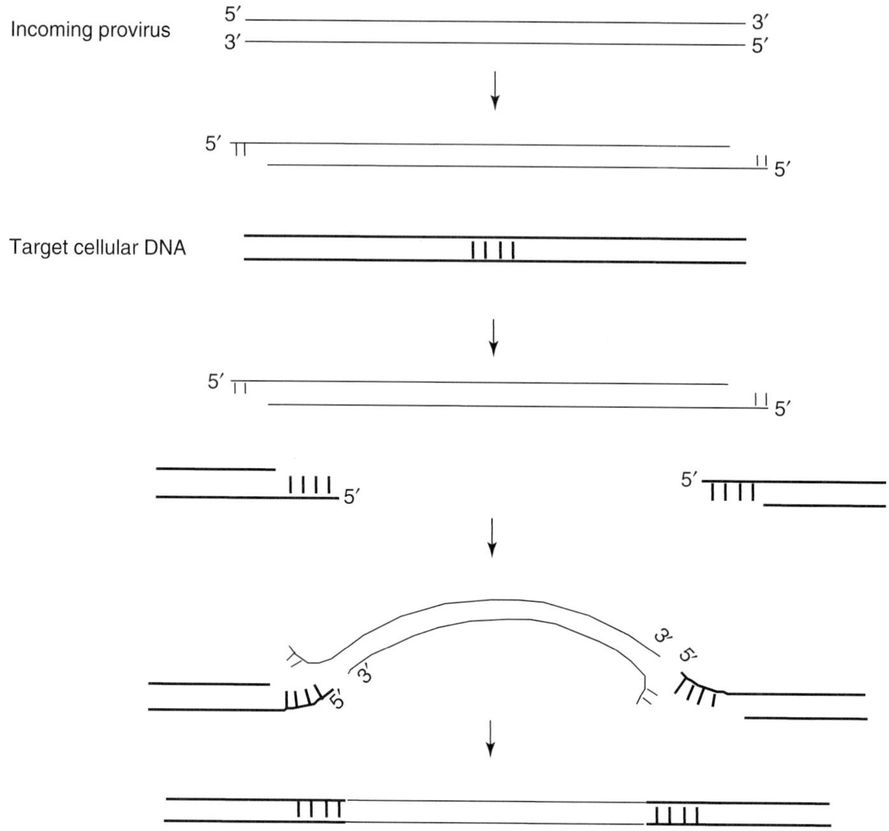


**Fig. 6** Integration process.

cleavage and the viral and cellular DNA ends are linked. Cellular enzymes fill in the defect (Fig. 6). The resulting sequence shows a reduplication of the cellular nucleotides at either end of the provirus.

The integrated provirus may remain transcriptionally silent. Transcriptional activation depends in part on the degree of stimulation of the cell. In T cells, for example, mitotic stimulation is associated with activation of the retroviral promoter. Viral latency is a useful strategy for this pathogen to escape immune recognition and studies suggest that, at any one time, probably less than 10% of infected cells have a transcriptionally active virus.

The 5′ long terminal repeat of the virus contains a promoter element with a conventional TATA box. When activated, this is transcribed by RNA polymerase II to produce a genome length transcript. Multiple small abortive transcripts are seen early on but sufficient full-length RNAs are produced to initiate gene expression. Instability sequences are found in the full-length RNA, which default the RNA transcript to be spliced, and intronic sequences encompassing the Gag and Pol regions and the central part of envelope are spliced out to produce multiply spliced RNAs that are then exported into the cytoplasm and translated on free cytoplasmic ribosomes. The earliest transcripts code for the regulatory proteins, Tat, Rev, and Nef (Fig. 2). Tat is a transcriptional activator that shuttles back into the nucleus and interacts with an RNA stem loop structure, the Tat Response Element (TAR) at the 5′ end of all viral transcripts (Fig. 2). Tat bound to TAR, recruits cyclin T and CDK9, which cause hyperphosphorylation of the C-terminal domain of the RNA polymerase II, massively enhancing its transcriptional efficiency. The sequence of transcription, export, and translation is replicated and amplified and more Tat protein is produced in what amounts to a positive feedback loop. The spliced RNA species encoding Rev is also translated and like the Tat protein, it shuttles back into the nucleus. It interacts with an RNA structure called the *Rev Responsive Element* (*RRE*) in the envelope intron (Fig. 2), enhancing export of both the full-length genomic RNA and the singly spliced RNA that encodes the envelope glycoprotein. Thus, the Tat protein causes an early amplification of transcription, and Rev subsequently leads to an early to late shift in the species of RNA, which are exported from the nucleus from those encoding regulatory factors to mRNAs for structural and enzymatic proteins. Unspliced full-length RNAs are translated on free cytoplasmic ribosomes to produce the Gag and Pol proteins. The singly spliced envelope encoding RNA binds to a ribosome and with the translation of its signal peptide, the complex is transported to the rough endoplasmic reticulum where envelope glycoproteins are translated and glycosylated using the conventional cellular glycoprotein secretory route.

The Gag and Pol proteins are encoded on the same transcript but in different reading frames. Translation of the polyprotein is interrupted by a frameshift sequence at the Gag/Pol overlap with an efficiency of about 5%. The ribosome shifts into the −1 frame, so that for approximately every 19 Gag proteins that are produced, 1 Gag/Pol fusion protein is synthesized. Gag and Gag/Pol proteins assemble at the plasma membrane in approximately the same stoichiometric ratio. At some stage, following transcription, the Gag protein has interacted with full-length genomic RNA transcripts and captured these through binding to a complex folded region of the RNA termed the *packaging signal* or Ψ region. This RNA is then transported into the assembling virion where it binds other Gag proteins through the zinc fingers of their nucleocapsid subfragments, the RNA acting to some extent as a scaffold for viral assembly. RNA–protein and protein–protein interactions contribute to the generation of a convex curved raft of protein underneath the cell membrane. Assembly continues with the viral bud bulging out of the cell and the Gag and Gag/Pol proteins accumulating to form a spherical viral core. This mode of assembly at the membrane defines lentiviruses as having a C-type morphology.

The viral core that forms is composed of Gag and Gag/Pol polyproteins, all aligned with the matrix region outermost and the NC region toward the center of the particle. At the extreme C-terminal of the Gag protein is a small region termed p6, which is intimately involved in the budding process. Mutations in p6 have been shown to lead to arrested budding in which stalked but otherwise intact particles remain attached to the plasma membrane of the cell. p6 has recently been shown to interact with a cellular protein, TSG101. In the normal cell, TSG101 is involved in the budding process of endosomes into multivesicular endosomes. It appears that the virus has hijacked this budding mechanism for an external budding process and uses the same cellular factors to facilitate this. The complete array of proteins involved is not yet elucidated.

After or during the process of budding, the protease component of the Gag/Pol polyprotein becomes activated. This enzyme is inactive in the Gag/Pol monomer and functions as a homodimeric molecule. An as yet unknown process appears to approximate two protease monomers leading to autolytic cleavage of the first protease enzyme, following which cleavage of further Gag/Pol molecules will lead to release of further protease subunits and a cascade of proteolytic cleavage of Gag and Gag/Pol can occur. This cleavage is accompanied by a morphological change in which the particle changes from a spherical one containing a "doughnut" ring of structural proteins into a typical retrovirus particle with matrix protein underlying the envelope and an apparent space between this and the pyramidal core, which is composed of capsid proteins.

## 2.3 Additional Accessory Proteins

These can be divided into those that are or are not incorporated into the viral particle.

### 2.3.1 Virion Associated Proteins

As noted previously, the Vpr protein is involved in the preintegration complex targeting the nucleus of the newly infected cell. There is evidence that Vpr binds to the nucleoporin hCG1. Vpr also appears to have other functions. It may be a transcriptional transactivator in its own right through binding to p300/CBP coactivators. It has also been implicated in G2 phase cell cycle arrest. This is suggested to be the phase of the cell cycle at which Tat transactivation is most efficient. Vpr may also have a role in pathogenesis of AIDS since it has clearly been associated with apoptosis in a number of different cell types including neurons. It appears to gain entry to the virus by binding to the p6 region of Gag.

In HIV-2 and SIV, a gene duplication appears to have occurred to produce a second regulatory protein, Vpx, with homology to Vpr. There is evidence that in these viruses, Vpx may have subsumed the specific nuclear entry role of Vpr.

Vif is another small accessory protein that is incorporated into the virus particle. It has been clear for many years that Vif is an essential protein for viral replication in some cell lines but not others. The Vif phenotype depends on virion associated Vif since expression of Vif in a target cell does not appear to allow replication of an incoming Vif negative virus. Recent work suggests that Vif may be interacting with an inhibitory cellular protein that has sequence homology to proteins of the cytidine deaminase family and it may be that the actions of this nucleotide-modifying enzyme must be neutralized for successful reverse transcription and integration to occur.

### 2.3.2 Nonvirion Associated Proteins

Vpu is a small protein found only in HIV-1 and $SIV_{cpz}$. It has a number of roles that enhance the efficiency of budding of the virus. The envelope glycoprotein, gp120, which on the virion particle interacts with the CD4 protein, is synthesized and exported in the infected cell through the same pathway as the cell's own CD4. Complexing of these two molecules in the ER has been documented. Vpu appears to facilitate disaggregation of these two by binding to and degrading CD4, enhancing the export of the gp120 SU protein.

Vpu also appears to have additional envelope-independent effect, enhancing virus export.

The Nef protein is found throughout the primate lentiviruses. It is a multifunctional molecule. It has been documented as being capable of downmodulating cell surface expression of certain cellular proteins including CD4 and the MHC proteins. Recent evidence suggests that Nef may be involved in recruiting lymphocytes to HIV-infected macrophage/monocytes. Specific effects on migration inhibition of lymphocytes have been documented and these properties seem to be a powerful functional argument for Nef having important effects on virus spread and infectivity. This is further supported by the observation that Nef-deleted or Nef-mutated viruses have been shown to be associated with a severely attenuated pathogenicity. Recipients of HIV-infected blood from a single donor in Australia remained well for many years and the virus was identified as having a defective Nef protein. In SIV infection, Nef mutant viruses show an attenuated phenotype and reversion of the Nef mutation to wild type is associated with a regain of virulence.

## 2.4 Envelope Variants

The envelope gene of HIV-1 encodes a polyprotein comprising the two components of the external viral receptor ligand. At the N-terminus is the surface (SU) or gp120 protein. At the C-terminal, the TM is found. This latter sequence contains a hydrophobic stretch of amino acids, which comprise the membrane anchor and a second hydrophobic region involved in binding to the target cell membrane and fusing it with the virus envelope. The SU region is divided into five constant (C1-5) and five variable (V1-5) regions. The SU protein of HIV is responsible for binding to target cells. The V3 loop region of SU has attracted much attention because of its critical importance in receptor binding. The CD4 protein found on lymphocytes and cells of the monocyte/macrophage series was identified early in the AIDS epidemic as being a major receptor for the virus. At the same time, it was clear that binding to CD4 was necessary but not sufficient for infection. Observations that activated lymphocytes released substances that block infection by HIV and the identification of these as chemokines and their receptors as belonging to the family of 7-transmembrane spanning chemokine receptors revealed the existence of a family of coreceptor molecules, which are required for successful entry and infection. The two major families of these are the so-called *CC chemokine receptors* and the *CXC receptors.* Two of these are the principal HIV receptors. The CC chemokine receptor CCR5 responds to the cytokines MIP1$\alpha$, MIP1$\beta$, and RANTES. The CXC chemokine receptor CXCR4 is bound by SDF1. There are a number of other receptors that have been identified as being involved in binding of HIV-1, HIV-2, and SIV. The identification of these receptors neatly mapped on the long-standing observation that there were at least two types of tropism identifiable in HIV. Some viruses appeared to be able to infect cells of the monocyte/macrophage series very efficiently but were very poor at infecting T lymphocytes. Others appear to be extremely efficient at infecting T lymphocyte cell lines *in vitro* but were very poor at infecting macrophages. These were broadly divided into "macrophage tropic" and "lymphocyte tropic" viruses (although this is a simplification). The chemokine receptor data link these tropisms with the fact that the lymphocyte tropic viruses have a surface envelope protein that preferentially binds to the CXCR4 receptor and the macrophage tropic to the CCR5 receptor. These tropisms are commonly abbreviated by describing viruses as X4 or R5, respectively. More recently, a mutation has been identified, which affects about 1% of Caucasians who are homozygous for a stop codon causing a truncation of the CCR5 protein, and renders these individuals uninfectable by macrophage tropic virus. This explained another observation that there were some individuals who appeared to be repeatedly exposed to HIV through high-risk sexual practices and yet did not become infected.

## 3 Routes of Transmission

Infection can occur via cell-free virus or through contact between infected cells and those of the recipient. SIV in monkeys is probably primarily transferred from host to host by blood contact. HIV is also infectious by blood contact but the predominant mode of transmission is by the sexual route. HIV positive mothers can transmit the virus to their newborn child at birth or through breast feeding. The fact that cell-free virus is infectious means that HIV was transmitted very efficiently in clotting factor concentrates in the 1980s, resulting in the infection of large cohorts of hemophiliac patients. In the developed world, blood donations are screened for HIV as they are in many countries in the developing world. There is a very small but finite risk that a blood donor in the "window" period before seroconversion may transmit an infection through blood donation but this is in the order of less than one in a million. Blood transmission occurs quite efficiently in the case of intravenous drug users who share needles. Sexual transmission can occur from both male to female and female to male with an estimated risk of 1 : 200. Mucous membrane transmission probably begins with the infection of sessile cells of the macrophage/monocyte system in the mucous membrane tissues with subsequent replication and spread of virus to other cells of the immune system. Recent evidence implicates Langerhans cells as the first target. These may bind the virus using the DC-sign cell surface lectin, which has a high affinity for gp120. Whether replication occurs within these cells or whether surface-bound virus infects migrating T cells and macrophages is not established.

## 4 Natural History of Infection

The characteristic natural history of infection is illustrated in Fig. 7. Following early replication in lymph nodes and within a period of approximately four to eight weeks after exposure to virus, viremia becomes detectable in the form of ELISA positivity to the p24 capsid antigen and detection of viral RNA in the serum. This viremia rises rapidly to a peak that may be around five million copies of the viruses per milliliter of blood. At this time, the beginnings of an immune response against the virus become detectable, including special antibodies directed against the viral proteins. Subsequently, the virus level in the blood falls to a "set point" and this correlates most closely with detection of a rise in the cytotoxic T lymphocyte cell mediated immunity. At the time of this seroconversion, between a third and a

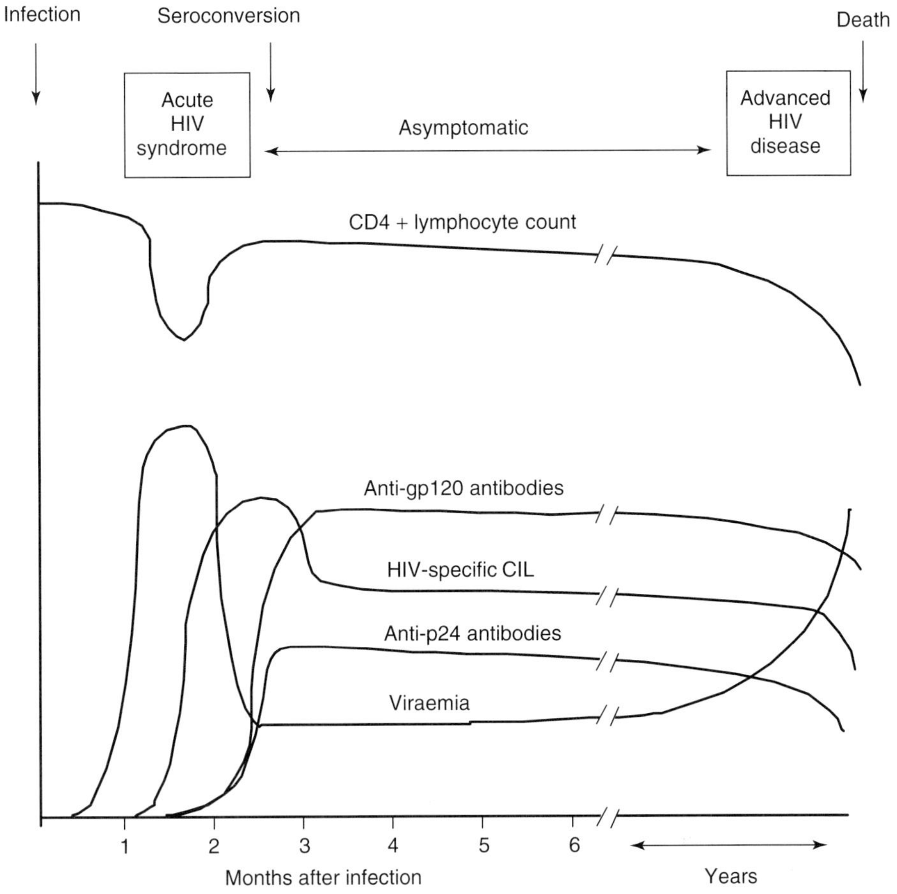


**Fig. 7** Natural history of HIV infection.

half of infected patients will have symptoms resembling a glandular feverlike illness with lymphadenopathy, transient petechial rashes, arthropathy, evidence of peripheral neuropathy, and occasionally a brief meningoencephalitic illness. However, virtually without exception, these symptoms resolve. There then follows a period of apparent stability in which the viral level or "viral load" remains stable and there is ongoing evidence of an immune response. In some individuals, persistent enlargement of lymph nodes occurs associated with weight loss and fevers. This so-called *persistent generalized lymphadenopathy* appears to be associated with a worse prognosis in terms of the speed of onset of immunodeficiency. Despite the asymptomatic nature of infection in most individuals at this stage, measurements of the CD4 lymphocyte count show that during the initial viral peak, there is a corresponding trough in lymphocyte numbers, implying a significant amount of direct virus mediated lymphocyte destruction. This rises again as the virus load is brought under control but does not achieve the preinfection level and from then on there is a very slow but progressive decline.

Calculations of the quantity of virus being produced during this time and the rate of lymphocyte destruction/regeneration have generated the remarkable figures that around $10^9$–$10^{10}$ viruses are produced and destroyed each day. This is probably an underestimate. The lymphocyte turnover is also around $10^9$ per day, but while the viral load may be constant or may rise, the lymphocyte count slowly falls such that there is a net loss of around $10^6$ lymphocytes per day. The fall in lymphocyte count correlates with a progressive decline in immunological competence and is a marker, if not the cause, of the eventual breakdown in immune competence. At this stage, the number of infected cells in the blood is very low, at around 1 in $10^5$. This might suggest that virus mediated lymphocyte destruction is less important at this stage, although localized destruction in lymph nodes would be difficult to detect. Immune clearance of infected cells, however, may be the dominant force affecting lymphocyte numbers. Depending on the success of the immune response in controlling the virus, the length of time between seroconversion and immunodeficiency may be anything from 2 up to 12+ years. However, as yet, although there are examples of very slow progressors, there is no evidence of complete control of viral infection by the immune system once the virus has established itself. At the late stages of the disease, the predominant type of virus appears to change with the emergence of faster replicating T-cell tropic (X4) variants.

Symptomatic immunodeficiency becomes apparent when the lymphocyte count falls below $200 \times 10^9\ L^{-1}$, at which point the individual is at very high risk of opportunistic infection and the development of unusual malignancies. The fall in lymphocyte count correlates with a loss of control over viral replication and a rise in detectable virus in the circulation in the absence of antiretroviral therapy. Death from an overwhelming infection or an uncontrolled malignancy usually occurs within two to three years of the onset of the severe immunodeficiency even with treatment of the individual infections and cancers.

As will be described later, the advent of antiretroviral therapy has transformed the natural history of disease in the western world. Pharmacological control of virus replication leads to a fall in viral load and a corresponding rise in the CD4 lymphocyte count. This leads to a reversal of the loss of immunocompetence, which, although it may not be complete, is enough to significantly reduce the risk of opportunistic infection.

## 5
## Immune Response to HIV

Following the initial peak of virus in the circulation after infection, the viral load declines to a set point, which, on average, is around 30 000 copies per milliliter. If it is higher than this, the onset of AIDS is quicker and vice versa. It appears that qualitative and quantative parameters of the immune response determine this set point. A broad antibody response develops within the first three months of infection with specificity for some of the structural proteins such as Gag. Antibodies to Envelope and Polymerase proteins appear swiftly thereafter, and the regulatory and accessory proteins also trigger an antibody response.

Antibodies interfering with interaction between the envelope protein and the cellular receptor may be neutralizing and the site at which they bind on the viral protein provides strong clues as to the important regions of SU involved in viral entry. In particular, the V3 loop is a major target for neutralizing antibody. A second region on gp120, which is involved in both CD4 and chemokine receptor binding, is another neutralization target. However, V3 loop neutralizing antibodies often appear to be more effective against laboratory strains of HIV than primary clinical isolates. It is possible that this relates to the different conformation of the envelope in the *in vitro* adapted strains in which the lack of immune selection pressure has allowed a more open structure that favors rapid cellular entry. *In vivo*, such a structure would predispose to the blockade by neutralizing antibodies and may be disadvantageous.

Other evidence suggesting that antibodies are of relatively low importance is the fact that some individuals in whom the illness appears to be held in check by the immune response do not have high levels of neutralizing antibody. In addition, passive protection using antibodies in animal models is only effective when they are present in very high titer. There is some evidence that high titers of neutralizing antibody may be associated with a lower "set point." The immense variability of the viral envelope facilitates immunological escape of the virus. This may occur either by amino acid variability or by modification of envelope glycosylation.

Cytotoxic T-cell responses can be detected before the antibody response and have long been considered one of the major immunological defenses against HIV. CD8+ T cells can kill virus infected cells by specific recognition of the HIV peptides presented on class I MHC proteins on the surface of the infected cell. *In vitro*, CTLs can eliminate infected cells even before new viruses have been produced. A second effect involves the release of $\beta$-chemokine mediators, which, as previously mentioned, are themselves able to bind to their natural ligand CCR5 and block the use of this protein by HIV for infection of the cell. The strong correlation between appearance of CD8+ T cells and the decline in viral replication in the early stages together with data on adoptive transfer in humans and CD8 depletion experiments in monkeys, which leads to a sustained increase in viral load, confirms the critical role played by this arm of the immune response in combating infection.

The signature of HIV infection is the decline in CD4 lymphocytes. The mechanisms of destruction of CD4 lymphocytes are probably many and varied as described earlier. Production of HIV with its gp120 within a cell that is already making CD4 may lead to protein aggregation within the cell and dysfunction leading to apoptosis. This can be observed *in vitro* as single cell killing. Expression of gp120 on the surface of the cell may induce fusion of that cell with a CD4+ cell. This latter phenomenon is also seen *in vitro* and leads to killing by syncitia formation. Passive binding of shed gp120 on to the CD4 proteins of immune competent cells may lead to them being a target for immunological clearance and the gp120 itself may have adverse effects. The cytotoxic T-cell response, clearing infected cells, is probably responsible for a very large proportion of the destruction of CD4+ infected cells. Thus, a number of direct and indirect mechanisms contribute to the loss of the CD4 cell population.

Individuals whose disease progresses very slowly (long-term nonprogressors – LTNPs) have clearly been shown to have powerful T-cell responses against HIV with the CD4 lymphoproliferative response correlating well with successful suppression of the viral load. Whether these cells act independently or through enhancement of CD8+ CTL killing is not clear and it may be a combination of both. Apart from specific genetic variants such as the CCR5 truncation mutation described earlier, there does appear to be some genetic influences in that certain HLA alleles HLAB35 and CW4 are associated with more rapid disease progression, and certain other polymorphisms in chemokines and chemokine receptor genes may influence responses to therapeutic strategies and vaccination.

The immune response is capable of a measure of control of HIV replication in all patients, and in some, there is very significant delay in the onset of AIDS. Recent evidence suggests that not only does the ability to mount an immune response decline as disease progresses but the ability to regenerate this when antiretroviral treatment is started also declines the longer the initiation of treatment is delayed. Indeed, treatment during the early stages of infection at seroconversion has been correlated with a much better long-term CD4 lymphoproliferative response. Thus, early treatment might lead to the ability to maintain a strong CD4 cell response and this might provide the possibility of powerful long-term control of viral replication. Against this is the risk of generating drug resistance early in infection and the loss of utility of the limited range of chemotherapeutic agents when they may be more useful at a later stage.

## 6 Drug Therapy of HIV

Antiviral drugs are commonly targeted at viral enzymes. This is the case with HIV in which the first two types of highly successful pharmacotherapies have been directed to inhibit the actions of the viral reverse transcriptase gene and the protease gene. Reverse transcriptase is clearly a virus-specific process since the cell does not have any functional RNA-dependent DNA polymerase activity. The reverse transcriptase enzyme does not have proof-reading capability and while, as mentioned later, this may be an advantage to the virus in generating variants, it also is an Achilles' heel for the virus in that nucleoside analogs that mimic the bases that the enzyme is attempting to incorporate into the DNA chain can substitute for these and once inserted cannot be excised. Pyrimidine analogs have been particularly effective and they appear to act both by inhibiting the enzyme and by chain termination. They require phosphorylation within the cell to form triphosphates. Another class of drug that inhibits RT is the nonnucleoside reverse transcriptase inhibitors (NNRTIs), which bind to reverse transcriptase away from the active site but are not incorporated into the transcript. A third major class of antiretroviral drugs are those inhibiting the viral protease. These take advantage of the target peptide sequence of the aspartyl protease of HIV, being slightly different from cellular homologs of this enzyme. Predictably, however, the specificity of these drugs is somewhat lower than RT inhibitors and they are associated with a greater number of side effects, probably relating to effects on cellular proteases. Newer classes of drugs are being developed including those that may inhibit viral entry, viral integration, and viral export. Injection of a peptide, which interferes with the formation of the fusogenic envelope of the virus and blocks cell entry, has recently been introduced into clinical practice.

Structured treatment interruptions (STI) is being tested as a strategy to reduce drug toxicity. As yet, the effect of this on overall prognosis and immune regeneration is not fully established.

## 7 Viral Escape

The previously described infidelity of the reverse transcriptase enzyme is responsible for considerable sequence change in the virus. Undoubtedly, many of the incorporated nucleotides will cause lethal mutations and a nonviable integrated provirus will result. The huge numbers of viruses being produced everyday, however, will ensure that a large number of viable variants will be produced, which may have minor changes in their amino acid sequence. Thus, it is easy to see that a virus entering a macrophage may mutate slightly during its reverse transcription and integration process such that the progeny virions from that cell may have a slightly greater predisposition to infect lymphocytes than the parental version. The other major mechanism of sequence variation comes from recombination. When a cell is infected with two different viruses, the diploid nature of the genome means that it is possible for an RNA strand from two different integrated proviruses to be copackaged. During reverse transcription in the target cell, the reverse transcriptase enzyme is known to dissociate from its template and may reassociate with the partner template. Thus, exchange of large blocks of genetic information between the two strands is quite common. Indeed, it is estimated that the enzyme skips between strands at least four times during every replication cycle. In this way, mixing and matching of genomes can occur with ease and, in this case, large blocks of functionally intact genome will be changed. Recombination is, therefore, believed to be as significant if not more significant than RT errors in generating diversity. Some of the major worldwide clades have clearly been formed from recombination. Genotype E in Thailand, for example, is a recombinant between a clade A Gag and a clade E envelope sequence.

Variation allows the virus to introduce point mutations, which will render the binding of antiviral drugs much less efficient, and strategies to combat viral resistance have become an integral part of clinical management. This is particularly the case since antiviral therapy in the western world became more widespread. The chances of acquiring a virus, which has already generated some antiviral resistance, has increased greatly. For this reason, where possible, it is recommended that antiviral resistance testing is undertaken before initiating antiretroviral therapy so that the combination of drugs chosen is one in which the virus is sensitive to each component. As well as this, it is vital that combinations of three or more drugs are used to prevent emergence of resistance. Both of these strategies are designed to avoid presenting the virus with a single drug against which it can develop resistance.

The genetic variation also contributes very largely to the difficulty the immune system has in neutralizing this virus. There are clearly documented examples of immunological escape where viral proteins, which have cytotoxic T-cell epitopes, mutate and thus escape cytotoxic T-cell recognition. In terms of envelope variation, the virus has the additional weapon of variable glycosylation. HIV envelope is extremely heavily glycosylated compared to a number of its close viral relatives and it is clear that glycosylation may significantly inhibit the function of neutralizing antibodies.

## 8 Vaccines

From the above, it is clear that the development of a vaccine against HIV is a monumental task compared to many other pathogens. The features that make it particularly challenging are as follows. Firstly, there is no evidence that complete sterilizing immunity occurs in the human race after infection. Thus, even the best immune response in the world does not appear to be capable of indefinitely checking virus replication. Secondly, the huge variability of the virus means that unlike, for example, polio or smallpox, vaccines containing a restricted number of variants will not successfully prevent infection and disease caused by HIV since the virus has such a vast repertoire of viable variants. Thirdly, the virus integrates into the genome of the cell. A number of the features of the life cycle of this and other retroviruses have been designed to specifically maintain cellular viability since death of the cell equates with death of the provirus. The action of many of the accessory genes of HIV in downmodulating cell surface proteins makes even a good immune response incapable of easily clearing infection. Over and above that, however, the virus has the capacity to become latent. Complete transcriptional arrest may occur (or a level of basal transcription insufficient to generate Tat), with transcriptional reactivation occurring under appropriate conditions such as triggering of the cell into mitosis. A transcriptionally inactive virus is completely invisible to the immune system and persistence of a small number of latently infected cells leaves the possibility of reactivation when the immune response wanes. Reactivation of other viruses such as *Varicella zoster* provides a comparable parallel. It is, thus, not surprising that progress toward a vaccine has been slow. To date, vaccines have been generated that are capable of preventing infection in SIV models. Most of these appear to be effective against homologous strains but with, as yet, relatively little evidence that effective cross-neutralization of heterologous strains occurs. Use of a live attenuated vaccine such as one deleted in *nef* and some of the other regulatory genes has been associated with protection against subsequent challenge. Attenuation, however, appears to be a finite phenomenon and mutation back to the wild-type sequence and full pathogenicity is a feature. Recent data using prime boost techniques in which carrier vaccine strains are used such as modified vaccinia Ankara have shown some promising evidence of good immune responses against HIV proteins. As yet, the duration of immune response using these techniques is not clear. A cohort of prostitutes in western Africa who were repeatedly exposed to HIV but remained uninfected originally, generated hope that a fully neutralizing immune response was possible with frequent low level viral challenge. The follow-up of these individuals showed that when sexual exposure to HIV ceases, the immunity wanes and the individuals become susceptible to infection again. This, again, is rather a concerning issue in terms of vaccine development since it would suggest that repeated boosts of vaccination will be required to maintain a level of immunity, which would otherwise be lost. There are strong parallels here with immunity to malaria.

## 9 Summary

The HIV epidemic is a global problem that has been controlled but not eliminated in the western world. The vast majority of infected individuals live in sub-Saharan Africa where access to treatment for the virus or its complications is extremely limited. Significant loss of life and disruption of family and society has occurred in this part of the world due to HIV. Recurrent suggestions that HIV may not be the causative agent of AIDS have seriously set back educational programs in a number of countries and are unfounded and irresponsible. Countries that have taken advice and help and approached the problem with a vigorous education campaign such as Uganda have reaped the rewards with the HIV positivity rate in the antenatal population halving over a period of less than 10 years. Until vaccines are developed and antiretroviral therapy becomes more widely available, these sort of approaches should be powerfully supported. In the western world, antiretroviral therapy has transformed the prognosis, and conceivably, a middle aged person acquiring HIV might now have a normal lifespan given the additive effects of the slow onset of clinical disease, the availability of antiretroviral drugs, and the combined time window these provide for the appearance of new therapies. While the death rate has fallen in the west, the incidence of HIV infection has continued to rise. This is partly because of increased longevity of infected individuals and also from increasing numbers of migrants from highly endemic areas who may seek access to western health resources. This trend is increasing and will provide an increasing burden on western healthcare services. There is still no reason why an uninfected person cannot be protected against HIV since transmission is preventable by barrier methods of contraception, vaginal virucidal agents, and so on. However, the key to the control of HIV in the near future in the developing and developed world is education.

## Acknowledgments

I would like to acknowledge S. Griffin for contributions to the figures.

## Bibliography

### Books and Reviews

British HIV Association Guidelines for Antiretroviral Treatment of HIV Seropositive Individuals. BHIVA Guidelines Co-ordinating Committee (1997), *Lancet* **349**, 1086–1092.

Carpenter, C.C., Cooper, D.A., Fischl, M.A., Gatell, J.M., Gazzard, B.G., Hammer, S.M., Hirsch, M.S., Jacobsen, D.M., Katzenstein, D.A., Montaner, J.S., Richman, D.D., Saag, M.S., Schechter, M., Schooley, R.T., Thompson, M.A., Vella, S., Yeni, P.G., Volberding, P.A. (2000) Antiretroviral therapy in adults: updated recommendations of the international AIDS society-USA panel, *JAMA* **283**, 381–390.

Coffin, J.M., Hughes, S.H., Varmus, H.E. (Eds.) (1997) *Retroviruses*, Cold Spring Harbor Laboratory Press, New York.

Cohen, O.J., Fauci, A.S. (2001) Pathogenesis and Medical Aspects of HIV-1 Infection, in: Knipe, D.M., Howley, P.M. (Eds.) *Fields Virology*, 4th edition, Lippincott Williams & Wilkins, Philadelphia, pp. 2043–2094.

Fackler, O.T., Peterlin, B.M., Weis, K. (2001) Lessons from HIV: movement of macromolecules inside the cell, *Curr. Mol. Med.* **1**, 1–7.

Freed, E.O., Martin, M.A. (2001) HIVs and their Replication, in: Knipe, D.M., Howley, P.M. (Eds.) *Fields Virology*, 4th edition,

Lippincott Williams & Wilkins, Philadelphia, pp. 1971–2042.
Jeang, K.-T. (Ed.) (2000) *Advances in Pharmacology: HIV-1: Molecular Biology and Pathogenesis – Viral Mechanisms (Vol 48) & Clinical Applications (Vol. 49)*, Academic Press, San Diego.
Persaud, D., Zhou, Y., Siliciano, J.M., Siliciano, R.F., (2003) Latency in human immunodeficiency virus type 1 infection: no easy answers, *J. Virol.* **77**, 1659–1665.
Weiss, R.A., Adler, M.W., Rowland-Jones, S.L. (Eds.) (2001) The changing face of HIV and AIDS, *Br. Med. Bull.* **58**(1).
Wyatt, R., Sodroski, J. (1998) The HIV-1 envelope glycoproteins: fusogens, antigens, and immunogens, *Science* **280**, 1884–1888.

### Primary Literature

Addo, M.M., Altfeld, M., Rosenberg, E.S., Eldridge, R.L., Philips, M.N., Habeeb, K., Khatri, A., Brander, C., Robbins, G.K., Mazzara, G.P., Goulder, P.J., Walker, B.D. (2001) The HIV-1 regulatory proteins Tat and Rev are frequently targeted by cytotoxic T lymphocytes derived from HIV-1-infected individuals, *Proc. Natl. Acad. Sci. U.S.A.* **98**, 1781–1786.
Adjorlolo-Johnson, G., De Cock, K.M., Ekpini, E., Vetter, K.M., Sibailly, T., Brattegaard, K., Yavo, D., Doorly, R., Whitaker, J.P., Kestens, L., Ou, C.Y., George, J.R., Gayle, H.D. (1994) Prospective comparison of mother-to-child transmission of HIV-1 and HIV-2 in Abidjan, Ivory Coast, *JAMA* **272**, 462–466.
Amara, R.R., Villinger, F., Altman, J.D., Lydy, S.L., O'Neil, S.P., Staprans, S.I., Montefiori, D.C., Xu, Y., Herndon, J.G., Wyatt, L.S., Candido, M.A., Kozyr, N.L., Earl, P.L., Smith, J.M., Ma, H.L., Grimm, B.D., Hulsey, M.L., Miller, J., McClure, H.M., McNicholl, J.M., Moss, B., Robinson, H.L. (2001) Control of a mucosal challenge and prevention of AIDS by a multiprotein DNA/MVA vaccine, *Science* **292**, 69–74.
Andreasson, P.A., Dias, F., Naucler, A., Andersson, S., Biberfeld, G. (1993) A prospective study of vertical transmission of HIV-2 in Bissau, Guinea-Bissau, *AIDS* **7**, 989–993.
Asjo, B., Morfeldt-Manson, L., Albert, J., Biberfeld, G., Karlsson, A., Lidman, K., Fenyo, E.M. (1986) Replicative capacity of human immunodeficiency virus from patients with varying severity of HIV infection, *Lancet* **2**, 660–662.
Baltimore, D. (1970) RNA-dependent DNA polymerase in virions of RNA tumour viruses, *Nature* **226**, 1209–1211.
Barouch, D.H., Kunstman, J., Kuroda, M.J., Schmitz, J.E., Santra, S., Peyerl, F.W., Krivulka, G.R., Beaudry, K., Lifton, M.A., Gorgone, D.A., Montefiori, D.C., Lewis, M.G., Wolinsky, S.M., Letvin, N.L. (2002) Eventual AIDS vaccine failure in a rhesus monkey by viral escape from cytotoxic T lymphocytes, *Nature* **415**, 335–339.
Barre-Sinoussi, F., Chermann, J.C., Rey, F., Nugeyre, M.T., Chamaret, S., Gruest, J., Dauguet, C., Axler-Blin, C., Vezinet-Brun, F., Rouzioux, C., Rozenbaum, W., Montagnier, L. (1983) Isolation of a T-lymphotropic retrovirus from a patient at risk for acquired immune deficiency syndrome (AIDS), *Science* **220**, 868–871.
Beaumont, T., van Nuenen, A., Broersen, S., Blattner, W.A., Lukashov, V.V., Schuitemaker, H. (2001) Reversal of human immunodeficiency virus type 1 IIIB to a neutralization-resistant phenotype in an accidentally infected laboratory worker with a progressive clinical course, *J. Virol.* **75**, 2246–2252.
Berry, N., Davis, C., Jenkins, A., Wood, D., Minor, P., Schild, G., Bottiger, M., Holmes, H., Almond, N. (2001) Vaccine safety. Analysis of oral polio vaccine CHAT stocks, *Nature* **410**, 1046–1047.
Borrow, P., Lewicki, H., Hahn, B.H., Shaw, G.M., Oldstone, M.B. (1994) Virus-specific CD8+ cytotoxic T-lymphocyte activity associated with control of viremia in primary human immunodeficiency virus type 1 infection, *J. Virol.* **68**, 6103–6110.
Brodie, S.J., Patterson, B.K., Lewinsohn, D.A., Diem, K., Spach, D., Greenberg, P.D., Riddell, S.R., Corey, L. (2000) HIV-specific cytotoxic T lymphocytes traffic to lymph nodes and localize at sites of HIV replication and cell death, *J. Clin. Invest.* **105**, 1407–1417.
Bukrinsky, M.I., Haggerty, S., Dempsey, M.P., Sharova, N., Adzhubel, A., Spitz, L., Lewis, P., Goldfarb, D., Emerman, M., Stevenson, M. (1993) A nuclear localization signal within HIV-1 matrix protein that governs infection of non-dividing cells, *Nature* **365**, 666–669.

Carrington, M., Nelson, G.W., Martin, M.P., Kissner, T., Vlahov, D., Goedert, J.J., Kaslow, R., Buchbinder, S., Hoots, K., O'Brien, S.J. (1999) HLA and HIV-1: heterozygote advantage and B*35-Cw*04 disadvantage, *Science* **283**, 1748–1752.

Chackerian, B., Rudensey, L.M., Overbaugh, J. (1997) Specific N-linked and O-linked glycosylation modifications in the envelope V1 domain of simian immunodeficiency virus variants that evolve in the host alter recognition by neutralizing antibodies, *J. Virol.* **71**, 7719–7727.

Champagne, P., Ogg, G.S., King, A.S., Knabenhans, C., Ellefsen, K., Nobile, M., Appay, V., Rizzardi, G.P., Fleury, S., Lipp, M., Forster, R., Rowland-Jones, S., Sekaly, R.P., McMichael, A.J., Pantaleo, G. (2001) Skewed maturation of memory HIV-specific CD8 T lymphocytes, *Nature* **410**, 106–111.

Chen, M.Y., Maldarelli, F., Karczewski, M.K., Willey, R.L., Strebel, K. (1993) Human immunodeficiency virus type 1 Vpu protein induces degradation of CD4 in vitro: the cytoplasmic domain of CD4 contributes to Vpu sensitivity, *J. Virol.* **67**, 3877–3884.

Cheng-Mayer, C., Seto, D., Levy, J. (1988) Biologic features of HIV-1 that correlate with virulence in the host, *Science* **240**, 80–82.

Cho, M.W., Kim, Y.B., Lee, M.K., Gupta, K.C., Ross, W., Plishka, R., Buckler-White, A., Igarashi, T., Theodore, T., Byrum, R., Kemp, C., Montefiori, D.C., Martin, M.A. (2001) Polyvalent envelope glycoprotein vaccine elicits a broader neutralizing antibody response but is unable to provide sterilizing protection against heterologous Simian/human immunodeficiency virus infection in pigtailed macaques, *J. Virol.* **75**, 2224–2234.

Chun, R.F., Jeang, K.T. (1996) Requirements for RNA polymerase II carboxyl-terminal domain for activated transcription of human retroviruses human T-cell lymphotropic virus I and HIV-1, *J. Biol. Chem.* **271**, 27888–27894.

Cochrane, A.W., Perkins, A., Rosen, C.A. (1990) Identification of sequences important in the nucleolar localization of human immunodeficiency virus Rev: relevance of nucleolar localization to function, *J. Virol.* **64**, 881–885.

Dalgleish, A.G., Beverley, P.C., Clapham, P.R., Crawford, D.H., Greaves, M.F., Weiss, R.A. (1984) The CD4 (T4) antigen is an essential component of the receptor for the AIDS retrovirus, *Nature* **312**, 763–767.

Deacon, N.J., Tsykin, A., Solomon, A., Smith, K., Ludford-Menting, M., Hooker, D.J., McPhee, D.A., Greenway, A.L., Ellett, A., Chatfield, C. (1995) Genomic structure of an attenuated quasi species of HIV-1 from a blood transfusion donor and recipients, *Science* **270**, 988–991.

De Cock, K.M., Zadi, F., Adjorlolo, G., Diallo, M.O., Sassan-Morokro, M., Ekpini, E., Sibailly, T., Doorly, R., Batter, V., Brattegaard, K. (1994) Retrospective study of maternal HIV-1 and HIV-2 infections and child survival in Abidjan, Cote d'Ivoire, *BMJ* **308**, 441–443.

Dittmar, M.T., Simmons, G., Hibbitts, S., O'Hare, M., Louisirirotchanakul, S., Beddows, S., Weber, J., Clapham, P.R., Weiss, R.A. (1997) Langerhans cell tropism of human immunodeficiency virus type 1 subtype A through F isolates derived from different transmission groups, *J. Virol.* **71**, 8008–8013.

Eckert, D.M., Kim, P.S. (2001) Mechanisms of viral membrane fusion and its inhibition, *Annu. Rev. Biochem.* **70**, 777–810.

Felber, B.K., Drysdale, C.M., Pavlakis, G.N. (1990) Feedback regulation of human immunodeficiency virus type 1 expression by the Rev protein, *J. Virol.* **64**, 3734–3741.

Finzi, D., Hermankova, M., Pierson, T., Carruth, L.M., Buck, C., Chaisson, R.E., Quinn, T.C., Chadwick, K., Margolick, J., Brookmeyer, R., Gallant, J., Markowitz, M., Ho, D.D., Richman, D.D., Siliciano, R.F. (1997) Identification of a reservoir for HIV-1 in patients on highly active antiretroviral therapy, *Science* **278**, 1295–1300.

Fischer, U., Huber, J., Boelens, W.C., Mattaj, I.W., Luhrmann, R. (1995) The HIV-1 Rev activation domain is a nuclear export signal that accesses an export pathway used by specific cellular RNAs, *Cell* **82**, 475–483.

Fouchier, R.A., Groenink, M., Kootstra, N.A., Tersmette, M., Huisman, H.G., Miedema, F., Schuitemaker, H. (1992) Phenotype-associated sequence variation in the third variable domain of the human immunodeficiency virus type 1 gp120 molecule, *J. Virol.* **66**, 3183–3187.

Gallay, P., Hope, T., Chin, D., Trono, D. (1997) HIV-1 infection of nondividing cells through the recognition of integrase by the importin/karyopherin pathway, *Proc. Natl. Acad. Sci. U.S.A.* **94**, 9825–9830.

Gallo, R.C., Sarin, P.S., Gelmann, E.P., Robert-Guroff, M., Richardson, E., Kalyanaraman, V.S., Mann, D., Sidhu, G.D., Stahl, R.E., Zolla-Pazner, S., Leibowitch, J., Popovic, M. (1983) Isolation of human T-cell leukemia virus in acquired immune deficiency syndrome (AIDS), *Science* **220**, 865–867.

Garcia, J.V., Miller, A.D. (1991) Serine phosphorylation-independent downregulation of cell-surface CD4 by nef, *Nature* **350**, 508–511.

Geijtenbeek, T.B., Kwon, D.S., Torensma, R., van Vliet, S.J., van Duijnhoven, G.C., Middel, J., Cornelissen, I.L., Nottet, H.S., KewalRamani, V.N., Littman, D.R., Figdor, C.G., van Kooyk, Y. (2000) DC-SIGN, a dendritic cell-specific HIV-1-binding protein that enhances trans-infection of T cells, *Cell* **100**, 587–597.

Gonzalez, E., Dhanda, R., Bamshad, M., Mummidi, S., Geevarghese, R., Catano, G., Anderson, S.A., Walter, E.A., Stephan, K.T., Hammer, M.F., Mangano, A., Sen, L., Clark, R.A., Ahuja, S.S., Dolan, M.J., Ahuja, S.K. (2001) Global survey of genetic variation in *CCR5*, *RANTES*, and *MIP-1α*: Impact on the epidemiology of the HIV-1 pandemic, *PNAS* **98**, 5199–5204.

Gottlieb, G.S., Sow, P.S., Hawes, S.E., Ndoye, I., Redman, M., Coll-Seck, A.M., Faye-Niang, M.A., Diop, A., Kuypers, J.M., Critchlow, C.W., Respess, R., Mullins, J.I., Kiviat, N.B. (2002) Equal plasma viral loads predict a similar rate of CD4+ T cell decline in human immunodeficiency virus (HIV) type 1- and HIV-2-infected individuals from Senegal, West Africa, *J. Infect. Dis.* **185**, 905–914.

Gottlinger, H.G., Dorfman, T., Sodroski, J.G., Haseltine, W.A. (1991) Effect of mutations affecting the p6 gag protein on human immunodeficiency virus particle release, *Proc. Natl. Acad. Sci. U.S.A.* **88**, 3195–3199.

Gray, R.H., Wawer, M.J., Brookmeyer, R., Sewankambo, N.K., Serwadda, D., Wabwire-Mangen, F., Lutalo, T., Li, X., vanCott, T., Quinn, T.C. (2001) Probability of HIV-1 transmission per coital act in monogamous, heterosexual, HIV-1-discordant couples in Rakai, Uganda, *Lancet* **357**, 1149–1153.

Greenberg, M.E., Iafrate, A.J., Skowronski, J. (1998) The SH3 domain-binding surface and an acidic motif in HIV-1 Nef regulate trafficking of class I MHC complexes, *EMBO J.* **17**, 2777–2789.

Gulick, R.M., Mellors, J.W., Havlir, D., Eron, J.J., Gonzalez, C., McMahon, D., Richman, D.D., Valentine, F.T., Jonas, L., Meibohm, A., Emini, E.A., Chodakewitz, J.A. (1997) Treatment with indinavir, zidovudine, and lamivudine in adults with human immunodeficiency virus infection and prior antiretroviral therapy, *N. Engl. J. Med.* **337**, 734–739.

Guy, B., Kieny, M.P., Riviere, Y., Le Peuch, C., Dott, K., Girard, M., Montagnier, L., Lecocq, J.P. (1987) HIV F/3′ orf encodes a phosphorylated GTP-binding protein resembling an oncogene product, *Nature* **330**, 266–269.

Hammer, S.M., Squires, K.E., Hughes, M.D., Grimes, J.M., Demeter, L.M., Currier, J.S., Eron, J.J. Jr., Feinberg, J.E., Balfour, H.H. Jr., Deyton, L.R., Chodakewitz, J.A., Fischl, M.A., AIDS Clinical Trials Group 320 Study Team. (1997) A controlled trial of two nucleoside analogues plus indinavir in persons with human immunodeficiency virus infection and CD4 cell counts of 200 per cubic millimeter or less, *N. Engl. J. Med.* **337**, 725–733.

Harrer, T., Harrer, E., Kalams, S.A., Elbeik, T., Staprans, S.I., Feinberg, M.B., Cao, Y., Ho, D.D., Yilma, T., Caliendo, A.M., Johnson, R.P., Buchbinder, S.P., Walker, B.D. (1996) Strong cytotoxic T cell and weak neutralizing antibody responses in a subset of persons with stable nonprogressing HIV type 1 infection, *AIDS Res. Hum. Retroviruses* **12**, 585–592.

Heinzinger, N.K., Bukinsky, M.I., Haggerty, S.A., Ragland, A.M., Kewalramani, V., Lee, M.A., Gendelman, H.E., Ratner, L., Stevenson, M., Emerman, M. (1994) The Vpr protein of human immunodeficiency virus type 1 influences nuclear localization of viral nucleic acids in nondividing host cells, *Proc. Natl. Acad. Sci. U.S.A.* **91**, 7311–7315.

Herrmann, C.H., Gold, M.O., Rice, A.P. (1996) Viral transactivators specifically target distinct cellular protein kinases that phosphorylate the RNA polymerase II C-terminal domain, *Nucleic Acids Res.* **24**, 501–508.

Ho, D.D., Neumann, A.U., Perelson, A.S., Chen, W., Leonard, J.M., Markowitz, M. (1995) Rapid turnover of plasma virions and CD4 lymphocytes in HIV-1 infection, *Nature* **373**, 123–126.

Huang, Y., Paxton, W.A., Wolinsky, S.M., Neumann, A.U., Zhang, L., He, T., Kang, S.,

Ceradini, D., Jin, Z., Yazdanbakhsh, K., Kunstman, K., Erickson, D., Dragon, E., Landau, N.R., Phair, J., Ho, D.D., Koup, R.A. (1996) The role of a mutant CCR5 allele in HIV-1 transmission and disease progression, *Nat. Med.* **2**, 1240–1243.

Isel, C., Karn, J. (1999) Direct evidence that HIV-1 Tat stimulates RNA polymerase II carboxyl-terminal domain hyperphosphorylation during transcriptional elongation, *J. Mol. Biol.* **290**, 929–941.

Jacks, T., Power, M.D., Masiarz, F.R., Luciw, P.A., Barr, P.J., Varmus, H.E. (1988) Characterization of ribosomal frameshifting in HIV-1 gag-pol expression, *Nature* **331**, 280–283.

Jost, S., Bernard, M.C., Kaiser, L., Yerly, S., Hirschel, B., Samri, A., Autran, B., Goh, L.E., Perrin, L. (2002) A patient with HIV-1 superinfection, *N. Engl. J. Med.* **347**, 731–736.

Jowett, J.B., Planelles, V., Poon, B., Shah, N.P., Chen, M.L., Chen, I.S. (1995) The human immunodeficiency virus type 1 vpr gene arrests infected T cells in the G2 + M phase of the cell cycle, *J. Virol.* **69**, 6304–6313.

Kestler, H.W. III, Ringler, D.J., Mori, K., Panicali, D.L., Sehgal, P.K., Daniel, M.D., Desrosiers, R.C. (1991) Importance of the nef gene for maintenance of high virus loads and for development of AIDS, *Cell* **65**, 651–662.

Kino, T., Gragerov, A., Slobodskaya, O., Tsopanomichalou, M., Chrousos, G.P., Pavlakis, G.N. (2002) Human immunodeficiency virus type 1 (HIV-1) accessory protein Vpr induces transcription of the HIV-1 and glucocorticoid-responsive promoters by binding directly to p300/CBP coactivators, *J. Virol.* **76**, 9724–9734.

Kirchhoff, F., Greenough, T.C., Brettler, D.B., Sullivan, J.L., Desrosiers, R.C. (1995) Brief report: absence of intact nef sequences in a long-term survivor with nonprogressive HIV-1 infection, *N. Engl. J. Med.* **332**, 228–232.

Kjems, J., Calnan, B.J., Frankel, A.D., Sharp, P.A. (1992) Specific binding of a basic peptide from HIV-1 Rev, *EMBO J.* **11**, 1119–1129.

Koup, R.A., Safrit, J.T., Cao, Y., Andrews, C.A., McLeod, G., Borkowsky, W., Farthing, C., Ho, D.D. (1994) Temporal association of cellular immune responses with the initial control of viremia in primary human immunodeficiency virus type 1 syndrome, *J. Virol.* **68**, 4650–4655.

Le Gall, S., Erdtmann, L., Benichou, S., Berlioz-Torrent, C., Liu, L., Benarous, R., Heard, J.M., Schwartz, O. (1998) Nef interacts with the mu subunit of clathrin adaptor complexes and reveals a cryptic sorting signal in MHC I molecules, *Immunity* **8**, 483–495.

Le Rouzic, E., Mousnier, A., Rustum, C., Stutz, F., Hallberg, E., Dargemont, C., Benichou, S. (2002) Docking of HIV-1 Vpr to the nuclear envelope is mediated by the interaction with the nucleoporin hCG1, *J. Biol. Chem.* **277**, 45091–45098.

Lever, A., Gottlinger, H., Haseltine, W., Sodroski, J. (1989) Identification of a sequence required for efficient packaging of human immunodeficiency virus type 1 RNA into virions, *J. Virol.* **63**, 4085–4087.

Liu, R., Paxton, W.A., Choe, S., Ceradini, D., Martin, S.R., Horuk, R., MacDonald, M.E., Stuhlmann, H., Koup, R.A., Landau, N.R. (1996) Homozygous defect in HIV-1 coreceptor accounts for resistance of some multiply-exposed individuals to HIV-1 infection, *Cell* **86**, 367–377.

Lyall, E.G., Blott, M., de Ruiter, A., Hawkins, D., Mercy, D., Mitchla, Z., Newell, M.L., O'Shea, S., Smith, J.R., Sunderland, J., Webb, R., Taylor, G.P. (2001) Guidelines for the management of HIV infection in pregnant women and the prevention of mother-to-child transmission, *HIV Med.* **2**, 314–334.

Lyles, R.H., Munoz, A., Yamashita, T.E., Bazmi, H., Detels, R., Rinaldo, C.R., Margolick, J.B., Phair, J.P., Mellors, J.W., Multicenter AIDS Cohort Study. (2000) Natural history of human immunodeficiency virus type 1 viremia after seroconversion and proximal to AIDS in a large cohort of homosexual men, *J. Infect. Dis.* **181**, 872–880.

Malim, M.H., Hauber, J., Le, S.Y., Maizel, J.V., Cullen, B.R. (1989) The HIV-1 rev trans-activator acts through a structured target sequence to activate nuclear export of unspliced viral mRNA, *Nature* **338**, 254–257.

Mariani, R., Kirchhoff, F., Greenough, T.C., Sullivan, J.L., Desrosiers, R.C., Skowronski, J. (1996) High frequency of defective nef alleles in a long-term survivor with nonprogressive human immunodeficiency virus type 1 infection, *J. Virol.* **70**, 7752–7764.

Mascola, J.R., Stiegler, G., VanCott, T.C., Katinger, H., Carpenter, C.B., Hanson, C.E., Beary, H., Hayes, D., Frankel, S.S., Birx, D.L.,

Lewis, M.G. (2000) Protection of macaques against vaginal transmission of a pathogenic HIV-1/SIV chimeric virus by passive infusion of neutralizing antibodies, *Nat. Med.* **6**, 207–210.

McCutchan, F.E., Artenstein, A.W., Sanders-Buell, E., Salminen, M.O., Carr, J.K., Mascola, J.R., Yu, X.F., Nelson, K.E., Khamboonruang, C., Schmitt, D., Kieny, M.P., McNeil, J.G., Burke, D.S. (1996) Diversity of the envelope glycoprotein among human immunodeficiency virus type 1 isolates of clade E from Asia and Africa, *J. Virol.* **70**, 3331–3338.

McKnight, A., Weiss, R.A., Shotton, C., Takeuchi, Y., Hoshino, H., Clapham, P.R. (1995) Change in tropism upon immune escape by human immunodeficiency virus, *J. Virol.* **69**, 3167–3170.

Mellors, J.W., Kingsley, L.A., Rinaldo, C.R. Jr., Todd, J.A., Hoo, B.S., Kokka, R.P., Gupta, P. (1995) Quantitation of HIV-1 RNA in plasma predicts outcome after seroconversion, *Ann. Int. Med.* **122**, 573–579.

Mellors, J.W., Rinaldo, C.R. Jr., Gupta, P., White, R.M., Todd, J.A., Kingsley, L.A. (1996) Prognosis in HIV-1 infection predicted by the quantity of virus in plasma, *Science* **272**, 1167–1170.

Migueles, S.A., Sabbaghian, M.S., Shupert, W.L., Bettinotti, M.P., Marincola, F.M., Martino, L., Hallahan, C.W., Selig, S.M., Schwartz, D., Sullivan, J., Connors, M. (2000) HLA B*5701 is highly associated with restriction of virus replication in a subgroup of HIV-infected long term nonprogressors, *Proc. Natl. Acad. Sci. U.S.A.* **97**, 2709–2714.

Myers, E.L., Allen, J.F. (2002) Tsg101, an inactive homologue of ubiquitin ligase e2, interacts specifically with human immunodeficiency virus type 2 gag polyprotein and results in increased levels of ubiquitinated gag, *J. Virol.* **76**, 11226–11235.

Narayan, O., Cork, L.C. (1985) Lentiviral diseases of sheep and goats: chronic pneumonia leukoencephalomyelitis and arthritis, *Rev. Infect. Dis.* **7**, 89–98.

Ossendorp, F., Eggers, M., Neisig, A., Ruppert, T., Groettrup, M., Sijts, A., Mengede, E., Kloetzel, P.M., Neefjes, J., Koszinowski, U., Melief, C. (1996) A single residue exchange within a viral CTL epitope alters proteasome-mediated degradation resulting in lack of antigen presentation, *Immunity* **5**, 115–124.

Oxenius, A., Price, D.A., Easterbrook, P.J., O'Callaghan, C.A., Kelleher, A.D., Whelan, J.A., Sontag, G., Sewell, A.K., Phillips, R.E. (2000) Early highly active antiretroviral therapy for acute HIV-1 infection preserves immune function of CD8+ and CD4+ T lymphocytes, *Proc. Natl. Acad. Sci. U.S.A.* **97**, 3382–3387.

Palella, F.J. Jr., Delaney, K.M., Moorman, A.C., Loveless, M.O., Fuhrer, J., Satten, G.A., Aschman, D.J., Holmberg, S.D. (1998) Declining morbidity and mortality among patients with advanced human immunodeficiency virus infection. HIV Outpatient Study Investigators, *N. Engl. J. Med.* **338**, 853–860.

Parren, P.W., Moore, J.P., Burton, D.R., Sattentau, Q.J. (1999) The neutralizing antibody response to HIV-1: viral evasion and escape from humoral immunity, *AIDS* **13**, S137–S162.

Patel, C.A., Mukhtar, M., Pomerantz, R.J. (2000) Human immunodeficiency virus type 1 Vpr induces apoptosis in human neuronal cells, *J. Virol.* **74**, 9717–9726.

Phillips, R.E., Rowland-Jones, S., Nixon, D.F., Gotch, F.M., Edwards, J.P., Ogunlesi, A.O., Elvin, J.G., Rothbard, J.A., Bangham, C.R., Rizza, C.R. (1001) Human immunodeficiency virus genetic variation that can escape cytotoxic T cell recognition, *Nature* **354**, 453–459.

Piguet, V., Chen, Y.L., Mangasarian, A., Foti, M., Carpentier, J.L., Trono D. (1998) Mechanism of Nef-induced CD4 endocytosis: Nef connects CD4 with the mu chain of adaptor complexes, *EMBO J.* **17**, 2472–2481.

Pircher, H., Moskophidis, D., Rohrer, U., Burki, K., Hengartner, H., Zinkernagel, R.M. (1990) Viral escape by selection of cytotoxic T cell-resistant virus variants in vivo, *Nature* **346**, 629–633.

Pornillos, O., Alam, S.L., Davis, D.R., Sundquist, W.I. (2002) Structure of the Tsg101 UEV domain in complex with the PTAP motif of the HIV-1 p6 protein, *Nat. Struct. Biol.* **9**, 812–817.

Quinn, T.C., Wawer, M.J., Sewankambo, N., Geberding, J.L. (1994) Incidence and prevalence of human immunodeficiency virus, hepatitis B virus, hepatitis C virus, and cytomegalovirus among health care personnel at risk for blood exposure: final report from a longitudinal study, *J. Infect. Dis.* **170**, 1410–1417.

Ratner, L., Haseltine, W., Patarca, R., Livak, K.J., Starcich, B., Josephs, S.F., Doran, E.R., Rafalski, J.A., Whitehorn, E.A., Baumeister, K.

(1985) Complete nucleotide sequence of the AIDS virus, HTLV-III, *Nature* **313**, 277–284.

Reitter, J.N., Means, R.E., Desrosiers, R.C. (1998) A role for carbohydrates in immune evasion in AIDS, *Nat. Med.* **4**, 679–684.

Rosenberg, E.S., Altfeld, M., Poon, S.H., Phillips, M.N., Wilkes, B.M., Eldridge, R.L., Robbins, G.K., D'Aquila, R.T., Goulder, P.J., Walker, B.D. (2002) Immune control of HIV-1 after early treatment of acute infection, *Nature* **407**, 523–526.

Rosenberg, E.S., Billingsley, J.M., Caliendo, A.M., Boswell, S.L., Sax, P.E., Kalams, S.A., Walker, B.D. (1997) Vigorous HIV-1-specific CD4+ T cell responses associated with control of viremia, *Science* **278**, 1447–1450.

Rowland-Jones, S., Sutton, J., Ariyoshi, K., Dong, T., Gotch, F., McAdam, S., Whitby, D., Sabally, S., Gallimore, A., Corrah, T. (1995) HIV-specific cytotoxic T-cells in HIV-exposed but uninfected Gambian women, *Nat. Med.* **1**, 59–64.

Schuitemaker, H., Kootstra, N.A., de Goede, R.E., de Wolf, F., Miedema, F., Tersmette, M. (1991) Monocytotropic human immunodeficiency virus type 1 (HIV-1) variants detectable in all stages of HIV-1 infection lack T-cell line tropism and syncytium-inducing ability in primary T-cell culture, *J. Virol.* **65**, 356–363.

Schwartz, D., Sharma, U., Busch, M., Weinhold, K., Matthews, T., Lieberman, J., Birx, D., Farzedagen, H., Margolick, J., Quinn T. (1994) Absence of recoverable infectious virus and unique immune responses in an asymptomatic HIV + long-term survivor, *AIDS Res. Hum. Retroviruses* **10**, 1703–1711.

Schwartz, O., Marechal, V., Le Gall, S., Lemonnier, F., Heard, J.M. (1996) Endocytosis of major histocompatibility complex class I molecules is induced by the HIV-1 Nef protein, *Nat. Med.* **2**, 338–342.

Sheehy, A.M., Gaddis, N.C., Choi, J.D., Malim, M.H. (2002) Isolation of a human gene that inhibits HIV-1 infection and is suppressed by the viral Vif protein, *Nature* **418**, 646–650.

Sodroski, J., Goh, W.C., Rosen, C., Dayton, A., Terwilliger, E., Haseltine, W. (1986) A second post-transcriptional trans-activator gene required for HTLV-III replication, *Nature* **321**, 412–417.

Sodroski, J., Rosen, C., Wong-Staal, F., Salahuddin, S.Z., Popovic, M., Arya, S., Gallo, R.C., Haseltine, W.A. (1985) Trans-acting transcriptional regulation of human T-cell leukemia virus type III long terminal repeat, *Science* **227**, 171–173.

Spira, A.I., Marx, P.A., Patterson, B.K., Mahoney, J., Koup, R.A., Wolinsky, S.M., Ho, D.D. (1996) Cellular targets of infection and route of viral dissemination after an intravaginal inoculation of simian immunodeficiency virus into rhesus macaques, *J. Exp. Med.* **183**, 215–225.

Swingler, S., Mann, A., Jacque, J., Brichacek, B., Sasseville, V.G., Williams, K., Lackner, A.A., Janoff, E.N., Wang, R., Fisher, D., Stevenson, M. (1999) HIV-1 Nef mediates lymphocyte chemotaxis and activation by infected macrophages, *Nat. Med.* **5**, 997–1003.

Temin, H.M., Mizutani, S. (1970) RNA-dependent DNA polymerase in virions of Rous sarcoma virus, *Nature* **226**, 1211–1213.

Tersmette, M., de Goede, R.E., Al, B.J., Winkel, I.N., Gruters, R.A., Cuypers, H.T., Huisman, H.G., Miedema, F. (1988) Differential syncytium-inducing capacity of human immunodeficiency virus isolates: frequent detection of syncytium-inducing isolates in patients with acquired immunodeficiency syndrome (AIDS) and AIDS-related complex, *J. Virol.* **62**, 2026–2032.

VerPlank, L., Bouamr, F., LaGrassa, T.J., Agresta, B., Kikonyogo, A., Leis, J., Carter, C.A. (2001) Tsg101, a homologue of ubiquitin-conjugating (E2) enzymes, binds the L domain in HIV type 1 Pr55(Gag), *Proc. Natl. Acad. Sci. U.S.A.* **98**, 7724–7729.

Wagner, L., Yang, O.O., Garcia-Zepeda, E.A., Ge, Y., Kalams, S.A., Walker, B.D., Pasternack, M.S., Luster, A.D. (1998) Beta-chemokines are released from HIV-1-specific cytolytic T-cell granules complexed to proteoglycans, *Nature* **391**, 908–911.

Walker, B.D., Chakrabarti, S., Moss, B., Paradis, T.J., Flynn, T., Durno, A.G., Blumberg, R.S., Kaplan, J.C., Hirsch, M.S., Schooley, R.T. (1987) HIV-specific cytotoxic T lymphocytes in seropositive individuals, *Nature* **328**, 345–348.

Walker, B.D., Flexner, C., Paradis, T.J., Fuller, T.C., Hirsch, M.S., Schooley, R.T., Moss, B. (1988) HIV-1 reverse transcriptase is a target for cytotoxic T lymphocytes in infected individuals, *Science* **240**, 64–66.

Wei, P., Garber, M.E., Fang, S.M., Fischer, W.H., Jones, K.A. (1998) A novel CDK9-associated C-type cyclin interacts directly with HIV-1 Tat and mediates its high-affinity, loop-specific binding to TAR RNA, *Cell* **92**, 451–462.

Wei, X., Ghosh, S.K., Taylor, M.E., Johnson, V.A., Emini, E.A., Deutsch, P., Lifson, J.D., Bonhoeffer, S., Nowak, M.A., Hahn, B.H., Saag, M.S., Shaw, G.M. (1995) Viral dynamics in human immunodeficiency virus type 1 infection, *Nature* **373**, 117–122.

Yang, O.O., Kalams, S.A., Rosenzweig, M., Trocha, A., Jones, N., Koziel, M., Walker, B.D., Johnson, R.P. (1996) Efficient lysis of human immunodeficiency virus type 1-infected cells by cytotoxic T lymphocytes, *J. Virol.* **70**, 5799–5806.

Yang, X., Herrmann, C.H., Rice, A.P. (1996) The human immunodeficiency virus Tat proteins specifically associate with TAK in vivo and require the carboxyl-terminal domain of RNA polymerase II for function, *J. Virol.* **70**, 4576–4584.

Yao, X.J., Gottlinger, H., Haseltine, W.A., Cohen, E.A. (1992) Envelope glycoprotein and CD4 independence of vpu-facilitated human immunodeficiency virus type 1 capsid export, *J. Virol.* **66**, 5119–5126.

Zaitseva, M., Blauvelt, A., Lee, S., Lapham, C.K., Klaus-Kovtun, V., Mostowski, H., Manischewitz, J., Golding, H. (1997) Expression and function of CCR5 and CXCR4 on human Langerhans cells and macrophages: implications for HIV primary infection, *Nat. Med.* **3**, 1369–1375.

# Index


*a*
2-5A synthetase-RNase L 225
accessory proteins 404f
acetylcholine (ACh) 150, 152
ACh, *see* acetylcholine
acinetobacter 160
Acquired Immune Deficiency Syndrome, *see* AIDS
acquired immunity 48
  *E. coli* lipopolysaccharide (LPS) 373
  toll-like (Toll) receptors 373
activation of T cells 61f
$\gamma$-activation sequences 222, *see also* GAS
acute renal failure 159
acute respiratory distress syndrome (ARDS) 202
acute-phase proteins 153
acute-phase response 154
AD, *see* Alzheimer's disease
adaptive immune response 114
adaptive immunity 5, 95, 102–107, 148
ADCC, *see* antibody-dependent cell-mediated cytotoxicity
adhesion molecules 86
  ICAM-1 49
  integrins 62
  LFA-3 49
  members of the Ig superfamily 62
  mucin-like molecules 62
  selectins 62
adrenaline 150, 152
adrenocorticotropin hormone 150, 152
affinity 143
affinity maturation 114, 128
AIDS (Acquired Immune Deficiency Syndrome) 396, 398
  cell biology of 395–412
  molecular biology of 395–412
allergic reactions 370
  degranulation
    histamine 369
    neurosensitizing molecules 369
    proinflammatory 369
allergy 159, 161
alloantigens 52
ALPS, *see* autoimmune lymphoproliferative syndrome
Alzheimer's disease (AD) 159
anaphylatoxin C5a 145
anchors 55
anergy 63
angiogenesis 150
animal models 348
antibody 50, 140, 154
  classes of 322–324
  engineering of 321–324
  genetic modifications of 330–336
  structure of 321–324
  subclasses of 322–324
antibody/antigen 320
antibody-dependent cell-mediated cytotoxicity (ADCC) 320, 322, 228
antibody fusion proteins 334–336
  expression systems 336
  genetic engineering of 319–336
antibody production 328–330
anticancer agent 208
antigen 75, 114, 118, 321
  biochemical nature 51
  entry sites 52
  foreign 76
  origin 52
  self 76
  types 50
antigen entry sites 52
antigen presentation 58, 75
  duration of 104
antigen presenting molecules
  CD1 molecules 56
  MHC class I 54
  MHC class II 56


*Immunology. From Cell Biology to Disease.* Edited by Robert A. Meyers.

ISBN: 978-3-527-31770-7

antigen processing 57, 85
MHC class I 58
MHC class II 60
antigen receptor 50
antigen recognition receptors 66
antigenic peptides 54, 76
antigen-presenting cell (APC) 29, 32, 48, 51ff, 103, 122
antigen-specific immunity
antibodies definition 47
antigens definition 47
anti-inflammatory cytokines 154
antimicrobial activity 226, 228
antimicrobial peptides 22
anti-nRNP 350
antiproliferative activity 226, 233
antiretroviral drugs 396, 410
anti-smith antigen (Sm) 350
anti-SS-A/Ro 350
anti-SS-B/La 350
antitumor activity 228
antiviral activity 208, 223, 226, 228
antiviral protein
2-5A synthetase 224
dsRNA-activatable protein kinase 224
APC, *see* antigen-presenting cell
APC, nonprofessional 49f
APC, professional 48f
apoptosis 80, 114, 123, 146, 150, 288
apoptotic bodies 80
apoptotic cell death 349
application 279
apposition system 153
ARDS, *see* acute respiratory distress syndrome
arthritis 369, 377f
ArthroSoft® 380
arthrosynovitis 371
asthma 159
autoantibodies 231, 346
autoimmune disease 161, 231
systemic lupus erythematosus (SLE) 52
autoimmune lymphoproliferative syndrome (ALPS) 363
autoimmunity 83, 89, 346
mast cells 372

***b***
bacterial superantigens
the immune response to 31
bacteriophage 320, 328
basophils 101
BBB, *see* blood-brain barrier
B-cell 53, 60, 71ff, 81, 94f, 348
activation 71
antigen presentation in the germinal center 74
antigen presentation outside the follicular center 72
B cell/follicular DC contacts 75
CD40–CD40L engagement 74
HLA-DO 75
memory 75
B-cell antigen receptor (BCR) 51, 71
B-cell memory 128f
BCR, *see* B-cell antigen receptor
biological activities associated with IFNs 223
biological response modifiers (BRMs) 233
biologically active peptides 373
bispecific antibodies 332
Blk 71
blood-brain barrier (BBB) 378
*Bordetella pertussis* 87
box C/D 360
brain 372
BrdUrd, *see* 5-bromodeoxyuridine
breast cancer 159
BRMs, *see* biological response modifiers
5-bromodeoxyuridine (BrdUrd) 284ff
BXSB 362
bystander activation 73

***c***
$\gamma$C receptor family
IL-2 261
IL-4 261
cajal body 352
calcium ion, intracellular 282
CAM, *see* cell adhesion molecule
cancer 162, 233f
capsid 399,
cardiolipin 51, 363
caspase 288
cathepsins 60
CCR, *see* cytokine receptor
CD1 51, 54, 56ff, 60, 80
structure 56
CD, *see* clusters of differentiation
CD2 64
CD3 64
CD4 65
CD9 61
CD19 72
CD28 63, 64
CD36 67, 80
CD40 63, 69, 74, 80, 82, 86
CD43 64
CD45 64
CD63 61, 85

CD80 50, 63, 69, 72, 74, 86
CD antigen 277, 279
CD4+ T-cell memory 127
  effector differentiation of 126
CD8+ T-cell memory
  maintenance of 122f
CD8+ T-cell memory generation
  models of 124, 126
    decreasing-potential model 125
    instructive model 125
    linear differentiation model 123
Cdk (cyclin-dependent kinase) 120
CD40L 68f, 74f, 82, 86
CEDIA 146, 149, 151
cell adhesion molecule (CAM) 98
cell cycle 280ff
cell death 287
cell-mediated immunity 347
  versus antibody mediated immunity 102f
cell-mediated immunity induction of
  T-cell activation
    costimulation 103f
    DC maturation 104f
    extravasation 105
cell-mediated inflammation 17f
cell membrane 289
cell proliferation 283
cell sorting 277
cellular effector mechanisms 17f
  oxygen-dependent killing 19f
central nervous system (CNS) 150
central supramolecular activation cluster (c-SMAC) 64
CGD, *see* chronic granulomatous disease
chemical-induced autoimmunity 348
chemoattractant
  microvascular leakage 371
  protease-activated receptor (PAR) 371
  tryptase
    gelatinase 371
    metalloproteinase 371
chemokine 4, 6, 8, 114, 127, 138, 141, 143, 150f, 155, 160, 217
  alternatively activated macrophages 70
  AMAC-1 (alternative macrophage activation-associated chemokine-1) 70
  MCP-1 (macrophage chemotactic protein) 70
  MIP-3$\beta$ 62
chimeric antibodies 326f
chromosome
  analysis sorting 290
chronic granulomatous disease (CGD) 227, 235
chronic inflammation
  lymphoid organ-like structures 106f
chronic inflammatory disease 161
chronic periodontitis 159
CLA, cutaneous lymphocyte-associated antigen
class I 232
class II MHC antigen 229, 232
ClearView® 152
clinical trials 208
clinical uses of IFNs 232
CLIP (class II MHC-associated Ii peptide) 60, 75
clusters of differentiation (CD) 114
CML 234
CNS, *see* central nervous system
colds 235
collagen matrix 79
colony stimulating factor (CSF) 143
colostrum 155
competence factors 186
complement receptor 75
constant region 320, 333
constitutive defenses 6–8
coronary disease 159
corticotropin-releasing factor (CRF) 150
costimulation 72
costimulatory molecules 74, 76
  4-1BBL 63
  CD40 63
  CD80 63
  CD86 63
  CD80 (B7.1) 49
  CD86 (B7.2) 49
CpG nucleotides 67
CpG oligonucleotides 86
C-reactive protein (CRP) 11
CRF, *see* corticotropin-releasing factor
CRH 378
  CRHR-1 377
  CRHR-2 377
  receptor 377
CRHR-1 antagonist
  Antalarmin 378
Crohn's disease 159, 161
cross-presentation
  MHC Class I 84
  proteasome-independent 84
  TAP-dependent 84
cross-priming 84
cross-reactants 154ff
cross-reacting molecules 153
CRP, *see* C-reactive protein
crystal 369
CSF, *see* colony stimulating factor

c-SMAC, *see* central supramolecular activation cluster
CTL (cytotoxic T-cell, cytotoxic T lymphocytes) 94, 102, 114, 227ff
CTLA-4 63
cutaneous lymphocyte-associated antigen (CLA) 79
cyclin-dependent kinase, *see* Cdk
cytogram 276
cytokine 4, 6, 8, 81, 114, 119, 121, 138, 207, 209–210, 215, 217, 228, 230, 233, 236, 241–264, 373
  historical perspective 170–173
  interleukin 167–202
cytokine genes 218
cytokine network 157
cytokine receptor (CCR5) 144f, 219
  hematopoietin family of 256–262
cytokine type II 256
cytomegalovirus 119
cytotoxic T cells (CD8+), *see also* CTL 108

### d

danger hypothesis 82
danger signals 82f
DCs (dendritic cells) 5, 51, 53f, 60, 62, 65, 76ff, 94, 100, 115, 262
  adhesion receptor DC-SIGN 81
  antigen uptake mechanisms 79
  $CD34^+$ bone marrow–derived precursors 78
  $CD14^+$ monocyte pathway 79
  $CD34^+$ precursor pathway 78
  follicular dendritic cells 75
  human DC subtypes 78
  immature 80
  immune evasion 87
  immunotherapy 88
  interdigitating dendritic cells (IDCs) 52
  interstitial 80
  macropinocytosis versus phagocytosis 80
  maturation 80f
  myeloid versus lymphoid 77
  plasmacytoid cell pathway 79
  plasticity 87
  receptors for endocytosis 80
  subsets 77
  T cell priming and polarization 85
  T helper cell polarization 86
DC-SIGN (DC-specific ICAM-3-grabbing nonintegrin) 62, 81
death domain 150
defense mechanisms of viruses 225
defensins 97, 148
delayed hypersensitivity (DH) 230
delayed-type hypersensitivity 149
delta ($\delta$) IFNs 209
dendritic cells, *see* DCs
DH, *see* delayed hypersensitivity
diabodies 331
diffusion 143
dimer linkage site (DLS) 399
DLS, *see* dimer linkage site
DNA 280ff
  degradation 288ff
DNA histogram 280
DNA polymerase $\zeta$ (pol$\zeta$) 357
DNase I 363
Doppler 378
double-stranded DNA (dsRNA) 217, 219, 222, 224f
drug-induced systemic autoimmunity 348
dsRNA, *see* double-stranded DNA
dsRNA-dependent protein kinase 225f

### e

EAE, *see* experimental allergic encephalomyelitis
EBV, *see* Eppstein–Barr virus
ECP, *see* eosinophil cationic protein
EDN, *see* eosinophil-derived neurotoxin
EGF (epidermal growth factor) 262
EIA 145
ELISPOT 151, 156
embryogenesis 153
EMIT 146, 149, 150ff
endogenous pyrogen 140
endometrium 151
endosomal compartments 60
endothelial cell 147, 150, 152, 157, 158
endothelin 373
endotoxin 146, 156, 160
  bivalent scFvs 330f
  monovalent scFvs 330f
  multivalent scFvs 330f
envelope variants 405f
enzyme kinetic 281
enzyme-linked immunosorbent assay (ELISA) 146f, 156, 352
eosinophil cationic protein (ECP) 100
eosinophil granulocytes 100f
eosinophil peroxidase (EPO) 100
eosinophil-derived neurotoxin (EDN) 100
epidermal growth factor, *see* EGF
epitope 56, 72, 142ff, 360
EPO, *see* eosinophil peroxidase
Eppstein–Barr virus (EBV) 74, 119, 160
ER-resident aminopeptidase ERAP-1 58
Evans Blue extravasation 378

exosomes 89
experimental allergic encephalomyelitis (EAE) 376
experimental inflammatory arthritis 371
expression systems 336

*f*
Fab 320, 336, 141
FACS, *see* fluorescence-activated cell sorter
Fas 362
Fc 320, 336
Fc receptor 75
Fc$\gamma$ receptors 67
FDCs, *see* follicular dendritic cells
fever 140, 148
fibrillarin 353
fibroblast IFN 209
flares 369
flavonoids 380
flow chamber 273ff
flow cytometer 273
  data processing 276
  fluidics 274
  optics 275
  signal processing 275
  applications to marine biology 291
  applications to microbiology 291
  clinical applications 283
  instrument 273
fluorescence 276
fluorescence microscopy 352
fluorescence-activated cell sorter (FACS) 148
flushing 378
follicular dendritic cells (FDCs) 75
foreign antigenic peptides 58
foreign antigens 52, 54
foreign peptides 55f, 83
FPIA 146, 149ff
FRET (fluorescence resonance energy transfer) 280
functional plasticity model 77
Fv 141
Fv/scFv 320
Fyn 64, 71

*g*
gamma activation site, *see* GAS
GAS (gamma activation site), *see also* $\gamma$-activation sequences 242, 248
gating 276
gene therapy 155
genes 210, 362
genetic loci 364
genetic polymorphism 156, 158
genetic predisposition 347
germinal center 74ff
gld 362
glucosamine 379
glucosamine sulfate or chondroitin sulfate 379f
glutathione
  intracellular 282
glycolipids 54
glycosylation 336
GM-CSF (granulocyte macrophage-colony-stimulating factor) 79, 100, 147–149, 152, 154, 159, 163, 218, 236, 262, 320
Gp130 cytokine family 143
gp130 receptor-like subfamily 260
gp130-related receptors
  IL-12 family of 260f
graft-versus-host disease (GvHD) 360
granules 376
granulocyte-macrophage colony-stimulating factor, *see* GM-CSF
granulocytes 53
granulomas 231
granzymes 108
Graves' disease 349
GvHD, *see* graft-versus-host disease

*h*
H-2 360
HACA, *see* human antichimeric antibody
HAMA, *see* human antimouse antibody
hairy cell leukemia (HCL) 162, 234
haptens 50f, 142
HCL, *see* hairy cell leukemia
HCV, *see* hepatitis C virus
heat stress 82
$\alpha$-helix 213ff
helper CD4+ T cell (Th cells) 115
hematopoiesis 147, 153, 169, 186
hematopoietic factors 143
hematopoietin 168, 174, 211, 219, 243, 249, 256
hematopoietin receptors (HR) 168, 174
hepatitis C vaccine 162, 236
hepatitis C virus (HCV) 119, 236
herpes simplex virus type 1 (HSV-1) 87
herpes simplex viruses (HSV) 122
histamine 378
HIV (human immunodeficiency virus) 67, 119, 160
  drug therapy of 410
  immune response to 408–410
  life cycle of 399–404

HIV (human immunodeficiency virus) (*continued*)
  molecular biology of 398–406
  origins of 396–398
  virus structure of 398
HIV-1 (human immunodeficiency virus 1) 81
HIV-1 replication 236
HLA-DM 75, 85
  molecular chaperone 61
  peptide editor 61
  peptide loading 61
  x-ray analysis 60
HLA-DO 75
  CLIP removal 76
  cochaperone 76
  HLA-DM–HLA-DO complexes 76
  lysosomal MIICs 76
H2-M 60
homeostasis 115, 129, 154f
homologous 169
homologs 193
hormone 142, 156
HR, *see* hematopoietin receptors
HSP (heat shock protein) 80, 83f, 156
HSV, *see* herpes simplex viruses
HTLV 160
HuIFN-$\alpha_2$ 211
HuIFN-$\gamma$ 219
HuIFN-ß 219
HuIFN-$\alpha$ subtypes 219
human antichimeric antibody (HACA) 327
human antimouse antibody (HAMA) 325
  in mice 327f
humanized antibodies 327
humoral effector mechanisms 20–25
humoral immunity 347
hybridoma 321, 324, 358
hypersensitivity reaction
  delayed type 105f
hypothalamus 378
hypothesis 83
hypoxia 152

### i

ibuprofen 379
ICAM-2 62f, 81
ICAM-3 62, 65, 81
ICAM-1 (intracellular adhesion molecule) 62–65, 81, 228
ICOS 63
IDDM, *see* insulin-dependent diabetes mellitus
idiopathic disease 362
IDO, *see* indoleamine 2,3-dioxygenase
IEMA 146, 151
IFMA 146
IFN (interferon) 139f, 207f, 217, 241ff, 264
IFN-$\alpha$ 79, 141, 143, 159, 162, 210, 213–217, 219, 230–232, 234–236
  $\alpha$-helical 211
  structure 211
  subtypes 208f, 211
IFN-$\alpha_2$ 233
IFN-$\beta$ 79, 162, 209, 211, 214f, 218f, 230, 232, 234, 236
IFN-$\gamma$ 68f, 147, 151f, 157, 159, 208f, 214f, 218, 222, 230–232, 234f, 363
IFN-$\delta$ 213
IFN-$\varepsilon$ 209, 213
IFN-$\kappa$ 209, 214, 218, 230
IFN-$\lambda$1 210, 215, 219–222
IFN-$\lambda$2 210, 215, 219–222
IFN-$\lambda$3 210
IFN-$\omega$ 209, 213f, 218f, 230, 232
IFN-$\tau$ 213f, 218, 230
IFN gene regulatory elements (IREs) 217
IFN producer cells (IPC) 216
IFN protein 211
IFN receptor family 256
IFN receptors
  accessory factor (AF) 219
  IFN binding proteins 219
  IFN-$\lambda$ receptor 258
  type II cytokine receptor family (CRF2) 219
IFNA 210, 218, 223
IFNA genes 217, 232
IFNB 223
IFNB genes 210, 217f
IFNE1 210
IFNG 218
IFNG gene 210
IFN-inducible genes 222
IFN-inducible proteins 222f
IFNK gene 210, 218
IFN-$\alpha_2, \ldots \alpha_n$ 230
IFN-response gene sequences 222
IFNs and pathogenesis 230
IFNs and pathophysiological phenomena 230
IFN-stimulated genes (ISGs) 246
IFN-stimulated response element (ISRE) 222, 243, 246
IFNT gene 210
IFNW 223
IFNW gene 210, 218
Ig, *see* immunoglobulin
Ig-like transcript (ILT)-3 80
IL, *see* interleukin
immediate hypersensitivity 149
immune defence, cell mediated 93–110

immune escape mechanisms 70
immune response 126
  silencing of 108f
immune system 5–8, 346, 377
immunity 115, 129, 230
immunoadhesins 335
immunodominant 328
immunofluorescence 279
immunoglobulin (antibody) synthesis 228
immunoglobulin (Ig) 115, 128, 140, 149, 321,
  IgA 74
  IgE 74
  IgG 74, 140
  IgM 74
immunoligands 334
immunologic memory 113–129
immunological synapse 74
  central supramolecular activation cluster (c-SMAC) 64
  peripheral supramolecular activation cluster (p-SMAC) 64
immunology 3–25
immunoproteasome 58, 84, 89
immunoreceptor tyrosine-based activation motifs (ITAMs) 71
immunoregulatory activity
  MHC antigens 227
include SP 373
indoleamine 2,3-dioxygenase (IDO) 227
inducible defenses 6–8
inducible nitric oxide synthase (iNOS) 229
  history of 406–408
inflammation 4, 8, 17, 151, 230
  carrageenan-induced aseptic 378
  chemoattraction
    endothelial molecule 374
  IL-6 374
  mast cells 380
  microvascular leakage 374
  proinflammatory neurosensitizing mediators 374
  TNF-$\alpha$ 374
inflammatory 377, 379
inflammatory arthritis 369, 371f, 378
inflammatory conditions 376f
inflammatory disorders 377
innate immune system 67
innate immunity 4, 5, 48, 99–102, 147
  affinity 9
  epithelial cells 97f
  evolution of 8f
  molecular basis of 8f
  recognition of 10–17
  specificity 9
iNOS, *see* inducible nitric oxide synthase
insulin sensitivity 379
insulin-dependent diabetes mellitus (IDDM) 232
integrin LFA-1 228
integrins 62, 67, 80
interference 154
interferon genes 210
interferon proteins 211
interferon-stimulated gene factor-3, *see* ISGF-3
interleukin (IL) 115, 139, 141, 143, 168
  biological activities of 185–197
  clinical uses of 200–202
  disease correlates 199f
  historical perspective 170–173
  pathophysiology of 199f
  physiology of 197–199
interleukin-1 69, 140f, 148, 151ff, 186f
interleukin-1$\alpha$ (IL-1$\alpha$) 144
interleukin-1$\beta$ (IL-1$\beta$ 82, 155
interleukin-2 (IL-2) 74
interleukin-2 (IL-2) 141, 149f, 187, 218
interleukin-3 (IL-3) 70
interleukin-3 (IL-3) 187–189
interleukin-4 (IL-4) 70, 74, 79
interleukin-4 (IL-4) 180f
interleukin-5 (IL-5) 74
interleukin-5 (IL-5) 190f, 218
interleukin-6 (IL-6) 69, 74
interleukin-6 (IL-6) 144, 147–151, 153ff, 157ff, 191
interleukin-7 191
interleukin-8 (IL-8) 144, 157, 192
interleukin-9 192
interleukin-10 (IL-10) 70, 144, 149, 152, 157ff, 192f, 215, 219f
interleukin-11 193
interleukin-12 (IL-12) 68f, 193
interleukin-13 194
interleukin-14 194
interleukin-15 194
interleukin-16 194
interleukin-17 194
interleukin-18 195
interleukin-19 195
interleukin-20 195
interleukin-21 195
interleukin-22 195
interleukin-23 196
interleukin-24 196
interleukin-25 196
interleukin-26 196
interleukin-27 196
interleukin-28 196f, 210

interleukin-29 197, 210
interleukin genes 175–179
interleukin proteins 173–179
interleukin receptors 79, 147, 152, 154, 219f, 258f, 262
  intracellular signaling pathways 183–185
  structure of 179–183
intracellular immunization 330
intracellular signaling pathways 183–185
intradermal 378
invariant chain 60
IPC, *see* IFN producer cells
IRE, *see* iron-responsive element
IRES (internal ribosomal entry site) 217
IRF (interferon regulatory factor) 217, 229, 247
IRFs (interferon response factors) 243
IRF-1 217, 222, 226, 229
IRF-2 217f, 226
IRF-3 217f, 236
IRF-5 218
IRF-7 217f
IRF-9 220–222
IRMA 146
iron-responsive element (IRE) 222
ischemia 152
ISGs, *see* IFN-stimulated genes
ISGF-3 220–222, 243, 246
ILT, *see* Ig-like transcript
isoforms 378
ISRE, *see* IFN-stimulated response element
ITAMs, *see* immunoreceptor tyrosine-based activation motifs

***j***

JAK (janus kinase) 243, 245, 229
  family members of 250–255
  family of 249–251
  Jak1 250
  Jak2 250
  Jak3 250
  JAK structure of 249f
  Tyk2 251
JAK1 220, 221
JAK2 220
JAK-STAT signaling
  regulation of 262–264
JAK-STAT signaling cascade 241–264
JLP, *see* juvenile laryngeal papilloma
joints 372
juvenile laryngeal papilloma (JLP) 235
juvenile RA 370
  proinflammatory molecules 381

***k***

Kaposi's sarcoma 236
K/BxN mouse 372
keratinocytes 53
knockout 362
knockout mice 153, 229
Kupffer cells 48

***l***

LAK, *see* lymphokine-activated killer cells
lymphokine-activated killer (LAK) cells
lamina propria 53
  Birbeck granules 48
  CD1a 79
  CD11c 79
  dendritic cell family 48
  E-cadherin 79
LAMs, *see* Lipoarabinomannans
LCMV, *see* lymphocytic choriomeningitis virus
langerin (CD207) 79f
laser 274
LCK 64
*Leishmania* 69
lentiviruses 396
leucine-rich repeat (LRR) 11
leukocyte IFN 209
leukocytes 4, 377
Leydig cell 151
LFA-1 63f, 69
license-to-kill model 63, 86
LICOS 63
ligand 144, 243, 251
light scattering 276
limitin 209
Lipoarabinomannans (LAMs) 56
lipopeptides 81
lipopolysaccharide, *see* LPS
*Listeria monocytogenes* 61, 155
lpr 362
LPS 10, 68, 70, 86
LPS recognition system 67
LRR, *see* leucine-rich repeat
lupus 363
lupus erythematosus 161
lupus nephritis 159
lyme arthritis 369
lymph nodes 52, 66, 77, 80
lymphatic system 115, 125
lymphocytes 9, 115, 118, 147–149, 157
  IL-2 377
lymphocytic choriomeningitis virus (LCMV) 121
lymphoid DCs 78
lymphoid tissues 52ff, 61, 80, 81

lymphokine-activated killer (LAK) cells 187, 162
lymphokines 140, 244
lymphoma 89
lymphotoxin 128, 144, 153
Lyn 71

*m*
mAb, *see* monoclonal antibody
M cell 53
Mac-1 (CD11b/CD18) 67
macrophage mannose receptor (MMR) 12
macrophage migration inhibitory factor (MIF) 140
macrophage scavenger receptors (MSR) 12
macrophages 48, 51, 53, 65ff, 67, 79, 81, 84, 99, 169, 178
  activation 68
  alveolar 54
  antigen recognition receptors 66
  bone marrow–derived monocytes 66
  cellular origin 65
  cross talk 65
  effector functions 69
  innate immune response 65
  Kupffer cells 66
macropinocytosis 80
major basic protein-2 (MBP-2) 100
malaria 159
malignant diseases 233
malignant melanoma 89
mannan-binding lectin (MBL) 11
mannose receptor (MR) 67, 80
MAP kinase (mitogen-activated protein kinase) 119, 146
margination 98
MAST® 152
mast cells 101
  activation 376
  bone marrow progenitor 372
  C3a and C5a 370
  chemokine receptor
    CXCR3 371
  chondroitin sulfate
    activation 380
  c-kit 372
  CRH 378
  cytokines 380
  degranulation 376
  disodium cromoglycate (cromolyn) 379
  epidermal growth factor (EGF) 371
  flavonoids 379
    quercetin 380
  histamine-releasing peptide (HRP) 373
  HPA
    IL-6 and IL-1 378
  HRP 373
  IL 371, 380
  immunoglobulin E (FcεRI) 372
  intragranular activation 376
  juvenile RA
    IL-6 370
  mediators 375f, 378
  metalloproteinases 370
  neurohormonal 377
  neurons 377f
  neuropeptides 372
  perivascular 377
  pheochromocytoma cells 372
  piecemeal degranulation 376
  platelet-derived growth factor (PDGF) 371
  proteolytic enzyme 372
  selective secretion 375
  stem cell factor 372
  TNF 371
  TNF-α 370, 380
  triggers 373f
  vascular endothelial growth factor (VEGF) 371
  vesicles 376
matrix metalloproteinases (MMP) 70
MBL, *see* mannan-binding lectin
MBP-2, *see* major basic protein-2
melanoma 162
membrane potential 281
memory CD4+ T cells
  generation of 120–122
  maintenance of 127
  memory generation of 127
  phenotype of 126f
  phenotyping 118f
  trafficking of 127
memory cd8+ T cells
  enhanced responsiveness of 119f
memory cells 118
mercury 360
metazoan organism 95
Metchnikoff, Elie 65
MG, *see* myasthenia gravis
MHC (major histocompatibility complex) 31, 49ff, 115, 122, 347
MHC antigen 233
MHC class I 58
  self-peptide repertoire of 56
  specificity pockets 55
  x-ray structure 55
MHC class I loading complex 59f

MHC class I molecules 54ff, 58, 84
MHC class I processing 58
MHC class II 32, 60, 80
  self-peptides 57
MHC class II molecules 50, 53, 56ff, 71, 73–75
MHC class II processing 60
MHC restriction 58, 73
microbead technology 291
$\beta_2$-microglobulin 54
MIF, *see* macrophage migration inhibitory factor *or* migration inhibitory factor
migration inhibitory factor (MIF) 155, 161
mitochondrial membranes 288
mixed connective tissue disease 350
MMR, *see* macrophage mannose receptor
molecular architecture 30
molecular chaperones 61
molecular characteristics 219
molecular mechanics 223
monoclonal antibody (mAb) 142, 358
monocytes 56, 66, 79, 80, 99, 158, 377
monokine 140, 244
  polymers of 333f
MR, *see* mannose receptor
MRL-lpr/lpr 362
MS, *see* multiple sclerosis
MSR, *see* macrophage scavenger receptors
Mtb, *see* mycobacterium tuberculosis
mucosa 53
multiple sclerosis (MS) 159, 162, 234
multiplexed arrays 291
multisystem autoimmune diseases 350
murine monoclonal antibodies 324–326
Mx 224
myasthenia gravis (MG) 349
*Mycobacteria leprae* 69
mycobacterial lipids 54, 56
mycobacterium tuberculosis (Mtb) 69
mycolic acids 56
*Mycoplasma arthritidis* 29
mycoplasma lipoproteins 81
myeloid cells 17
myeloid DCs 78
myeloma 89
myxoma virus 160

***n***

natural killing 99
necrosis 80, 82, 287
nerve growth factor (NGF) 129, 371
neuromediators 148
neuron 378ff
neurotensin (NT) 373
neutropenia 163
neutrophils 70, 98f
NES, *see* nuclear export sequence
New Zealand Black (NZB) mice 231
NF-$\kappa$B 71, 74, 217f
NGF, *see* nerve growth factor
nitric oxide 152
nitric oxide radicals 69
nitric oxide synthesis 231
NK (natural killer) cells 5, 69, 99, 227, 229, 233, 243, 250, 255
NO synthetase (iNOS), *see also* inducible nitric oxide synthase 69
nonallergic triggers 370
noncytokine receptors 262
nonvirion associated proteins 404f
noradrenaline 150
NT, *see* neurotensin
nuclear export 263
nuclear export sequence (NES) 263
nucleolus 352
nucleus 352
NZB, *see* New Zealand Black
(NZB × NZW) F1 362

***o***

OA
  WOMAC 379
"Open sandwich" (OS) immunoassay 153
opsonins 67
opsonization 96
organogenesis 153
osteoarthritis 378
ovulation 151
oxidative species 282
oxygen radicals 69

***p***

PAF, *see* platelet-activating factor
pain 369
PAMP, *see* pathogen-associated molecular patterns
paralogs 169, 195
paratope 142ff
Parkinson's disease (PD) 159
passive immunization 117
pathogen 96, 115, 126
  *Leishmania* 61
  *Listeria monocytogenes* 61
  oxygen-independent destruction of 18f
pathogen-associated molecular patterns (PAMP) 66, 95, 10f, 146, 156
pattern-recognition receptors (PRR) 11–13, 66, 88, 95, 156
PD, *see* Parkinson's disease

PDGF (platelet-derived growth factor) 70, 262, 371
peptide
antifungal 67
peptide loading 59
peptide receptors 54
peptide-binding cleft 54
peptidoglycan (PGN) 11, 81
peripheral supramolecular activation cluster (p-SMAC) 64
persistent generalized lymphadenopathy 407
Peyer's patches 53
PG, *see* proteoglycans
PGN, *see* peptidoglycan
pH 282
phage libraries 328
phagocytes 49
phagocytosis 4, 11, 18, 80
phagolysosomes 69
phosphatases 263
PIC, *see* preintegration complex
pituitary adenylate cyclase-activating polypeptide 150
PGE2, *see* prostaglandin E2
PKC (protein kinase C) 64
PKR 224, 226
placebo 379
placenta 151
plant defense
general elicitors of 8
plasma cells 71, 116, 128
plasmacytoid DCs 78
platelet-activating factor (PAF) 99, 152
platelets 102
pockets 56ff
poly $\zeta$, *see* DNA polymerase $\zeta$
polyclonal antibodies 142, 324
polymorphism 54
polymorphonuclear neutrophils 18
polysaccharides 379
potential physiological and pathophysiological roles of IFNs 228
PPRs 81
pre-IFN-$\alpha$ 211
preintegration complex (PIC) 401
pre-mRNA splicing 357
pristane 362
progression factors 186
proliferating cell nuclear antigen 353
prostaglandin E2 (PGE2) 82, 150
prostaglandins 152
prostate 89
proteasome 58
protein kinase C 222
protein tyrosine kinase (PTK) 184, 243
proteins induced by IFNs 222
proteoglycan chondroitin sulfate 379
proteoglycans (PG) 380
chondroitin 378
glucosamine sulfate 378
provirus 396, 401
PRR, *see* pattern-recognition receptors
*Pseudomonas aeruginosa* 160
p-SMAC, *see* peripheral supramolecular activation cluster
PTK, *see* protein tyrosine kinase

***q***

quercetin 381
cytokine secretion 380
the glycoside 380

***r***

RA, *see* rheumatoid arthritis
rapid test device (RTD) 151
reactive oxygen intermediates (ROI) 227, 229, 231, 235
receptor 139, 243, 249, 258
receptor extended gp130 259
receptor signaling 220
recombinase-activating genes (RAGs) 9
regulation of cellular gene expression by IFNs 219
regulation of IFN Synthesis 215
remissions 369
renal cancer 89
reproduction 150
respiratory tract 54
responsive genes 217
retrovirus 396, 398
Rev Responsive Element (RRE) 403
reverse transcriptase polymerase chain reaction (RT-PCR) 156
rheumatoid arthritis (RA) 159, 161, 369, 378f, 381
rhinoviruses 235
RIA 145, 149
ring of Waldeyer 54
risk factors 369
RNase L 224
ROI, *see* reactive oxygen intermediates
RRE, *see* Rev Responsive Element
RTD, *see* rapid test device
RT-PCR, *see* reverse transcriptase polymerase chain reaction

**S**

SAgs, *see* superantigens
salicylate 380
saliva 155
SAP, *see* serum amyloid P component *or* SLAM-associated protein
SCF 371
SCID, *see* severe combined immunodeficiency disease
schizophrenia 159
scleroderma 351, 377
secretion 380
selective antibody system 153
selective release
  cytokines 369
self-antigens 52, 60, 75f, 85
self-peptide 83
self-peptide repertoire 56
self-proteins 66
seminal fluid 155
sepsis 159, 161f
septic shock 155, 160
Sertoli cells 151
serum 150
serum amyloid P component (SAP) 11
SEs, *see* staphylococcal enterotoxins
SFP, *see* staphylococcal food poisoning
severe combined immunodeficiency disease (SCID) 250
SH2 (Src homology 2) 243, 247
side effects 230, 233f
signal transducer and activator of transcription 184, 220, 245
signaling 145
signaling lymphocyte activation molecule (SLAM) 128
signaling pathway 217
single-chain receptor family 262
SIRS, *see* systemic inflammatory response syndrome
Sjogrens syndrome 350
SLAM, *see* signaling lymphocyte activation molecule
SLAM-associated protein (SAP) 128
SLE, *see* systemic lupus erythematosus
Sm, *see* anti-smith antigen
small nuclear ribonucleoprotein 346
SNP (single-nucleotide polymorphisms) 158
SOCS (suppressors of cytokine signaling) 244, 246, 264
  feedback loop 263
soluble receptors 146
SP 377
SPEs, *see* streptococcal pyrogenic exotoxins
specialized lineage model 77
species 69
specificity 153
specificity may 154ff
spermatogenesis 151
spleen 377
spleen cell 378
standardization 154
staphylococcal enterotoxins (SEs) 29, 160
staphylococcal food poisoning (SFP) 38
*Staphylococcus aureus* 160
STAT 244
  evolution of 251f
  family members of 254
    stat2 255
    stat3 255
    stat4 255
    stat6 256
    stat5a 255f
    stat5b 255f
  stat1 254–255
  structure of
    coiled-coil domain 253
    DNA binding domain 253
    linker domain 253
    $NH_2$ domain 252f
    SH2 and tyrosine activation domain 253f
    transcription activation domain 254
STAT1 220f
STAT2 220f
STI, *see* structured treatment interruptions
stomatitis vesicular virus 155
streptococcal pyrogenic exotoxins (SPEs) 29
*Streptococcus pneumoniae* 88
stress
  acute 377
  contact dermatitis 378
  corticotropin-releasing hormone or CRH 377
  CRH 371
  hormones 369
  HPA axis 377
  hypothalamic-pituitary adrenal (HPA) axis 377
  IL-6 levels 371
  immune system 377
  parathyroid hormone 373
  somatostatin 373
  urocortin (Ucn) 377
structured treatment interruptions (STI) 411
studies
  absorption 379
  bioavailability 379

double-blind 379
placebo-controlled trial 379
substance P 150
subtypes 211, 213ff
sulfonamide antibiotics 380
superantigens (SAgs) 29
superoxide anion 152
synergistic actions 380
synergistic effects 380
synergy 157
systemic inflammatory response syndrome (SIRS) 10
systemic lupus erythematosus (SLE) 159, 231, 350, 363

*t*

TAP, *see* transporter associated with antigen processing
TAR, *see* Tat Response Element
T cell activation 61, 75
effector T cell 63
T cell priming 85
T cells 29, 62, 71, 75, 82, 94f, 244, 250, 348
$CD4^+$ 61, 65, 69, 73, 85, 86
$CD8^+$ 61, 65, 68, 84–86
cytotoxic 49, 57
helper 49
naive 48
priming 85
TCGF, *see* T-cell growth factor
T helper cell polarization 86f
T helper cells 94, 100
$T_{H2}$ 69
$T_{H1}$ cells 69
T lymphocytes 170f, 378
Tapasin 60
Tat Response Element (TAR) 403
T-cell effector cytokine 259
T-cell growth factor (TCGF) 171
T-cell mediated immunity
mechanisms of 107f
T-cell memory 109
CD8+ 118–126
T-cell receptor (TCR) 5, 29, 51, 63, 64, 99, 116, 123
binding to the 35
T-cell zones 72
TCR, *see* T-cell receptor
TCR binding sites 35
tears 155
telomeres 116
tetraspan microdomains 85
tetraspan network 61
tetraspanin 61, 72, 79, 85
TfR, *see* transferrin receptor
TGF$\beta$, *see* transforming growth factor $\beta$
Th cells (CD4+) 107f
Th1 and Th2 cytokines 149
Th1 cells 69f, 94, 105, 149, 161
Th2 cells 69f, 94, 105, 149, 161
therapy
cyclooxygenase 2 (COX-2) inhibitors 378
immune modifying drugs 378
immunosuppressant 378
nonsteroidal anti-inflammatory drugs (NSAIDs) 378
$T_{H1}$ phenotype 87
$T_{H2}$ phenotype 87
thymic leukemia (TL) antigen 120
thymus 61, 77, 377
TIL, *see* tumor-infiltrating lymphocytes
tissue cells 101
tissue remodeling 70
TNF, *see* tumor necrosis factor, *see also* tumor necrosis factor receptor
TNF$\alpha$ 68f, 82, 151, 155
TNF blockers
etanercept or infliximab 378
TNF receptor 69
TNF superfamily 143
toll-like receptors (TLRs) 12ff, 66, 76, 83, 86, 146, 156
TLR2 67, 81
TLR3 68, 81
TLR4 67, 81
TLR5 68
TLR9 68, 81
toxic shock syndrome (TSS) 29
Tr1 149
transcription factor 217
trans-endothelial migration 81
transferrin receptor (TfR) 334
transforming growth factor $\beta$ (TGF$\beta$) 70, 144
transgene 116
transgenic 362
routes of 406
transplantation 159, 162
transporter associated with antigen processing (TAP) 58
trastuzumab (Herceptin) 327
trauma 152
*Treponema pallidum* 51
trigeminal system
submucosal neurons
calcitonin gene-related peptide (CGRP)-containing neurons 371
triggers 372
trilisate 380

trophoblasts 209
TSS, *see* toxic shock syndrome
tuberculosis 162
tumor antigens 84
tumor cells 80
tumor necrosis factor (TNF) 120, 148, 150, 153, 157–159, 161
tumor necrosis factor (TNF) receptor 335
tumor rejection antigens 89
tumor-infiltrating lymphocytes (TIL) 162
TYK2 220, 221
type I IFN receptor 256–258
type I IFNs 209, 215, 217, 222f, 228f, 231f, 236
type II IFN receptor 222, 258
type II IFNs 209, 218, 221f, 223f, 227–229, 235
type III IFNs 210f, 219, 221f, 224
tyrosine kinase
  nonreceptor tyrosine kinases 220
tyrosine kinases 145
tyrosine motif 244, 254, 259

***u***
urocortin 377

***v***
vaccination 88, 116f
vaccine 412
vaccinia virus 69, 87, 155, 160
vagal nerve 150
van der Waals' interactions 24
variable region 321, 324
vascular-endothelial growth factor (VEGF) 70
vasoactive intestinal peptide 150
vasodilation 378
veiled cells 52
VEGF, *see* vascular-endothelial growth factor
vesicular stomatitis virus 104
vesicular stomatitis virus (VSV) infection 121
viral escape 411f
viral hepatitis
  hepatitis-B and -C virus infections 231
virion associated proteins 404
virus 140
virus infections 208
VSV, *see* vesicular stomatitis virus

***w***
ways of studying IFN "physiology" 229
Western blot 148

***x***
X-linked lymphoproliferative syndrome (XLP) 128

***y***
Yaa 362
*Yersinia enterocolitica* 160
*Yersinia pestis* 29
*Yersinia pseudotuberculosis* 29

## Related Titles

Falus, A. (ed.)

**Immunogenomics and Human Disease**

2006

ISBN 978-0-470-01530-8

Pollard, K. M. (ed.)

**Autoantibodies and Autoimmunity**

Molecular Mechanisms in Health and Disease

2006

ISBN 978-3-527-31141-5

Kropshofer, H., Vogt, A. B. (eds.)

**Antigen Presenting Cells**

From Mechanisms to Drug Development

2005

ISBN 978-3-527-31108-8

Coligan, J. E.

**Short Protocols in Immunology**

2005

ISBN 978-0-471-71578-8

Coico, R., Sunshine, G., Benjamini, E.

**Immunology**

A Short Course

2004

ISBN 978-0-471-22689-5

Eales, L.-J.

**Immunology for Life Scientists**

2003

ISBN 978-0-470-84524-0

Stewart, C. C., Nicholson, J. K. A. (eds.)

**Immunophenotyping**

2000

ISBN 978-0-471-23957-4